Finite Element Method

AF510901

To my colleague Dr. Sri Radhakrishan

[signature]
December 2022

Finite Element Method

Physics and Solution Methods

Sinan Müftü
Professor, Department of Mechanical and Industrial Engineering
Northeastern University, Boston, United States

ELSEVIER

ACADEMIC PRESS
An imprint of Elsevier
elsevier.com/books-and-journals

Academic Press is an imprint of Elsevier

125 London Wall, London EC2Y 5AS, United Kingdom
525 B Street, Suite 1650, San Diego, CA 92101, United States
50 Hampshire Street, 5th Floor, Cambridge, MA 02139, United States
The Boulevard, Langford Lane, Kidlington, Oxford OX5 1GB, United Kingdom

Copyright © 2022 Elsevier Inc. All rights reserved.

No part of this publication may be reproduced or transmitted in any form or by any means, electronic or mechanical, including photocopying, recording, or any information storage and retrieval system, without permission in writing from the publisher. Details on how to seek permission, further information about the Publisher's permissions policies and our arrangements with organizations such as the Copyright Clearance Center and the Copyright Licensing Agency, can be found at our website: www.elsevier.com/permissions.

This book and the individual contributions contained in it are protected under copyright by the Publisher (other than as may be noted herein).

Notices

Knowledge and best practice in this field are constantly changing. As new research and experience broaden our understanding, changes in research methods, professional practices, or medical treatment may become necessary.

Practitioners and researchers must always rely on their own experience and knowledge in evaluating and using any information, methods, compounds, or experiments described herein. In using such information or methods they should be mindful of their own safety and the safety of others, including parties for whom they have a professional responsibility.

To the fullest extent of the law, neither the Publisher nor the authors, contributors, or editors, assume any liability for any injury and/or damage to persons or property as a matter of products liability, negligence or otherwise, or from any use or operation of any methods, products, instructions, or ideas contained in the material herein.

ISBN: 978-0-12-821127-4

For Information on all Academic Press publications visit our website at https://www.elsevier.com/books-and-journals

Publisher: Matthew Deans
Acquisitions Editor: Brian Guerin
Editorial Project Manager: Rafael G. Trombaco
Production Project Manager: Sojan P. Pazhayattil
Cover Designer: Victoria Pearson Esser

Working together to grow libraries in developing countries

www.elsevier.com • www.bookaid.org

Typeset by Aptara, New Delhi, India

This book is dedicated to my family

Lynne, Serra, and Emre
Yunus, Gülgün, and Ali

Contents

Preface xv
Acknowledgment xix

1. Introduction **1**

1.1 Modeling and simulation **1**
1.1.1 Boundary and initial value problems 1
1.1.2 Boundary value problems 3
1.2 Solution methods **5**

2. Mathematical modeling of physical systems **9**

2.1 Introduction **9**
2.2 Governing equations of structural mechanics **9**
2.2.1 External forces, internal forces, and stress 9
2.2.2 Stress transformations 12
2.2.3 Deformation and strain 17
2.2.4 Strain compatibility conditions 21
2.2.5 Generalized Hooke's law 22
2.2.6 Two-dimensional problems 27
2.2.7 Balance laws 30
2.2.8 Boundary conditions 32
2.2.9 Total potential energy of conservative systems 36
2.3 Mechanics of a flexible beam **41**
2.3.1 Equation of motion of a beam 43
2.3.2 Kinematics of the Euler–Bernoulli beam 45
2.3.3 Stresses in an Euler-Bernoulli beam 46
2.3.4 Kinematics of the Timoshenko beam 47
2.3.5 Stresses in a Timoshenko Beam 48
2.3.6 Governing equations of the Euler–Bernoulli beam theory 48
2.3.7 Governing equations of the Timoshenko beam theory 49
2.4 Heat transfer **50**
2.4.1 Conduction heat transfer 50
2.4.2 Convection heat transfer 51
2.4.3 Radiation heat transfer 51
2.4.4 Heat transfer equation in a one-dimensional solid 53
2.4.5 Heat transfer in a three-dimensional solid 56

2.5 Problems 58
References 59

3. Integral formulations and variational methods 61
3.1 Introduction 61
3.2 Mathematical background 63
3.2.1 Divergence theorem 63
3.2.2 Green-Gauss theorem 63
3.2.3 Integration by parts 64
3.2.4 Fundamental lemma of calculus of variations 64
3.2.5 Adjoint and self-adjoint operators 64
3.3 Calculus of variations 65
3.3.1 Variation of a functional 66
3.3.2 Functional derivative 67
3.3.3 Properties of functionals 68
3.3.4 Properties of the variational derivative 69
3.3.5 Euler–Lagrange equations and boundary conditions 69
3.4 Weighted residual integral and the weak form of the boundary value problems 74
3.4.1 Weighted residual integral 74
3.4.2 Boundary conditions 75
3.4.3 The weak form 76
3.4.4 Relationship between the weak form and functionals 77
3.5 Method of weighted residuals 85
3.5.1 Rayleigh–Ritz method 86
3.5.2 Galerkin method 87
3.5.3 Polynomials as basis functions for Rayleigh–Ritz and Galerkin methods 87
3.6 Problems 104
References 109

4. Finite element formulation of one-dimensional boundary value problems 111
4.1 Introduction 111
4.1.1 Boundary value problem 111
4.1.2 Spatial discretization 112
4.2 A second order, nonconstant coefficient ordinary differential equation over an element 113
4.2.1 Deflection of a one-dimensional bar 113
4.2.2 Heat transfer in a one-dimensional domain 114
4.3 One-dimensional interpolation for finite element method and shape functions 114
4.3.1 C^0 continuous, linear shape functions 115
4.3.2 C^0 continuous, quadratic shape functions 117

4.3.3 General form of C^0 shape functions 118
4.3.4 One-dimensional, Lagrange interpolation functions 119
4.4 Equilibrium equations in finite element form 120
4.4.1 Element stiffness matrix for constant problem parameters 123
4.4.2 Element stiffness matrix for linearly varying problem parameters *a*, *p*, and *q* 124
4.5 Recovering specific physics from the general finite element form 124
4.6 Element assembly 125
4.7 Boundary conditions 128
4.7.1 Natural boundary conditions 128
4.7.2 Essential boundary conditions 129
4.8 Computer implementation 130
4.8.1 Main-code 130
4.8.2 Element connectivity table 130
4.8.3 Element assembly 130
4.8.4 Boundary conditions 131
4.9 Example problem 131
4.10 Problems 133

5. Finite element analysis of planar bars and trusses 139

5.1 Introduction 139
5.2 Element equilibrium equation for a planar bar 141
5.2.1 Problem definition 141
5.2.2 Weak form of the boundary value problem 141
5.2.3 Total potential energy of the system 142
5.2.4 Finite element form of the equilibrium equations of an elastic bar 143
5.3 Finite element equations for torsion of a bar 147
5.4 Coordinate transformations 148
5.4.1 Transformation of unit vectors between orthogonal coordinate systems 148
5.4.2 Transformation of equilibrium equations for the one-dimensional bar element 149
5.5 Assembly of elements 151
5.6 Boundary conditions 155
5.6.1 A formal definition 156
5.6.2 Direct assembly of the active degrees of freedom 158
5.6.3 Numerical implementation of the boundary conditions 158
5.7 Effects of initial stress or initial strain 165
5.7.1 Thermal stresses 165
5.7.2 Initial stresses 166
5.8 Postprocessing: Computation of stresses and reaction forces 166
5.8.1 Computation of stresses in members 166

5.8.2 Reaction forces 167
5.9 **Error and convergence in finite element analysis** 172
Problems 177
Reference 185

6. Euler–Bernoulli beam element 187

6.1 **Introduction** 187
6.2 **C^1-Continuous interpolation function** 187
6.3 **Element equilibrium equation** 189
6.3.1 Problem definition 189
6.3.2 Weak form of the boundary value problem 189
6.3.3 Total potential energy of a beam element 191
6.3.4 Finite element form of the equilibrium equations of an Euler–Bernoulli beam 192
6.4 **General beam element with membrane and bending capabilities** 195
6.5 **Coordinate transformations** 196
6.5.1 Vector transformation between orthogonal coordinate systems in a two-dimensional plane 196
6.5.2 Transformation of equilibrium equations for the Euler-Bernoulli beam element with axial deformation 197
6.6 **Assembly, boundary conditions, and reaction forces** 198
6.7 **Postprocessing and computation of stresses in members** 198
Problems 206
Reference 209

7. Isoparametric elements for two-dimensional elastic solids 211

7.1 **Introduction** 211
7.2 **Solution domain and its boundary** 213
7.2.1 Outward unit normal and tangent vectors along the boundary 213
7.3 **Equations of equilibrium for two-dimensional elastic solids** 215
7.4 **General finite element form of equilibrium equations for a two-dimensional element** 216
7.4.1 Variational form of the equation of equilibrium 217
7.4.2 Finite element form of the equation of equilibrium 218
7.5 **Interpolation across a two-dimensional domain** 221
7.5.1 Two-dimensional polynomials 221
7.5.2 Two-dimensional shape functions 223
7.6 **Mapping between general quadrilateral and rectangular domains** 229
7.6.1 Jacobian matrix and Jacobian determinant 230
7.6.2 Differential area in curvilinear coordinates 233
7.7 **Mapped isoparametric elements** 239
7.7.1 Strain–displacement operator matrix, [B] 238

7.7.2 Finite element form of the element equilibrium equations for a Q4-element 240

7.8 Numerical integration using Gauss quadrature 245

7.8.1 Coordinate transformation 245

7.8.2 Derivation of second-order Gauss quadrature 246

7.8.3 Integration of two-dimensional functions by Gauss quadrature 248

7.9 Numerical evaluation of the element equilibrium equations 250

7.10 Global equilibrium equations and boundary conditions 252

7.10.1 Assembly of global equilibrium equation 252

7.10.2 General treatment of the boundary conditions 253

7.10.3 Numerical implementation of the boundary conditions 254

7.11 Postprocessing of the solution 254

References 255

8. Rectangular and triangular elements for two-dimensional elastic solids 257

8.1 Introduction 257

8.1.1 Total potential energy of an element for a two-dimensional elasticity problem 259

8.1.2 High-level derivation of the element equilibrium equations 260

8.2 Two-dimensional interpolation functions 262

8.2.1 Interpolation and shape functions in plane quadrilateral elements 262

8.2.2 Interpolation and shape functions in plane triangular elements 263

8.3 Bilinear rectangular element (Q4) 264

8.3.1 Element stiffness matrix 264

8.3.2 Consistent nodal force vector 267

8.4 Constant strain triangle (CST) element 273

8.5 Element defects 275

8.5.1 Constant strain triangle element 275

8.5.2 Bilinear rectangle (Q4) 277

8.6 Higher order elements 277

8.6.1 Quadratic triangle (linear strain triangle) 277

8.6.2 Q8 quadratic rectangle 278

8.6.3 Q9 quadratic rectangle 279

8.6.4 Q6 quadratic rectangle 280

8.7 Assembly, boundary conditions, solution, and postprocessing 282

References 290

9. Finite element analysis of one-dimensional heat transfer problems 291

9.1 Introduction 291

9.2 One-dimensional heat transfer 291

9.2.1 Boundary conditions for one-dimensional heat transfer 292

9.3 Finite element formulation of the one-dimensional, steady state, heat transfer problem 293

9.3.1 Element equilibrium equations for a generic one-dimensional element 294

9.3.2 Finite element form with linear interpolation 299

9.4 Element equilibrium equations: general ordinary differential equation 300

9.5 Element assembly 302

9.6 Boundary conditions 305

9.6.1 Natural boundary conditions 305

9.6.2 Essential boundary conditions 306

9.7 Computer implementation 307

Problems 309

10. Heat transfer problems in two-dimensions 313

10.1 Introduction 313

10.2 Solution domain and its boundary 313

10.3 The heat equation and its boundary conditions 315

10.3.1 Boundary conditions for heat transfer in two-dimensional domain 315

10.4 The weak form of heat transfer equation in two dimensions 316

10.5 The finite element form of the two-dimensional heat transfer problem 318

10.5.1 Finite element form with linear, quadrilateral (Q4) element 320

10.6 Natural boundary conditions 323

10.6.1 Internal edges 324

10.6.2 External edges subjected to prescribed heat flux 324

10.6.3 External edges subjected to convection 326

10.6.4 External edges subjected to radiation 327

10.7 Summary of finite element form of the heat equation and natural boundary conditions 328

10.8 Numerical integration of element equilibrium equations 329

10.9 Element assembly 331

10.10 Imposing the essential boundary conditions 332

10.10.1 Symbolic representation of essential boundary conditions 333

10.10.2 Numerical implementation of essential boundary conditions 333

Problems 340

Reference 341

11. Transient thermal analysis 343

11.1 Introduction 343

11.2 Transient heat transfer equation 344

11.2.1 Boundary/initial value problem 344
11.2.2 Element equilibrium equation of one-dimensional, transient heat transfer 345
11.2.3 Global equilibrium equation of one-dimensional, transient heat transfer 348
11.2.4 Global boundary conditions 348
11.3 Finite difference approximations to derivatives 348
11.3.1 Temporal discretization of a continuous function 348
11.3.2 Taylor series expansion 349
11.3.3 Approximations to the first derivative of a function 349
11.4 Direct time integration of the heat transfer equation 351
11.4.1 Forward difference or Euler method 351
11.4.2 Backward difference method 352
11.4.3 Central difference or Crank–Nicholson method 353
11.4.4 Generalized trapezoidal method 353
11.5 Solution algorithm 355
11.5.1 Explicit and implicit time integration methods 356
11.6 Convergence, stability, and accuracy of time integration methods 358
11.6.1 Modal expansion of the semidiscrete first-order equation 359
11.6.2 Stability of the semidiscretized first-order equation 360
11.6.3 Modal expansion of the generalized trapezoidal algorithm 361
11.6.4 Stability of the generalized trapezoidal algorithm 361
11.6.5 Fourier–von Neumann stability analysis of the generalized trapezoidal method 362
11.6.6 Consistency and rate of convergence 366
References 376

12. Transient analysis of solids and structures 377
12.1 Introduction 377
12.2 Vibration of single degree of freedom systems 378
12.2.1 Free vibrations: complementary solution 379
12.2.2 Response to harmonic excitations: particular solution 382
12.2.3 Combined response: complimentary and particular solutions 384
12.2.4 Transient vibration 385
12.3 Initial/boundary value problems for deformable solids 386
12.3.1 Two-dimensional deformable solid 386
12.3.2 One-dimensional bar 387
12.3.3 Euler–Bernoulli beam 387
12.4 Vibration response of an Euler–Bernoulli beam 388
12.4.1 Eigenvalue problem 389
12.4.2 Free vibration problem 391
12.5 Semidiscrete equations of motion 394

12.5.1 Two-dimensional deformable element 394
12.5.2 One-dimensional elastic bar element 396
12.5.3 Euler–Bernoulli beam element 398
12.6 Mass matrix 400
12.6.1 Consistent mass matrices 400
12.6.2 Lumped mass matrix 402
12.7 Damping matrix 403
12.8 Global equation of motion 405
12.9 Analytical analysis of vibration of semidiscrete systems 405
12.9.1 Eigenvalue problem for the semidiscrete equation of motion 406
12.9.2 Orthogonality of the eigenvectors 414
12.9.3 Response to initial excitations by modal analysis 415
12.10 Direct time integration of the equation of motion of a solid 421
12.10.1 Central finite difference approximations: explicit time integration 423
12.10.2 Linear and average acceleration methods: implicit time integration 424
12.10.3 Newmark's method for direct time integration 426
12.10.4 α-Method for direct time integration 428
12.10.5 Initial conditions 430
12.10.6 Solution algorithm 430
12.11 Convergence, stability, and accuracy of time integration methods 431
12.11.1 Stability of the explicit method 431
12.11.2 Stability and consistency of the Newmark and α-methods 435
Problems 440
References 444

Appendix A MATLAB 445
Appendix B Guidelines for writing a finite element code in MATLAB 459
Appendix C Finite element analysis with ANSYS 475
Appendix D ANSYS tutorial: beam and bar elements 483
Appendix E ANSYS tutorial: two-dimensional linear elastic analysis 499
Appendix F ANSYS tutorial: thermomechanical deformation 503
Index 515

Preface

This book comes out of teaching graduate level finite element method and applied mechanics courses over the last two decades at Northeastern University. One of the major goals in writing this book is to convey to the reader that the physical models, the solution methods, and the results are inseparable parts of analysis. The knowledge in these areas is spread over many excellent textbooks including references [1–9] in the finite element literature, [10–18] in the mechanics literature, [19–23] in the engineering mathematics literature, and [24–25] in the heat transfer literature, among others. I have drawn upon these textbooks and my own derivations in preparing this book which aims for a unified presentation between the two covers.

Having a good grasp of the underlying physics of a problem, its mathematical representation, and the capabilities and the limitations of the solution methods is necessary to setup the problems and to interpret the results effectively. I hope that this book will help guide the users of the finite element codes to have a holistic view in their analysis. This book also aims to provide detailed background information on the development of the finite element method with a target audience who is interested to develop their own codes for engineering research.

All physical systems in the realm addressed by finite element analysis can be represented by boundary/initial value problems. In this book, we primarily focus on the solution of one- and two-dimensional linear elasticity and heat transfer problems. Extension of these solution techniques to three-dimensional analysis is straight forward, therefore it is not addressed here to preserve the clarity of presentation and avoid repetition. A chapter is dedicated at the beginning to the detailed derivation of the boundary/initial value problems in heat transfer and elasticity. Chapter 2 is intended to serve as a reference for not only this book, but also for those who want a concise review of these topics.

The finite element method falls under the general umbrella of *variational methods*. The connections between the classical variational techniques and the finite element method are introduced in Chapters 3 and 4. The former is dedicated to a brief introduction to variational calculus followed by two classical variational methods: the *Rayleigh–Ritz* and *Galerkin*. In Chapter 4, we develop the Rayleigh–Ritz-based finite element method for the solution of the boundary value problems governed by *self-adjoint* ordinary differential equations. Here, and in the rest of the book, the finite element method is developed as a natural extension of the Rayleigh–Ritz method and the boundary value problem is defined over a subsegment (element) of the solution domain.

Chapters 5 and 6 develop the finite element form of the boundary value problems that represent the deformation of *bars* and *Euler–Bernoulli beams*, respectively. These chapters are extended to analysis of two-dimensional trusses and frames, respectively. Chapters 4–6 are foundational in all aspects of solving an engineering problem by using finite element method, with topics ranging from the physical model to element derivation, element assembly, imposing boundary conditions, and postprocessing.

Chapter 7 gives a detailed account of solution of two-dimensional linear elasticity problems by using isoparametric elements. In Chapter 8, an analysis of two-dimensional finite element method is provided for rectangular and simple triangular elements much like in reference [3]. Chapter 9 introduces the finite element method for the solution of one-dimensional, steady state heat transfer problems. This is extended to two-dimensional heat transfer problems by using isoparametric formulation in Chapter 10.

Chapters 4–10 of the book are involved with setup and solution of boundary value problems at steady state. Solution methods for problems in which transient effects are significant are introduced in Chapters 11 and 12 for parabolic and hyperbolic systems for heat transfer and solid mechanics problems, respectively. Derivation of the finite difference approximations to the derivative operators; direct time integration methods for the heat transfer (parabolic) problems; and their convergence, stability, and accuracy characteristics are presented in Chapter 11. Transient analysis methods for solids and structures are presented in Chapter 12. In addition to the derivation of the semidiscrete equation of motion by finite element method, this chapter includes comprehensive treatments of vibration of single- and multidegree of freedom systems and solution approaches to the transient problems with modal, as well as direct time integration methods.

This book is written for an audience who wants to write their own finite element codes, who wants to use commercially available finite element programs, and who wants to make informed choices on element types, solution convergence, and in-depth understanding of physical models. The target audience for the first part of the book is the first-year graduate students. A one semester course can be delivered by using Chapters 1–6 and 8. The target audience for the second part of the book is graduate students in their second-year or higher. A one semester course can be delivered for this group by using Chapters 7 and 9–12.

Examples in the book have been prepared by using Mathematica and by finite element programs written in MATLAB. Appendix A provides a summary of the MATLAB functions I find useful to write a finite element program. Two of the MATLAB-based finite element programs used in the book are provided in Appendix B. We also provide extensive instructions on using the commercially available finite element analysis program *ANSYS Mechanical APDL* for one- and two-dimensional solid mechanics and heat transfer problems. These include step-by-step instructions to use the program from the graphical user interface and a summary of how to prepare input macros with the *ANSYS Parametric Design Language* (APDL). Several APDL macros are also provided.

Acknowledgments

I would like to take this opportunity to thank the many people who helped me along the way and who were influential in writing of this book in one fashion or another. Many of the students who were in ME 5657 and ME 7275 at Northeastern University asked many deep, penetrating questions to help me focus on the details and pointed out typos and early mistakes. I had many enjoyable discussions about the topic and the course with my former graduate students Dr. Dinçer Bozkaya, Dr. Ernesto Lopez, Dr. Hsuan-Yu Chou, Dr. Qian Sheng, Dr. Hankang Yang, Dr. Soroush Irandoust, and Dr. Runyang Zhang who were teaching assistants during different periods. Dr. Tuğçe Kaşıkçı was the one who pointed out the obvious by asking "why don't these notes have any examples?" This book was written during trying times for humanity when the Covid-19 epidemic was raging around the World and for me personally. I would like to thank the editors at Elsevier, Mr. Brian Guerin and Mr. Rafael Guilherme Trombaco, for their patience and understanding with the delays. Mr. Sojan Pazhayattil for the excellent typesetting. Ms. Berin Üçyiğit was my eighth and ninth grade mathematics teacher at Tevfik Fikret Lisesi in Ankara, Turkey. I am grateful to her for the love of mathematics and geometry I developed under her tutelage.

Finally, I would like to thank my family. My father instilled the sense of right and wrong in me and my mother provided for me and my brother after he passed away. She showed both of us the value of never giving up. I am grateful to both for their love and for the opportunities they provided us. Lastly, I would like to thank my wife Lynne and our children Serra and Emre for the patience and encouragement they afforded me during this project.

Chapter 1

Introduction

One of the major aims of natural sciences is to identify the *cause–effect* relationships in nature. Grasping these relationships is not only useful to satisfy inherent curiosity, but it is also helpful in building and designing systems that improve our daily lives.

Engineers and scientists interested in developing predictive capabilities for natural processes are generally involved in two major tasks: *modeling* and *simulation*. Our understanding of the term *model* refers to developing a mathematical representation of a physical process that is observed in nature. To this end, partial differential equations (PDEs) are typically used to formulate mathematical representations of physical processes. The term *simulation*, on the other hand, refers to solving the mathematical model and analyzing the results to gain insight into the cause–effects relationships.

1.1 Modeling and simulation

1.1.1 Boundary and initial value problems

Consider a system with *dependent variables* u, v, and w defined over a domain Ω, which itself occupies a subsection of space (Fig. 1.1). In general, each variable can take different values at different points in the domain and these values can also vary in time. Spatial position of a point P in the domain Ω can be identified with respect to a spatial reference system (e.g., (x, y, z)). If the position of point P also varies in time, the position of point P is said to be time dependent. Thus, for example, if u is a function of space and time $u = u(x, y, z, t)$. In these notes, we will consider *boundary value problems* (BVPs) and *initial value problems* that are formulated by using PDEs. A very general representation of such a problem can be given as follows:

$$\mathcal{L}(u, v, w) = f \text{ in } \Omega, \text{ for } 0 \leq t \leq \tau \tag{1.1}$$

where $\mathcal{L}(\cdot)$ is a differential operator of independent spatial variables x, y, z and time t, $f = f(x, y, z, t)$ is typically a function that represents the internal effects that act on the system, and τ is the duration of interest.

The dependent variables interact with the outside of the domain Ω through the *boundary* Γ of the domain, and typically experience changes as a result

Finite Element Method: Physics and Solution Methods. DOI: https://doi.org/10.1016/B978-0-12-821127-4.00001-3
Copyright © 2022 Elsevier Inc. All rights reserved.

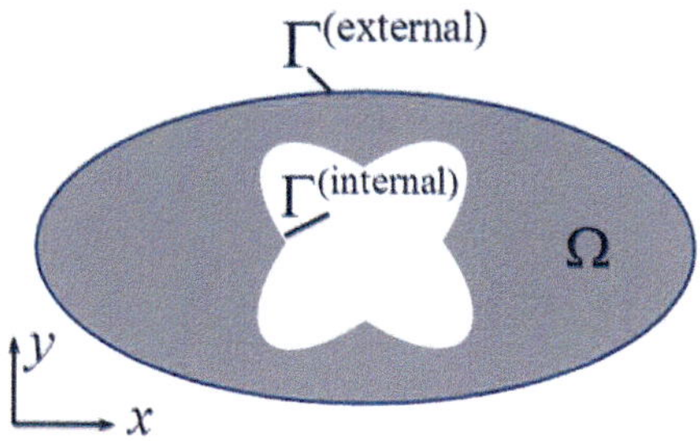

FIGURE 1.1 A domain Ω with internal and external boundaries. Note that boundary conditions can be applied on both internal and boundaries.

of the external effects that are imposed on the boundary. These external effects are known as the *boundary conditions* which depend on the physics of the problem.

The *Dirichlet boundary condition* represents a prescribed value for a dependent variable,

$$u = u_b(t)\,\text{on}\,\Gamma_E \tag{1.2}$$

Here the variable u of the solution domain is prescribed to u_b on a segment of the boundary Γ_E. In general, this prescribed variable can be a function of time t. The Dirichlet boundary condition is also known as the *essential boundary condition.*

The *von Neumann boundary condition* typically describes the external effects that cause a change in the system. Such effects include external forces, heat flow, etc. As we will demonstrate later in the notes, the von Neumann boundary conditions are typically represented as follows:

$$\mathcal{B}(u, v, w) = g(t)\,\text{on}\,\Gamma_N \tag{1.3}$$

where $\mathcal{B}(\cdot)$ is another differential operator, g is a given function, and Γ_N represents the segment of the boundary over which the von Neumann boundary condition is applied. The von Neumann boundary condition, also known as the *natural boundary condition* or the *nonessential boundary condition*, can also vary in time.

The *initial states* of the dependent variables are also required to describe the problem *properly*. For example, the *initial values* of the variable u and its temporal rate of change $u_{,t}$ and $u_{,tt}$ can prescribed in the solution domain as *initial conditions* as follows:

$$u\,(x, y, z, 0) = u^{(0)},\;\; u_{,t}(x, y, z, 0) = \dot{u}^{(0)},\;\; u_{,tt}(x, y, z, 0) = \ddot{u}^{(0)} \text{ in } \Omega \tag{1.4}$$

where $(\cdot)_{,t}$ and $(\dot{\cdot})$ indicate differentiation with respect to time. Similar initial conditions have to be defined for v and w as well. Number of initial conditions that are sufficient to describe the problem properly depends on the physical system, and this topic is addressed in later chapters.

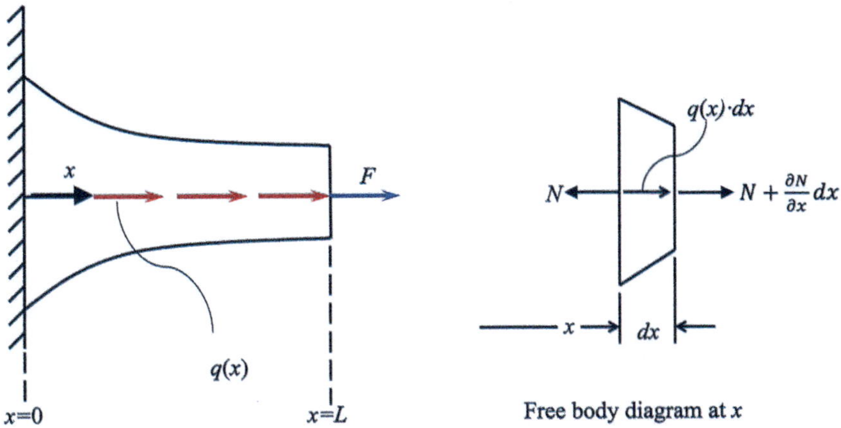

FIGURE 1.2 A bar with variable cross-sectional area, subjected to an axially distributed load $q(x)$, and the free body diagram of a small segment of this bar at position x.

1.1.2 Boundary value problems

In some problems, only the *steady state* of the dependent variables is of interest and the temporal variation is neglected (or negligible). Thus, for example, u becomes only a function of the spatial dimensions $u = u(x, y, z)$. A *steady state boundary value problem* can be formulated by dropping the time dependence as follows:

$$\mathcal{L}\,(u, v, w) = f \text{ in } \Omega \tag{1.5}$$

where for a boundary value problem $\mathcal{L}(\cdot)$ is a differential operator of the independent spatial variables (x, y, z) and $f = f(x, y, z)$. A steady state boundary value problem is also subject to the Dirichlet and/or von Neumann conditions on the boundary of the domain.

Example 1.1 Equation of motion of a solid bar

a) Derive the equation of motion of an elastic bar in terms of its deflection $u(x,t)$. Initially, assume that the bar has a variable cross-sectional area $A(x)$ and that it is subjected to distributed axial load $q(x,t)$ and a concentrated force F at its free end as shown in Fig. 1.2. Also assume small deflections, linear elastic material behavior with constant elastic modulus E, and constant mass density ρ.

b) Obtain the steady state solution for the case of constant cross-section and zero distributed force.

Solution 1.1a: The solution domain Ω for this problem spans $0 < x < L$. The boundaries Γ of the solution domain are located at $x = 0$ and $x = L$. Internal forces develop in the bar in response to external loading. The internal normal force $N(x)$ at the cross-section x can be defined as follows:

$$N(x) = \bar{\sigma}(x)A(x) \tag{1.6}$$

where the average normal stress $\bar{\sigma}$ is defined as follows:

$$\bar{\sigma}(x) = \frac{1}{A(x)} \int_{A(x)} \sigma dA$$

and where σ is the internal normal stress, A is the cross-sectional area of the bar. The equation of motion of the bar can be obtained by using Newton's second law on a small segment of the bar (Fig. 1.2). The balance of internal and inertial forces gives,

$$\begin{aligned} &\sum F_x = \rho A dx \frac{\partial^2 u}{\partial t^2} && \text{(a)} \\ &-N + q\,dx + \left(N + \frac{\partial N}{\partial x} dx\right) = \rho A dx \frac{\partial^2 u}{\partial t^2} && \text{(b)} \\ &\frac{\partial N}{\partial x} = -q + \rho A \frac{\partial^2 u}{\partial t^2} && \text{(c)} \end{aligned} \tag{1.7}$$

Hooke's law defines the constitutive relationship between the internal stress and strain for linear, elastic materials. For a slender bar, the Hooke's law can be given as follows:

$$\bar{\sigma} = E\varepsilon \tag{1.8}$$

where E is the elastic (Young's) modulus of the material. The strain–displacement, ε–u, relationship is given as follows:

$$\varepsilon = \frac{\partial u}{\partial x} \tag{1.9}$$

Combining Eqs. (1.6–1.9), we find the internal force resultant as follows:

$$N = \bar{\sigma}A = EA\frac{\partial u}{\partial x} \tag{1.10}$$

The equation of motion can then be found by combining Eqs. (1.7c) and (1.10),

$$\frac{\partial}{\partial x}\left[EA\frac{\partial u}{\partial x}\right] = -q + \rho A \frac{\partial^2 u}{\partial t^2} \tag{1.11}$$

This is a PDE that governs the dynamics of axial deflection $u(x,t)$ along the bar. Its solution requires two boundary conditions and two initial conditions. The boundaries of this bar are located at $x = 0, L$. At the $x = L$ boundary, the force resultant should be equal to the applied load, i.e., $N(L) = F$. By using Eq. (1.10), this condition can be expressed in terms of the bar deflection. The boundary conditions for this problem then become,

$$\text{Boundary conditions: } u(0) = 0 \quad \text{(a)}$$
$$\left.\frac{\partial u}{\partial x}\right|_{x=L} = \frac{F}{EA(L)} \quad \text{(b)} \tag{1.12}$$

The initial conditions represent the state of deflection and velocity of the entire bar at $t = 0$. In general, these conditions can be represented as follows:

$$\text{Initial conditions: } u(x,0) = u^{(0)}(x) \quad \text{(a)}$$
$$\left.\frac{\partial u}{\partial t}\right|_{t=0} = \dot{u}^{(0)}(x) \quad \text{(b)} \tag{1.13}$$

where $u^{(0)}(x)$ and $\dot{u}^{(0)}(x)$ are known functions.

Solution 1.1b: Let us find the steady state deflection of a bar with constant cross-sectional area, A_0, subjected only to a concentrated force F at its free end. Eqs. (1.11) and (1.12) become,

$$\text{Governing equation, (in } \Omega\text{): } EA_0\frac{d^2u}{dx^2} = 0 \text{ in } 0 < x < L \tag{1.14}$$

$$\text{Boundary conditions, (on } \Gamma\text{): } u(0) = 0 \quad \text{(a)}$$
$$\left.\frac{du}{dx}\right|_{x=L} = \frac{F}{EA_0} \quad \text{(b)} \tag{1.15}$$

Solution of Eq. (1.14) gives,

$$u(x) = \frac{C_1}{EA_0}x + C_2 \tag{1.16}$$

where C_1 and C_2 are integration constants that can be found by using the boundary conditions as follows:

$$C_1 = F \text{ and } C_2 = 0$$

And the variation of $u(x)$ becomes,

$$u(x) = \frac{Fx}{EA_0} \tag{1.17}$$

Note that this relationship shows that the displacement varies linearly for a bar with cross-section that is fixed on one end and pulled by a force on the other.

1.2 Solution methods

In this work, *modeling* refers to mathematical formulation of a physical process. This requires background in the related subjects, certain mathematical tools, and experimental observations. In Chapter 2, we present the formulation of models for *deformation* of elastic solids and *transfer and storage of thermal energy* in solids and fluids. Solution of the mathematical model can be a challenging task and forms the general background of this work. *Analytical solutions* which can be expressed as relatively straight forward relationships between the dependent and independent variables exist only for a relatively small number of situations where the geometry and the physical nature of the problem can be simplified. Numerical methods are used otherwise. Among the numerical solution methods for solving PDEs are the *finite difference, variational*, and *finite element* methods.

The finite difference method (FDM) is implemented on the differential form of the BVP. The derivative operators of the PDE are approximated by finite difference operators. The solution domain is discretized in to a grid, and the unknowns are the values of the dependent variable at the nodes. The discretized version of the PDE is evaluated at each grid point. This results in a set of algebraic equations which can be represented in matrix form,

$$[K]\{D\} = \{R\} \tag{1.18}$$

where $[K]$ is the stiffness matrix representing the discretized form of the partial derivatives, $\{D\}$ is the vector of unknown nodal values of the dependent variable, and $\{R\}$ is the loading vector representing the external effects. The boundary conditions often require specialized treatment of the finite difference operators and modify the $[K]$ matrix. The FDM is effective over relatively simple shapes such as rectangular and cylindrical domains in two-dimensional problems and parallelepiped or spherical domains in three-dimensional problems.

Solution of PDEs by *variational methods* involves use of a weighted-residual integral, e.g.,

$$\int_{\Omega} w(A(u) - f)d\Omega = 0 \tag{1.19}$$

where w is a weight function. One way to distinguish the well-known techniques such as the *Rayleigh–Ritz, Galerkin*, and *collocation methods,* which fall under the umbrella of the variational methods, is the choice of the weight functions. Nevertheless, in these traditional methods, an approximation function is chosen for the dependent variable over the entire solution domain. Therefore, traditional variational approaches can be impractical when the shape of the domain is not one of the simple shapes mentioned above. A more complete discussion of the variational methods is given in Chapter 3.

The *finite element method* (FEM) is based on the variational methods. However, the approximation function for the dependent variable is developed for small regions, known as elements, with regular shapes such as rectangles, triangles, tetrahedra, and hexahedra. As most irregular solution domains can be approximated as a collection of smaller regions (Figs. 1.3 and 1.4) the finite element method presents a powerful technique to solve BVPs over very complex shapes. In the FEM, the domain over which the problem is defined is divided into *elements* interconnected at nodal points and along edges. An *element* can take

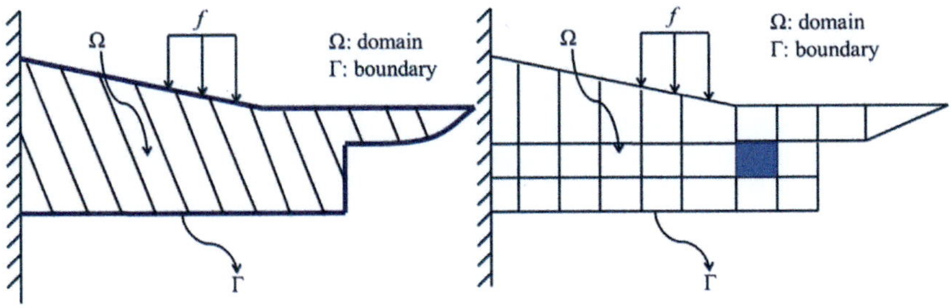

FIGURE 1.3 Definitions of domain and boundary on a two-dimensional object subjected to distributed external forces, *f*. Finite element analysis requires discretization of the solution domain to subdomains called elements.

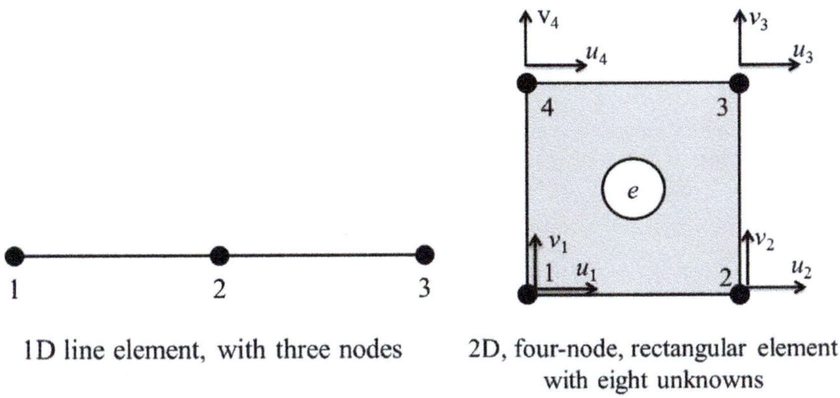

FIGURE 1.4 Elements communicate with the other elements on their boundaries. A line element with three nodes and a 2D element with four nodes are shown in this figure.

various shapes and can have a number of *nodes* associated with it. The dependent variable (i.e., the unknown quantity) is determined *at the nodes*.

FEM overcomes the disadvantage of the traditional variational methods by providing a systematic procedure for the derivation of the approximation functions over subregions of the domain. As the solution domain is represented by a collection of elements, it is relatively straight forward to implement nonhomogeneous, and or discontinuous material properties.

As problems with complex geometry, material properties and boundary conditions cannot be effectively solved with analytical techniques, use of FEM provides must be accompanied with experimental verification. In addition, if possible, analytical forms of the problems should also be sought for simplified conditions as a check to the finite element solution. Finally, the importance of a systematic mesh convergence study for the solution cannot be understated.

Chapter 2

Mathematical modeling of physical systems

2.1 Introduction

The goal of this chapter is to give brief descriptions to modeling of deformation of linear elastic solids and thermal energy transfer and storage in physical systems. More detailed discussion of these topics can be found in the specialized references provided at the end of this chapter. Our goal is to demonstrate how to obtain mathematical models (representations) of physical systems by using the fundamental laws of physics. Thus, we will show that deformation of elastic solids can be described by using Newton's laws of motion. This will result in equations of motion represented as partial differential equations. Vibration of a long and slender bar (Section 2.1), deflection of a general deformable body (Section 2.2), and deflection of beams (Section 2.3) constitute examples of such systems. The principle of conservation of energy will be used to describe effects of heat transfer in a continuum (Section 2.4).

2.2 Governing equations of structural mechanics

When a deformable body is subjected to external effects such as external forces and/or imposed displacements on its boundary, its shape will change and internal forces will develop throughout its volume. The level of deformation for given external effects depends on the material of the deformable body. In this section, the *equations of motion* for small deflections of linear, elastic materials are presented. In particular, we are interested in small deformations of linear, elastic solids. To this end, following are discussed: *i*) concepts of external and internal forces and the concept of stress, *ii*) elastic deformations and the concept of small strain, *iii*) linear elastic constitutive relations, *iv*) balance laws, and *v*) total potential energy of a deformable body.

2.2.1 External forces, internal forces, and stress

External forces acting on continua fall under two categories: *body force* and *surface traction* as depicted in Fig. 2.1.

Finite Element Method: Physics and Solution Methods. DOI: https://doi.org/10.1016/B978-0-12-821127-4.00008-6
Copyright © 2022 Elsevier Inc. All rights reserved.

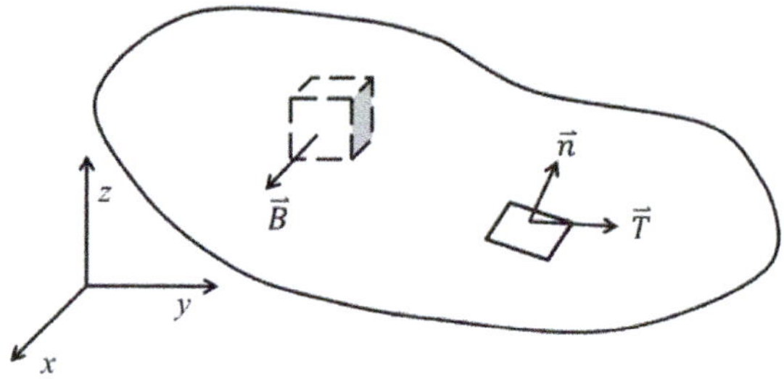

FIGURE 2.1 Body force $\vec{B}$, surface traction $\vec{T}_n$ and surface normal $\vec{n}$.

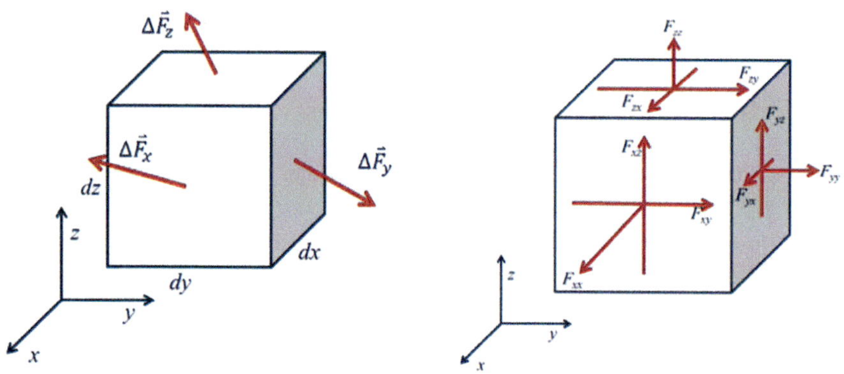

FIGURE 2.2 Infinitesimally small volume dV of sides dx, dy, and dz. Faces are subjected to $\Delta\vec{F}_x$, $\Delta\vec{F}_y$, and $\Delta\vec{F}_z$, which can be expressed in terms of their Cartesian components as given in Eq. (2.1).

The *body force*, $\vec{F}_B$ acts on all material points contained in the deformable solid (domain) and it is typically due to presence of an external field that affects the solid at-a-distance. For example, gravitational and magnetic fields result in body forces. The unit of body force is force/volume $[F/L^3]$.[1]

In contrast, the *surface tractions* $\vec{T}$ act across the surface (boundary) of the solid body. External tractions are typically due to direct contact of one body with another. Examples include the contact stresses and wind pressure. The unit of surface traction is force/area $[F/L^2]$. *Concentrated force* or *distributed force* (over a line) are special cases of surface tractions, with units $[F]$ and force/length $[F/L]$, respectively.

Internal forces develop in a deformable solid when it is subjected to external forces, and help keep the material intact. In general, the internal forces vary from point to point within the body. If we consider an infinitesimally small hexahedral volume $\Delta V (= dx.dy.dz)$ (Fig. 2.2) with the corresponding surface areas $\Delta A_x = dy\,dz$, $\Delta A_y = dx\,dz$, $\Delta A_z = dx\,dy$, we will see that each face of

1. F and L represent force and length, respectively. The $[\circ]$ notation is used to indicate the units of the variable.

this small segment will be subjected to a resultant internal force, $\Delta \vec{F}_x$, $\Delta \vec{F}_y$, and $\Delta \vec{F}_z$. Note that the subscripts x, y, and z in the force vector designations indicate the direction of the unit outward normal of the face on which the force is acting. The same convention is also used in designating the faces of the hexahedron.

The force vectors on each face can then be written with their Cartesian components,

$$\begin{aligned}\Delta \vec{F}_x &= \Delta F_{xx}\hat{i} + \Delta F_{xy}\hat{j} + \Delta F_{xz}\hat{k}\\ \Delta \vec{F}_y &= \Delta F_{yx}\hat{i} + \Delta F_{yy}\hat{j} + \Delta F_{yz}\hat{k}\\ \Delta \vec{F}_z &= \Delta F_{zx}\hat{i} + \Delta F_{zy}\hat{j} + \Delta F_{zz}\hat{k}\end{aligned} \tag{2.1}$$

where $\hat{i}$, $\hat{j}$ and $\hat{k}$ are the unit normal vectors along the x, y, and z axes, respectively. If we represent a generic force component ΔF_{ij} by using the subscripts i and j, we note that the first subscript ($i = x, y, z$) represents the vector $\Delta \vec{F}_i$, hence the surface on which the force is acting. The second subscript ($j = x, y, z$) represents the direction of the force component acting along one of the coordinate axes.

As we consider smaller volumes ΔV inside the material, the force to area ratio ($\Delta \vec{F}_i / \Delta A_i$) approaches a constant value even though the volume approaches zero. This ratio is known as *stress*. It is simpler to define stress in terms of its components acting along the three major axes, as the effects caused by these stress components are physically observable.

Normal stress represents the force intensity acting normal to ΔA_i ($i = x, y, z$) and it is defined as follows:

$$\sigma_{xx} = \lim_{\Delta A \to 0} \frac{\Delta F_{xx}}{\Delta A_x}, \quad \sigma_{yy} = \lim_{\Delta A \to 0} \frac{\Delta F_{yy}}{\Delta A_y}, \quad \text{and } \sigma_{zz} = \lim_{\Delta A \to 0} \frac{\Delta F_{zz}}{\Delta A_z} \tag{2.2}$$

When viewed one at a time, it can be seen that the normal stress causes elongation/compression along the axis that it is associated with.

Shear stress represents the force intensity in the plane of ΔA and it is defined as follows:

$$\begin{aligned}\sigma_{xy} &= \lim_{\Delta A \to 0} \frac{\Delta F_{xy}}{\Delta A_x}, \quad \sigma_{xz} = \lim_{\Delta A \to 0} \frac{\Delta F_{xz}}{\Delta A_x}\\ \sigma_{yx} &= \lim_{\Delta A \to 0} \frac{\Delta F_{yx}}{\Delta A_y}, \quad \sigma_{yz} = \lim_{\Delta A \to 0} \frac{\Delta F_{yz}}{\Delta A_y}\\ \sigma_{zx} &= \lim_{\Delta A \to 0} \frac{\Delta F_{zx}}{\Delta A_z}, \quad \sigma_{zy} = \lim_{\Delta A \to 0} \frac{\Delta F_{zy}}{\Delta A_z}\end{aligned} \tag{2.3}$$

Shear stresses cause distortion of the small volume segment, ΔV.

The stress components are often represented by using the index notation, σ_{ij} where the subscripts $i, j = x, y, z$. As mentioned above, the indices in the stress are not arbitrary: i, the first subscript, indicates the direction of the outward unit normal of the plane on which the stress component acts; and, j, the second subscript, indicates the direction of the stress component.

The symbol τ_{ij} is often used to indicate shear stress, and used interchangeably with the symbol σ_{ij}.

Stress at a point in the continuum is represented by nine independent stress components. In indicial notation, this is represented as follows:

$$\sigma_{ij} \text{ for } (i, j = x, y, z) \tag{2.4}$$

The stress components at a point can be represented in a *tensor* as follows:

$$[\sigma] = \begin{bmatrix} \sigma_{xx} & \tau_{xy} & \tau_{xz} \\ \tau_{yx} & \sigma_{yy} & \tau_{yz} \\ \tau_{zx} & \tau_{zy} & \sigma_{zz} \end{bmatrix} \tag{2.5}$$

It can be shown that the shear stress components have the following complimentary property, $\tau_{xy} = \tau_{yx}$, $\tau_{yz} = \tau_{zy}$, $\tau_{xz} = \tau_{zx}$. The *stress tensor* then becomes symmetrical.

In finite element formulation, the following vector representation will be used,

$$\{\sigma\} = \begin{Bmatrix} \sigma_{xx} & \sigma_{yy} & \sigma_{zz} & \tau_{xy} & \tau_{yz} & \tau_{zx} \end{Bmatrix}^T \tag{2.6}$$

2.2.2 Stress transformations

The state of stress of the point *P'* is expressed by the stress tensor in the global, Cartesian, coordinates (*x, y, z*) by Eq. (2.5). However, the choice of this coordinate system and the small volume ($dV = dx.dy.dz$) is arbitrary. We would like to be able to transform the stress state between different orientations.

In order to develop these transformations, we consider an oblique cross-section of the hexahedron by a plane of arbitrary, but known orientation. This results in the tetrahedral volume on one side of the plane as shown in Fig. 2.3. The traction components $\vec{T}_x$, $\vec{T}_y$, $\vec{T}_z$, and $\vec{T}_n$, acting on the small triangular areas ΔA_x, ΔA_y, ΔA_z and ΔA_n, respectively, must be in static equilibrium in order to keep the continuum whole. Let us look into the arbitrary plane in more detail before we state this equilibrium condition.

The orientation of the plane is identified by its *outward unit normal* $\vec{n}$, as shown in the figure. The unit normal is defined as follows:

$$\vec{n} = n_x\hat{i} + n_y\hat{j} + n_z\hat{k} \tag{2.7}$$

in Cartesian coordinate system. In vector notation, the unit normal is expressed as follows:

$$\{n\} = \begin{Bmatrix} n_x & n_y & n_z \end{Bmatrix}^T \tag{2.8}$$

The components n_x, n_y, and n_z are the direction cosines of $\vec{n}$ with respect to the (*x, y, z*) axes, and have the property,

$$n_x^2 + n_y^2 + n_z^2 = 1 \tag{2.9}$$

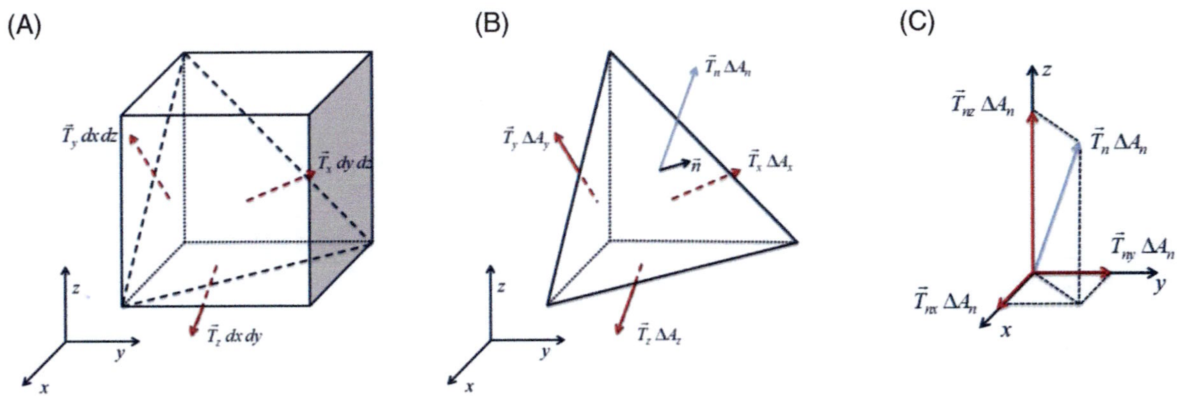

FIGURE 2.3 A tetrahedral volume formed by an oblique plane with unit normal $\vec{n}$ and the (x, y), (y, z), and (z, y) planes. The area of the oblique plane is ΔA_n, and the other three faces of the tetrahedron are ΔA_x, ΔA_y, and ΔA_z. Note that traction $\vec{T}_n$ acts on the plane $\vec{n}$, and tractions $\vec{T}_x$, $\vec{T}_y$ and $\vec{T}_z$ act on the other faces, as shown. These tractions are not necessarily normal to the surface on which they act. The $\vec{T}_n$and its components along the Cartesian coordinate axes can be represented as shown.

It can be shown that the *direction cosines* can be expressed as the ratios ΔA_i to ΔA_n as follows [7]:

$$n_i = \frac{\Delta A_i}{\Delta A_n} \text{ for } i = x, y, z \tag{2.10}$$

Let's consider traction $\vec{T}_n$ acting on the oblique plane in Fig. 2.3. Since the traction represents the intensity of force per unit area, it can be expressed by the vector representation as follows:

$$\vec{T}_n = T_{nx}\hat{\imath} + T_{ny}\hat{\jmath} + T_{nz}\hat{k} \tag{2.11}$$

Next, consider the static equilibrium of the forces acting on the pyramid, as follows:

$$\vec{T}_n \Delta A_n = \vec{T}_x \Delta A_x + \vec{T}_y \Delta A_y + \vec{T}_z \Delta A_z \tag{2.12}$$

where $\vec{T}_x$, $\vec{T}_y$ and $\vec{T}_z$ are the tractions acting on the other faces (ΔA_x, ΔA_y, and ΔA_z) of the tetrahedron. By using Eq. (2.10), this expression becomes,

$$\vec{T}_n = \vec{T}_x n_x + \vec{T}_y n_y + \vec{T}_z n_z \tag{2.13}$$

On each lateral face of the tetrahedron, we can express the tractions in terms of their components, by using Eqs. (2.1) and (2.2) as follows:

$$\begin{aligned} \vec{T}_x &= \sigma_{xx}\hat{\imath} + \tau_{xy}\hat{\jmath} + \tau_{xz}\hat{k} \\ \vec{T}_y &= \tau_{yx}\hat{\imath} + \sigma_{yy}\hat{\jmath} + \tau_{yz}\hat{k} \\ \vec{T}_z &= \tau_{zx}\hat{\imath} + \tau_{zy}\hat{\jmath} + \sigma_{zz}\hat{k} \end{aligned} \tag{2.14}$$

Combining this relationship with Eq. (2.13), we find that equilibrium dictates that,

$$\vec{T}_n = \left(\sigma_{xx}\hat{\imath} + \tau_{xy}\hat{\jmath} + \tau_{xz}\hat{k}\right)n_x + \left(\tau_{yx}\hat{\imath} + \sigma_{yy}\hat{\jmath} + \tau_{yz}\hat{k}\right)n_y + \left(\tau_{zx}\hat{\imath} + \tau_{zy}\hat{\jmath} + \sigma_{zz}\hat{k}\right)n_z \tag{2.15}$$

This relationship can be rearranged as follows:

$$\begin{aligned} \vec{T}_n = {} & \left(\sigma_{xx}n_x + \tau_{yx}n_y + \tau_{zx}n_z\right)\hat{\imath} + \left(\tau_{xy}n_x + \sigma_{yy}n_y + \tau_{zy}n_z\right)\hat{\jmath} \\ & + \left(\tau_{xz}n_x + \tau_{yz}n_y + \sigma_{zz}n_z\right)\hat{k} \end{aligned} \tag{2.16}$$

This equation can be represented in tensor and matrix forms as follows:

$$\vec{T}_n = \sigma \cdot \vec{n} \tag{2.17a}$$

$$\{T_n\} = [\sigma]^T \cdot \{n\} \tag{2.17b}$$

respectively. Note that for the case where the stress tensor is symmetric, we will have $[\sigma]^T = [\sigma]$ in Eq. (2.17a). By using Eqs. (2.11) and (2.16), we can find the components of $\vec{T}_n$ acting on the oblique plane as follows:

$$\begin{aligned} T_{nx} &= \left(\sigma_{xx}n_x + \tau_{yx}n_y + \tau_{zx}n_z\right) \\ T_{ny} &= \left(\tau_{xy}n_x + \sigma_{yy}n_y + \tau_{zy}n_z\right) \\ T_{nz} &= \left(\tau_{xz}n_x + \tau_{yz}n_y + \sigma_{zz}n_z\right) \end{aligned} \tag{2.18}$$

2.2.2.1 *Normal and shear components of tractions*

Traction $\vec{T}_n$ on an oblique plane $\vec{n}$ can be expressed by using the normal and shear components on the plane as follows:

$$\vec{T}_n = \sigma_{nn}\cdot\vec{n} + \sigma_{nt}\cdot\vec{t} \tag{2.19}$$

where the normal and tangential traction components are σ_{nn} and σ_{nt}, respectively. These components can be found as follows:

$$\sigma_{nn} = \vec{T}_n\cdot\vec{n} \quad \text{and} \quad \sigma_{nt} = \vec{T}_n\cdot\vec{t} \tag{2.20}$$

where $\vec{t}$ is a unit vector in the oblique the plane, which has the unit normal $\vec{n}$. The vector $\vec{t}$ has following components expressed in the global Cartesian system,

$$\vec{t} = t_x\hat{i} + t_y\hat{j} + t_z\hat{k} \quad \text{or} \quad \{t\} = \left\{t_x \quad t_y \quad t_z\right\}^T \tag{2.21}$$

The normal component of the traction $\vec{T}_n$ is found by using Eqs. (2.16) and (2.20) as follows:

$$\begin{aligned} \sigma_{nn} &= \vec{T}_n\cdot\vec{n} = \left([\sigma]^T\cdot\{n\}\right)\cdot\{n\} \\ &= \left[\left(\sigma_{xx}n_x + \tau_{yx}n_y + \tau_{zx}n_z\right)\hat{i} + \left(\tau_{xy}n_x + \sigma_{yy}n_y + \tau_{zy}n_z\right)\hat{j} \right. \\ &\quad \left. + \left(\tau_{xz}n_x + \tau_{yz}n_y + \sigma_{zz}n_z\right)\hat{k}\right]\cdot\left(n_x\hat{i} + n_y\hat{j} + n_z\hat{k}\right) \end{aligned}$$

or, after rearranging,

$$\begin{aligned} \sigma_{nn} = &\left(\sigma_{xx}n_x + \tau_{yx}n_y + \tau_{zx}n_z\right)n_x + \left(\tau_{xy}n_x + \sigma_{yy}n_y + \tau_{zy}n_z\right)n_y \\ &+ \left(\tau_{xz}n_x + \tau_{yz}n_y + \sigma_{zz}n_z\right)n_z \end{aligned} \tag{2.22}$$

Similarly, the tangential component of the traction $\vec{T}_n$ is found as follows:

$$\begin{aligned} \sigma_{nt} = &\left(\sigma_{xx}n_x + \tau_{yx}n_y + \tau_{zx}n_z\right)t_x + \left(\tau_{xy}n_x + \sigma_{yy}n_y + \tau_{zy}n_z\right)t_y \\ &+ \left(\tau_{xz}n_x + \tau_{yz}n_y + \sigma_{zz}n_z\right)t_z \end{aligned} \tag{2.23}$$

In fact, Eqs. (2.22) and (2.23) can be used to transform the stresses to any orientation $\vec{n}$ and $\vec{t}$.

2.2.2.2 *Principal stresses and directions*

Infinitely many planes can pass through a given point in a continuum, such as the plane that gives rise to the tetrahedron shown in Fig. 2.3. The normal and shear traction components σ_{nn} and σ_{nt} will vary according to the orientation of the cutting-plane. It is reasonable to assume that among these planes, there is one on which the shear tractions vanish, i.e., $\sigma_{nt} = 0$. This plane is called the *principal plane*, and its orientation is called the *principal orientation* $\vec{n}_p$. The normal component of the traction σ_{nn} acting on the principle plane has the magnitude λ and it is named the *principal stress*.

The traction acting on the principal plane can be expressed by a simple vector argument and by using Eq. (2.17a) as follows:

$$\vec{T}_{n_p} = \lambda \vec{n}_p \tag{2.24a}$$

$$\vec{T}_{n_p} = [\sigma]^T \{n_p\} \tag{2.24b}$$

By combining these two relationships, we find,

$$\begin{aligned}\lambda\left(n_{p_x}\hat{i} + n_{p_y}\hat{j} + n_{p_z}\hat{k}\right) = \big[&\left(\sigma_{xx}n_{p_x} + \tau_{yx}n_{p_y} + \tau_{zx}n_{p_z}\right)\hat{i} + \left(\tau_{xy}n_{p_x} + \sigma_{yy}n_{p_y} + \tau_{zy}n_{p_z}\right)\hat{j} \\ &+ \left(\tau_{xz}n_{p_x} + \tau_{yz}n_{p_y} + \sigma_{zz}n_{p_z}\right)\hat{k}\big]\end{aligned} \tag{2.25}$$

The following system of equations can be obtained for the unknown vector components of the normal vector of the principle plane from this relationship,

$$\begin{aligned}\left(\sigma_{xx}n_{p_x} + \tau_{yx}n_{p_y} + \tau_{zx}n_{p_z}\right) &= \lambda n_{p_x} \\ \left(\tau_{xy}n_{p_x} + \sigma_{yy}n_{p_y} + \tau_{zy}n_{p_z}\right) &= \lambda n_{p_y} \\ \left(\tau_{xz}n_{p_x} + \tau_{yz}n_{p_y} + \sigma_{zz}n_{p_z}\right) &= \lambda n_{p_z}\end{aligned} \tag{2.26}$$

This equation can be represented in matrix form as follows:

$$\begin{bmatrix} (\sigma_{xx} - \lambda) & \tau_{yx} & \tau_{zx} \\ \tau_{xy} & (\sigma_{yy} - \lambda) & \tau_{zy} \\ \tau_{xz} & \tau_{yz} & (\sigma_{zz} - \lambda) \end{bmatrix} \begin{Bmatrix} n_{p_x} \\ n_{p_y} \\ n_{p_z} \end{Bmatrix} = 0 \tag{2.27a}$$

or,

$$([\sigma] - \lambda[I])\{n_p\} = 0 \tag{2.27b}$$

where $[I]$ is the identity matrix. This relationship represents the equilibrium of the principal stress λ acting on the principal plane $\vec{n}_p$.

A nontrivial solution of Eq. (2.27a) is possible if the determinant of the term in the brackets is zero,

$$\det\left([\sigma]^T - \lambda[I]\right) = 0 \tag{2.28}$$

In general, this determinant can be expressed as the following polynomial equation in terms of the unknown principle stress magnitude,

$$\lambda^3 - I_1\lambda^2 - I_2\lambda - I_3 = 0 \tag{2.29}$$

This relationship is also known as the *characteristic equation*. The coefficients of this third order polynomial equation I_1, I_2, and I_3 are called the *stress invariants* and they represent the state of stress in the material as follows:

$$\begin{aligned}I_1 &= \sigma_{xx} + \sigma_{yy} + \sigma_{zz} \\ I_2 &= -\left(\sigma_{xx}\sigma_{yy} + \sigma_{yy}\sigma_{zz} + \sigma_{zz}\sigma_{xx}\right) + \tau_{xy}^2 + \tau_{yz}^2 + \tau_{zx}^2 \\ I_3 &= 2\tau_{xy}\tau_{yz}\tau_{zx} + \sigma_{xx}\sigma_{yy}\sigma_{zz} - \sigma_{xx}\tau_{yz}^2 - \sigma_{yy}\tau_{xz}^2 - \sigma_{zz}\tau_{xy}^2\end{aligned} \tag{2.30}$$

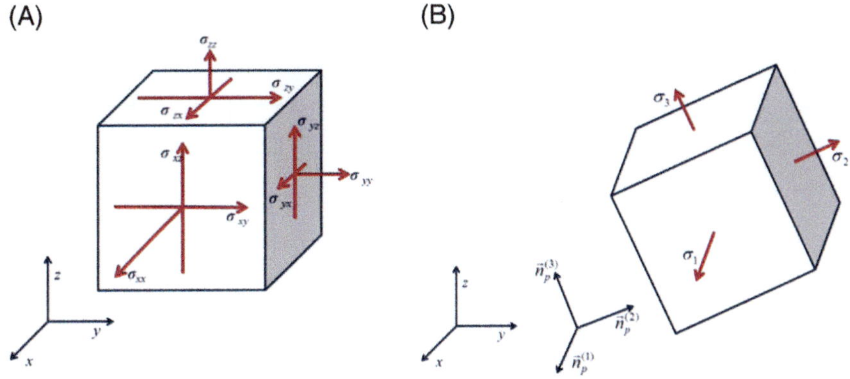

FIGURE 2.4 The stress state at point expressed in the Cartesian coordinate system and the corresponding principal stresses and directions.

The characteristic equation has three roots, for a given state of stress. λ_1, λ_2, and λ_3 correspond to three different planes $\vec{n}_p^{(1)}$, $\vec{n}_p^{(2)}$, and $\vec{n}_p^{(3)}$. The three principle planes are mutually orthogonal [7], and the state of principle stress can be represented by a hexahedron as shown in Fig. 2.4.

The *principle stresses* are assigned as $\sigma_1 = \lambda_1$, $\sigma_2 = \lambda_2$, and $\sigma_3 = \lambda_3$ and ordered from highest to lowest (algebraically) as follows:

$$\sigma_1 \geq \sigma_2 \geq \sigma_3 \tag{2.31}$$

The principal direction $\vec{n}_p^{(i)}$ for each principle stress σ_i is obtained by substituting σ_i into Eq. (2.27a) and by using the identity,

$$\left(n_{p_x}^{(i)}\right)^2 + \left(n_{p_y}^{(i)}\right)^2 + \left(n_{p_z}^{(i)}\right)^2 = 1 \tag{2.32}$$

2.2.3 Deformation and strain

When subjected to external forces, an internal material point located, for example, at position P before the loading, moves to point P' as depicted in two dimensions (a two-dimensional solid) in Fig. 2.6. The position of all material points in this solid domain are referred to a fixed Cartesian reference frame, and the position of point P' is found as follows (Fig. 2.6):

$$\vec{r}_{p'} = \vec{r}_p + \vec{u} \tag{2.33}$$

where $\vec{u}$ is the *deformation vector*. For a general deformation in three-dimensional space, the deformation vector is represented as follows:

$$\vec{u} = u_x\hat{i} + u_y\hat{j} + u_z\hat{k} \tag{2.34}$$

where u_x, u_y, and u_z are the projections of $\vec{u}$ onto the x, y, and z axes, respectively.

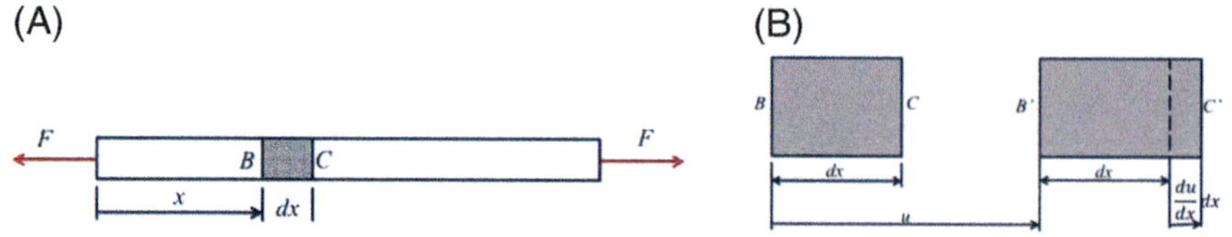

FIGURE 2.5 One-dimensional deformation of a bar.

2.2.3.1 *Uniaxial deformation*

Uniaxial deformation, also referred to as the *simple loading,* is the type of deformation that can be described with respect to a single orientation (e.g., x). Typical examples of this type of loading are the elongation/compression of a bar by a tensile/compressive force (Fig. 2.5).

In order to describe the uniaxial deformation in a slender bar, consider an infinitesimally small segment BC of length dx before deformation as shown in Fig. 2.19B. Once the external load is applied in the x direction, the bar deforms. Point B moves to B' where its position change u is described by the deformation vector as follows:

$$\vec{u} = u\,\hat{\imath} \tag{2.35}$$

Note that $u = u(x)$. As the point B moves to the position B', the point C moves to position C' and undergoes a position change of $u + (du/dx)dx$. Thus, the distance between points CB changes from dx to $dx + (du/dx)dx$.

Elongation/contraction along a given direction is measured by using the *normal strain* ε which is defined as the ratio of the change in length along a given direction to the original length as follows:

$$\varepsilon = \frac{1}{dx}\left[\left(1 + \frac{du}{dx}\right)dx - dx\right] = \frac{du}{dx} \tag{2.36}$$

2.2.3.2 *Planar deformation*

In some problems deformation is confined into a plane. A material point P in the undeformed material will move to point P' after deformation as shown in Fig. 2.6. This deformation can be represented by a *deformation vector* $\vec{u}$. In case of the planar deformation depicted in Fig. 2.6, the deformation vector can be expressed in a Cartesian coordinate system as follows:

$$\vec{u} = u_x\,\hat{\imath} + u_y\,\hat{\jmath} \tag{2.37}$$

where u_x and u_y are the projections of $\vec{u}$ on the x and y axes, respectively. In vector form, the deformation vector is represented as follows:

$$\{u\} = \left\{u_x \quad u_y\right\}^T \tag{2.38}$$

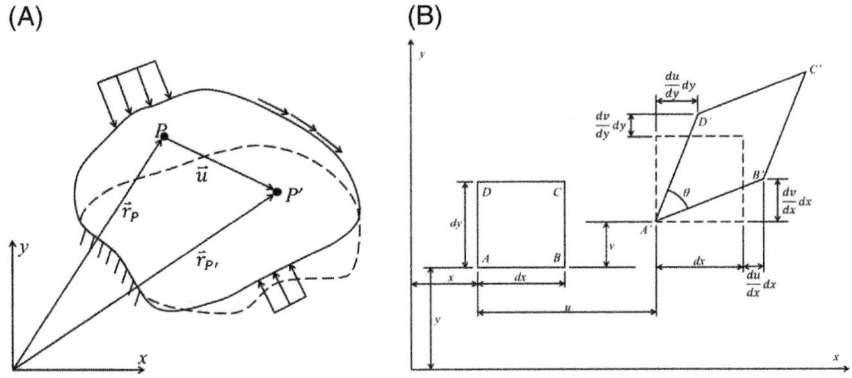

FIGURE 2.6 (A) As a result of the external forces and having a fixed boundary, an internal material point P of the deformable solid deforms to point P'. The deformation vector is $\vec{u}$ and the position vectors for points P and P' are $\vec{r}_P$ and $\vec{r}_{P'}$, respectively. (B) Undeformed (ABCD) and deformed (A'B'C'D') states of a small segment ($dx.dy.1$) and the displacement components.

Note that, in general, the deflection components u_x and u_y vary from point to point inside the deforming solid. Therefore, these variables are functions of the x and y coordinates, $u_x = u_x(x,y)$ and $u_y = u_y(x,y)$.

Deformation of a small rectangle around the point P with side lengths dx, dy and unit depth is shown Fig. 2.6. As a result of deformation, point A in the undeformed configuration moves to point A'. Similarly points B, C, and D move to B', C', and D'.

For the two-dimensional deformation depicted in Fig. 2.6, the original length of the side AB is dx. The length of the projection of the deformed line $A'B'$ on the x axis is $(dx + (\partial u_x/\partial x)\, dx$. Thus, by using Eq. (2.36), we can define the normal strain along the x direction as follows:

$$\varepsilon_{xx} = \frac{\partial u_x}{\partial x} \tag{2.39a}$$

It can be similarly shown that the normal strain along the y direction is defined as follows:

$$\varepsilon_{yy} = \frac{\partial u_y}{\partial y} \tag{2.39b}$$

Change in the relative orientation of two mutually perpendicular lines is defined as the *shear strain.* For the x and y directions, this is defined as $\gamma_{xy} = \pi/2 - \theta$. For small deformations, this can be approximated, and results in the following definition of the shear strain,

$$\gamma_{xy} = \frac{\partial u_x}{\partial y} + \frac{\partial u_y}{\partial x} \tag{2.40}$$

The strain state at a point in planar deformation can be expressed by the *strain tensor*,

$$[\varepsilon] = \begin{bmatrix} \varepsilon_{xx} & \gamma_{xy} \\ \gamma_{yx} & \varepsilon_{yy} \end{bmatrix} \tag{2.41}$$

In general, the strain tensor is symmetric, i.e., $\gamma_{xy} = \gamma_{yx}$. In developing the finite element method, it is notationally advantageous to express the strain tensor in vector form,

$$\{\varepsilon\} = \{\varepsilon_{xx} \quad \varepsilon_{yy} \quad \gamma_{xy}\}^T \tag{2.42}$$

Eqs. (2.39a) and (2.40) can also be represented in a derivative operator format as follows:

$$\{\varepsilon\} = \begin{bmatrix} \frac{\partial}{\partial x} & 0 \\ 0 & \frac{\partial}{\partial y} \\ \frac{\partial}{\partial y} & \frac{\partial}{\partial x} \end{bmatrix} \begin{Bmatrix} u_x \\ u_y \end{Bmatrix} \tag{2.43}$$

This format can be further simplified to the following form,

$$\{\varepsilon\} = [\partial]\{u\} \tag{2.44}$$

Use of this notation is helpful in developing the finite element method.

2.2.3.3 Three-dimensional deformation

In general, material points in a deformable body are displaced in the three-dimensional space when subjected to external effects. The deformation vector is expressed as follows:

$$\vec{u} = u_x\hat{i} + u_y\hat{j} + u_z\hat{k} \tag{2.45}$$

where u_x, u_y and u_z are the projections of $\vec{u}$ onto the x, y, and z axes, respectively. Note that in case of three-dimensional deformation, these variables are functions of all three coordinate directions, i.e. $u_x = u_x(x,y,z)$, $u_y = u_y(x,y,z)$ and $u_z = u_z(x,y,z)$. In vector notation, the deformation vector is given as follows:

$$\{u\} = \{u_x \quad u_y \quad u_z\}^T \tag{2.46}$$

It can be shown that, in three-dimensions, strain components are defined by the following relationships,

$$\begin{aligned} &\varepsilon_{xx} = \frac{\partial u_x}{\partial x}, \quad \varepsilon_{yy} = \frac{\partial u_y}{\partial y}, \quad \varepsilon_{zz} = \frac{\partial u_z}{\partial z} \\ &\gamma_{xy} = \frac{\partial u_x}{\partial y} + \frac{\partial u_y}{\partial x}, \quad \gamma_{yz} = \frac{\partial u_y}{\partial z} + \frac{\partial u_z}{\partial y}, \quad \gamma_{zx} = \frac{\partial u_z}{\partial x} + \frac{\partial u_x}{\partial z} \end{aligned} \tag{2.47}$$

In general, the strain state at a point can be represented by nine independent strain components. However, in most problems the shear strains display the following symmetry $\gamma_{xy} = \gamma_{yx}$, $\gamma_{yz} = \gamma_{zy}$, and $\gamma_{xz} = \gamma_{zx}$ [7]. Thus the strain tensor is given as follows:

$$[\varepsilon] = \begin{bmatrix} \varepsilon_{xx} & \gamma_{xy} & \gamma_{xz} \\ \gamma_{yx} & \varepsilon_{yy} & \gamma_{yz} \\ \gamma_{zx} & \gamma_{zy} & \varepsilon_{zz} \end{bmatrix} \tag{2.48}$$

Note that γ_{ij} given here is the engineering strain. The mathematical definition of shear strain has a factor of $\frac{1}{2}$ [7]. It is advantageous, in the finite element method to use the following vector representation of the strain components,

$$\{\varepsilon\} = \left\{\varepsilon_{xx} \quad \varepsilon_{yy} \quad \varepsilon_{zz} \quad \gamma_{xy} \quad \gamma_{yz} \quad \gamma_{zx}\right\}^T \tag{2.49}$$

Symmetry of the shear strain was assumed in this notation. Thus the strain state in three-dimensions is represented by six strain components as shown in Eq. (2.49). Similar to Eq. (2.43) in matrix operator format, Eq. (2.47) can be expressed as follows:

$$\begin{Bmatrix} \varepsilon_{xx} \\ \varepsilon_{yy} \\ \varepsilon_{zz} \\ \gamma_{xy} \\ \gamma_{yz} \\ \gamma_{zx} \end{Bmatrix} = \begin{bmatrix} \frac{\partial}{\partial x} & 0 & 0 \\ 0 & \frac{\partial}{\partial y} & 0 \\ 0 & 0 & \frac{\partial}{\partial z} \\ \frac{\partial}{\partial y} & \frac{\partial}{\partial x} & 0 \\ 0 & \frac{\partial}{\partial z} & \frac{\partial}{\partial y} \\ \frac{\partial}{\partial z} & 0 & \frac{\partial}{\partial x} \end{bmatrix} \begin{Bmatrix} u_x \\ u_y \\ u_z \end{Bmatrix} \tag{2.50}$$

Symbolically Eq. (2.50) is expressed as follows:

$$\{\varepsilon\} = [\partial]\{u\} \tag{2.51}$$

2.2.4 Strain compatibility conditions

An *elastic deformation* should not cause holes in a deformable body that does not have any holes before deformation. Moreover, no material overlap should be predicted by the displacement field. The strain compatibility conditions ensure that these constraints are satisfied [7].

In a planar deformation, where $u_x = u_x(x,y)$, $u_y = u_y(x,y)$ and $u_z = 0$, consider the following combination of the strains,

$$\frac{\partial^2 \varepsilon_{yy}}{\partial x^2} + \frac{\partial^2 \varepsilon_{xx}}{\partial y^2} - \frac{\partial^2 \gamma_{xy}}{\partial x \partial y} \tag{a}$$

Using the definitions given in Eq. (2.47) we will find,

$$\frac{\partial^3 u_y}{\partial x^2 \partial y} + \frac{\partial^3 u_x}{\partial y^2 \partial x} - \frac{\partial^2}{\partial x \partial y}\left(\frac{\partial u_y}{\partial x} + \frac{\partial u_x}{\partial y}\right) = 0 \tag{b}$$

Thus we see that the relationship (a) must be equal to zero. This is the strain compatibility equation for a two-dimensional deformation in the x, y plane, which imposes a specific relationship between the strains and the strain–displacement relationships.

For three-dimensional deformations where $u_x = u_x(x,y,z)$, $u_y = u_y(x,y,z)$ and $u_z = u_z(x,y,z)$ there are a total of six strain compatibility conditions. These can be found as follows:

$$\frac{\partial^2 \varepsilon_{yy}}{\partial x^2} + \frac{\partial^2 \varepsilon_{xx}}{\partial y^2} - \frac{\partial^2 \gamma_{xy}}{\partial x \partial y} = 0 \tag{2.52a}$$

$$\frac{\partial^2 \varepsilon_{yy}}{\partial z^2} + \frac{\partial^2 \varepsilon_{zz}}{\partial y^2} - \frac{\partial^2 \gamma_{yz}}{\partial z \partial y} = 0 \tag{2.52b}$$

$$\frac{\partial^2 \varepsilon_{zz}}{\partial x^2} + \frac{\partial^2 \varepsilon_{xx}}{\partial z^2} - \frac{\partial^2 \gamma_{zy}}{\partial x \partial z} = 0 \tag{2.52c}$$

$$2\frac{\partial^2 \varepsilon_{xx}}{\partial y \partial z} = \frac{\partial}{\partial x}\left(-\frac{\partial \gamma_{yz}}{\partial x} + \frac{\partial \gamma_{zx}}{\partial y} + \frac{\partial \gamma_{xy}}{\partial z}\right) \tag{2.52d}$$

$$2\frac{\partial^2 \varepsilon_{yy}}{\partial z \partial x} = \frac{\partial}{\partial y}\left(-\frac{\partial \gamma_{zx}}{\partial y} + \frac{\partial \gamma_{xy}}{\partial z} + \frac{\partial \gamma_{yz}}{\partial x}\right) \tag{2.52e}$$

$$2\frac{\partial^2 \varepsilon_{zz}}{\partial x \partial y} = \frac{\partial}{\partial z}\left(-\frac{\partial \gamma_{xy}}{\partial z} + \frac{\partial \gamma_{yz}}{\partial x} + \frac{\partial \gamma_{zx}}{\partial y}\right) \tag{2.52f}$$

In three-dimensional problems, all six of the strain compatibility conditions must be satisfied at each point. In two-dimensional problems, the relevant condition, i.e., only one of (Eq. 2.52a–c), must be satisfied at each point of the domain.

2.2.5 Generalized Hooke's law

As stated in the introduction to Section 2.2, when a deformable body is subjected to external effects such as external forces and/or imposed displacements on its boundary, its shape will change and internal forces will develop throughout its volume. The level of deformation for given external effects depends on the material of the deformable body. *Constitutive relations* are empirically obtained, material specific relationships between the stress and the strain in the body. Here we are primarily interested in linear elastic relationships.

2.2.5.1 Hooke's law

The deformation behavior of a specific material is determined experimentally. These experiments are designed such that only one of the stress components and the corresponding strain dominates the problem. This state is known as a *simple-loading* state.

For linear, isotropic materials tensile loading of a slender test specimen, i.e., the *simple-tension test*, reveals two fundamental material properties. The relationship between the normal stress and the normal strain is found by conducting a simple-tension test, as follows:

$$\sigma_{ii} = E\varepsilon_{ii} \quad \text{for} \quad i = x, y, z \tag{2.53}$$

where E is the *elastic modulus* of the material, also referred to as the *Young's modulus*. The relationship between the longitudinal strain ε_l and the transverse strain ε_t represents the *Poisson's ratio*, the second material property,

$$\nu = -\frac{\varepsilon_t}{\varepsilon_l} \tag{2.54}$$

The *simple-shear* test reveals the relationship between the shear strain and the shear stress,

$$\tau_{ij} = G\gamma_{ij} \quad \text{for} \quad i, j = x, y, z \quad \text{and} \quad i \neq j \tag{2.55}$$

where G is the *shear modulus*, or *modulus of rigidity*. For a linear, elastic, isotropic material the following relationship holds:

$$G = \frac{E}{2(1+\nu)} \tag{2.56}$$

2.2.5.2 Generalized Hooke's law

In previous sections it was indicated that, in general, the stress and strain tensors at a point have nine independent components each, if we do not take into account the symmetries. Therefore, the possibility exists for all of these 18 components to be interrelated. In it most general form, the linear elastic constitutive law, also known as *generalized Hooke's law*, can be expressed as follows:

$$\sigma_{ij} = c_{ijrs}\varepsilon_{rs} \tag{2.57}$$

where the subscripts $i, j, r, s = x, y, z$ and the coefficients c_{ijrs} are empirically determined. Note that the tensor notation is used in expressing Eq. (2.57) where σ and ε are second order tensors and c_{ijrs} is a fourth order tensor [7]. Repeated indices imply summation, such that for σ_{xx} the most general form of the Hooke's law would be,

$$\begin{aligned}\sigma_{xx} = {} & c_{xxxx}\varepsilon_{xx} + c_{xxxy}\gamma_{xy} + c_{xxxz}\gamma_{xz} + c_{xxyx}\gamma_{yx} + c_{xxyy}\varepsilon_{yy} + c_{xxyz}\gamma_{yz} + c_{xxzx}\gamma_{zx} \\ & + c_{xxzy}\gamma_{zy} + c_{xxzz}\varepsilon_{zz}\end{aligned} \tag{2.58}$$

It can easily be deduced that 81 material properties would be required in case of an anisotropic material with no-symmetries in the strain and stress tensors. In matrix notation, Eq. (2.57) can be expressed as follows:

$$\{\sigma\} = [E]\{\varepsilon\} \tag{2.59}$$

where $[E]$ is an 81×81 elasticity matrix and $\{\sigma\}$ and $\{\varepsilon\}$ are 9×1 vectors.

2.2.5.3 Generalized Hooke's law for anisotropic materials with symmetric stress and strain tensors

For an anisotropic material with symmetric stress and strain tensors (i.e. $\sigma_{ij} = \sigma_{ji}$ and $\varepsilon_{ij} = \varepsilon_{ji}$) the number of independent material constants is reduced from 81 to 36, as each one of $\{\sigma\}$ and $\{\varepsilon\}$ has 6 independent variables. The generalized Hooke's law in this case can be expressed in matrix notation as follows:

$$\begin{Bmatrix} \sigma_{xx} \\ \sigma_{yy} \\ \sigma_{zz} \\ \tau_{xy} \\ \tau_{yz} \\ \tau_{zx} \end{Bmatrix} = \begin{bmatrix} c_{11} & c_{12} & c_{13} c_{14} & c_{15} & c_{16} \\ c_{21} & c_{22} & c_{23} c_{24} & c_{25} & c_{26} \\ c_{31} & c_{32} & c_{33} c_{34} & c_{35} & c_{36} \\ c_{41} & c_{42} & c_{43} c_{44} & c_{45} & c_{46} \\ c_{51} & c_{52} & c_{53} c_{54} & c_{55} & c_{56} \\ c_{61} & c_{62} & c_{63} c_{64} & c_{65} & c_{66} \end{bmatrix} \begin{Bmatrix} \varepsilon_{xx} \\ \varepsilon_{yy} \\ \varepsilon_{zz} \\ \gamma_{xy} \\ \gamma_{yz} \\ \gamma_{zx} \end{Bmatrix} \tag{2.60}$$

The elasticity matrix $[E]$ has 36 independent constants that need to be determined experimentally.

2.2.5.4 Generalized Hooke's law for isotropic materials with symmetric stress and strain tensors

In case the material is elastically isotropic and the stress and strain tensors are symmetric the material behavior can be characterized with two material constants,

E: Elastic modulus or Young's modulus

v: Poisson's ratio

For a three-dimensional problem, it can be shown that the following relationships exist between the stresses and strains,

$$\varepsilon_{xx} = \frac{1}{E}\left[\sigma_{xx} - \nu\left(\sigma_{yy} + \sigma_{zz}\right)\right] \tag{2.61a}$$

$$\varepsilon_{yy} = \frac{1}{E}\left[\sigma_{yy} - \nu(\sigma_{zz} + \sigma_{xx})\right] \tag{2.61b}$$

$$\varepsilon_{zz} = \frac{1}{E}\left[\sigma_{zz} - \nu\left(\sigma_{xx} + \sigma_{yy}\right)\right] \tag{2.61c}$$

$$\tau_{xy} = G\gamma_{xy} \tag{2.61d}$$

$$\tau_{yz} = G\gamma_{yz} \tag{2.61e}$$

$$\tau_{zx} = G\gamma_{zx} \tag{2.61f}$$

where shear modulus $G = E/2(1 + \nu)$.

Note that Eq. (2.61a) can be inverted and expressed as follows:

$$\sigma_{xx} = \lambda(\varepsilon_{xx} + \varepsilon_{yy} + \varepsilon_{zz}) + 2\mu\varepsilon_{xx} \tag{2.62a}$$

$$\sigma_{yy} = \lambda(\varepsilon_{xx} + \varepsilon_{yy} + \varepsilon_{zz}) + 2\mu\varepsilon_{yy} \tag{2.62b}$$

$$\sigma_{zz} = \lambda(\varepsilon_{xx} + \varepsilon_{yy} + \varepsilon_{zz}) + 2\mu\varepsilon_{zz} \tag{2.62c}$$

$$\tau_{xy} = \mu\gamma_{xy} \tag{2.62d}$$

$$\tau_{yz} = \mu\gamma_{yz} \tag{2.62e}$$

$$\tau_{zx} = \mu\gamma_{zx} \tag{2.62f}$$

where, the *Lamé constants* are defined as follows:

$$\lambda = \frac{\nu E}{(1+\nu)(1-2\nu)} \tag{2.63a}$$

$$\mu = G \tag{2.63b}$$

For a linear elastic, isotropic material, it is sufficient to determine E and ν from uniaxial - loading tests. The elasticity matrix for an isotropic, elastic material can be found from Eq. (2.62a) as follows:

$$[E] = \begin{bmatrix} \lambda+2\mu & \lambda & \lambda & 0 & 0 & 0 \\ \lambda & \lambda+2\mu & \lambda & 0 & 0 & 0 \\ \lambda & \lambda & \lambda+2\mu & 0 & 0 & 0 \\ 0 & 0 & 0 & \mu & 0 & 0 \\ 0 & 0 & 0 & 0 & \mu & 0 \\ 0 & 0 & 0 & 0 & 0 & \mu \end{bmatrix} \tag{2.64}$$

2.2.5.5 Effects of initial stress/strain and thermal strain

Thermal stress in a one-dimensional problem: Consider a long and slender bar of length L and initial temperature $T^{(0)}$. If the temperature of the bar is changed by ΔT, material points in the bar would experience *thermal strain* proportional to the temperature change,

$$\varepsilon^{(th)} = \alpha\,\Delta T \tag{2.65}$$

the proportionality constant α is a material property known as the *coefficient of thermal expansion* with units of K^{-1} or $(°C)^{-1}$. If the bar is not constrained on its ends, its length will change by an amount,

$$\Delta L = \int_0^L \alpha\,\Delta T\,dx \tag{2.66}$$

but no internal stress will develop.

On the other hand if both ends of the bar are constrained, internal forces and hence stress will develop in the bar. If such constraint conditions exist, the *thermal stress* in the bar can be found from Hooke's law as follows:

$$\sigma^{(th)} = E\,\alpha\,\Delta T \tag{2.67}$$

Next, consider a constrained bar subjected to external forces and change of temperature. The *total strain* in this bar can be found by using the superposition of the mechanical component of the strain and the thermal strain,

$$\varepsilon = \frac{\sigma}{E} + \varepsilon^{(th)} = \frac{\sigma}{E} + \alpha\,\Delta T \tag{2.68}$$

The inverse of this relation gives the corresponding *total stress*,

$$\sigma = E(\varepsilon - \alpha\,\Delta T) \tag{2.69}$$

Generalized stress–strain relations with thermal effects: For materials with isotropic material properties temperature change only causes normal strain in the material. The stress–strain relations for a three-dimensional isotropic material subjected to a temperature change ΔT are expressed as follows [8]:

$$\varepsilon_{xx} - \alpha\,\Delta T = \frac{1}{E}\left[\sigma_{xx} - \nu\left(\sigma_{yy} + \sigma_{zz}\right)\right] \tag{2.70a}$$

$$\varepsilon_{yy} - \alpha\,\Delta T = \frac{1}{E}\left[\sigma_{yy} - \nu\left(\sigma_{zz} + \sigma_{xx}\right)\right] \tag{2.70b}$$

$$\varepsilon_{zz} - \alpha\,\Delta T = \frac{1}{E}\left[\sigma_{zz} - \nu\left(\sigma_{xx} + \sigma_{yy}\right)\right] \tag{2.70c}$$

$$\gamma_{xy} = \frac{\tau_{xy}}{G} \tag{2.70d}$$

$$\gamma_{yz} = \frac{\tau_{yz}}{G} \tag{2.70e}$$

$$\gamma_{zx} = \frac{\tau_{zx}}{G} \tag{2.70f}$$

This relationship can be inverted and expressed in the following form that is suitable for finite element implementation, [2],

$$\begin{Bmatrix} \sigma_{xx} \\ \sigma_{yy} \\ \sigma_{zz} \\ \tau_{xy} \\ \tau_{yz} \\ \tau_{zx} \end{Bmatrix} = \begin{bmatrix} \lambda + 2\mu & \lambda & \lambda & 0 & 0 & 0 \\ \lambda & \lambda + 2\mu & \lambda & 0 & 0 & 0 \\ \lambda & \lambda & \lambda + 2\mu & 0 & 0 & 0 \\ 0 & 0 & 0 & \mu & 0 & 0 \\ 0 & 0 & 0 & 0 & \mu & 0 \\ 0 & 0 & 0 & 0 & 0 & \mu \end{bmatrix} \begin{Bmatrix} \varepsilon_{xx} \\ \varepsilon_{yy} \\ \varepsilon_{zz} \\ \gamma_{xy} \\ \gamma_{yz} \\ \gamma_{zx} \end{Bmatrix} - \frac{E\alpha\Delta T}{1 - 2\nu} \begin{Bmatrix} 1 \\ 1 \\ 1 \\ 0 \\ 0 \\ 0 \end{Bmatrix} \tag{2.71}$$

General case of initial stress/strain: It is conceivable for a material to have an initial strain/stress state with all six strain/stress components resulting in,

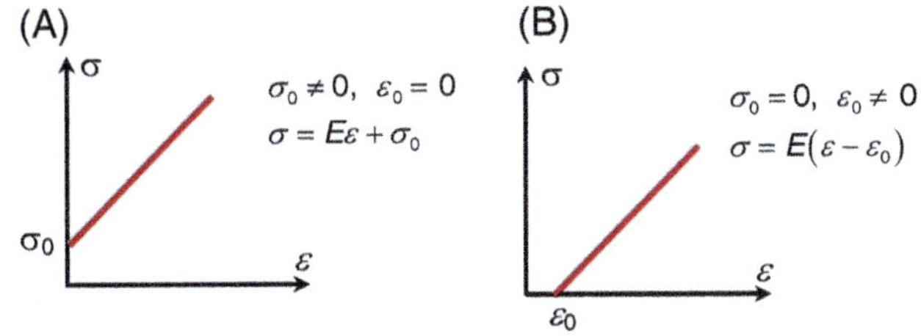

FIGURE 2.7 One-dimensional Hooke's law in case of (A) initial stress and (B) initial strain.

$$\left\{\varepsilon^{(0)}\right\} = \left\{\varepsilon_{xx}^{(0)} \quad \varepsilon_{yy}^{(0)} \quad \varepsilon_{zz}^{(0)} \quad \gamma_{xy}^{(0)} \quad \gamma_{yz}^{(0)} \quad \gamma_{zx}^{(0)}\right\}^T \tag{2.72a}$$

$$\left\{\sigma^{(0)}\right\} = \left\{\sigma_{xx}^{(0)} \quad \sigma_{yy}^{(0)} \quad \sigma_{zz}^{(0)} \quad \tau_{xy}^{(0)} \quad \tau_{yz}^{(0)} \quad \tau_{zx}^{(0)}\right\}^T \tag{2.72b}$$

where the superscript $^{(0)}$ indicates the initial state. In the case where the material has initial stress $\{\sigma^{(0)}\}$, before external loading, Hooke's law is modified to accommodate the initial stress as follows:

$$\{\sigma\} = [E]\{\varepsilon\} + \left\{\sigma^{(0)}\right\} \tag{2.73}$$

One-dimensional depiction of the initial stress is shown in Fig. 2.7A. In case the material has initial strain $\{\varepsilon^{(0)}\}$ prior to loading, the generalized Hooke's law is expressed as follows:

$$\{\sigma\} = [E]\left(\{\varepsilon\} - \left\{\varepsilon^{(0)}\right\}\right) \tag{2.74}$$

One-dimensional depiction of the effects of initial strain on the stress–strain relationship are plotted in Fig. 2.7B. Note that, in a three-dimensional problem, it is conceivable that the effects of initial stress and initial strain exist in different parts of the material, and both equations may need to be used albeit at different locations.

2.2.6 Two-dimensional problems

Most elasticity problems are three dimensional as dictated by their geometry. However, in some cases, approximation of a problem in two dimensions could be feasible, and have solutions that are reasonably close to the solution of the three-dimensional problem. This reduction in dimensionality of the problem also renders some relatively simple solutions feasible, in particular, for the *plane strain* and *plane stress* cases. Before discussing these approximations, we first introduce the generalized Hooke's law for a two-dimensional problem for an anisotropic material.

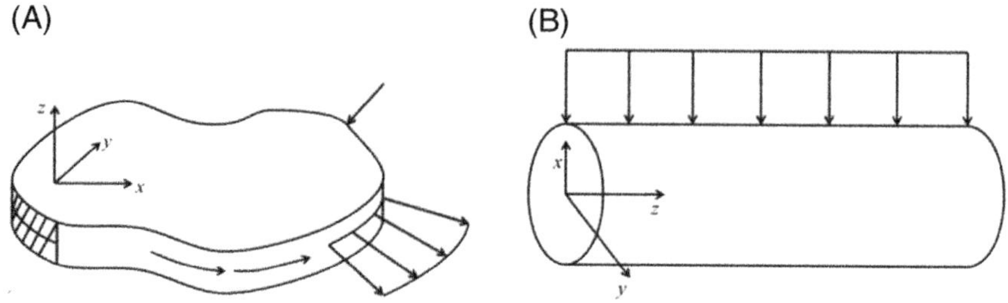

FIGURE 2.8 (A) Plane stress and (B) plane strain configurations. Problem is defined in the (x, y) plane.

2.2.6.1 Generalized Hooke's law for two-dimensional problems

In a 2D problem, the elastic material property matrix given in Eq. (2.60) becomes,

$$[E] = \begin{bmatrix} c_{11} & c_{12} & c_{13} \\ c_{21} & c_{22} & c_{23} \\ c_{31} & c_{32} & c_{33} \end{bmatrix} \tag{2.75}$$

where c_{ij} $(i, j = 1, 2, 3)$ are material properties obtained experimentally. For isotropic materials, where the material properties are independent of the material direction, c_{ij} are simplified.

2.2.6.2 Plane stress problem

If the thickness of the domain is small in comparison with the other two dimensions, and if the top and bottom surfaces are traction free (Fig. 2.8A), the stress components acting on the z-planes can be neglected,

$$\sigma_{zz} = 0,\ \tau_{zx} = 0,\ \tau_{zy} = 0 \tag{2.76}$$

for all x, y, z-locations. By using the generalized Hooke's law the above equation implies that the transverse shear strain components are zero,

$$\gamma_{zx} = 0,\ \gamma_{zy} = 0 \tag{2.77}$$

Moreover, by using Eq. (2.61a) with the condition $\sigma_{zz} = 0$ we find,

$$\varepsilon_{zz} = -\frac{\nu}{E}\left(\sigma_{xx} + \sigma_{yy}\right) \tag{2.78}$$

Thus by using Eqs. (2.61a), it can be shown that the plane stress problem is defined by the following constitutive relationship,

$$\begin{aligned} \varepsilon_{xx} &= \frac{1}{E}\left(\sigma_{xx} - \nu\sigma_{yy}\right) \\ \varepsilon_{yy} &= \frac{1}{E}\left(\sigma_{yy} - \nu\sigma_{xx}\right) \\ \gamma_{xy} &= \frac{\tau_{xy}}{G} \end{aligned} \tag{2.79}$$

The elasticity matrix for the plane stress problem can be given as follows:

$$[E] = \frac{E}{1-\nu^2}\begin{bmatrix} 1 & \nu & 0 \\ \nu & 1 & 0 \\ 0 & 0 & \frac{1}{2}(1-\nu) \end{bmatrix} \tag{2.80}$$

2.2.6.3 *Plane strain problem*

Consider a long cylindrical body whose axis is parallel to the z axis (Fig. 2.8B). If the external loading does not vary along the z-axis, such a geometry renders itself to be analyzed by using the assumption that strain variation is confined to (x, y) plane. The plane-strain conditions are described as follows:

i. $u_x = u_x(x,y)$ and $u_y = u_y(x,y)$
ii. $u_z = 0$ (no extension in the z-direction).

Considering the strain–displacement relations in Eq. (2.47) in light of these assumptions one finds,

$$\varepsilon_{zz} = 0, \ \gamma_{zx} = 0, \ \gamma_{zy} = 0. \tag{2.81}$$

An equivalent statement can be made in terms of stresses by using the constitutive relations given in Eq. (2.62a), as follows:

$$\tau_{zy} = 0, \tau_{zx} = 0 \tag{2.82}$$

Moreover, by using Eqs. (2.81) ($\varepsilon_{zz} = 0$) and (2.61) we get,

$$\sigma_{zz} = \nu\left(\sigma_{xx} + \sigma_{yy}\right) \tag{2.83}$$

Using this relationship, it can then be shown that the constitutive relations for a plane-strain problem are,

$$\begin{aligned} \varepsilon_{xx} &= \frac{1-\nu^2}{E}\sigma_{xx} - \frac{\nu(1+\nu)}{E}\sigma_{yy} \\ \varepsilon_{yy} &= \frac{1-\nu^2}{E}\sigma_{yy} - \frac{\nu(1+\nu)}{E}\sigma_{xx} \\ \gamma_{xy} &= \frac{\tau_{xy}}{G} \end{aligned} \tag{2.84}$$

The elasticity matrix can then be expressed by using the Lamé constants as follows:

$$[E] = \begin{bmatrix} (\lambda+2G) & \lambda & 0 \\ \lambda & (\lambda+2G) & 0 \\ 0 & 0 & G \end{bmatrix} \tag{2.85}$$

Plane-strain relations with thermal effects: Using the plane-strain conditions given in Eqs. (2.81)–(2.83) the stress–strain relationships with the thermal effects become [2],

$$\begin{Bmatrix} \sigma_{xx} \\ \sigma_{yy} \\ \tau_{xy} \end{Bmatrix} = \frac{E}{(1+\nu)(1-2\nu)} \begin{bmatrix} 1-\nu & \nu & 0 \\ \nu & 1-\nu & 0 \\ 0 & 0 & \frac{1}{2}(1-2\nu) \end{bmatrix} \begin{Bmatrix} \varepsilon_{xx} \\ \varepsilon_{yy} \\ \gamma_{xy} \end{Bmatrix} - \frac{E\alpha\Delta T}{1-2\nu} \begin{Bmatrix} 1 \\ 1 \\ 0 \end{Bmatrix} \tag{2.86a}$$

$$\sigma_{zz} = \frac{E\nu}{(1+\nu)(1-2\nu)}\left(\varepsilon_{xx}+\varepsilon_{yy}\right) + \frac{E\alpha\Delta T}{1-2\nu} \tag{2.86b}$$

$$\tau_{zy} = \tau_{xz} = 0 \tag{2.86c}$$

Plane stress relations with thermal effects: Using the plane stress conditions given in Eqs. (2.76)–(2.78), the stress–strain relationships with the thermal effects become [2],

$$\begin{Bmatrix} \sigma_{xx} \\ \sigma_{yy} \\ \tau_{xy} \end{Bmatrix} = \frac{E}{\left(1-\nu^2\right)} \begin{bmatrix} 1 & \nu & 0 \\ \nu & 1 & 0 \\ 0 & 0 & \frac{1}{2}(1-\nu) \end{bmatrix} \begin{Bmatrix} \varepsilon_{xx} \\ \varepsilon_{yy} \\ \gamma_{xy} \end{Bmatrix} - \frac{E\alpha\Delta T}{1-\nu} \begin{Bmatrix} 1 \\ 1 \\ 0 \end{Bmatrix} \tag{2.87a}$$

$$\varepsilon_{zz} = -\frac{\nu}{E}\left(\sigma_{xx}+\sigma_{yy}\right) + \alpha\Delta T \tag{2.87b}$$

$$\tau_{zy} = \tau_{xz} = 0 \tag{2.87c}$$

2.2.7 Balance laws

2.2.7.1 Balance law for the one-dimensional problem

The equation of motion for a one-dimensional problem depends on the geometry that is under consideration. If the geometry is a one-dimensional slab, extending to infinity in two directions (e.g., y and z) there is no possibility of a surface force influence. In this case, equilibrium of an infinitesimally small section of the body will result in the following relationship,

$$\sigma_{xx}A - \left(\sigma_{xx} + \frac{\partial \sigma_{xx}}{\partial x}dx\right)A + F_{Bx}A\,dx = \rho A\frac{\partial^2 u_x}{\partial t^2}dx \tag{2.88}$$

where F_{Bx} is body force vector, ρ is the mass density per unit volume, A is the area of the slab in the yz-plane and $\partial^2 u_x/\partial t^2$ is the local acceleration of the material points. The equation of motion of an infinitely wide one-dimensional medium is then found as follows:

$$\frac{\partial \sigma_{xx}}{\partial x} + F_{Bx} = \rho\frac{\partial^2 u_x}{\partial t^2} \tag{2.89}$$

More often, the one-dimensional geometry is a slender member with cross-sectional area A. In this case, we need to consider the possibility of surface traction acting along the peripheral surface (Fig. 2.9). Applying Newton's second

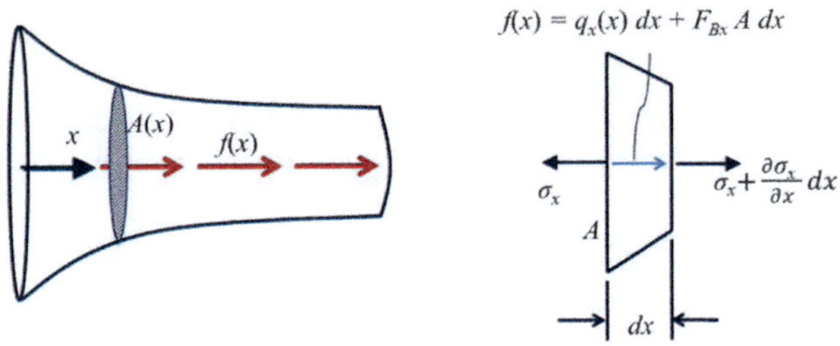

FIGURE 2.9 One-dimensional bar with cross-sectional area $A(x)$ subjected to surface traction q_x and body force F_{Bx} and the free body diagram of a small segment in the bar.

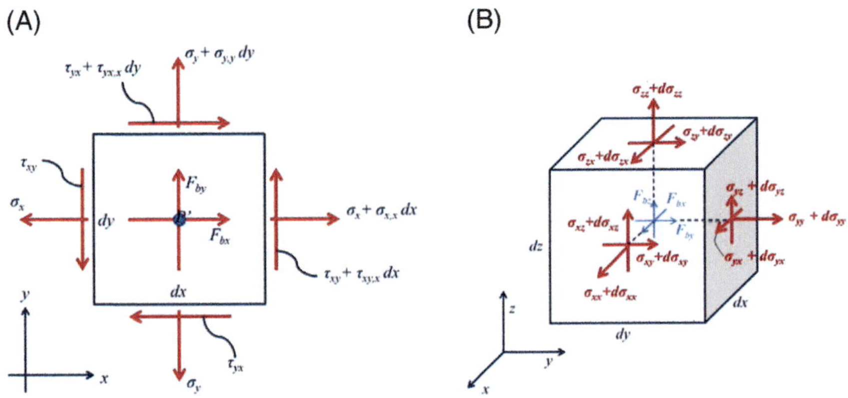

FIGURE 2.10 Free body diagram around an infinitesimally small segment of a continuum for (A) two two-dimensional and (B) three-dimensional analysis.

law on an infinitesimally small section of the body results in the following relationship,

$$\sigma_{xx}A - \left(\sigma_{xx}A + \frac{\partial \sigma_{xx}A}{\partial x}dx\right) + q_x dx + F_{Bx}A\,dx = \rho A\frac{\partial^2 u_x}{\partial t^2}dx \tag{2.90}$$

where q_x is the force per unit length acting on the outer periphery of the body along the x-direction. The equation of motion of slender, one-dimensional body is then found as follows:

$$\frac{\partial \sigma_{xx}A}{\partial x} + q_x + F_{Bx}A = \rho A\frac{\partial^2 u_x}{\partial t^2} \tag{2.91}$$

2.2.7.2 Balance laws for two-dimensional problems

Consider the internal stresses around point P' of a two-dimensional deformable body. An infinitesimally small rectangular segment of this body of size $(dx.dy)$ with unit depth, is shown in Fig. 2.10A. By considering the Newton's second law and adding up the forces acting along the x-axis we find,

$$\sigma_{xx}dy - \left(\sigma_{xx} + \frac{\partial \sigma_{xx}}{\partial x}dx\right)dy - \tau_{yx}dx + \left(\tau_{xy} + \frac{\partial \tau_{xy}}{\partial y}dy\right)dx + F_{Bx}dx\,dy = \rho\,dx\,dy\frac{\partial^2 u_x}{\partial t^2} \tag{2.92}$$

where F_{Bx} is the projection of the body force vector onto the x-axis. Note that the body force is defined as force per unit volume. Dividing this equation by $(dx.dy)$ the equation of motion in the x-direction is found as follows:

$$\frac{\partial \sigma_{xx}}{\partial x} + \frac{\partial \tau_{yz}}{\partial y} + F_{Bx} = \rho\frac{\partial^2 u_x}{\partial t^2} \tag{2.93a}$$

Similarly, the equation of motion in the y-direction can be stated as follows:

$$\frac{\partial \tau_{xy}}{\partial x} + \frac{\partial \sigma_{yy}}{\partial y} + F_{By} = \rho\frac{\partial^2 u_y}{\partial t^2} \tag{2.93b}$$

where F_{By} is the projection of the body force vector onto the y-axis.

2.2.7.3 Balance laws for three-dimensional problems

Fig. 2.10B shows the stresses acting on the sides of an infinitesimally small segment $(dx\,dy\,dz)$ of a three-dimensional deformable body under. The details of the derivation for the three-dimensional equations of motion are similar to the two-dimensional derivation just presented. Therefore, we directly give the equations of motion for a three-dimensional solid as follows:

$$\frac{\partial \sigma_{xx}}{\partial x} + \frac{\partial \tau_{yx}}{\partial y} + \frac{\partial \tau_{zx}}{\partial z} + F_{Bx} = \rho\frac{\partial^2 u_x}{\partial t^2} \tag{2.94a}$$

$$\frac{\partial \tau_{xy}}{\partial x} + \frac{\partial \sigma_{yy}}{\partial y} + \frac{\partial \tau_{zy}}{\partial z} + F_{By} = \rho\frac{\partial^2 u_y}{\partial t^2} \tag{2.94b}$$

$$\frac{\partial \tau_{xz}}{\partial x} + \frac{\partial \tau_{yz}}{\partial y} + \frac{\partial \sigma_{zz}}{\partial z} + F_{Bz} = \rho\frac{\partial^2 u_z}{\partial t^2} \tag{2.94c}$$

2.2.8 Boundary conditions

The geometric positions of all material points of the deformable solid are described as a domain Ω over which the deformation, strain and stress are sought. The solution domain is connected to the outside through its boundaries, Γ. The *external tractions* or the *imposed displacements* acting on the boundary are known as the *boundary conditions*; along with body forces, they are ultimately responsible for the deformation of the domain.

2.2.8.1 Boundary

The solution domain is bounded by one or multiple *boundaries*. Points on the boundary are identified by their unit outward normal $\vec{n}^{(b)}$.

2.2.8.2 Displacement and traction boundaries

At a given position on the boundary either the displacements or the external tractions can be specified, but not both at the same time. This is explained in detail in Chapter 3. The boundary Γ of the solid Ω can be thought of as the union of a subsection of the boundary where displacements are defined Γ_E and another subsection where the tractions are prescribed Γ_N[2]. These can also be called the *displacement* (Γ_E) and *traction boundaries* (Γ_N). These boundaries have the following properties:

$$i)\,\Gamma = \Gamma_E \cup \Gamma_N \text{ and } ii)\,\Gamma_E \cap \Gamma_N = 0 \tag{2.95}$$

The state of each material point on the boundary should be identified before a solution to the governing equations can be sought. Identification of the exact boundary conditions can sometimes be tricky. Nevertheless, the following conditions should be defined on the boundary.

2.2.8.3 Boundary conditions

On a displacement boundary the displacement vector $\vec{u}^{(b)}$is prescribed,

$$\vec{u} = \vec{u}^{(b)} \text{ on } \Gamma_E \tag{2.96}$$

Note that if the deformation on Γ_E is restricted, i.e. $\vec{u}^{(b)} = 0$ on Γ_E, the boundary is said to be *fixed*. Otherwise this vector represents the displacements imposed on the boundary.

On a traction boundary the stress on the body should be equal to the external traction $\vec{T}_n^{(b)}$,

$$[\sigma]\{n^{(b)}\} = \{T_n^{(b)}\} \text{ on } \Gamma_N \tag{2.97}$$

In case the boundary is free of tractions, the *free-traction boundary condition* is $\vec{T}_n^{(b)} = 0$. Otherwise, it is natural to specify the traction relative to the local unit normal and unit tangent vectors, i.e.,

$$\vec{T}_n^{(b)} = \sigma_{nn}^{(b)}\vec{n}^{(b)} + \sigma_{nt}^{(b)}\vec{t}^{(b)} \tag{2.98}$$

where the normal and tangential traction components are $\sigma_{nn}^{(b)}$ and $\sigma_{nt}^{(b)}$, are given by Eqs. (2.22) and (2.23). Note that we will use the symbols q_n and q_t for $\sigma_{nn}^{(b)}$ and$\sigma_{nt}^{(b)}$, respectively to indicate tractions imposed on the boundary. For a two-dimensional problem Eq. (2.97) becomes,

$$q_n = (\sigma_{xx}n_x + \tau_{yx}n_y)n_x + (\tau_{xy}n_x + \sigma_{yy}n_y)n_y \tag{2.99a}$$

$$q_t = (\sigma_{xx}n_x + \tau_{yx}n_y)t_x + (\tau_{xy}n_x + \sigma_{yy}n_y)t_y \tag{2.99b}$$

The boundary Γ of a domain in two dimensions is generally represented by the vector $\vec{r}(s)$, where s is the coordinate running along the curve (i.e. $\Gamma = \Gamma(s)$) as shown in Fig. 2.11. In general, it is helpful to express this vector with respect

2. The subscripts $_E$ and $_N$ refer to the essential and natural boundary conditions, respectively.

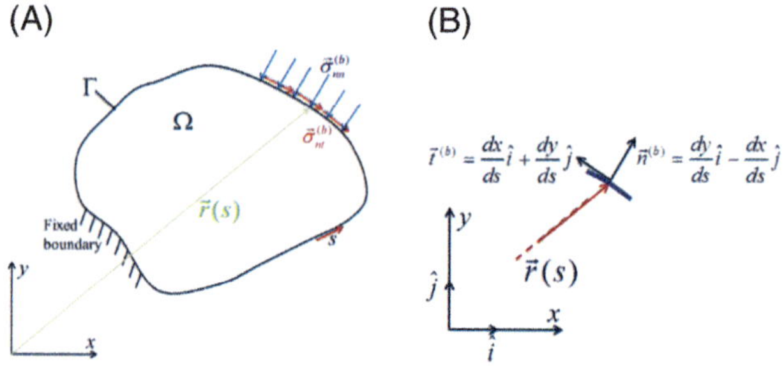

FIGURE 2.11 (A) Position vector $\vec{r}(s)$ of the boundary Γ. (B) The normal $\vec{n}$ and tangent $\vec{t}$ vectors of the boundary.

to a global Cartesian coordinate system as follows:

$$\vec{r}(s) = x(s)\vec{i} + y(s)\vec{j} \tag{2.100}$$

The tangential and normal unit vectors along the boundary are defined as follows [9] (Fig. 2.12),

$$\vec{t}^{(b)}(s) = \frac{d\vec{r}}{ds} \text{ and } \vec{n}^{(b)}(s) = \frac{1}{\left|d\vec{t}^{(b)}/ds\right|}\frac{d\vec{t}^{(b)}}{ds} \tag{2.101}$$

It can then be shown that the tangent and outward normal vectors become [7],

$$\vec{t}^{(b)} = \frac{dx}{ds}\hat{i} + \frac{dy}{ds}\hat{j} \text{ and } \vec{n}^{(b)} = \frac{dy}{ds}\hat{i} - \frac{dx}{ds}\hat{j} \tag{2.102}$$

We can then express the traction vector that is defined with respect to the local $\left(\vec{n}^{(b)}, \vec{t}^{(b)}\right)$ system in the global coordinate system as follows:

$$q_n = \left(\sigma_{xx}\frac{dy}{ds} - \tau_{yx}\frac{dx}{ds}\right)\frac{dy}{ds} - \left(\tau_{xy}\frac{dy}{ds} - \sigma_{yy}\frac{dx}{ds}\right)\frac{dx}{ds} \tag{2.103a}$$

$$q_t = \left(\sigma_{xx}\frac{dy}{ds} - \tau_{yx}\frac{dx}{ds}\right)\frac{dx}{ds} + \left(\tau_{xy}\frac{dy}{ds} - \sigma_{yy}\frac{dx}{ds}\right)\frac{dy}{ds} \tag{2.103b}$$

Example 2.1 Traction boundary conditions acting on a rectangular domain

Consider a rectangular domain subjected to the following external traction components:

$$\text{Edge}-1: x=a, \quad 0 \le y \le b: \quad \frac{dy}{ds}=1, \quad \frac{dx}{ds}=0: \quad \vec{n}^{(1)}=\hat{i}-0\hat{j}, \quad \vec{t}^{(1)}=0\hat{i}+\hat{j}$$

$$\text{Edge}-2: a \le x \le 0, \quad y=b, : \quad \frac{dy}{ds}=0, \quad \frac{dx}{ds}=-1: \quad \vec{n}^{(2)}=0\hat{i}+\hat{j}, \quad \vec{t}=-\hat{i}+0\hat{j}$$

$$\text{Edge}-3: x=0, \quad b \le y \le 0: \quad \frac{dy}{ds}=-1, \quad \frac{dx}{ds}=0: \quad \vec{n}^{(3)}=-\hat{i}+0\hat{j}, \quad \vec{t}^{(3)}=0\hat{i}-\hat{j}$$

$$\text{Edge}-4: 0 \le x \le a, \quad y=0, : \quad \frac{dy}{ds}=0, \quad \frac{dx}{ds}=1: \quad \vec{n}^{(4)}=0\hat{i}-\hat{j}, \quad \vec{t}^{(4)}=\hat{i}+0\hat{j}$$

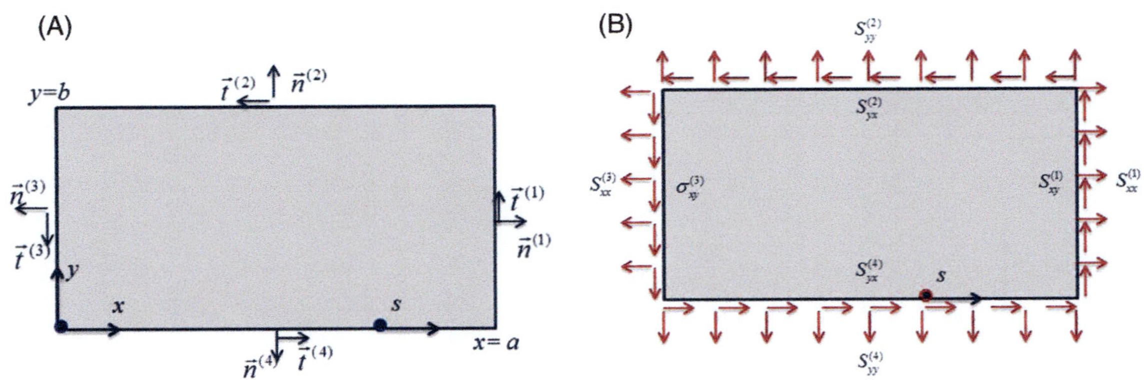

FIGURE 2.12 (A) The unit normal and tangent vectors on the boundary of a rectangular domain. (B) The boundary tractions for Example 2.1.

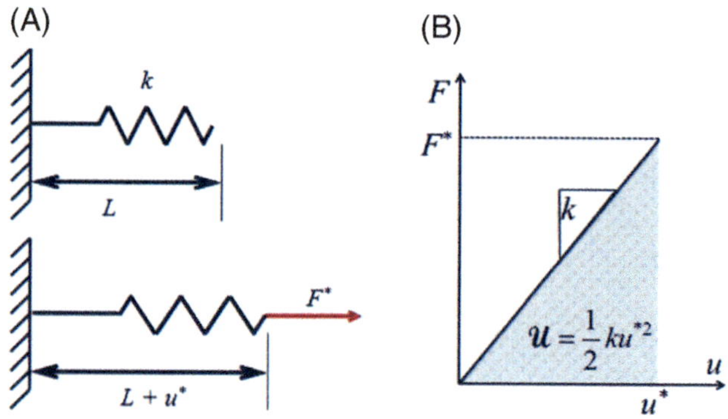

FIGURE 2.13 (A) Unstretched and stretched configurations of a linear spring. (B,C) The load-displacement curve of the spring with the strain energy $\mathcal{U}$.

The traction boundary conditions can be found by using Eq. (2.103a) as follows:

$$\text{Edge} - 1: \sigma_{xx} = S_{nn}^{(1)}, \quad \tau_{xy} = S_{nt}^{(1)} \tag{2.104a}$$

$$\text{Edge} - 2: \sigma_{yy} = S_{nn}^{(2)}, \quad \tau_{xy} = -S_{nt}^{(2)} \tag{2.104b}$$

$$\text{Edge} - 3: \sigma_{xx} = S_{nn}^{(3)}, \quad \tau_{xy} = S_{nt}^{(3)} \tag{2.104c}$$

$$\text{Edge} - 4: \sigma_{yy} = S_{nn}^{(4)}, \quad \tau_{xy} = -S_{nt}^{(4)} \tag{2.104d}$$

2.2.9 Total potential energy of conservative systems

A system is conservative if the work done by internal and external forces are both independent of the path taken between initial and final states. Here we consider conservative systems. Under this assumption, the *total potential energy* π_p of a deformable body subjected to external forces is given by the following relationship,

$$\pi_p = U + \Omega \tag{2.105}$$

where U is the strain energy of elastic distortion and Ω is the potential possessed by applied loads (by virtue of having the capacity to do work if displaced through a distance).

2.2.9.1 *Total potential energy of a linear spring*

In order to motivate the concept of strain energy let's recall the familiar linear spring subjected to an axial load. The force-displacement relationship (Fig. 2.13) for a linear spring is given by the familiar relationship,

$$F = ku \tag{2.106}$$

where k is the spring constant, and u is the end displacement of spring.

Let us assume that for an external force of magnitude F^* the string stretches by an amount u^*. Energy stored in the spring in the stretched configuration is found by computing the area under the force-displacement curve, and for a linear spring it is given as follows:

$$\mathcal{U} = \frac{1}{2}ku^{*2} \tag{2.107}$$

The factor 1/2 takes into account that the force F increases linearly and slowly during deformation.

The work W done by the force to stretch the spring by this amount is $W = F^*u^*$. By doing this work, the load loses the same amount in potential. The potential of the applied load is expressed as follows:

$$\Omega = -W = -F^*u^* \tag{2.108}$$

Note that here we assume that the load is regarded as acting at its full value, F^*. The total potential energy of the system then becomes,

$$\pi_p = \frac{1}{2}ku^{*2} - F^*u^* \tag{2.109}$$

2.2.9.2 *Total potential energy of an elastic body*

We will next consider a deformable continuum, and the energy stored in it when it is loaded by external forces. In an elastic material, *work done by internal forces is equal in magnitude to the change in strain energy.*

Strain energy for one-dimensional deformation

Let's consider a bar under uniaxial tension (Fig. 2.14A). This bar is in a state of pure tensile stress. If we take an infinitesimally small volume from this bar as shown in Fig. 2.14B, we will see that it experiences a constant stress σ and constant strain ε, along the x-axis.

The internal force in this state can be found as follows $dF = \sigma(dy\, dz)$ and the corresponding displacement from the relationship $\varepsilon = \Delta(dx)/dx$ as $d(dx) = \varepsilon\, dx$. If the material behaves in a linear elastic manner we can assume that the internal work is given by the familiar relationship:

$$\frac{1}{2}dF\, d(dx) = \frac{1}{2}\sigma\varepsilon(dx\, dy\, dz) = \frac{1}{2}\sigma\varepsilon dV \tag{2.110}$$

where $dV = (dx\, dy\, dz)$ represents the small volume of the element. Shortly, we will drop the assumption of linear elastic material, and generalize this relationship. For now, let's state that the incremental *internal work* is stored in the material as incremental *strain energy*$d\mathcal{U}$,

$$d\mathcal{U} = \frac{1}{2}\sigma\varepsilon dV \tag{2.111}$$

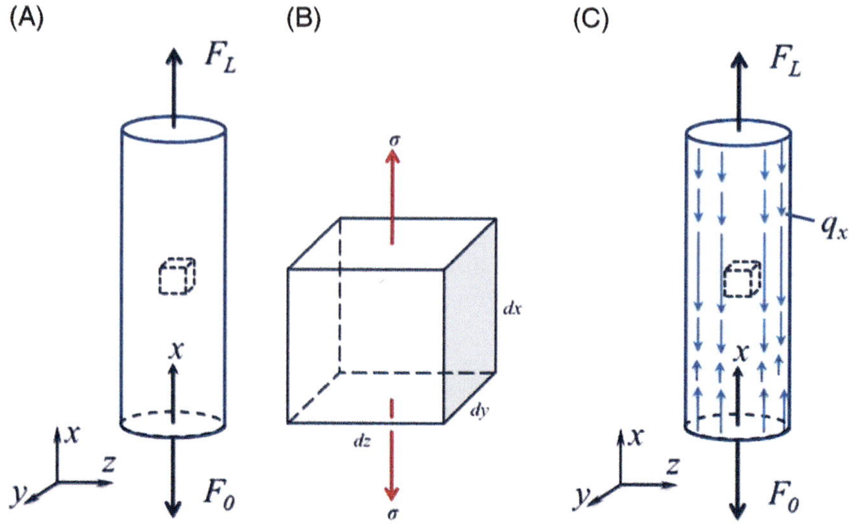

FIGURE 2.14 (A) Uniaxial loading state and (B) an infinitesimally small volume from this material. (C) Uniaxial loading state with surface force q_x.

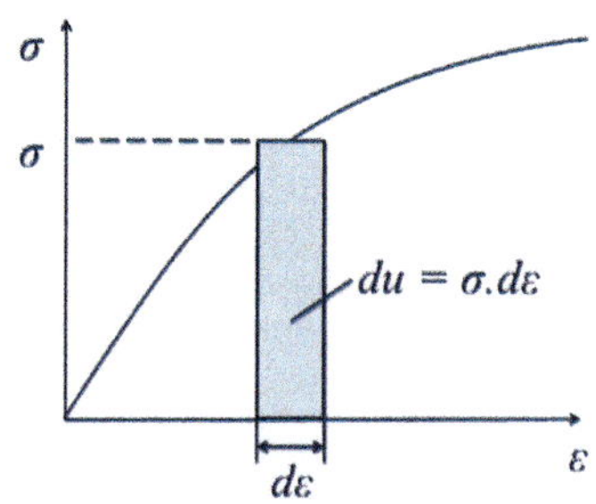

FIGURE 2.15 The increment of the strain energy density, du is computed for a given stress σ level assuming that strain is incremented by $d\varepsilon$.

The ratio of the strain energy to the volume of the material is the *strain energy density u*,

$$u = \frac{d\mathcal{U}}{dV} = \frac{1}{2}\sigma\varepsilon \tag{2.112}$$

Let's next consider a material with nonlinear stress–strain relationship as shown in the Fig. 2.15. The increment of strain energy density du for a constant stress can be approximated as the area under the stress–strain curve for the strain increment $d\varepsilon$. This is depicted in Fig. 2.15 and computed as follows:

$$du = \sigma d\varepsilon \tag{2.113}$$

The incremental strain energy $d\mathcal{U}$ in the bar, presented in Fig. 2.14, can be found as follows:

$$d\mathcal{U} = \sigma \, d\varepsilon \, dV$$

For a linear elastic material the strain energy is found by using the Hooke's law as follows:

$$\mathcal{U} = \frac{1}{2}\int_0^L \sigma \, \varepsilon \, A dx = \frac{1}{2}\int_0^L E\left(\frac{du}{dx}\right)^2 A dx \tag{2.114}$$

Work by external forces

The bar shown in Fig. 2.14C is subjected to two types of external forces: a distributed force $q_x(x)$ per unit length, acting along its outer surface and two concentrated forces F_0 and F_L acting at its ends $x = 0$ and L. The work due to these external forces can be found by considering the *local deformation* of the bar, and it is given as follows:

$$W = \int_0^L q_x \, u \, dx + F_0 u(0) + F_L u(L) \tag{2.115}$$

Total potential energy of an elastic bar

The total potential energy $\pi_p = \mathcal{U} - W(= \mathcal{U} + \Omega)$ of a bar subjected to two end forces and distributed surface force q_x as shown in Fig. 2.14C can be found as follows:

$$\pi_p = \frac{1}{2}\int_0^L E\left(\frac{du}{dx}\right)^2 A dx - \int_0^L q_x \, u \, dx - F_0 u(0) - F_L u(L) \tag{2.116}$$

where the first term represents the total strain energy in the system, the second term represents the external work done on the system due to distributed force q_x, and the last two terms represent the external work done on the system by the two forces F_0 and F_L that are applied at the ends of the bar.

Strain energy for three-dimensional deformation

The strain energy density expression given in Eq. (2.113) can be generalized for a nonlinearly elastic, three-dimensional do domain as follows:

$$du = \sigma_{xx} d\varepsilon_{xx} + \sigma_{yy} d\varepsilon_{yy} + \sigma_{zz} d\varepsilon_{zz} + \tau_{xy} d\gamma_{xy} + \tau_{yz} d\gamma_{yz} + \tau_{zx} d\gamma_{zx} \tag{2.117}$$

Using the vector representation of stress $\{\sigma\}$ and strain $\{\varepsilon\}$ the strain energy density is expressed as follows:

$$du = \{\sigma\}^T \{d\varepsilon\} \tag{2.118}$$

Recall that the generalized Hooke's law, $\{\sigma\} = [E]\{\varepsilon\}$, given in Eq. (2.59) expresses the stress–strain relationship for a linear, elastic material. The transpose of the stress vector can be expressed by using the following expression from matrix algebra,

$$([E]\{\varepsilon\})^T = \{\varepsilon\}^T[E]^T \tag{2.119}$$

Thus, using this in Eq. (2.118), the increment of strain energy density is found as follows:

$$du = \{\sigma\}^T\{d\varepsilon\} = \{\varepsilon\}^T[E]^T\{d\varepsilon\} \tag{2.120}$$

Integrating this expression with respect to $\{d\varepsilon\}$, we obtain the following relationship,

$$u = \frac{1}{2}\{\varepsilon\}^T[E]^T\{\varepsilon\} \tag{2.121}$$

Since the energy density is a scalar we have $u = u^T$. This results in the following strain energy density expression for a generic linear elastic material,

$$u = \frac{1}{2}\{\varepsilon\}^T[E]\{\varepsilon\} \tag{2.122}$$

The total strain energy $\mathcal{U}$ is obtained by integrating u through the volume of the elastic body, as follows:

$$\mathcal{U} = \int_{\forall} u dV \tag{2.123}$$

If the material has initial stress $\{\sigma_0\}$ and initial strain $\{\varepsilon_0\}$, the *strain energy density* expression for a generic linear elastic material becomes,

$$u = \frac{1}{2}\{\varepsilon\}^T[E]\{\varepsilon\} - \{\varepsilon\}^T[E]\{\varepsilon_0\} \tag{2.124a}$$

or

$$u = \frac{1}{2}\{\varepsilon\}^T[E]\{\varepsilon\} + \{\varepsilon\}^T\{\sigma_0\} \tag{2.124b}$$

and the *strain energy* expression becomes,

$$\mathcal{U} = \int_V \{\varepsilon\}^T\left(\frac{1}{2}[E]\{\varepsilon\} - [E]\{\varepsilon_0\} + \{\sigma_0\}\right)dV \tag{2.125}$$

Note that the strain energy expression includes both the initial strain and the initial stress because the integral is taken over the entire volume of the material that can be subjected to different conditions.

2.2.9.3 *Potential possessed by applied loads*

In general, the external forces that act on a given deformable volume can be classified as *body forces* and *surface forces*. In vector notation these are expressed as follows:

Body force $\{F_B\} = \{F_{Bx} \quad F_{By} \quad F_{Bz}\}^T$
Surface traction $\{T\} = \{q_x \quad q_y \quad q_z\}^T$
Concentrated force $\{F\}$

The deformation vector $\vec{u}$ is expressed by using the vector representation as follows:

$$\{u\} = \{u_x \quad u_y \quad u_z\}^T$$

In general, the work required to move a force $\vec{F}$ through a displacement $\vec{u}$ is expressed as follows:

$$W = \vec{F}.\vec{u} \text{ or } W = \{u\}^T\{F\} \tag{2.126}$$

Considering that the body force acts per unit volume of the material, and the surface traction acts over the outer surface $S(=\Gamma)$ of the body, the total work done by the external forces is given as follows:

$$W = \int_V \{u\}^T\{F_b\}dV + \int_S \{u\}^T\{T\}dS + \sum \{u_F\}^T\{F\} \tag{2.127}$$

where $\{u_F\}$ represents the deflections where the concentrated loads $\{F\}$ are applied. The potential due this work is $\Omega = -W$ as before.

2.2.9.4 Total potential energy of a deformable body

The total potential energy of a linear elastic material that is subjected to conservative loads is expressed as follows:

$$\pi_p = \mathcal{U} - W$$
$$\pi_p = \frac{1}{2}\int_V \{\varepsilon\}^T[E]\{\varepsilon\}dV - \int_V \{u\}^T\{F_B\}dV - \int_S \{u\}^T\{T\}dS - \sum \{u_F\}^T\{F\} \tag{2.128}$$

In case, there are initial stresses and initial strains in the material, the total potential energy becomes,

$$\begin{aligned}\pi_p = &\int_V \{\varepsilon\}^T\left(\frac{1}{2}[E]\{\varepsilon\} - [E]\{\varepsilon_0\} + \{\sigma_0\}\right)dV - \int_V \{u\}^T\{F_B\}dV \\ &- \int_S \{u\}^T\{T\}dS - \sum \{u_F\}^T\{F\}\end{aligned} \tag{2.129}$$

2.3 Mechanics of a flexible beam

Mechanics of a deformable medium that describe the three-dimensional loading and deformation states are described in the previous sections. If the plane stress and plane strain conditions can be satisfied by the geometry and the loading conditions, the analysis can be reduced to two spatial dimensions. Beam theory

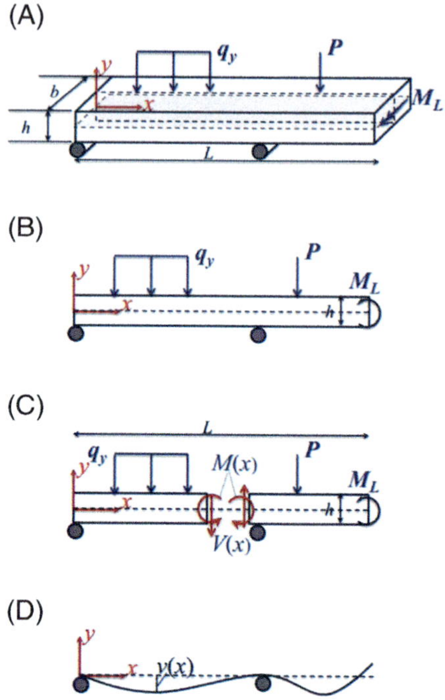

FIGURE 2.16 (A,B) Beam mechanics can be represented by variations along a single space dimension: in this case x. (C) The beam resists external effects primarily by developing internal bending moments $M(x)$ and shear forces $V(x)$. (D) The transverse deflection of the beam can be represented by the deflection of its longitudinal (neutral) axis.

is another form of simplification that typically applies to long and slender deformable members that are typically loaded normal to their longitudinal axes [1,5]. For such structural members the analysis can be simplified to one spatial dimension.

A long (L) and slender ($h \times b$) beam subjected to a transverse distributed load q_y, transverse concentrated force F and a bending moment M_L is shown in Fig. 2.16. Such structural members resist external loads by developing internal bending moments $M(x)$ and transverse shear forces $V(x)$. The transverse deflection of the beam is u_y and the longitudinal deflection is u_x. However, the customary notation for beam theory is $u = u_x$ and $v = u_y$, as depicted in Fig. 2.17. Both of these variables are functions of the longitudinal position x. If the external loading depends on time, then the beam deflection will also vary in time, t.

The following assumptions are made in order to develop the beam theory: (1) transverse deflection v is smaller than thickness h of the beam ($v < h$); (2) material is isotropic and behaves linearly and elastically; and (3) transverse normal stress σ_y is negligible. The nature of the transverse shear strain gives rise to the two theories, as described below.

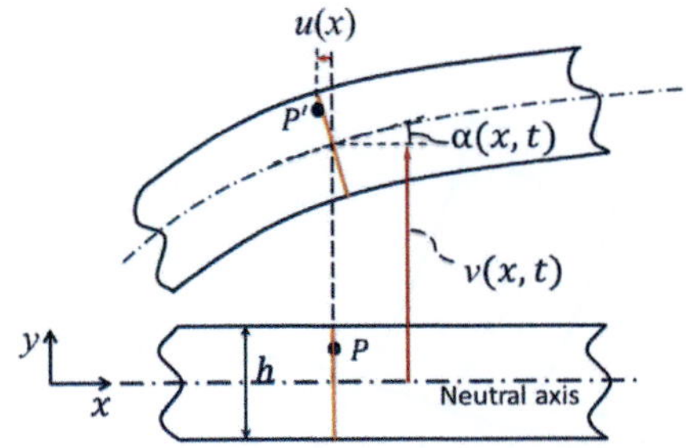

FIGURE 2.17 Definitions of the transverse $v(x)$ and longitudinal $u(x)$ deflection components and the slope $\alpha(x)$ of the neutral axis.

Euler–Bernoulli beam theory: in this theory the transverse shear strain γ_{xy} is considered negligible. This indicates that a line normal to neutral axis before deformation remains normal after deformation (Kirchhoff-Love assumption). Line *a'b'* which is originally normal to the neutral axis of the beam is assumed to remain normal after deformation (Fig. 2.18). Thus, the slope $\alpha(x)$ of the deflected beam becomes,

$$\alpha = \frac{dv}{dx} \tag{2.130}$$

Timoshenko beam theory: in this theory the transverse shear strain γx_y is not negligible. The slope of the deflected beam has contributions from bending and shear deformation. This introduces another variable into the problem. When the shear deformation is included in analysis the slope of the beam becomes,

$$\alpha = \frac{dv}{dx} = \gamma_{xy} + \theta \tag{2.131}$$

where the slope component due to bending is θ.

2.3.1 Equation of motion of a beam

We will next develop the equations of motion for a beam subjected to external loading. First these equations will be developed in terms of the internal bending moment M and shear force resultant V. In this form, these equations apply to both EB and Timoshenko beam theories. Let's consider the deflection of a beam subjected to distributed force $q_y(x,t)$. Free body diagram of an infinitesimally small segment of a beam subjected to q_y is shown in Fig. 2.19. The balance of forces acting in the transverse direction should be equal to the acceleration of the beam segment. This is expressed as follows:

$$V - \left(V + \frac{\partial V}{\partial x}dx\right) - q_y\,dx = \rho A dx \frac{\partial^2 v}{\partial t^2} \tag{2.132}$$

$$q_y + \frac{\partial V}{\partial x} = \rho A \frac{\partial^2 v}{\partial t^2} \tag{2.133}$$

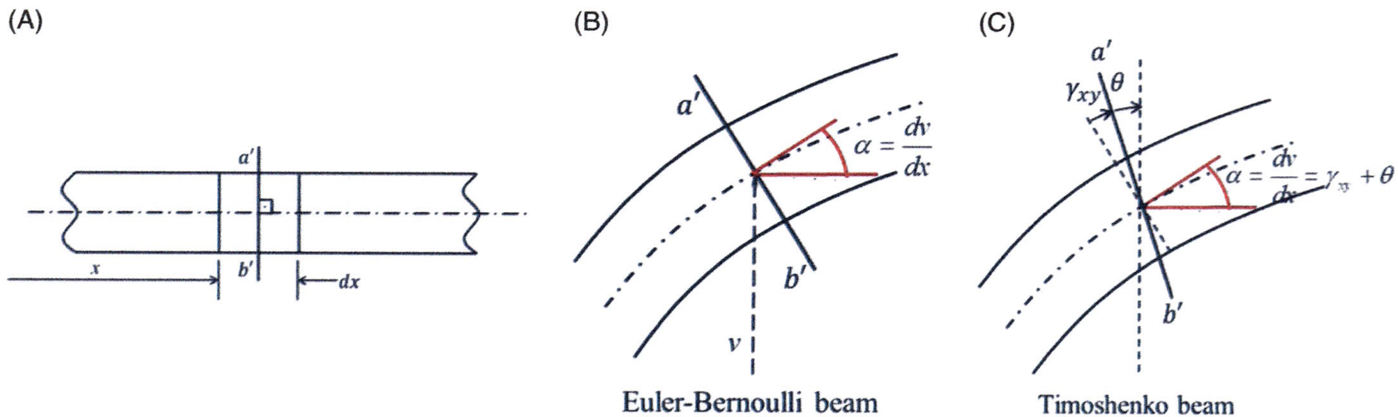

FIGURE 2.18 Geometric definition of the Kirchhoff–Love assumption that leads to the EB and Timoshenko beam theories. (A) Line *a'b'* represents a normal to the neutral axis before deformation. (B) In EB beam theory this line is assumed to remain normal to the neutral axis after deformation. (C) In Timoshenko beam theory the shear deformation becomes significant part of the theory.

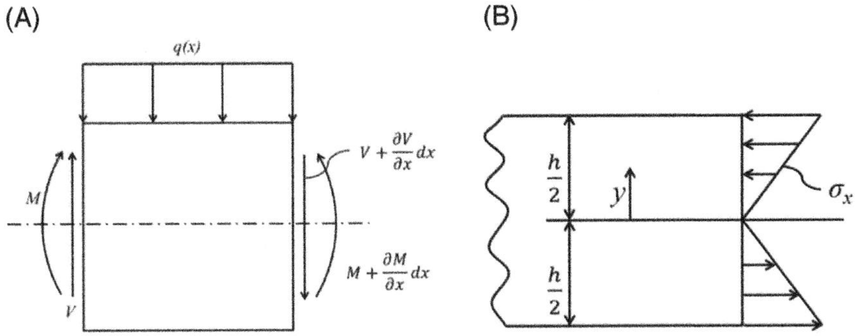

FIGURE 2.19 (A) Free body diagram of an infinitesimally small segment of a beam subjected to distributed force q_y. The width of the small segment is dx. (B) Variation of the bending stress through the thickness of the beam.

The balance of moments acting on the small beam segment, about the z-axis gives,

$$\rho I \frac{\partial^2 \alpha}{\partial t^2} = -M + \left(M + \frac{\partial M}{\partial x} dx \right) - V dx + q_y \frac{dx^2}{2} \tag{2.134}$$

$$\rho I \frac{\partial^2 \alpha}{\partial t^2} = \frac{\partial M}{\partial x} - V + q_y \frac{dx}{2} \tag{2.135}$$

where I is the second moment of area of the beam. Taking the limit as $dx \to 0$ we find

$$\frac{\partial M}{\partial x} = V + \rho I \frac{\partial^2 \alpha}{\partial t^2} \tag{2.136}$$

Thus we see that equation of motion of the beam is represented by two coupled Eqs. (2.133) and (2.136), where M, V, v and α are the unknown variables. The kinematics of deformation allow further simplifications.

2.3.2 Kinematics of the Euler–Bernoulli beam

When a beam is subjected to *pure bending* (i.e., two end moments) its top fibers are compressed in the longitudinal direction while its bottom fibers are stretched. It can therefore be surmised that a fiber, known as *neutral axis* (NA), exists between the top and bottom fibers, where the beam neither elongates nor is compressed in bending. The stretching or compression can be visualized as shown in Fig. 2.17.

A point P located a distance z from the neutral axis will experience longitudinal displacement u due to bending. This displacement component is given by the following relationship,

$$u = -y \frac{dv}{dx} \tag{2.137}$$

The bending strain is then computed as follows:

$$\varepsilon_x = -y\frac{d^2v}{dx^2} \tag{2.138}$$

2.3.3 Stresses in an Euler–Bernoulli beam

Using Hooke's law, the longitudinal stress due to bending can be derived as follows:

$$\sigma_x = -E\left(y\frac{d^2v}{dx^2}\right) \tag{2.139}$$

This is also known as the *bending stress*. Eq. (2.139) shows that the bending stress is zero on the neutral axis ($y = 0$) and otherwise varies linearly through the thickness of the beam (Fig. 2.19). Close inspection of this figure shows that the bending stress can be represented by a resultant bending moment about the neutral axis. This gives rise to the *moment–curvature* relationship as follows:

$$M = \int_{-h/2}^{h/2} \sigma_x y b dy = -E\frac{d^2v}{dx^2}\int_{-h/2}^{h/2} y^2 b dy \tag{2.140a}$$

$$M = -EI\frac{d^2v}{dx^2} \tag{2.140b}$$

where the second moment of area of the cross-section of the beam is defined as follows:

$$I = \int_{-h/2}^{h/2} y^2 b dy \tag{2.141}$$

Note that for small beam deflections the *curvature* of the deflected beam is approximately given by the term, d^2v/dx^2. By combining Eqs. (2.139) and (2.140a), a relationship between the bending moment and the bending stress can be obtained as follows:

$$\sigma_x = -\frac{My}{I} \tag{2.142}$$

In case the beam deflection is due to the combined action of transverse loads and external bending moments, then internal shear force will develop in the beam. It is shown a bit later in this document that the shear force resultant is given by the following relationship:

$$V = \frac{dM}{dx} = -\frac{d}{dx}\left(EI\frac{d^2v}{dx^2}\right) \tag{2.143}$$

The stress-displacement relationships for the Euler-Bernoulli beam are summarized in Table 2.1.

TABLE 2.1 Slope, moment shear force, bending stress and shear stress relationships used in Euler–Bernoulli and Timoshenko beam theories [1].

	Euler–Bernoulli	Timoshenko
Slope: $\alpha(x)$	dv/dx	$\gamma_{xy} + \theta$
Bending moment resultant: M	$-EI\frac{d^2v}{dx^2}$	$-EI\frac{d\theta}{dx}$
Shear force resultant: V	$-\frac{d}{dx}\left(EI\frac{d^2v}{dx^2}\right)$	$k(GA\gamma_{xy})$
Bending stress: σ_x	$-Ey\frac{d^2v}{dx^2}$	$-Ey\frac{d\theta}{dx}$
Shear stress: τ_{xy}	$\frac{VQ}{It}$	$\frac{V}{A} = kG\left(\frac{dv}{dx} - \theta\right)$

TABLE 2.2 Boundary conditions of the Euler-Bernoulli and Timoshenko beam theories for three common types of boundaries found in analysis of beams.

	Constrained variables	Euler-Bernoulli beam theory	Timoshenko beam theory
Fixed (clamped)	v, α	$v=0,\ v_{,x}=0$	$v=0,\ \theta=0$
Simply supported (pinned)	v, M	$v=0,\ v_{,xx}=0$	$v=0,\ \theta_{,x}=0$
Free	V, M	$v_{,xx}=0,\ v_{,xxx}=0$	$\theta_{,x}=0,\ w_{,x}-\theta=0$

2.3.4 Kinematics of the Timoshenko beam

Timoshenko beam theory should be used in cases where the transverse the shear strain γ_{xy} is not negligible. In this theory, the Kirchhoff-Love assumption about the behavior of a line normal to the neutral axis no longer applies. Consequently, the slope of the deflected beam has a component due to bending θ and another component due to shear strain as shown in Fig. 2.18, and it is given as follows:

$$\frac{dv}{dx} = \gamma_{xy} + \theta \tag{2.144}$$

The longitudinal displacement u due to bending is found by the following relationship:

$$u = -y\theta \tag{2.145}$$

where θ is the bending angle. The longitudinal strain becomes,

$$\varepsilon_x = -y\frac{d\theta}{dx} \tag{2.146}$$

2.3.5 Stresses in a Timoshenko beam

As in the case of the Euler–Bernoulli beam, by using the Hooke's law, the moment curvature relationship is expressed as follows:

$$M = \int_{-h/2}^{h/2} \sigma_x y b dy = -E\frac{d\theta}{dx}\int_{-h/2}^{h/2} y^2 b dy \tag{2.147a}$$

$$M = -EI\frac{d\theta}{dx} \tag{2.147b}$$

Also as in the case of the Euler–Bernoulli beam an internal shear stress develops in the beam due to transverse loading. The shear stress is expressed by using the Hooke's law, as $\tau_{xy} = G\gamma_{xy}$, where G is the shear modulus of the material. The shear force resultant V is obtained by integrating the shear stress τ_{xy} through the thickness ($-h/2 \leq z \leq h/2$) of the beam, as follows:

$$V = \int_{-h/2}^{h/2} \tau_{xy} b dz = \int_{-c/2}^{c/2} (G\gamma_{xy}) b dz = k(GA\gamma_{xy}) \tag{2.148}$$

where A is the cross-sectional area of the beam (Fig. 2.16). Note that, in this derivation it is assumed that the shear strain γ_{xy} is constant thought the beam thickness. The factor k is the *shear correction factor* which compensates for assuming that γ_{xy} does not vary in y-direction.

Using the geometric relation given in Eq. (2.131) the shear force resultant V can be expressed as follows:

$$V = kGA\left(\frac{dv}{dx} - \theta\right) \tag{2.149}$$

Thus the number of unknowns for the Timoshenko beam theory can be reduced to two (v, θ). Once the shear force resultant is known the shear stress τ_{xy} can be found as follows:

$$\tau_{xy} = \frac{V}{A} = kG\left(\frac{dv}{dx} - \theta\right) \tag{2.150}$$

The stress-displacement relationships for the Timoshenko beam are summarized in Table 2.1.

2.3.6 Governing equations of the Euler–Bernoulli beam theory

Using the equilibrium Eq. (2.136), the shear force resultant V can be expressed as follows:

$$V = \frac{\partial M}{\partial x} - \rho I\frac{\partial^2 \alpha}{\partial t^2} \tag{2.151}$$

This expression is then used to eliminate V from Eq. (2.133) as follows:

$$\rho A\frac{\partial^2 v}{\partial t^2}+\frac{\partial}{\partial x}\left(\rho I\frac{\partial^2 \alpha}{\partial t^2}\right)-\frac{\partial^2 M}{\partial x^2}=q_y \tag{2.152}$$

By combining the moment–curvature relation, Eq. (2.140a), the Kirchhoff–Love assumption and Eq. (2.152), we find the following equation of motion,

$$\rho A\frac{\partial^2 v}{\partial t^2}+\rho I\frac{\partial^4 v}{\partial x^2\partial t^2}+\frac{\partial^2}{\partial x^2}\left(EI\frac{\partial^2 v}{\partial x^2}\right)=q_y \tag{2.153}$$

Typically the rotational acceleration term is negligible. Therefore, the equation of motion of an Euler–Bernoulli beam is expressed as follows:

$$\rho A\frac{\partial^2 v}{\partial t^2}+\frac{\partial^2}{\partial x^2}\left(EI\frac{\partial^2 v}{\partial x^2}\right)=q_y \tag{2.154}$$

For steady-state problems the inertial effects are omitted and the equation of equilibrium becomes,

$$\frac{\partial^2}{\partial x^2}\left(EI\frac{\partial^2 v}{\partial x^2}\right)=q_y \tag{2.155}$$

For the sake of completeness let's summarize the variables and their dimensions,

$q_y(x)$: Distributed transverse force [F/L]
I: Second moment of beam cross-sectional area [L^4]
E: Elastic modulus [F/L^2]
A: Cross-sectional area of the beam [L^2]
ρ: Volumetric mass density of the beam [mass/Vol]

2.3.7 Governing equations of the Timoshenko beam theory

In Timoshenko beam theory, the transverse deflection v and bending slope θ are the dependent variables. The steps of derivation are similar to that for the Euler–Bernoulli beam. Recall the relationships for the bending moment, transverse shear force resultant and slope of the deflection curve for the Timoshenko beam theory,

$$M=-EI\frac{\partial\theta}{\partial x},\quad V=kGA\left(\frac{dv}{dx}-\theta\right),\quad \alpha=\gamma_{xy}+\theta \tag{2.156}$$

from Eqs. (2.147a), (2.150), and (2.131), respectively, and the equilibrium Eqs. (2.133) and (2.136),

$$q+\frac{\partial V}{\partial x}=\rho A\frac{\partial^2 v}{\partial t^2}$$

$$\frac{\partial M}{\partial x}-V=\rho I\frac{\partial^2 \theta}{\partial t^2}$$

Combining these relationships gives,

$$\frac{\partial}{\partial x}\left[kGA\left(\frac{\partial v}{\partial x}-\theta\right)\right]=\rho A\frac{\partial^2 v}{\partial t^2}-q_y \tag{2.157a}$$

$$\frac{\partial}{\partial x}\left[EI\frac{\partial \theta}{\partial x}\right]+kGA\left(\frac{\partial v}{\partial x}-\theta\right)=-\rho I\frac{\partial^2 \theta}{\partial t^2} \tag{2.157b}$$

If G, A, E, I do not vary with x these equations reduce to the following form:

$$kGA\left(\frac{\partial^2 v}{\partial x^2}-\frac{\partial \theta}{\partial x}\right)=\rho A\frac{\partial^2 v}{\partial t^2}-q_y \tag{2.158a}$$

$$\frac{\partial}{\partial x}\left[EI\frac{\partial \theta}{\partial x}\right]+kGA\left(\frac{\partial v}{\partial x}-\theta\right)=-\rho I\frac{\partial^2 \theta}{\partial t^2} \tag{2.158b}$$

The variables of the Timoshenko beam theory are summarized as follows:

$v(x)$: Transverse deflection [L]
$\theta(x,t)$: Rotation of transverse cross-section due to bending
E, G: Elastic and shear moduli [F/L^2]
k: Shear correction factor
A,I: Area and second moment of area of beam's cross-section [L^2]
$q_y(x)$: Distributed force acting in the transverse direction [F/L]

2.4 Heat transfer

Heat flow within a body or on its boundaries takes place with different physical mechanisms. Inside the body, heat flows by *conduction* described in the next section. Thermal conduction within solids takes place by lattice vibration and transport of free electrons (Section 2.4.1). Good electrical and thermal conductors have a large number of free electrons. On the boundaries the heat is transferred to or from the body by *convection* to and from an adjacent fluid (Section 2.4.2) and/or by *radiation* (Section 2.4.3). Inside the body heat may be generated due to such effects as radioactivity, resistance to electric current, absorption of microwave radiation.

In what follows the basic governing equations of heat transfer will be derived. This is based on discussions in [3] and [6].

2.4.1 Conduction heat transfer

When a temperature gradient exists inside a body, energy is transferred from the high-temperature region to the low temperature region. The rate of *heat-transfer*, q^h, per unit area, A, is known as the *heat flux*, $f = q^h/A$. Heat flux is proportional to the temperature gradient as follows:

$$f=\frac{q^h}{A}\propto\frac{\partial T}{\partial x} \tag{2.159}$$

where T is the temperature. In SI system the unit of *heat flow, q*, is Watt, which is also equal to Joule/s, and the unit of *heat flux, f*, is Watt/m^2. A proportionality constant known as the *thermal conductivity, k*, can be implemented in the above relationship,

$$q^h = -kA\frac{\partial T}{\partial x} \tag{2.160}$$

This relationship is known as *Fourier's law of heat conduction*. The thermal conductivity is a material property with units of W/m.K. The minus sign in this equation ensures that heat flows from a high temperature region to a low temperature region in the body. This is graphically depicted in Fig. 2.20.

2.4.2 Convection heat transfer

It is well known that a hot plate of metal will cool faster when placed in front of a fan, than when it is exposed to still air. Consider the hot plate shown in Fig. 2.21. The temperature of the plate is T and temperature of the fluid away from the wall is T_∞ at the free stream. In case $T > T_\infty$, the plate is cooled by the fluid with a physical mechanism known as the *convection heat transfer*. In this process, similar to a fluid boundary layer, the temperature changes from T to T_∞ in a *thermal boundary layer*, and a temperature gradient in the y-direction exists.

Newton's law of cooling is used to express the overall effect of *convection* where rate of heat transfer becomes:

$$q^h = hA_w(T - T_\infty) \tag{2.161}$$

where h is the *convection heat-transfer coefficient* with units of [W/m^{2}K] and A_w is the area of the wall exposed to the fluid, over which the convection heat transfer taking place. The *convection coefficient* which is related to the viscosity of the fluid, can be calculated for some simple systems, but can also be measured for complex situations.

2.4.3 Radiation heat transfer

In contrast to the mechanisms of conduction and convection, where energy transfer through a material medium is involved, energy is also be transferred through vacuum. The mechanism that allows this is electromagnetic radiation. In particular, we are interested in the *thermal radiation*.

An ideal radiator, or a black-body, will emit energy at a rate proportional to the fourth power of the absolute temperature of the body. The net heat transfer rate is the difference between the energy absorbed by the material and the energy emitted by the material,

$$q^h = \sigma A_w\left(T_s^4 - T^4\right) \tag{2.162}$$

where A_w is the surface area of body exposed to the radiation, T_s is the temperature of the emitting source, and σ ($= 5.669 \times 10^{-8}$ W/m^2/K^4) is the *Stefan-Boltzmann constant*. This equation is called the *Stefan-Boltzmann law of*

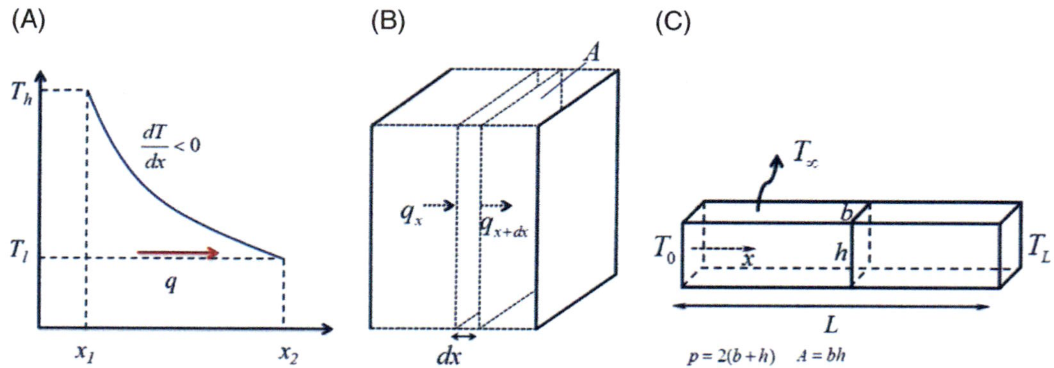

FIGURE 2.20 (A) Heat flow from the high temperature region to the low temperature region creates a negative temperature gradient $dT/dx < 0$ as shown. In order to define a positive heat flow for these conditions a negative sign is used in the Fourier's law of heat conduction. (B) Heat transfer in a one-dimensional domain where the periphery of the object does not participate in energy exchange. (C) Heat transfer from a fin, can be considered one-dimensional. The base temperature of the fin is T_0, the end temperature is T_L and the fin is located in a fluid with temperature T_∞.

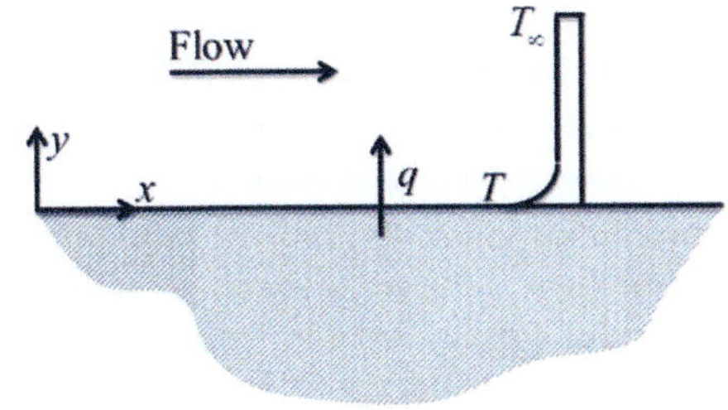

FIGURE 2.21 Convective flow q on the boundary of a solid and a fluid increases with flow velocity. The temperature of the fluid changes from that of the wall T to the free stream temperature T_∞ in a boundary layer.

thermal radiation, and it only applies to black-bodies. The idealized emission conditions of black-bodies may not be realized in some cases. The complicating factors include:

- *Emissivity* (ε) of the two surfaces. This property is influenced by the surface roughness, degree of oxidization of the surface, surface temperature and the wavelength of the emitted energy.
- *View factor* (F) accounts for the fact that the surfaces are not always parallel with respect to each other. View factor between two surfaces F_{1-2} may be defined as the fraction of radiation energy received by one of the surfaces when the energy ordinates from the other one.

The effects of emissivity and view factor can be combined into a factor ψ and the modified Stefan–Boltzmann law can be given as,

$$q^h = \psi\, \sigma\, A_w\left(T_s^4 - T^4\right) \tag{2.163}$$

Note that this equation can be factored as,

$$q^h = \psi\, \sigma\, A_w\left(T_s^2 + T^2\right)(T_s + T)(T_s - T) \tag{2.164}$$

where a nonlinear, temperature-dependent *radiation coefficient* h_{rad} can be defined as,

$$h_{rad} = \psi\, \sigma\left(T_s^2 + T^2\right)(T_s + T) \tag{2.165}$$

and heat transfer rate becomes,

$$q^h = h_{rad}\, A_w(T_s - T) \tag{2.166}$$

Even though this equation has the same form as the convection heat transfer Eq. (2.161), it is actually strongly nonlinear in temperature T.

2.4.4 Heat transfer equation in a one-dimensional solid

Consider a general one-dimensional problem where the temperature inside the body varies with time and there is a heat source/sink within a body.

Fig. 2.20B shows a small volume of such a body with width dx in the direction of conduction. Heat flows in the direction normal to the area A, as shown in the figure. The body is considered infinitely wide in all other directions, so that the heat transfer may be considered one dimensional.

The *energy balance* requires that the *rate of change of heat energy in this volume to be equal to the sum of the heat energy flowing across the boundary of the volume per unit time and the heat energy generated inside the volume per unit time.* Mathematically, this is stated as follows:

$$q_x^h - q_{x+dx}^h + QAdx = \rho cAdx\frac{\partial T}{\partial t} \tag{2.167}$$

The conducted components of the heat flow are expressed by using the Fourier's law,

$$\begin{aligned} &\text{Energy conducted in left face:} && q_x^h = -k_x A\frac{\partial T}{\partial x} \\ &\text{Energy conducted out right face:} && q_{x+dx}^h = -\left[k_x A\frac{\partial T}{\partial x} + \frac{\partial}{\partial x}\left(k_x A\frac{\partial T}{\partial x}\right)dx\right] \end{aligned} \tag{2.168}$$

The energy is generated within the solid due to a heat source Q which generates energy per unit volume. The unit of Q is W/m^3. The total energy generated in the given volume then becomes,

$$\text{Energy generated within unit volume:} \quad QAdx \tag{2.169}$$

The balance of the conducted and generated energy should be zero, when the system is at steady state. But, it is possible for the energy to have a transient condition before a steady state is established. The excess energy is stored inside the body, and it is proportional to the transient gradient of temperature,

$$\text{Energy stored internally:} \quad \rho c(Adx)\frac{\partial T}{\partial t} \tag{2.170}$$

where c is the *heat capacity* of the material. The unit of heat capacity is J/kg K. Substituting Eqs. (2.168) – (2.170) in to (2.167) we find the conduction equation for a one-dimensional medium that spans the range $x_0 < x < x_L$ as follows:

$$AQ + \frac{\partial}{\partial x}\left(k_x A\frac{\partial T}{\partial x}\right) = \rho cA\frac{\partial T}{\partial t} \quad \text{in}\, x_0 < x < x_L \tag{2.171}$$

Heat transfer in a fin is influenced by two fundamental mechanisms. Inside the fin heat moves according to Fourier's law of heat conduction, while on its peripheral boundary heat is transferred through convection to a surrounding fluid considered to be at temperature T_∞ (Fig. 2.20C). In this case the energy balance equation can be derived by using Eq. (2.167) as follows:

$$q_x^h - q_{x+dx}^h - q_{out}^{convection} + QAdx = \rho cAdx\frac{\partial T}{\partial t} \tag{2.172}$$

The conduction heat transfer is governed by Eq. (2.160) and the convection heat transfer is governed by Eq. (2.161). Substitution of these relationships into Eq. (2.172) gives,

$$-k_xA\frac{\partial T}{\partial x} + QAdx - \left(-k_xA\frac{\partial T}{\partial x} - \frac{\partial}{\partial x}\left(k_xA\frac{\partial T}{\partial x}\right)dx\right) - hA_w(T - T_\infty)$$
$$= \rho cAdx\frac{\partial T}{\partial t} \tag{2.173}$$

Thus the energy balance for a thin fin can be found by simplifying Eq. (2.173) as follows:

$$AQ + \frac{\partial}{\partial x}\left(k_xA\frac{\partial T}{\partial x}\right) + hp(T_\infty - T) = \rho cA\frac{\partial T}{\partial t} \quad \text{in}\, x_0 < x < x_L \tag{2.174}$$

Note that in order for the heat to be transferred from the fin to the surrounding fluid, the term that involves $(T - T_\infty)$ is used in this equation. The area over which convection takes place is the *peripheral area* of the fin which is given as $A_w = p\,dx$, where p is the *perimeter* of the fin (Fig. 2.20C). A is the cross sectional area of the fin.

2.4.4.1 Boundary conditions for 1D heat transfer

The physics of heat transfer dictate that the temperature T, or the heat flux f to take on specified values on the boundary of the body. In Chapter 3, it is shown that only T or f can be specified at a given position on the boundary; but, they cannot be defined simultaneously at the same position. The boundary conditions of a 1D heat transfer problem are defined as follows:

$$\text{at } x = x_0 \text{ either } T = T_b \text{ or } \left(-k_x\frac{\partial T}{\partial x}\right)n = f_B \tag{2.175a}$$

$$\text{at } x = x_L \text{ either } T = T_b \text{ or } \left(-k_x\frac{\partial T}{\partial x}\right)n = f_B \tag{2.175b}$$

Note that n represents the unit outward normal of the boundary, $n = -1, +1$ at $x = x_0, x_L$ respectively, T_b is the specified (prescribed) value of the temperature on the boundary and f_B is the prescribed heat flux. Depending on the conditions the prescribed heat flux is dictated by one (or combination) of the following

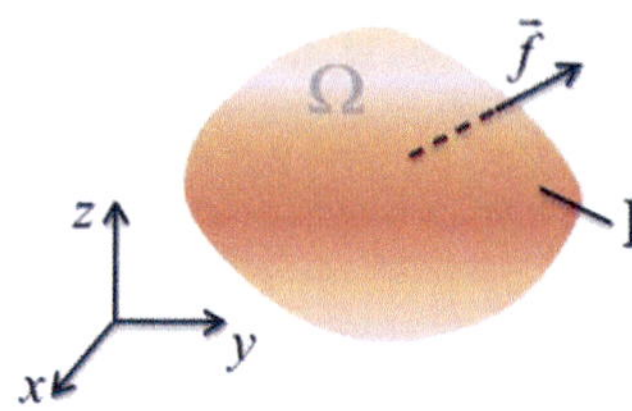

FIGURE 2.22 Positive heat flux $\vec{f}$ crossing the boundary Γ of a three-dimensional domain Ω.

physical mechanisms[3],

$$\begin{array}{l}\text{Insulated Boundary: } f_B = f_B(t)\\ \text{Convected Boundary: } f_B = h(T - T_\infty)\\ \text{Radiated Boundary: } f_B = \sigma\,\psi\left(T^4 - T_s^4\right)\end{array} \tag{2.176}$$

In addition, to the boundary conditions, the time dependent nature of the problem requires the *initial condition* of the body temperature to be specified,

$$T(x,0) = T^{(0)}(x) \tag{2.177}$$

2.4.5 Heat transfer in a three-dimensional solid

Consider an arbitrary domain Ω enclosed by a surface Γ as shown in Fig. 2.22. *Conservation of (heat) energy* in such a domain requires the following balance statement to be satisfied [4]: *rate of change of heat energy in this volume should be equal to the sum of the heat energy flowing across the boundary of the volume per unit time and the heat energy generated inside the volume per unit time.*

Flow of thermal energy per unit area and per unit time in a continuous medium is characterized by the *heat flux* vector $\vec{f}$. The heat flux that contributes to the energy balance is the normal component of $\vec{f}$ with respect to the boundary it is crossing. This is expressed by $\vec{f}.\vec{n}$ where $\vec{n}$ is the unit outward normal of the boundary. Thus a positive value for $\vec{f}.\vec{n}$ indicates an outward flow of energy from the boundary.

The thermal energy stored within the volume is expressed by,

$$\text{thermal energy} = \int_\Omega c\rho T d\Omega \tag{2.178}$$

where c is the heat capacity, ρ is the mass density and V is the volume of the material. The conservation of thermal energy can then be expressed mathematically

3. A note on the signs: The signs of Eq. (2.4.18) can become confusing. The following notes attempt to reduce the uncertainty. Note that f_B for convection and radiation given in Eq. (2.4.18) are considered positive. Rate of heat transfer q is considered positive when heat flows into the body. In addition, the negative sign in the bracket of Eq. (2.4.17b) ensures that the heat flows from high temperature region to the low temperature region in the body. If one considers all of the possible combinations one can show that these two conditions are satisfied when Eq. (2.4.18) is used.

as follows:

$$\frac{\partial}{\partial t}\int_{\Omega} c\rho T d\Omega = -\int_{\Gamma} \vec{f}.\vec{n} d\Gamma + \int_{\Omega} Q d\Omega \tag{2.179}$$

where Q is the internal energy generated per unit volume and unit time. The − sign in this equation indicates that thermal energy is lost from the volume.

Recall that for any vectorial variable $\vec{v}$, the *divergence theorem* [9] states the integral of the gradient of the vector field inside the volume is equal to the surface integral of the normal component of the vector field as follows:

$$\int_{\Omega} \nabla \cdot \vec{v} d\Omega = \int_{\Gamma} \vec{v} \cdot \vec{n} d\Gamma \tag{2.180}$$

where ∇ is the *gradient operator*, defined in Cartesian coordinates, as follows:

$$\nabla(\cdot) = \frac{\partial(\cdot)}{\partial x}\hat{i} + \frac{\partial(\cdot)}{\partial y}\hat{j} + \frac{\partial(\cdot)}{\partial z}\hat{k} \tag{2.181}$$

where $\hat{i}$, $\hat{j}$ and $\hat{k}$ are the unit vectors along the x-, y-, and z-axes, respectively. Therefore, using the divergence theorem Eq. (2.180), the energy conservation Eq. (2.179) becomes,

$$\frac{\partial}{\partial t}\int_{\Omega} c\rho T d\Omega = -\int_{\Omega} \nabla\cdot\vec{f} d\Omega + \int_{\Omega} Q d\Omega \tag{2.182a}$$

or

$$\int_{\Omega} \left(c\rho\frac{\partial T}{\partial t} + \nabla\cdot\vec{f} - Q \right) d\Omega = 0 \tag{2.182b}$$

or

$$c\rho\frac{\partial T}{\partial t} + \nabla\cdot\vec{f} - Q = 0 \tag{2.182c}$$

The final step in this derivation is to recall *Fourier's law of conduction,*

$$\vec{f} = -k\nabla T \tag{2.183}$$

which combined with Eq. (2.182a) gives the conduction equation for a general three-dimensional domain as follows:

$$c\rho\frac{\partial T}{\partial t} = \nabla\cdot(k\nabla T) + Q \tag{2.184}$$

The initial condition and the boundary conditions described in Section 2.4.4.1 are generalized here for the three-dimensional problem.

$$\text{Initial condition: } T(x, y, z, 0) = T_0(x, y, z) \tag{2.185}$$

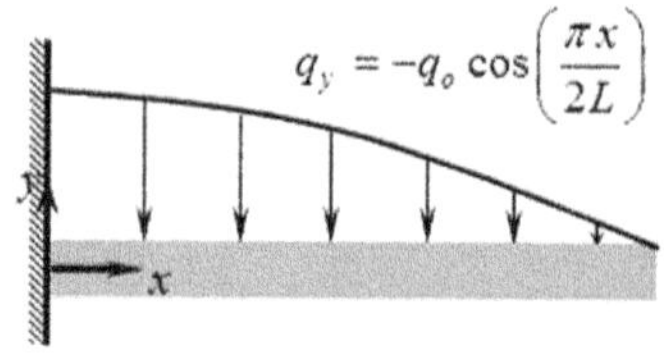

FIGURE 2.23 Problem 7.

Boundary conditions:

$$\text{Prescribed temperature}: T(x, y, z, t) = T_B(x, y, z, t) \text{ on } \Gamma_E \tag{2.186}$$

$$\text{Heat flux on the boundary:} \ -k\nabla T \cdot \vec{n} = f_B \text{ on } \Gamma_N \tag{2.187}$$

The heat flux on the boundary can be one of the following defined positive as follows:

$$\begin{aligned} &\text{Insulated Boundary:} \quad f_B = f_B(t) \\ &\text{Convected Boundary:} \ f_B = h(T - T_\infty) \\ &\text{Radiated Boundary:} \quad f_B = \sigma\psi\left(T^4 - T_s^4\right) \end{aligned} \tag{2.188}$$

2.5 Problems

2.1 Imagine that stresses in the xy plane are reported to be $\sigma_x = -6a_1x^2$, $\sigma_y = 12a_1x^2$ and $\tau_{xy} = 12a_1y^2$ where a_1 is a constant.
  a) Consider the square region $0 \le x \le b$, $0 \le y \le b$. Write expressions for tractions Φ_x and Φ_y on each side of this square, in terms of x, y, b and a_1.
  b) If body forces are zero, is this state of stress in fact possible? Explain.

2.2 Determine if the following stress field is a valid solution of a plane elasticity problem: $\sigma_x = 3a_1x^2y, \sigma_y = a_1y^3$, and $\tau_{xy} = -3a_1xy^2$, where a_1 is a constant. The body is isotropic and linearly elastic, and body forces are zero.

2.3 Let displacements in an isotropic body in a state of plane stress be as stated by

$$\begin{aligned} u &= a_1 + a_2x + a_3y + a_4x^2 + a_5xy + a_6y^2 \\ v &= a_7 + a_8x + a_9y + a_{10}x^2 + a_{11}xy + a_{12}y^2 \end{aligned}$$

What relation among the constants a_i is needed to satisfy the strain compatibility condition?

2.4 Show that for a plane stress problem the principal stress analysis described for 3D problems reduces to Mohr's circle approach.

2.5 A cantilevered I-beam is loaded with a distributed load, as shown in Fig. 2.23. The deflection, v, of the centerline of the beam as a function of the position, x, is given by the equation:

$$v = -\frac{q_0 L}{3\pi^4 EI}\left(48L^3\left(\cos\frac{\pi x}{2L} - 1\right) + \pi^3 x^2(3L - x)\right)$$

where $L = 3$ m is the length, $E = 70$ GPa is the elastic modulus, $I = 52.9 \times 10^{-6}$ m^4 is the moment of area, and $q_0 = 15$ kN/m.

a) What are the boundary conditions of this problem?
b) Show that the given solution satisfies the boundary conditions.
c) Plot the variation of deflection along the length of the beam.
d) Plot the variation of internal bending moment along the length of the beam.

References

[1] A.P. Boressi, R.J. Schmidt, O.M. Sidebottom, Advanced Mechanics of Materials, John Wiley and Sons, New York, 1993.

[2] A.F. Bower, Applied Mechanics of Solids, CRC Press, Florida, 2010.

[3] R.D. Cook, D.S. Malkus, M.E. Plesha, R.J. Witt, Concepts and Applications of Finite Element Analysis, John Wiley and Sons, New York, 2011.

[4] R. Haberman, Elementary Applied Partial Differential Equations, Prentice-Hall, New Jersey, 1987.

[5] R.C. Hibbeler, Mechanics of Materials, Pearson, New Jersey, 2016.

[6] J.P. Holman, Heat Transfer, McGraw Hill, Singapore, 1984.

[7] L.E. Malvern, Introduction to the Mechanics of a Continuous Medium, Prentice-Hall, New Jersey, 1969.

[8] S.P. Timoshenko, Theory of Elasticity, McGraw-Hill, New York, 1951.

[9] C.R. Wylie, L.C. Barrett, Advanced Engineering Mathematics, McGraw-Hill, New York, 1995.

Chapter 3

Integral formulations and variational methods

3.1 Introduction

Mathematical models of many physical systems are expressed as boundary value problems (BVPs) defined over a domain Ω such as,

$$\mathcal{L}(u) = f \text{ in } \Omega \tag{3.1}$$

where u is a *dependent variable*, f is a known function, $\mathcal{L}$ is a differential operator.

A *domain* is a collection of points in space with the property that if P is a point in the domain then all points sufficiently close to P belong to the domain. The *boundary* of a domain is the set of points such that in any neighborhood of each of these points, there are points that belong to the domain as well as points that do not (Fig. 3.1). If the dependent variable (e.g., temperature) is a function of one independent variable (e.g., x), the dependence can be expressed as $T = T(x)$. In this case, the domain is one-dimensional (1D) and the boundary consists of the two end points. If the dependent variable is a function of two independent variables (e.g., x and y) the functional dependence can be expressed as $T = T(x, y)$. In this case, the domain is an area, and the boundary is a closed curve. In general a two-dimensional or three-dimensional domain can have multiple internal boundaries.

The dependent variable is subjected to some conditions imposed on the boundary Γ. In particular:

The *essential* or *Dirichlet boundary conditions* (BCs) specify the value of the dependent variable on a part of the boundary, Γ_E. Examples include prescribed displacement, temperature, concentration.

The *Natural* or *von Neumann*, or *nonessential boundary conditions* specify the value of the physical effects that make a change in the primary variable on a part of the boundary Γ_N. Examples include prescribed force, heat flux, concentration flux.

Note that the boundary can be of either essential or a natural boundary, i.e. $\Gamma = \Gamma_E \cup \Gamma_N$ and $\Gamma_E \cap \Gamma_N = 0$.

The problem described by Eq. (3.1) can also be described by an integral equation. This form of the problem allows the solution to be obtained by using

Finite Element Method: Physics and Solution Methods. DOI: https://doi.org/10.1016/B978-0-12-821127-4.00012-8
Copyright © 2022 Elsevier Inc. All rights reserved.

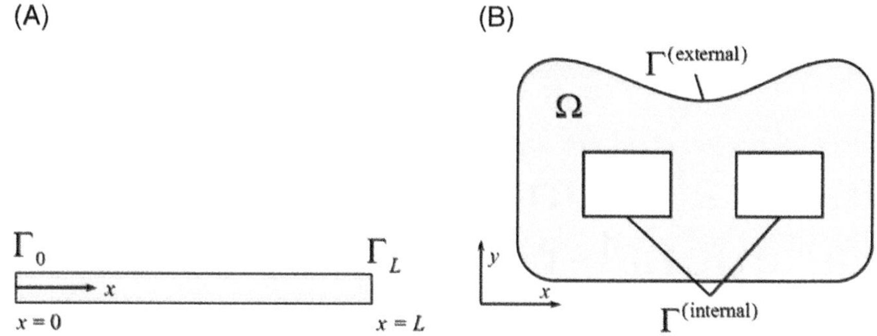

FIGURE 3.1 (A) A one-dimensional domain with two end points as the boundary. (B) An two-dimensional domain Ω in with internal $\Gamma^{(\text{internal})}$ and external $\Gamma^{(\text{external})}$ boundaries.

variational methods. The integral form $I(u)$ of the boundary value problem, typically called a *functional*, is of the form:

$$I(u) = \int_{\Omega} F(x, u, u_{,x}, \ldots)d\Omega \tag{3.2}$$

where u is the dependent variable, $u_{,x} = du/dx$, x is the coordinate axis. In solid mechanics, the integral $I(u)$ represents the total potential energy Π_p of the system. Therefore, it can be obtained in a fairly straightforward manner. However, in other disciplines of applied mechanics and applied mathematics, such as heat transfer, fluid mechanics, and electromagnetic theory, among others, obtaining the functionals directly may not be straightforward. In this chapter, we will address this, and show that a functional can be obtained from the governing partial differential equation of the boundary value problem.

We will also introduce various solution techniques to solve the boundary value problems. In general all of these techniques stem from the *method of weighted residuals*. Some of the weighted residual methods are the *Galerkin* method, *Petrov–Galerkin* method, *Least squares* method and the *collocation* method. The *Rayleigh–Ritz* method is a special form of the weighted residual methods and we will see how it differs from the other methods.

Example 3.1 The following boundary value problem: will be used as the sample problem in the rest of this chapter

$$\text{BVP: } \frac{d}{dx}\left(a\frac{du}{dx}\right) + c\,u = -q(x) \text{ in } x_0 < x < x_L \tag{3.3a}$$

$$\text{BCs: } u(x_0) = u_0, \tag{3.3b}$$

and

$$\left(a\frac{du}{dx}\right)_{x=x_L} = Q_L \tag{3.3c}$$

where $a = a(x)$, $c = c(x)$ and $q(x)$ are given functions, u_0 and Q_L are known scalars, the one-dimensional solution domain spans in the range is $x_0 < x < x_L$. Note that if the specified values u_0 and Q_L are zero boundary conditions are said to be *homogenous*.

3.2 Mathematical background

3.2.1 Divergence theorem

Consider a vector field $\vec{v}$ acting over a domain Ω surrounded by a boundary Γ. The divergence of the vector field is given by,

$$div(\vec{v}) = \nabla \cdot \vec{v} \tag{3.4}$$

and it represents the quantity of the vector field variable emanating from any point of the vector field. In a two-dimensional Cartesian reference frame the *del* operator is defined as follows:

$$\nabla = \frac{\partial}{\partial x}\,\hat{i} + \frac{\partial}{\partial y}\,\hat{j} \tag{3.5}$$

where $\hat{i}$ and $\hat{j}$ are the unit normals along the x and y-axes, respectively.

The *divergence theorem* (or *Gauss theorem*) states that *the volume/area integral of the divergence of any continuously differentiable vector is the closed surface/contour integral of the outward normal component of the vector*. The planar form of the theorem is expressed as follows [4],

$$\int_{\Omega} \nabla \cdot \vec{v}\, d\Omega = \oint_{\Gamma} \vec{n} \cdot \vec{v}\, d\Gamma \tag{3.6}$$

3.2.2 Green-Gauss theorem

The *Green-Gauss theorem* states that for a continuously differentiable scalar function $w(x, y)$ and a vector $\vec{v}$ the following relationship holds [4],

$$\int_{\Omega} \nabla \cdot \left(w\, \vec{v}\right) d\Omega = \oint_{\Gamma} w\, \vec{v} \cdot \vec{n}\, d\Gamma \tag{3.7}$$

By using the product rule of differentiation the following relationship can be established,

$$\nabla \cdot \left(w\, \vec{v}\right) = (\nabla w) \cdot \vec{v} + w \nabla \cdot \vec{v} \tag{3.8}$$

Then the Green-Gauss theorem in Eq. (3.7) can be expressed as follows [4]:

$$\int_{\Omega} w \nabla \cdot \vec{v}\, d\Omega = \oint_{\Gamma} w \vec{v}.\vec{n}\, d\Gamma - \int_{\Omega} (\nabla w) \cdot \vec{v}\, d\Omega \tag{3.9}$$

3.2.3 Integration by parts

Consider the one-dimensional functions on $u(x)$ and $v(x)$. The derivative of the product $u\,v$ with respect to x is found by using the product rule of differentiation as follows:

$$\frac{d(u\,v)}{dx} = u\frac{dv}{dx} + v\frac{du}{dx} \tag{3.10}$$

Integrating this equation with respect to x between x_0 and x_L gives,

$$[u\,v]_{x_0}^{x_L} = \int_{x_0}^{x_L} \left(u\frac{dv}{dx} + v\frac{du}{dx}\right)dx \tag{3.11}$$

The *integration by parts formula* is found by rearranging this equation as follows:

$$\int_{x_0}^{x_L} \left(v\frac{du}{dx}\right)dx = [v\,u]_{x_0}^{x_L} - \int_{x_0}^{x_L} \left(u\frac{dv}{dx}\right)dx \tag{3.12}$$

3.2.4 Fundamental lemma of calculus of variations

This lemma is given without proof but used extensively in variational calculus and weighted residual methods [2].

Consider a continuous function $g(x)$ defined in $x_0 \leq x \leq x_L$, and an *arbitrary function* $w(x)$ defined in the same interval with homogenous end conditions such that, $w(x_0) = w(x_L) = 0$. If

$$\int_{x_0}^{x_L} w(x)g(x)dx = 0$$

then the *fundamental lemma of calculus of variations* states that $g(x)$ *has to be zero in this interval*, i.e., $g(x) = 0$ in $x_0 \leq x \leq x_L$.

3.2.5 Adjoint and self-adjoint operators

Consider a differential operator $\mathcal{L}$ and two arbitrary functions $u(x)$ and $v(x)$ with homogeneous end conditions, $u(x_0) = u(x_L) = 0$ and $v(x_0) = v(x_L) = 0$. The *inner product* of $u(x)$ with $\mathcal{L}(v)$ is defined as follows:

$$\langle u, \mathcal{L}(v)\rangle = \int_{x_0}^{x_L} u\,\mathcal{L}(v)\,dx \tag{3.13}$$

An *adjoint differential operator* $\mathcal{L}^*$ satisfies the following relationship:

$$\langle u, \mathcal{L}(v)\rangle = \langle v, \mathcal{L}^*(u)\rangle \tag{3.14}$$

Consider the following second order differential equation with variable coefficients $a = a(x)$, $b = b(x)$ and $c = c(x)$,

$$\mathcal{L}(u) = a\frac{d^2u}{dx^2} + b\frac{du}{dx} + cu = 0 \text{ in } x_0 \leq x \leq x_L \tag{3.15}$$

By using integration by parts and the end conditions stated above, it can be shown that the adjoint differential operator $\mathcal{L}^*$ associated with $\mathcal{L}$ for Eq. (3.15) is given as follows [2],

$$\mathcal{L}^*(v) = \frac{d^2}{dx^2}(a\,v) - \frac{d}{dx}(b\,v) + c\,v \tag{3.16}$$

Differential operators for which the following condition holds,

$$\mathcal{L}^* = \mathcal{L} \tag{3.17}$$

are called *self-adjoint differential operators*.

By using Eqs. (3.15) and (3.16) it can be shown that a second order, linear differential equation is self-adjoint if and only if it is of the form [2],

$$\mathcal{L}(\cdot) = \frac{d}{dx}\left(a\frac{d(\cdot)}{dx}\right) + c(\cdot) \tag{3.18}$$

where $(\cdot)$ represents any arbitrary function of the type defined in this section. This differential operator is called the *Sturm–Liouville differential operator*.

Similarly, the following fourth order differential operator,

$$\mathcal{L}(\cdot) = \frac{d}{dx^2}\left(s\frac{d(\cdot)}{dx^2}\right) + \frac{d}{dx}\left(a\frac{d(\cdot)}{dx}\right) + c(\cdot) \tag{3.19}$$

where $a = a(x)$, $c = c(x)$ and $s = s(x)$, is self-adjoint, if the boundary conditions are homogenous and of the form [2],

$$u = \frac{du}{dx} = 0 \text{ or } u = s\frac{d^2u}{dx^2} = 0 \text{ or } \frac{du}{dx} = \frac{d}{dx}\left(s\frac{d^2u}{dx^2}\right) = 0 \tag{3.20}$$

on the boundaries.

3.3 Calculus of variations

A functional is a *function of functions*. A typical functional we encounter in the context of finite element method is an integral expression. For a function $u = u(x)$ a functional $I(u)$ may be of the form,

$$I(u) = \int_{x_0}^{x_L} F\left(x, u, u_{,x}\right)dx \tag{3.21}$$

where $u = u(x)$ and $u_{,x} = du/dx$. Note that the value of the functional is a scalar. Functionals of more than one function (e.g. u, w) and spatial variable (e.g. x, y)

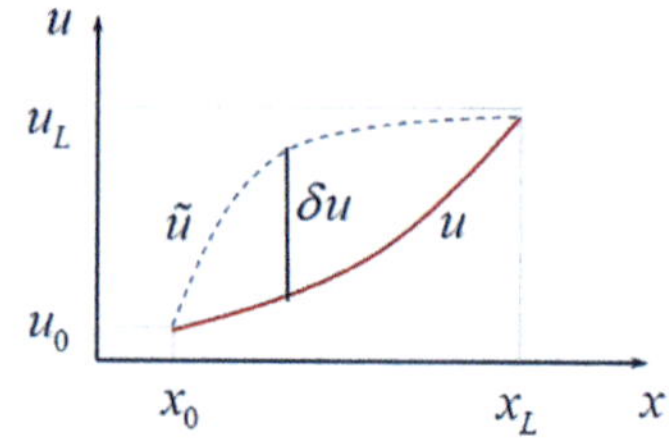

FIGURE 3.2 Graphical depiction of the function u, varied function $\tilde{u}$ and the variation δu.

can also be encountered in models of physical systems,

$$B(u, w) = \int_{\Omega} F(x, y, u, w, u_{,x}, u_{,y}, w_{,x}, w_{,y}, u_{,xx}, u_{,yy}, w_{,xx}, w_{,yy}, \ldots) d\Omega \tag{3.22}$$

A functional develops naturally when the total potential energy of a system is expressed as shown in Example 3.2, or in weighted residual formulations of boundary value problems, as will be introduced later in the chapter.

Example 3.2 The total potential energy (see Section 2.2.9) of a bar fixed at one end, subjected to an axial load F_L on the other end ($x = L$) and a distributed force $q(x)$ along its entire length is a functional:

$$\Pi_p = \mathcal{U} - W = \int_0^L \frac{1}{2} \sigma \, \varepsilon \, A dx - \left(\int_0^L q u dx + F_L u(L) \right) \tag{3.23}$$

By using the Hooke's law $\sigma = E\varepsilon$ and recalling the definition of normal strain $\varepsilon = u_{,x}$ we find:

$$\Pi_p(u, u_{,x}) = \int_0^L \left(\frac{1}{2} E A u_{,x}^2 - q u \right) dx - F_L u(L) \tag{3.24}$$

3.3.1 Variation of a functional

Consider the functional given in Eq. (3.21). Let's assume first, that the values of the independent variable u on the boundaries are given as,

$$u(x_0) = u_0 \text{ and } u(x_L) = u_L \tag{3.25}$$

where u_0 and u_L are prescribed values. Thus, we see that *essential boundary conditions* are specified on both ends of the domain. Recall that $I(u)$ is a scalar, therefore its value between these points will depend on the function $u(x)$ or the chosen path between the end points (Fig. 3.2).

Let's assume that a path $u(x)$ which extremizes the functional I exists. We will see that this path, called the *extremal path*, is the function that represents the solution of the problem. Let's also call all of the other paths between these points, $\tilde{u}(x)$, *varied paths*. In Fig. 3.2, we can see that the varied paths can be defined as follows:

$$\tilde{u} = u + \varepsilon\, v \tag{3.26}$$

where ε is an arbitrarily small parameter and $v(x)$ is any differentiable function. The varied function $\tilde{u}$ and $v(x)$, should have the same values as the function u at the end points. Thus we see that the varied function $v(x)$ should have the following property,

$$v(x_0) = 0 \text{ and } v(x_L) = 0 \tag{3.27}$$

In other words, v(x) should satisfy the homogenous form of the essential boundary conditions.

The difference between the extremal path and one of the varied paths is defined as follows:

$$\delta u = \tilde{u} - u \tag{3.28}$$

By using Eq. (3.26), the *variation of u, δu* can be found as follows:

$$\delta u = \tilde{u} - u = \varepsilon\, v \tag{3.29}$$

The function $\delta u = \delta u(x)$ represents the *variation* of $\tilde{u}(x)$ from $u(x)$. The symbol δ is called the *del-operator* or the *variational operator*. The variational operator represents variation of the function rather than a pointwise difference between the functions. Comparing Eqs. (3.29) with (3.27), we see that $\delta u(x)$ should have the same property (3.27) as $v(x)$. Thus the boundary conditions for the del-operator can be expressed as follows.

$$\delta u(x_0) = 0 \text{ and } \delta u(x_L) = 0 \tag{3.30}$$

This holds on a boundaries where essential boundary conditions are specified. In other words, on the boundaries where the value of the function is specified, the variation of $u(x)$ should be zero.

3.3.2 Functional derivative

While the dependent variable changes from u to $u + \delta u$ the functional changes from $I(u)$ to $I(u) + \delta I(u)$. The variation represents a change from a stationary function to a function close to it. Therefore, in order to find the stationary function the variation of the functional would have to be zero. Let's consider the variation of a functional that is a function of u and $u_{,x}$, such as Eq. (3.21). Let's also assume that this functional does not vary on its boundaries. The stationary value of such a function, can be found by setting the variation of the functional to zero,

$$\delta I(u) = \delta \int_{x_0}^{x_L} F(x, u, u_{,x})dx = \int_{x_0}^{x_L} \delta F(x, u, u_{,x})dx = 0 \tag{3.31}$$

The variation of the function F inside the functional can be expressed as follows [2],

$$\delta F = \frac{\partial F}{\partial x}\delta x + \frac{\partial F}{\partial u}\delta u + \frac{\partial F}{\partial u_{,x}}\delta u_{,x} \tag{3.32}$$

However, the variational operator δ measures the change of the value of the functional as the function u is varied. In fact it is not concerned with the effects of point-to-point changes in u. Thus the first term in Eq. (3.32) is zero, and the *variational derivative* of $F(x, u, u_{,x})$ becomes,

$$\delta F = \frac{\partial F}{\partial u}\delta u + \frac{\partial F}{\partial u_{,x}}\delta u_{,x} \tag{3.33}$$

It is instructive to note the similarities and the differences between the variational derivative operator and the total differential operator. Recall the total differential df of a function $f = f(x, y, z)$ is given as follows:

$$df = \frac{\partial f}{\partial x}dx + \frac{\partial f}{\partial y}dy + \frac{\partial f}{\partial z}dz \tag{3.34}$$

and, the total differential dF of the function $F(x, u, u_{,x})$ is given as follows:

$$dF = \frac{\partial F}{\partial x}dx + \frac{\partial F}{\partial u}du + \frac{\partial F}{\partial u_{,x}}du_{,x} \tag{3.35}$$

As stated above the variation sought by the variational operator is on the effects of the value of the function but not on the effects of the point-to-point variations of the function. This is sometimes stated as *the variational operator δ seeks a variation of F at a fixed x location.*

3.3.3 Properties of functionals

A generic functional $I(u)$ is *linear* if it satisfies the following condition:

$$I(\alpha u + \beta v) = \alpha I(u) + \beta I(v) \tag{3.36}$$

where α and β two arbitrary scalars and u and v are two arbitrary functions. A generic functional $B(u, v)$ of two dependent variables is *bilinear* if both dependent variables can be shown to be linear separately. The linearity of the first argument can be shown if the following statement can be satisfied,

$$B(\alpha u_1 + \beta u_2, v) = \alpha B(u_1, v) + \beta B(u_2, v) \tag{3.37}$$

where u_1 and u_2 are two arbitrary functions. Similarly the linearity of the second argument can be shown if,

$$B(u, \alpha v_1 + \beta v_2) = \alpha B(u, v_1) + \beta B(u, v_2) \tag{3.38}$$

can be satisfied for any dependent variables v_1 and v_2. A bilinear functional is *symmetric* if the following statement can be demonstrated:

$$B(u, v) = B(v, u) \tag{3.39}$$

3.3.4 Properties of the variational derivative

The following properties of the variational derivative are analogous to differentiation and they are stated without proof. For $F_1 = F_1(u)$ and $F_2 = F_2(u)$ it can be shown that the following properties of the derivative operator hold for the variational derivative operator as well [3]:

$$\text{Variation of the sums:} \quad \delta(F_1 \pm F_2) = \delta F_1 \pm \delta F_2 \tag{3.40a}$$

$$\text{Variation of the products:} \quad \delta(F_1 F_2) = F_2 \delta F_1 + F_1 \delta F_2 \tag{3.40b}$$

$$\text{Variation of the ratios:} \quad \delta\left(\frac{F_1}{F_2}\right) = \frac{F_2 \delta F_1 - F_1 \delta F_2}{F_2^2} \tag{3.40c}$$

$$\text{Variation of the powers:} \quad \delta\left[(F_1)^n\right] = n(F_1)^{n-1} \delta F_1 \tag{3.40d}$$

$$\text{Commutation with differential operator:} \quad \delta\left(\frac{du}{dx}\right) = \frac{d}{dx}(\delta u) \tag{3.40e}$$

$$\text{Commutation with integral operator:} \quad \delta \int_0^b u(x)dx = \int_a^b \delta u(x)dx \tag{3.40f}$$

Example 3.3 :

$$\text{Linear functional: } l(v) = \int_0^L v\, f\, dx + \frac{dv}{dx}(L) M_0$$

$$\text{Bilinear functional: } B(v, w) = \int_0^L a \frac{dv}{dx} \frac{dw}{dx} dx$$

where $f(x)$, M_0, $a = a(x)$ are given.

3.3.5 Euler–Lagrange equations and boundary conditions

We indicated that the value of $I(u)$ depends on the chosen varied path $\tilde{u}$. It can be stated that as the chosen path approaches the extremal path, $\tilde{u} \to u$, the variation of the functional will approach zero, $\delta I \to 0$. Thus, for the extremal path we have the condition, $\delta I = 0$ stated in Eq. (3.31). Let us evaluate the condition $\delta I = 0$ by using the definition of the variational derivative given in Eq. (3.33). Note that we will be using many of the properties of the variational derivative operator stated in Eqs. (3.40a)–(3.40f) without explicitly pointing to them.

$$\delta I(u) = \delta \int_{x_0}^{x_L} F(x, u, u_{,x}) dx = 0$$

$$= \int_{x_0}^{x_L} \delta F(x, u, u_{,x}) dx = \int_{x_0}^{x_L} \left(\frac{\partial F}{\partial u} \delta u + \frac{\partial F}{\partial u_{,x}} \delta u_{,x} \right) dx = 0 \tag{3.41}$$

Note that the term $\delta u_{,x} = \delta(\partial u/\partial x)$ or $\delta u_{,x} = \partial(\delta u)/\partial x$. Thus, Eq. (3.41) can be stated as,

$$\delta I(u) = \int_{x_0}^{x_L} \left(\frac{\partial F}{\partial u} \delta u + \frac{\partial F}{\partial u_{,x}} \frac{\partial(\delta u)}{\partial x} \right) dx = 0 \tag{3.42}$$

By using integration by parts on the second term of this equation we can obtain an equivalent expression that only has δu in the integral,

$$\delta I(u) = \int_{x_0}^{x_L} \frac{\partial F}{\partial u} \delta u dx + \int_{x_0}^{x_L} \frac{\partial F}{\partial u_{,x}} \frac{\partial(\delta u)}{\partial x} dx$$

$$= \int_{x_0}^{x_L} \frac{\partial F}{\partial u} \delta u dx + \left[\delta u \left(\frac{\partial F}{\partial u_{,x}} \right) \right]_{x_0}^{x_L} - \int_{x_0}^{x_L} \delta u \frac{\partial}{\partial x} \left(\frac{\partial F}{\partial u_{,x}} \right) dx = 0 \tag{3.43}$$

After rearranging we find the following final form:

$$\delta I(u) = \int_{x_0}^{x_L} \left[\frac{\partial F}{\partial u} - \frac{\partial}{\partial x} \left(\frac{\partial F}{\partial u_{,x}} \right) \right] \delta u \, dx + \delta u(x_L) \left(\frac{\partial F}{\partial u_{,x}} \right)_{x_L} - \delta u(x_0) \left(\frac{\partial F}{\partial u_{,x}} \right)_{x_0} = 0 \tag{3.44}$$

Note that we initially assumed that δu is an arbitrary variation. Therefore, in order for this equation to be satisfied, each one of the terms must vanish separately.

Considering the integral term in Eq. (3.44) we note that the term in the square brackets must always be zero as stated by the fundamental lemma of variational calculus stated in Section 3.2.4. We can also reach the same conclusion by considering that the variation δu is not always zero; therefore, the integral can only vanish if the term in the square brackets is equal to zero. We thus find the following equation from this condition:

$$\frac{\partial F}{\partial u} - \frac{\partial}{\partial x} \left(\frac{\partial F}{\partial u_{,x}} \right) = 0 \tag{3.45}$$

Eq. (3.45) is known as the ***Euler–Lagrange equation*** for a functional of the type $F = F(u, u_{,x})$ as the one given in Eq. (3.41). This equation represents a differential equation that must be satisfied by the unknown function $u(x)$. In fact, it represents an equilibrium equation depending on the physics of the problem.

The second and third terms of Eq. (3.44) are evaluated on the boundaries of the integral. They represent the conditions that must be satisfied on the boundaries. Thus, we find,

$$\text{at } x = x_0 : \text{ either } \delta u(x_0) = 0 \tag{3.46a1}$$

or

$$\frac{\partial F}{\partial u_{,x}} = 0 \tag{3.46a2}$$

and

$$\text{at } x = x_L : \text{ either } \delta u(x_L) = 0 \tag{3.46b1}$$

or

$$\frac{\partial F}{\partial u_{,x}} = 0 \tag{3.46b2}$$

In particular, the condition that $\delta u = 0$ on a given boundary in equations (a1) and (b1) states existence of a specified value for u at that boundary, such as $u(x_0) = u_0$ at $x = x_0$ or $u(x_0) = u_L$ at $x = x_L$. Thus we conclude that the second and third terms of (3.44) give the proper form of the boundary conditions that must be applied on the boundaries to be of the following form,

$$\text{at } x = x_0 : \text{ either } u \text{ is specified} \tag{3.47a1}$$

or

$$\frac{\partial F}{\partial u_{,x}} = 0 \tag{3.47a2}$$

and

$$\text{at } x = x_L : \text{ either } u \text{ is specified} \tag{3.47b1}$$

or

$$\frac{\partial F}{\partial u_{,x}} = 0 \tag{3.47b2}$$

Clearly, there are limited choices for the boundary conditions that can be applied on a boundary. Eq. (3.47a1)–(3.47b2) provide the ***admissible*** (or the proper form of the) ***boundary conditions***.

The condition is said to be an ***essential boundary condition*** (EBC) if the value of the dependent variable is specified. Eqs. (3.47a1) and (b1) represent the essential boundary conditions.

On the other hand, the type of condition represented by (3.47)-(a2) and (b2) is said to be the ***natural boundary condition*** (NBC). Thus the condition

TABLE 3.1 Combinations of admissible boundary conditions for the second order equation given in Example 3.1.

x_0	x_L
u	u
u	$a\frac{du}{dx}$
$a\frac{du}{dx}$	u
$a\frac{du}{dx}$	$a\frac{du}{dx}$

$\partial F/\partial u_{,x} = 0$ is a natural boundary condition for this functional. It will become apparent later in the text that the natural boundary conditions represent the physical drivers (effects) on the boundaries.

It is also very important to note that Eq. (3.44) admits only one boundary condition on a given boundary. That is *on a given boundary either an essential or a natural boundary condition can be prescribed, but, both type of boundary conditions may not be specified on the same boundary.*

It is left as an excercise to the reader to show that, for the second order equation given in Eq. (3.3a) the admissible boundary conditions are those summarized in Table 3.1.

Example 3.4a Consider the functional,

$$I(u) = \int_0^L \left[\frac{1}{2}EAu_{,x}^2 - qu\right]dx.$$

Find the Euler–Lagrange equation and the admissible boundary conditions for this functional.

Solution: After recognizing that $F = \frac{1}{2}EAu_{,x}^2 - qu$ we can find the Euler–Lagrange equation and the admissible form of the boundary conditions as follows:

The Euler-Lagrange-equation: $\frac{\partial F}{\partial u} - \frac{\partial}{\partial x}\left(\frac{\partial F}{\partial u_{,x}}\right) = 0$ leads to $-q - \frac{\partial}{\partial x}(EAu_{,x}) = 0$

This can be written in a more familiar form as follows: $\frac{\partial}{\partial x}\left(EA\frac{\partial u}{\partial x}\right) = -q$ which coincides with the equilibrium equation of an elastic bar subjected to distributed forces along its length.

The general form of the boundary conditions are found as follows:

EBCs: at $x = 0$: $u(0)$ is specified and at $x = L$: $u(L)$ is specified.

NBCs: at $x = 0, L$: $\frac{\partial F}{\partial u_{,x}} = \frac{\partial}{\partial u_{,x}}\left(\frac{1}{2}EAu_{,x}^2 - qu\right) = 0$, which leads to

$$EA\frac{\partial u}{\partial x} = 0$$

PACKING SLIP

1 4 3 9 6 4 8

Via

Market ID: 76926298
Order ID: 1439648
Order Date: 09/16/2026
Shipping Method:

Ship To

Alibris APEX DC 76926298-95
APEX
300 AVONDALE AVE.
GRANDVIEW HEIGHTS, OH 43212-3473, US

SKU	QTY	TITLE	TOTAL
BSM.184AI	1	Finite Element Method: Physics and Solution Methods	93.83

Order Total: $93.83

Your order has been packaged and created with great care. Enjoy! Bay State Books

#1439648

Example 3.4b Consider the functional,

$$I(u) = \int_0^L \left[\frac{1}{2} EAu_{,x}^2 - qu \right] dx - u(L)Q_L.$$

Find the equation of equilibrium associated with this functional and the admissible boundary conditions for this functional.

Solution: Note that in the development up to this point the funtionals we considered did not involve a term on the boundary, $u(L)Q_L$. The Euler-Lagrange equations and particularly the boundary conditions will be affected by the presence of this term in the functional. Therefore, let's start by minimizing the functional using the variational derivative to answer this question.

$$\delta I(u) = \delta \left[\int_0^L \left[\frac{1}{2} EAu_{,x}^2 - qu \right] dx - u(L)Q_L \right] = 0$$

Using the commutation property this expression becomes,

$$\delta I(u) = \int_0^L \delta \left[\frac{1}{2} EAu_{,x}^2 - qu \right] dx - \delta u(L)Q_L = 0$$

The variational derivative operation gives,

$$\delta I(u) = \int_0^L \left(EAu_{,x}\delta u_{,x} - q\delta u \right) dx - \delta u(L)Q_L = 0$$

Recognize that $\delta u_{,x} = \delta(\partial u/\partial x) = \partial(\delta u)/\partial x$. Using this relationship in the above we find,

$$\delta I(u) = \int_0^L \left[EA\frac{\partial u}{\partial x}\frac{\partial \delta u}{\partial x} - q\delta u \right] dx - \delta u(L)Q_L = 0$$

Using integration by parts on the first term,

$$\delta I(u) = \delta u(L)\left(EA\frac{\partial u}{\partial x} \right)_{x=L} - \delta u(0)\left(EA\frac{\partial u}{\partial x} \right)_{x=0}$$
$$- \int_0^L \left[\delta u \frac{\partial}{\partial x}\left(EA\frac{\partial u}{\partial x} \right) + q\delta u \right] dx - \delta u(L)Q_L = 0$$

Rearranging this equation we find the final form of the equation,

$$\delta I(u) = \delta u(L)\left[\left(EA\frac{\partial u}{\partial x}\right)_{x=L} - Q_L\right] - \delta u(0)\left(EA\frac{\partial u}{\partial x}\right)_{x=0}$$
$$- \int_0^L \left[\frac{\partial}{\partial x}\left(EA\frac{\partial u}{\partial x}\right) + q\right]\delta u dx = 0$$

In order for this equation to be valid each one of its terms must vanish separately. We find the equilibrium equation by using the fundamental lemma of variational calculus on the integral term,

$$\frac{\partial}{\partial x}\left(EA\frac{\partial u}{\partial x}\right) + q = 0$$

Setting the first and the second terms to zero, we find the admissible boundary conditions as follows:

$$\text{at } x = 0: \text{ either } u(0) \text{ is specified, or } \left(EA\frac{\partial u}{\partial x}\right)_{x=0} = 0,$$
$$\text{at } x = L: \text{ either } u(L) \text{ is specified, or } \left(EA\frac{\partial u}{\partial x}\right)_{x=L} - Q_L = 0.$$

3.4 Weighted residual integral and the weak form of boundary value problems

In general, integral formulations of the boundary value problems are based on weighted integral statements of the governing equations. In this section, we will define the *weak form* of a *boundary value problem* as a weighted-integral statement of the differential equation in which differentiation is distributed evenly between the dependent variable u and a weight function w. We will also see that an advantage of the weak-form, among others, is that it includes boundary condition in the formulation.

3.4.1 Weighted residual integral

Let us first define the *residual* of the governing equation. For example, the residual of Eq. (3.1) is defined as follows:

$$R(u) = \mathcal{L}(u) - f = 0 \text{ in } \Omega \tag{3.48}$$

Note that the residual, R is zero, if the *exact solution* u of the boundary value problem is known. Otherwise, if we have an approximate solution, the residual will have an error ε which can be represented as follows:

$$R(\tilde{u}) = \varepsilon \neq 0 \tag{3.49}$$

The weighted-residual statement of the original differential equation or simply the *weighted-residual integral* (WRI) is a bilinear functional defined as follows:

$$I(u, w) = \int_{\Omega} w\ R\ d\Omega = 0 \tag{3.50}$$

where w is a *weight function*. This can be any nonzero integrable function that helps to find an approximate solution to the problem.

For Eq. (3.3a), the residual becomes,

$$R = \left[\frac{d}{dx}\left(a\frac{du}{dx}\right) + cu + q\right] = 0 \text{ in } x_0 < x < x_L \tag{3.51}$$

and the corresponding weighted residual integral is,

$$I(u, w) = \int_{x_0}^{x_L} wRdx = \int_{x_0}^{x_L} w\left[\frac{d}{dx}\left(a\frac{du}{dx}\right) + cu + q\right]dx = 0 \tag{3.52}$$

3.4.2 Boundary conditions

The order of differentiation of the dependent variable u in the weighted residual statement can be reduced by using *integration by parts* until the differentiation is distributed evenly between the dependent variable u and the weight function w. *The boundary terms that results from integration by parts represent the admissible boundary conditions of the problem.*

Let's demonstrate this on the example boundary value problem whose weighted residual integral form is given by Eq. (3.52). In order to distribute the differentiation operator evenly between the weight function w and the independent variable u let's use integration by parts on the first term,

$$\int_{x_0}^{x_L} \left[w\frac{d}{dx}\left(a\frac{du}{dx}\right) + c\,w\,u + w\,q\right]dx = 0 \tag{3.53a}$$

$$w(x_L)\left(a\frac{du}{dx}\right)_{x_L} - w(x_0)\left(a\frac{du}{dx}\right)_{x_0} - \int_{x_0}^{x_L}\left[a\frac{du}{dx}\frac{dw}{dx} - c\,w\,u - w\,q\right]dx = 0 \tag{3.53b}$$

Note that in order for Eq. (3.53a) to be satisfied, each one of its terms must vanish separately. Thus it can be easily shown that on the boundaries the following conditions must be satisfied,

$$\text{at } x = x_0 : \text{ either } w(x_0) = 0 \tag{3.54a1}$$

or

$$a(x_0)\frac{du}{dx}\bigg|_{x_0} = 0 \tag{3.54a2}$$

and

$$\text{at } x = x_L : \text{ either } w(x_L) = 0 \tag{3.54b1}$$

or

$$a(x_L)\left.\frac{du}{dx}\right|_{x_L} = 0 \tag{3.54b2}$$

Close inspection shows that conditions (a1) and (b1), i.e., $w(x_0) = 0$ and $w(x_L) = 0$, are in the same form as the conditions (3.46a1) and (3.46b1), i.e., $\delta u(x_0) = 0$ and $\delta u(x_L) = 0$, obtained for the variation δu on the boundaries. Thus we note that w is an arbitrary function, but its value becomes zero on a boundary where u is specified. As both w and δu have the same property on the boundaries, but otherwise arbitrary functions, we can (and will) interchange w with δu. By replacing $w(x_0)$ with $\delta u(x_0)$ and $w(x_L)$ with $\delta u(x_L)$ we can rewrite Eq. (3.54a1) as follows:

$$\text{at } x = x_0 : \text{either } u \text{ is specified} \tag{3.55a1}$$

or

$$a(x_0)\left.\frac{du}{dx}\right|_{x_0} = 0 \tag{3.55a2}$$

and

$$\text{at } x = x_L : \text{either } u \text{ is specified} \tag{3.55b1}$$

or

$$a(x_L)\left.\frac{du}{dx}\right|_{x_L} = 0 \tag{3.55b2}$$

Thus we find the admissible boundary conditions for the BVP given in Example 3.1 as follows:

Essential boundary condition u is specified on a boundary

Natural boundary condition $a(x)\frac{du}{dx} = 0$ on a boundary

Observation: The foregoing discussion can be generalized to find the admissible boundary conditions of a Sturm-Liouville differential equation.

The type and form of the admissible boundary conditions can be obtained by investigating the *boundary terms*, once the weighted residual integral of the differential equation is integrated by parts. This will show that,

- The EBCs are given by the terms that involve the weight function w, and
- The NBCs are given by the terms that multiply the weight function w.

3.4.3 The weak form

The weak form of a differential equation is obtained by applying the prescribed boundary conditions to the weighted residual integral, following the integration by parts. For example, recall the boundary conditions of the example boundary value problem in Eqs. (3.3b), (3.3c),

$$\text{at } x = x_0: \qquad u(x_0) = u_0$$

$$\text{at } x = x_L: \quad a(x_L)\left.\frac{du}{dx}\right|_{x_L} = Q_L$$

and the weighted residual integral for this example is given in Eq. (3.53b). Note that an essential boundary condition is specified at $x = x_0$. This implies that $\delta u(x_0) = 0$ and therefore we should use $w(x_0) = 0$. A natural boundary condition is specified on the other boundary, $x = x_L$. This implies that the quantity $a(x_L)\ u_{,x}(x_L)$ must take the value Q_L. These observations are implemented in Eq. (3.53b) which becomes,

$$\underbrace{w(x_L)\left(a(x_L)\left.\frac{du}{dx}\right|_{x_L}\right)}_{=Q_L} - \underbrace{w(x_0)}_{=0}\left(a(x_0)\left.\frac{du}{dx}\right|_{x_0}\right) - \int_{x_0}^{x_L}\left[a\frac{du}{dx}\frac{dw}{dx} - c\,w\,u - w\,q\right]dx = 0$$

The *weak form* of the boundary value problem described in Example 3.1 then becomes,

$$w(x_L)Q_L - \int_{x_0}^{x_L}\left[a\frac{du}{dx}\frac{dw}{dx} - c\,w\,u - w\,q\right]dx = 0 \tag{3.56}$$

It should be noted that the natural boundary condition is now incorporated in the integral formulation. This will be helpful later when we develop solution methods.

3.4.4 Relationship between the weak form and functionals

We showed in the previous section that the weight function w and the variation of the dependent variable δu behave in the same way. Therefore, as w is an arbitrary function, we can use δu in its place. The weak form of the boundary value problem given in Example 3.1 becomes,

$$-\int_{x_0}^{x_L}\left[a\frac{du}{dx}\frac{d\delta u}{dx} - c\,u\,\delta u\right]dx + \int_{x_0}^{x_L} q\,\delta u dx + Q_L\delta u(x_L) = 0 \tag{3.57}$$

Noting that $\frac{1}{2}\delta\left(u^2\right) = u\delta u$ and $\frac{1}{2}\delta\left(u_{,x}^2\right) = u_{,x}\delta u_{,x}$ the first term is modified as follows:

$$-\int_{x_0}^{x_L}\frac{1}{2}\delta\left[a\left(\frac{du}{dx}\right)^2 - c\,u^2\right]dx + \int_{x_0}^{x_L} q\,\delta u\,dx + \delta u(x_L)\,Q_L = 0$$

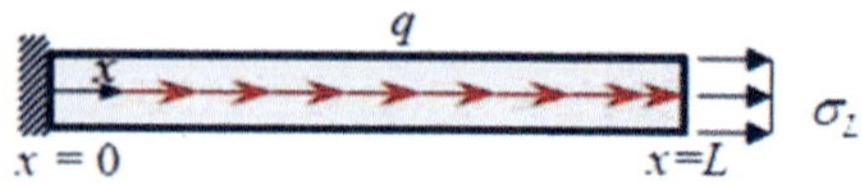

FIGURE 3.3 A slender bar fixed on one end and subjected to distributed force q along its surface and traction σ_0 on the other end.

Using the commutation property of the variational derivative the following relationship is obtained,

$$\delta\left[-\frac{1}{2}\int_{x_0}^{x_L}\left(a\left(\frac{du}{dx}\right)^2 - cu^2\right)dx + \int_{x_0}^{x_L} q\,u\,dx + u(x_L)Q_L\right] = 0 \tag{3.58}$$

Note that the term in the square brackets is a functional $I(u,\ u_{,x})$ and Eq. (3.58) represents the variation of the functional δI. This result indicates that the weak-integral-form of the boundary value problem, e.g. Eq. (3.57), is equal to mimimum of the variation of the functional I associated with the problem.

For solid mechanics problems $I(u,\ u_{,x})$ is the total potential energy of the system. The statement

$$\delta I(u, u_{,x}) = 0 \tag{3.59}$$

represents the *minimum total potential energy principle: of all admissible functions u, that which makes the total potential energy I(u) a minimum, also satisfies the differential equation of equilibrium and natural boundary conditions.*

Example 3.5 Stretching of a slender bar

Consider a uniform elastic bar loaded by a distributed axial load q and normal stress σ_L applied at its end as shown in Fig. 3.3. Find the weak form of this problem, identify the associated functional, linear and bilinear forms.

Solution: The axial displacement of this bar can be found from the following BVP:

$$\text{BVP:}\quad AE\frac{d^2u}{dx^2} = -q(x),\quad \text{for } 0 < x < L, \tag{3.60a}$$

$$\text{EBC:}\quad u(0) = 0, \tag{3.60b}$$

$$\text{NBC:}\quad Eu_{,x}|_{x=L} = \sigma_L. \tag{3.60c}$$

Compare these equations with Eq. (3.3a) given in Example 3.1. We see that in this case, $a(x) = AE;\ Q_L = \sigma_L A;\ u_0 = 0$. We can use the weak-form of the weighted-residual integral developed in (3.56) to get,

$$\int_0^L\left[AE\frac{dw}{dx}\frac{du}{dx} - wq\right]dx - w(L)(\sigma_L A) = 0 \tag{3.61}$$

We identify bilinear and linear functionals in Eq. (3.61) as follows:

$$B(w, u) = AE\int_0^L \frac{dw}{dx}\frac{du}{dx}dx \tag{3.62a}$$

$$\ell(w) = \int_0^L wqdx + w(L)(\sigma_L A) \tag{3.62b}$$

Making the substitution $\delta u \rightarrow w$ and using the properties of the variational derivative, the bilinear form can also be expressed as follows:

$$B(w, u) = B(\delta u, u) = \frac{1}{2}\delta B(u, u) = \frac{1}{2}\delta\left[\int_0^L AE\left(\frac{du}{dx}\right)^2 dx\right] \tag{3.63.a}$$

Similarly, for $\ell(w)$,

$$\ell(w) = \ell(\delta u) = \delta(\ell(u)) = \delta\left(\int_0^L uqdx + w(L)(\sigma_L A)\right) \tag{3.63.b}$$

Combining these two equations we can identify the variation of the functional δI,

$$\delta I = \delta\left(\frac{1}{2}B(u, u) - l(u)\right) = 0$$

where the functional $I(u)$ can be expressed as follows:

$$I(u) = \frac{1}{2}\int_0^L AE\left(\frac{du}{dx}\right)^2 dx - \int_0^L qudx - u(L)(\sigma_L A) \tag{3.64}$$

This functional is the total potential energy of the system Π_p, where the terms are identified as follows:

$\frac{1}{2}\int_0^L AE\left(\frac{du}{dx}\right)^2 dx$: Total strain energy

$\int_0^L qudx$: Work done by external distributed load (traction).

$u(L)(\sigma_L A)$: Work done by the end load $Q_L = \sigma_L A$.

This example demonstrates that for a given boundary value problem, we can obtain a functional by finding its weak-weighted-residual form. In this case the functional represents the total potential energy of the system. Conversely, if we know the total potential energy of the system, π_p, by minimizing the $\delta\pi_p = 0$ we can find the governing equation of equilibrium and the boundary conditions associated with the problem (strong form).

Example 3.6: Consider the BVP given below where a and c are known functions.

a) Obtain the weak form of the BVP

b) Express the weak form as a functional of dependent variable u.

$$\text{BVP:} \ -\frac{d}{dx}\left(a\frac{du}{dx}\right) - cu + x^2 = 0 \ \text{for} \ 0 < x < 1 \tag{3.65a}$$

$$\text{BCs:} \ u(0) = 0, \quad \left(a\frac{du}{dx}\right)\Big|_{x=1} = 1 \tag{3.65b}$$

Solution:

$$\text{Let's first define the residual: } R = -\frac{d}{dx}\left(a\frac{du}{dx}\right) - cu + x^2 \tag{3.66}$$

a) Weighted residual integral:

$$\int_0^1 w\,R\,dx = -\int_0^1 w\,\frac{d}{dx}\left(a\frac{du}{dx}\right)dx + \int_0^1 \left(wx^2 - cwu\right)dx = 0 \tag{3.67}$$

After integrating the first term by parts we find,

$$-\left\{\left[a\frac{du}{dx}w\right]_0^1 - \int_0^1 a\frac{dw}{dx}\frac{du}{dx}dx\right\} + \int_0^1 (x^2w - cwu)dx = 0 \tag{3.68}$$

$$\int_0^1 a\frac{dw}{dx}\frac{du}{dx}dx - \int_0^1 (cwu - x^2w)dx - \left[a\frac{du}{dx}w\right]_0^1 = 0 \tag{3.69}$$

Using the boundary conditions (3.65b) given for this problem we get

$$u(0) = 0 \quad \Rightarrow \quad w(0) = 0$$

$$a\frac{du}{dx}\bigg|_{x=1} = 1 \quad \Rightarrow \quad w(1)a\frac{du}{dx} = w(1)$$

Thus we find that the boundary term is reduced to:

$$\left[wa\frac{du}{dx}\right]_0^1 = w(1) \tag{3.70}$$

b) The linear and bilinear components of the functional given in Eq. (3.69) are identified as follows:

$$B(u, w) = \int_0^1 \left(a\frac{dw}{dx}\frac{du}{dx} - cwu\right)dx \tag{3.71a}$$

$$\ell(w) = -\int_0^1 wx^2dx + w(1) \tag{3.71b}$$

Consider Eq. (3.71a) and replace w with δu,

$$B(\delta u, u) = \int_0^1 \left(a \frac{d(\delta u)}{dx}\frac{du}{dx} - c\, u\, \delta u \right) dx \tag{3.72}$$

We can use the properties of the variational derivative operator as follows:

$$\delta\left(\frac{du}{dx}\frac{du}{dx}\right) = \frac{d(\delta u)}{dx}\frac{du}{dx} + \frac{du}{dx}\frac{d(\delta u)}{dx} = 2\frac{d(\delta u)}{dx}\frac{du}{dx}$$

and find,

$$\frac{d(\delta u)}{dx}\frac{du}{dx} = \frac{1}{2}\delta\left(\frac{du}{dx}\frac{du}{dx}\right) \tag{3.73}$$

Similarly we show the following relationship,

$$\delta\left(cu^2\right) = 2cu\delta u \quad \Rightarrow c\, u\, \delta u = \frac{1}{2}\delta\left(cu^2\right) \tag{3.74}$$

Then $B(\delta u, u)$ given in Eq. (3.72) becomes,

$$\frac{1}{2}\delta(B(u, u)) = \frac{1}{2}\delta \int_0^L \left(a\left(\frac{du}{dx}\right)^2 - cu^2 \right) dx \tag{3.75}$$

Next, we replace w with δu in Eq. (3.71b),

$$\ell(\delta u) = -\int_0^1 (\delta u) x^2 dx + \delta(u(1)) \tag{3.76}$$

It is straightforward to show that,

$$\ell(\delta u) = \delta(\ell(u)) = \delta\left[-\int_0^1 ux^2 dx + u(1) \right] \tag{3.77}$$

Combining Eqs. (3.75) and (3.77), we find:

$$\delta(B(u, u) - \ell(u)) = \delta\left[\frac{1}{2}\int_0^1 \left[a\left(\frac{du}{dx}\right)^2 - cu^2 + 2ux^2 \right] dx - u(1) \right] = 0 \tag{3.78}$$

And, finally the functional is found as follows:

$$I(u) = \frac{1}{2}\int_0^1 \left[a\left(\frac{du}{dx}\right)^2 - cu^2 + 2ux^2 \right] dx - u(1) \tag{3.79}$$

Example 3.7 Weak form of the Euler–Bernoulli beam equation

Consider the problem of finding the solution $v(x)$ to the differential equation:

$$\frac{d^2}{dx^2}\left(b\frac{d^2v}{dx^2}\right) = f_y \text{ in } 0 < x < L \tag{3.80}$$

where $f_y(x)$ and $b(x)$ are known functions. This differential equation represents the transverse deflection of an Euler–Bernoulli beam with bending rigidity $b(x)$ and subjected to an external force per unit width $f_y(x)$.

a) Identify the general forms of the essential and natural boundary conditions for Eq. (3.80) by using the weighted residual formulation.

b) Obtain the weak form of the system for the following prescribed boundary conditions:

$$v(0) = 0 \tag{3.81a}$$

and

$$\left.\frac{dv}{dx}\right|_{x=0} = 0 \tag{3.81b}$$

$$\left.b\frac{d^2v}{dx^2}\right|_{x=L} = M_0 \tag{3.81c}$$

and

$$\left.\frac{d}{dx}\left(b\frac{d^2v}{dx^2}\right)\right|_{x=L} = 0 \tag{3.81d}$$

where M_0 is a known independent variable.

c) Obtain the functional form of this equation.

Solution:

a)

$$\text{Residual: } R = \frac{d^2}{dx^2}\left(b\frac{d^2v}{dx^2}\right) - f_y = 0 \tag{3.82}$$

$$\text{Weighted residual form: } \int_0^L w\left[\frac{d^2}{dx^2}\left(b\frac{d^2v}{dx^2}\right) - f_y\right]dx = 0 \tag{3.83}$$

where w is an arbitrary weight function. In order to distribute differentiation of w and u equally, the first part of this integral is integrated by parts, twice. To accomplish this, let's define the following temporary variable,

$$\bar{v} = \frac{d}{dx}\left(b\frac{d^2v}{dx^2}\right)$$

and substitute $\bar{v}$ in Eq. (3.83),

$$\int_0^L w\left[\frac{d^2}{dx^2}\left(b\frac{d^2v}{dx^2}\right) - f_y\right]dx = \int_0^L \left[w\frac{d}{dx}(\bar{v}) - f_y w\right]dx$$

$$= [w\bar{v}]_0^L - \int_0^L \left[\bar{v}\frac{dw}{dx} + f_y w\right]dx$$

Substituting $\bar{v}$,

$$\int_0^L w\left[\frac{d^2}{dx^2}\left(b\frac{d^2v}{dx^2}\right) - f_y\right]dx$$

$$= \left[w\frac{d}{dx}\left(b\frac{d^2v}{dx^2}\right)\right]_0^L - \int_0^L \left[\frac{dw}{dx}\frac{d}{dx}\left(b\frac{d^2v}{dx^2}\right) + wf_y\right]dx$$

Next, the first term of the integral is integrated by parts one more time. To accomplish this let's define another temporary variable,

$$\bar{\bar{v}} = b\frac{d^2v}{dx^2}$$

The integral equation then becomes,

$$\int_0^L w\left[\frac{d^2}{dx^2}\left(b\frac{d^2v}{dx^2}\right) - f\right]dx = \left[w\frac{d}{dx}\left(b\frac{d^2v}{dx^2}\right)\right]_0^L - \int_0^L \left[\frac{dw}{dx}\frac{d}{dx}(\bar{\bar{v}}) + w\,f_y\right]dx$$

$$= \left[w\frac{d}{dx}\left(b\frac{d^2v}{dx^2}\right)\right]_0^L - \left[\frac{dw}{dx}\bar{\bar{v}}\right]_0^L + \int_0^L \left[\bar{\bar{v}}\frac{d^2w}{dx^2} + w\,f_y\right]dx$$

Substituting $\bar{\bar{v}}$, the weak form of Eq. (3.80) is found as follows:

$$\int_0^L \left[b\frac{d^2w}{dx^2}\frac{d^2v}{dx^2} - w\,f_y\right]dx + \left[w\frac{d}{dx}\left(b\frac{d^2v}{dx^2}\right)\right]_0^L - \left[\frac{dw}{dx}\left(b\frac{d^2v}{dx^2}\right)\right]_0^L = 0 \tag{3.84}$$

The terms in the square brackets represent the form of the admissible boundary conditions for Eq. (3.80). Recall that the weight function w is an arbitrary function that can be replaced by δv. Similarly, dw/dx can be replaced with $d(\delta v)/dx$ which is equivalent to $\delta(dv/dx)$. Eqs. (2.140a) and (2.143) show that the bending moment and shear force resultants are of the following form:

$$M = -b\frac{d^2v}{dx^2} \text{ and } V = -\frac{d}{dx}\left(b\frac{d^2u}{dx^2}\right)$$

We can then conclude that first boundary term gives the following boundary conditions:

At $x = 0$ and L either v is specified or $V = 0$

The second boundary terms gives the following boundary conditions:

At $x = 0$ and L either dv/dx is specified or $M = 0$.

Thus we see that at each boundary, two boundary conditions must be specified, resulting in a total of four conditions. This is not surprising as we note that the governing equation is a fourth order differential equation. We also conclude that the essential boundary conditions of this problem are specified on the variables,

$$v \text{ and } \frac{dv}{dx}$$

and the natural boundary conditions are specified on the variables,

$$b\frac{d^2v}{dx^2} \text{ and } \frac{d}{dx}\left(b\frac{d^2v}{dx^2}\right).$$

b) Weak-form for the given boundary conditions: The boundary conditions for this problem are given in Eq. (3.81a)–(3.81d) and are repeated below:

$$\text{Essential BCs: } v(0) = \left.\frac{dv}{dx}\right|_{x=0} = 0$$

implies weight function to be $w(0) = \left.\frac{dw}{dx}\right|_{x=0} = 0$

$$\text{Natural BCs: } \left(b\frac{d^2v}{dx^2}\right)_{x=L} = M_0 \text{ and } \left(\frac{d}{dx}\left(b\frac{d^2v}{dx^2}\right)\right)_{x=L} = 0$$

Thus the boundary-term of Eq. (3.84) will simplify considerably. The weak form of the boundary value problem defined by Eqs. (3.80, 3.81) becomes:

$$\int_0^L \left[b\frac{d^2w}{dx^2}\frac{d^2v}{dx^2} - wf_y\right]dx - M_0\left.\frac{dw}{dx}\right|_{x=L} = 0 \tag{3.85}$$

c) Functional form: The bilinear and linear functionals of Eq. (3.85), are identified as follows:

$$B(w, v) - \ell(w) = 0 \tag{3.86a}$$

$$B(w, v) = \int_0^L b\frac{d^2w}{dx^2}\frac{d^2v}{dx^2}dx \tag{3.86b}$$

$$\ell(w) = \int_0^L w\, f dx + M_0\left.\frac{dv}{dx}\right|_{x=L} \tag{3.86c}$$

It is now straight forward to replace the weight function with the variational operator, as demonstrated in Example 3.6.

$$B(w, v) \rightarrow B(\delta v, v) \rightarrow \frac{1}{2}\delta[B(v, v)]$$

$$\ell(w) \rightarrow \ell(\delta v) \rightarrow \delta[\ell(v)]$$

Then the functional $I(u)$ for this problem becomes:

$$I(v) = \int_0^L \left[\frac{1}{2}b\left(\frac{d^2v}{dx^2}\right) - f_y v\right]dx - M_0 \frac{dv}{dx}\bigg|_{x=L} \tag{3.87}$$

It can be easily verified that this functional is the total potential energy $\pi_p(v)$ of a cantilevered beam subjected to a distributed force f and end moment M_0.

Strain energy: $\frac{1}{2}\int_0^L b\left(\frac{d^2v}{dx^2}\right)^2 dx$

Work done by distributed load: $\int_0^L f_y\ u\ dx$

Work done by end moment: $M_0 \frac{dv}{dx}|_{x=L} = M_0\theta_{x=L}$

where θ is the slope of the deflected beam.

3.5 Method of weighted residuals

We now turn our attention to finding approximate solutions to boundary value problems. In particular, *the method of weighted residuals describes a group of solution methods to obtain approximate solutions to an integral equation expressed in the form of* Eq. (3.50).

We will seek an approximate solution $\tilde{u}$ for the problem. We chose this approximate solution as a superposition of certain *basis functions* $\phi_j(x)$, as follows:

$$\tilde{u}(x) = \phi_0(x) + c_1\phi_1(x) + c_2\phi_2(x) + \ldots + c_N\phi_N(x) = \phi_0(x) + \sum_{j=1}^{N} c_j\phi_j(x) \tag{3.88}$$

where c_j are unknown coefficients. Note that Eq. (3.88) represents a finite series with $N + 1$ basis functions. In order to find the N unknown coefficients Eq. (3.50) is generalized by using N *weight functions* w_i,

$$\int_\Omega w_i R(\tilde{u}) d\Omega = 0 \text{ for } i = 1, \ldots N \tag{3.89}$$

The choices for the basis and weight functions are not arbitrary. The following conditions should be followed, in choosing the basis and weight functions.

Basis functions: The basis functions ϕ_0, ϕ_i, should be chosen such that:

(i) The approximate solution $\tilde{u}$ satisfies the boundary conditions of the problem. (C1)
(ii) The approximate solution $\tilde{u}$ is as many times differentiable as required by the original differential equation. (C2)
(iii) The basis functions ϕ_0, ϕ_j are linearly independent. (C3)

Weight functions: The choice of the weight functions, w_i, depends on the method used.

These methods are the *Rayleigh–Ritz, Galerkin, Petrov–Galerkin, least-squares*, and *collocation* methods. The finite element method also uses a variational (weighted residual) approach. As we will see later, the choice of basis functions in the FEM are different from that of classical variational methods. In these notes we only introduce the Rayleigh-Ritz and the Galerkin methods. Detailed discussion of the other methods can be found for example in reference [1].

3.5.1 Rayleigh–Ritz method

The Rayleigh–Ritz method seeks an approximate solution of the form given in Eq. (3.88) *to the weak form of the boundary value problem.* For example, the weak form of Example-3.1 is approximated as follows:

$$w_i(x_L)Q_L - \int_{x_0}^{x_L} \left[a\frac{d\tilde{u}}{dx}\frac{dw_i}{dx} - cw_i\tilde{u} - w_i q \right] dx = 0 \text{ for } i = 1..N \tag{3.90}$$

where $w_i(x)$ are the weight functions. An approximate solution of the form,

$$\tilde{u}(x) = \phi_0(x) + \sum_{j=1}^{N} C_j\phi_j(x) \tag{3.91}$$

is sought, where C_j are the coefficients to be determined, $\phi_0(x)$ and $\phi_j(x)$ are the basis functions.

The weak form of the boundary value problem has certain advantages that are helpful in the selection of the basis functions. First, the natural boundary conditions of the problem are included in the weak form. This means that the natural boundary conditions can be approximately satisfied, if we happen to pick an approximate solution satisfying only the essential boundary conditions of the problem. Therefore, the condition (C1) can be relaxed. Second, the weak form has a reduced order of differentiation. Therefore, the continuity requirement stated in condition (C2) is reduced. With these caveats, the following conditions should be followed to pick the basis functions for the Rayleigh–Ritz method:

Basis functions: The basis functions ϕ_0, ϕ_j must be chosen such that:

i. The approximate solution $\tilde{u}$ satisfies at least the essential boundary conditions of the problem (C4)
ii. The approximate solution $\tilde{u}$ is as many times differentiable as required by the weak form integral (C5)
iii. The basis functions must be ϕ_0, ϕ_j linearly independent (C3)

Weight functions: In the Rayleigh–Ritz method, the weight functions are chosen as the same functions as the basis functions (i.e., $w_i = \phi_i$).

3.5.2 Galerkin method

The Galerkin method seeks an approximate solution of the form given in Eq. (3.88) *to the weighted residual form of the boundary value problem.* For example, the weighted residual form of Example 3.1 is approximated as follows:

$$\int_{x_0}^{x_L} w_i \left[\frac{d}{dx}\left(a\frac{d\tilde{u}}{dx} \right) + c\tilde{u} + q(x) \right] dx = 0 \text{ for } i = 1..N \tag{3.92}$$

The boundary conditions of this example were

$$u(x_0) = u_0, \text{ and } \left(a\frac{du}{dx} \right)_{x=x_L} = Q_L$$

and they play a prominent role in the choice of basis functions.

Basis functions: The basis functions ϕ_0, ϕ_j must be chosen such that:

i. The approximate solution $\tilde{u}$ satisfies the boundary conditions of the problem (C1)
ii. The approximate solution $\tilde{u}$ is as many times differentiable as required by the original differential equation (C2)
iii. The basis functions ϕ_0, ϕ_j are linearly independent (C3)

Weight functions: In Galerkin's method, the weight functions are chosen as the same functions as the basis functions (i.e., $w_i = \phi_i$).

3.5.3 Polynomials as basis functions for Rayleigh–Ritz and Galerkin methods

Neither one of Galerkin or Rayleigh–Ritz methods place any restrictions on the basis functions as long as the conditions C1–C5 are met. Therefore, transcendental and polynomial functions can be used as basis functions. In this chapter, we focus on the using polynomials.

The following condition is helpful when choosing polynomial basis functions:

i. Basis functions ϕ_0, ϕ_j must be chosen such that $\tilde{u}$ is a complete polynomial, (C6)

$$\tilde{u} = \tilde{u}(x, x^2, x^3 \ldots, x^N)$$

This condition will ensure to find the exact solution through variational techniques, in case the exact solution is a complete polynomial.

It should be noted that there is no unique procedure to determine polynomial basis functions to satisfy all the requirements mentioned above. We could easily find different polynomial functions that satisfy the boundary conditions of the problem as well as the requirements C1–C6. Here we focus on one approach.

For the approximate solution to satisfy an *essential boundary condition*, we require ϕ_0 to satisfy the nonzero part of the essential boundary condition. For example, if the boundary condition is of the following form:

$$\text{at } x = x_0 : u(x_0) - u_0$$

we could pick the basis functions such that

$$\phi_0(x_0) = u_0 \text{ and } \phi_j(x_0) = 0. \tag{3.93}$$

For the approximate solution to satisfy a *natural boundary condition*, it is usually helpful to start with,

$$\phi_0 = A + Bx + Cx^2 \text{ and } \phi_j = A + Bx^{j+1} + Cx^{j+2} \text{ for } j = 1 - N \tag{3.94}$$

and find values of the constant A, B and C from the boundary conditions. Examples 3.8 and 3.9 demonstrate the use of this approach to find the approximate functions for solutions with Rayleigh–Ritz and Galerkin methods, respectively.

Example 3.8 Rayleigh–Ritz solution of the model boundary value problem

Find an approximate solution of the boundary value problem given in Eq. (3.3a) by using the Rayleigh–Ritz method with polynomial basis functions. Assume that $a(x) = a$ is constant, $c = 0$, and $q(x) = q$ constant [1]. The solution domain (x_0, x_L) becomes $(0, L)$.

$$\text{BVP: } \frac{d}{dx}\left[a\frac{du}{dx}\right] = -q \text{ for } 0 < x < L \tag{3.95}$$

$$\text{BCs: } u(0) = u_0 \text{ and } a(L)\left.\frac{du}{dx}\right|_L = Q_L \text{ (a), (b)} \tag{3.96}$$

The weak form of this equation is given by (3.90),

$$\int_0^L a\frac{d\tilde{u}}{dx}\frac{dw_i}{dx}dx - \int_0^L w_i q dx - w_i(L)Q_L = 0 \text{ for } i = 1..N, \tag{3.97}$$

The approximate solution is given by (3.91),

$$\tilde{u}(x) = \phi_0(x) + \sum_{j=1}^{N} c_j\phi_j(x) \tag{3.98}$$

Implementation of the Rayleigh–Ritz solution:

1. $\phi_0 = u_0$ (3.99)
2. Choose ϕ_j by taking the following into consideration:
   a) Differentiable at least once, as the functional involves the first order derivative, $\frac{du}{dx}$.
   b) ϕ_j satisfies homogeneous part of the essential boundary conditions.

   The following choice of approximation function will satisfy these criteria:

   $$\phi_j = x^j \ \left(\text{i.e. } \phi_1 = x;\ \phi_2 = x^2;\ \phi_3 = x^3\right) \tag{3.100}$$

   Thus for this problem the approximate solution is given as follows:

   $$\tilde{u}(x) = u_0 + \sum_{j=1}^{N} c_j x^j \tag{3.101}$$

3. Weight function should be chosen as,

   $$w_i = \phi_i \tag{3.102}$$

4. Let's calculate the derivatives of $\tilde{u}$ and w_i

   $$\frac{d\tilde{u}}{dx} = \sum_{j=1}^{N} c_j\frac{d\phi_j}{dx} = \sum_{j=1}^{N} c_j jx^{j-1} \tag{3.103a}$$

   $$\frac{dw}{dx} = \frac{d\phi_i}{dx} = ix^{i-1} \tag{3.103b}$$

   Use Eqs. (3.101) and (3.103a) in the weak form (3.97). This gives,

   $$\int_0^L \left[air^{i-1}\sum_{j=1}^{N} c_j jx^{j-1} - x^i q\right]dx - L^i Q_L = 0 \text{ for } i = 1, \ldots, N$$

   or after rearranging,

   $$\int_0^L \left[a\sum_{j=1}^{N} c_j ijx^{j+i-2} - x^i q\right]dx = L^i Q_L \text{ for } i = 1, \ldots, N \tag{3.104}$$

we get upon integration of the polynomials;

$$a\sum_{j=1}^{N} c_j \frac{ij}{j+i-1} x^{j+i-1}\Big|_0^L = L^i Q_L + \frac{1}{i+1} x^{i+1} q\Big|_0^L \quad \text{for } i = 1, \ldots, N \tag{3.105}$$

This is a recursive formula from which the coefficients for c_j can be obtained by matrix methods. Let's show this for $N = 3$.

$$\begin{aligned}
i = 1 &\Rightarrow a\left[\left(c_1 \frac{1.1}{1+1-1} L^{1+1-1} + c_2 \frac{1.2}{2+1-1} L^{2+1-1} + c_3 \frac{1.3}{3+1-1} L^{3+1-1}\right)\right] \\
&= \frac{1}{1+1} L^{1+1} q + L Q_L \\
i = 2 &\Rightarrow a\left[\left(c_1 \frac{2.1}{1+2-1} L^{1+2-1} + c_2 \frac{2.2}{2+2-1} L^{2+2-1} + c_3 \frac{2.3}{3+2-1} L^{3+2-1}\right)\right] \\
&= \frac{1}{2+1} L^{2+1} q + L^2 Q_L \\
i = 3 &\Rightarrow a\left[\left(c_1 \frac{3.1}{1+3-1} L^{1+3-1} + c_2 \frac{3.2}{2+3-1} L^{2+3-1} + c_3 \frac{3.3}{3+3-1} L^{3+3-1}\right)\right] \\
&= \frac{1}{3+1} L^{3+1} q + L^3 Q_L
\end{aligned} \tag{3.106}$$

In matrix form, these algebraic equations can be represented as:

$$p\begin{bmatrix} L & L^2 & L^3 \\ L^2 & \frac{4}{3}L^3 & \frac{6}{4}L^4 \\ L^3 & \frac{6}{4}L^4 & \frac{9}{5}L^5 \end{bmatrix} \begin{Bmatrix} c_1 \\ c_2 \\ c_3 \end{Bmatrix} = \begin{Bmatrix} \frac{1}{2}L^2 q + L Q_L \\ \frac{1}{3}L^3 q + L^2 Q_L \\ \frac{1}{4}L^4 q + L^3 Q_L \end{Bmatrix} \tag{3.107}$$

$$\text{Solution gives: } c_1 = \frac{1}{a}(Q_L + qL),\ c_2 = -\frac{q}{2a},\ c_3 = 0 \tag{3.108}$$

Thus the approximate solution for $N = 3$ is found as follows:

$$\tilde{u}_{N=3} = u_0 + \frac{1}{a}(Q_L + qL)x - \frac{q}{2a}x^2 + 0x^3 \tag{3.109}$$

This constitutes the Rayleigh–Ritz solution of the boundary value problems. It can be shown that for $a(x) = a$ and $q(x) = q$ where a and q are constant real scalars, the exact solution of the BVP (3.95)–(3.96) is:

$$u = u_0 + \frac{1}{a}(Q_L + qL)x - \frac{q}{2a}x^2 \tag{3.110}$$

Thus we see that the Rayleigh–Ritz method can find the exact solution if the exact solution is of the polynomial form and if ϕ_j's are chosen as complete polynomials. Note that in this example, choosing $N = 2$ would have been sufficient to find the exact solution, and any N value greater than 2 would give $c_3, c_4 \ldots$ as zeros.

Example 3.9 Galerkin's method

Find an approximate solution for the following BVP by using Galerkin's method,

$$\text{BVP: } \frac{d}{dx}\left(a(x)\frac{du}{dx}\right) + u = q(x) \text{ in } 0 < x < L \tag{3.111}$$

$$\text{BCs: } u(0) = 0, \; a\frac{du}{dx}\bigg|_{x=1=Q_0} \tag{3.112}$$

Consider this problem for $L = a(x) = Q_0 = 1$; $q(x) = -x^2$ and use a two term polynomial approximation (i.e. $N = 2$).

Solution:

For the given L, a, Q_0 and $q(x)$ the problem becomes,

$$\text{BVP: } -\frac{d^2u}{dx^2} - u + x^2 = 0 \text{ in } 0 < x < 1 \tag{3.113}$$

$$\text{BCs: } u(0) = 0 \text{ and } \frac{du}{dx}\bigg|_{x=1} = 1 \tag{3.114}$$

$$\text{Approximate solution: } \tilde{u} = \phi_0 + \sum_{j=1}^{2} C_j\phi \tag{3.115}$$

$$\text{Residual: } \qquad R = -\frac{d^2u}{dx^2} - u + x^2 \tag{3.116a}$$

$$\text{Approximate residual: } R = -\frac{d^2\tilde{u}}{dx^2} - \tilde{u} + x^2 \tag{3.116b}$$

Finding the approximation functions: The requirements on ϕ_0 and ϕ_j are as follows:

$$\left.\begin{aligned} \phi_0(0) &= 0 \\ \frac{d\phi_0}{dx}(1) &= 1 \end{aligned}\right\} \text{ Satisfy non-homogeneous parts of the boundary conditions} \tag{3.117}$$

$$\left.\begin{aligned} \phi_j(0) &= 0 \\ \frac{d\phi_j}{dx}(1) &= 0 \end{aligned}\right\} \text{ Satisfy the homogeneous part of the boundary conditions} \tag{3.118}$$

Verify that the following function satisfies the conditions for the ϕ_0 approximation,

$$\phi_0 = x. \tag{3.119}$$

As ϕ_1 should be twice differentiable, the smallest polynomial we can try is a parabola,

$$\phi_1(x) = A + Bx + Cx^2 \tag{3.120}$$

The coefficients can be determined from the boundary conditions: $\phi_1(0) = 0$ and $\phi_{1,x}(1) = 0$

$$\phi_1 = A + Bx + Cx^2 \text{ and } \frac{d\phi_1}{dx} = B + 2Cx \tag{3.121}$$

by using the boundary conditions we find,

$$A = 0 \text{ and } B = -2C \tag{3.122}$$

Thus ϕ_1 becomes,

$$\phi_1(x) = -Cx(2 - x) \tag{3.123}$$

To find ϕ_2 use a third order polynomial,

$$\phi_2(x) = A + Bx^2 + Dx^3 \tag{3.124}$$

consider the boundary conditions: $\phi_2(0) = 0$ and $\phi_{2,x}(1) = 0$,

$$\phi_2 = A + Bx^2 + Dx^3 \text{ and } \frac{d\phi_2}{dx} = 2Bx + 3Dx^2 \tag{3.125}$$

by using the boundary conditions we find,

$$A = 0 \text{ and } 2B = -3D \tag{3.126}$$

Thus ϕ_2 becomes,

$$\phi_2(x) = Bx^2\left(1 - \frac{2}{3}x\right) \tag{3.127}$$

We can drop C and B, because they will be absorbed in to coefficients, C_j, of the approximate solution. The basis functions can then be expressed as follows:

$$\phi_1(x) = x(2 - x) \text{ and } \phi_2(x) = x^2\left(1 - \frac{2}{3}x\right) \tag{3.128}$$

The approximate solution is thus found as follows:

$$\tilde{u} = x + C_1 x(2 - x) + C_2 x^2\left(1 - \frac{2}{3}x\right) \tag{3.129}$$

Note that even though $\phi_1(x)$ and $\phi_2(x)$ are incomplete polynomials, $\tilde{u}$ is a complete third-order polynomial. With the approximation functions determined, we can now evaluate the residual by substituting $\tilde{u}$ into R,

$$R = -\frac{d^2}{dx^2}\left[\phi_0 + \sum_{j=1}^{N} C_j\phi_j\right] - \left[\phi_0 + \sum_{j=1}^{N} C_j\phi_j\right] + x^2 = -\sum_{j=1}^{N} C_j\frac{d^2\phi_j}{dx^2} - \phi_0 - \sum_{j=1}^{N} C_j\phi_j + x^2 \tag{3.130}$$

Using the following derivatives,

$$\frac{d\phi_1}{dx} = 2 - 2x; \ \frac{d^2\phi_1}{dx^2} = -2; \ \frac{d\phi_2}{dx} = 2x - 2x^2; \ \frac{d^2\phi_2}{dx^2} = 2 - 4x \tag{3.131}$$

the residual becomes,

$$R = -[-C_1 2 + C_2(2-4x)] - x - \left[C_1(2x - x^2) + C_2\left(x^2 - \frac{2}{3}x^3\right)\right] + x^2$$

Rearranging this equation we find,

$$R = C_1(2 - 2x + x^2) + C_2\left(-2 + 4x - x^2 + \frac{2}{3}x^3\right) - x + x^2 \tag{3.132}$$

Coefficients C_1 and C_2 can be determined by using *Galerkin's method*, which requires the following integral equations to be evaluated,

$$\int_0^1 w_i R dx = 0 \quad i = 1, 2 \tag{3.133}$$

where $w_1 = \phi_1$ and $w_2 = \phi_2$. By using Eqs. (3.128), and (3.132) and rearranging the terms, we can set up Eq. (3.133) as follows:

for $i = 1$,

$$\int_0^1 \left[C_1(2x - 2x^2 + x^3) + C_2\left(-2x + 4x^2 - x^3 + \frac{2}{3}x^4\right) - x^2 + x^3\right] dx = 0 \tag{3.134a}$$

for $i = 2$,

$$\int_0^1 \left[C_1(+2x^2 - 2x^3 + x^4) + C_2\left(-2x^2 + 4x^3 - x^4 + \frac{2}{3}x^5\right) - x^3 + x^4\right] dx = 0 \tag{3.134b}$$

These integrals result in the following set of algebraic equations:

$$\begin{aligned} \frac{4}{5}C_1 + \frac{17}{90}C_2 &= \frac{7}{60} \\ \frac{238}{1260}C_1 + \frac{116}{1260}C_2 &= \frac{35}{1260} \end{aligned} \tag{3.135}$$

The coefficients are found as follows:

$$C_1 = \frac{623}{4306}; \ C_2 = \frac{21}{4306} \tag{3.136}$$

Then the *approximate solution* with the Galerkin method becomes,

$$\tilde{u} = -\frac{x}{2153}(-2776 + 301x + 7x^2) \tag{3.137}$$

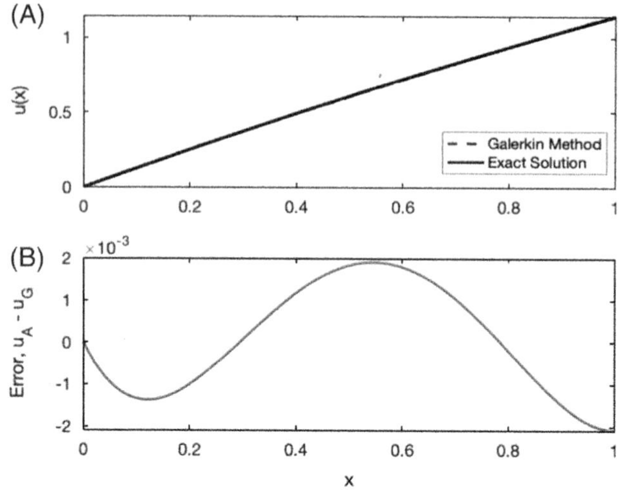

FIGURE 3.4 (A) The approximate and the exact solutions and (B) the corresponding error plotted in the solution domain.

Convergence of the solution

Is the solution presented in Eq. (3.137) accurate? If we know the analytical solution u, we can compare the numerical approximation $\tilde{u}$ to the analytical solution and answer this question easily. But, if we know the analytical solution why would we even need an approximate solution? This is an approach taken in most numerical analysis to establish the validity of the numerical methods. Otherwise, we are always better off using the analytical solution. In this case the *analytical solution* of the problem can be shown to be,

$$u = \frac{2\cos(1-x) - \sin x}{\cos 1} + x^2 - 2 \tag{3.138}$$

In Fig. 3.4, we compare the results predicted by Eqs. (3.137) and (3.138). This figure shows that the two solutions are indistinguishable to the naked eye. Moreover, when we plot the difference $u - \tilde{u}$ we note that the error is bound by $|2\times10^{-3}|$.

If this level of error is insufficient for our goals or if we want to see if the approximate solution can be further improved, we can increase the level of polynomial approximation by including more terms ($N > 2$) in the solution. Close inspection of the procedure outlined above will show that the following basis functions are appropriate for $\tilde{u}$ to satisfy the boundary conditions of the problem.

$$\phi_0 = x,\ \phi_j = x^j\left(1 - \frac{j}{j+1}x\right),\ \tilde{u} = x + \sum_{j=1}^{N} C_j\phi_j \tag{3.139}$$

TABLE 3.2 Convergence of the approximate solutions presented in Eqs. (3.134a–3.135) for the differential equation and the boundary conditions given in Eqs. (3.111–3.112).

N	$(\tilde{u})_{x=1/2}$	$\left.\left\lvert\frac{\tilde{u}_{N+1}-\tilde{u}_N}{\tilde{u}_N}\right\rvert\right\rvert_{x=1/2}$	ε_t
1	0.609375		-2.8×10^{-2}
2	0.609324	8.3×10^{-5}	2.5×10^{-2}
3	0.611157	3.0×10^{-3}	1.7×10^{-4}
4	0.611156	1.1×10^{-6}	9.2×10^{-5}
5	0.611159	4.9×10^{-6}	-3.7×10^{-7}
6	0.611159	1.3×10^{-9}	-1.5×10^{-7}

The problem can then be solved by using,

$$\int_0^1 w_i R dx = 0 \quad i = 1 - N \tag{3.140}$$

with $w_i = \phi_i$

The results of this are presented for N values between 1 and 6 in Table 3.1. Note that finding the coefficients for the higher order approximations is a considerable effort, and therefore, the details are not given here. In order to evaluate the level of convergence without making a comparison to the exact solution we use two metrics. The first one is the value of the function at the mid-point, $x = 1/2$, and the second one is the integral of the error in the residual,

$$\varepsilon_t = \int_0^1 R(\tilde{u})dx \tag{3.141}$$

The results of the convergence analysis are provided in Table 3.2. We see that five digit accuracy is obtained with $N = 3$.

Example 3.10 Galerkin's method

Find an approximate solution for the following boundary value problem by using Galerkin's method,

$$\text{BVP: } 4\frac{d^2u}{dx^2} + 3u = 2x^3 \text{ in } 0 < x < 1 \tag{3.142}$$

$$\text{BCs: } \left.\frac{du}{dx}\right|_{x=0} = 1, \quad u(1) = 2 \tag{3.143}$$

Use a two term polynomial approximation (i.e. $N = 2$).

Solution:

$$\text{Residual: } R = 4\frac{d^2u}{dx^2} + 3u - 2x^3 \tag{3.144a}$$

$$\text{Approximate residual: } R = 4\frac{d^2\tilde{u}}{dx^2} + 3\tilde{u} - 2x^3 \tag{3.144b}$$

Find an approximate solution,

$$\tilde{u} = \phi_0 + C_1\phi_1 + C_2\phi_2 \tag{3.145}$$

that satisfies the boundary conditions (3.143). This can be accomplished if ϕ_0 satisfies the nonhomogeneous parts of the boundary conditions,

$$\left.\frac{d\phi_0}{dx}\right|_{x=0} = 1 \text{ and } \phi_0(1) = 2 \tag{3.146}$$

and ϕ_j ($j = 1, 2$) satisfy the homogeneous part of the boundary conditions

$$\left.\frac{d\phi_j}{dx}\right|_{x=1} = 0 \text{ and } \phi_j(1) = 0 \tag{3.147}$$

It can be shown that the following function satisfies the conditions for ϕ_0,

$$\phi_0 = 2 + (x - 1) \tag{3.148}$$

By inspection we can find, for a generic length of the solution domain L, the following function satisfies the conditions in Eq. (3.147),

$$\phi_j(x) = x^{j+1} - L^{j+1} \tag{3.149}$$

Here, $L = 1$. Thus we have,

$$\phi_1(x) = x^2 - 1, \text{ and } \phi_2(x) = x^3 - 1 \tag{3.150}$$

Note that even though $\phi_1(x)$ and $\phi_2(x)$ are incomplete polynomials, $\tilde{u}$ is a complete third-order polynomial,

$$\tilde{u} = 2 + (x - 1) + C_1\left(x^2 - 1\right) + C_2\left(x^3 - 1\right) \tag{3.151}$$

and as expected it satisfies the boundary conditions,

$$\left.\frac{d\tilde{u}}{dx}\right|_{x=0} = 1, \ \tilde{u}(1) = 2 \tag{3.152}$$

The approximate residual becomes,

$$R = 4\frac{d^2}{dx^2}\left[\phi_0 + \sum_{j=1}^{2} c_j\phi_j\right] + 3\left[\phi_0 + \sum_{j=1}^{2} C_j\phi_j\right] - 2x^3 \tag{3.153}$$

Using the following derivatives,

$$\frac{d^2\phi_1}{dx^2} = 2; \text{ and } \frac{d^2\phi_2}{dx^2} = 6x$$

the residual becomes,

$$R = 4(2C_1 + 6C_2x) + 3\left(x + 1 + C_1\left(x^2 - 1\right) + C_2\left(x^3 - 1\right)\right) - 2x^3 \tag{3.154}$$

Rearranging this equation we find,

$$R = 3 - 2x^3 + 3x + C_1\left(3x^2 + 5\right) + 3C_2\left(x^3 + 8x - 1\right) \tag{3.155}$$

Coefficients C_1 and C_2 can be determined by using *Galerkin's method*, which requires the following integral equations to be evaluated,

$$\int_0^1 w_i R d_x = 0 \quad i = 1, 2 \tag{3.156}$$

where $w_1 = \phi_1$ and $w_2 = \phi_2$. By using Eqs. (3.150), and (3.155) we can set up Eq. (3.156) as follows:

for $i = 1$,

$$\int_0^1 \left(x^2 - 1\right)\left[3 - 2x^3 + 3x + C_1\left(3x^2 + 5\right) + 3C_2\left(x^3 + 8x - 1\right)\right]dx = 0 \tag{3.157a}$$

for $i = 2$,

$$\int_0^1 \left(x^3 - 1\right)\left[3 - 2x^3 + 3x + C_1\left(3x^2 + 5\right) + 3C_2\left(x^3 + 8x - 1\right)\right]dx = 0 \tag{3.157b}$$

These integrals result in the following set of algebraic equations:

$$\begin{aligned} \frac{56}{15}C_1 + \frac{17}{4}C_2 &= -\frac{31}{12} \\ \frac{17}{4}C_1 + \frac{369}{70}C_2 &= -\frac{411}{140} \end{aligned} \tag{3.158}$$

The coefficients are found as follows:

$$C_1 = -\frac{3195}{4529}; \quad C_2 = \frac{23}{1941} \tag{3.159}$$

Then the *approximate solution* with the Galerkin method becomes,

$$\tilde{u} = 1 + x - \frac{3195}{4529}\left(x^2 - 1\right) + \frac{23}{1941}\left(x^3 - 1\right) \tag{3.160}$$

Convergence of the solution

Close inspection of the procedure outlined above will show that the following basis functions are appropriate for $\tilde{u}$ to satisfy the boundary conditions of the problem.

$$\phi_0 = 2 + (x - 1), \quad \phi_j(x) = x^{j+1} - 1, \tag{3.161}$$

TABLE 3.3 Convergence of the approximate solution for the differential equation and the boundary conditions given in Eqs. (3.142–3.143).

N	$(\tilde{u})_{x=1/2}$	$\left\lvert\dfrac{\tilde{u}_{N+1}-\tilde{u}_N}{\tilde{u}_N}\right\rvert_{x=1/2}$	ε_t
1	2.01897		1.15
2	2.01872	1.2×10^{-4}	-1.7×10^{-1}
3	2.02064	9.5×10^{-4}	-3.2×10^{-2}
4	2.02066	9.99×10^{-6}	8.3×10^{-4}
5	2.02066	1.21×10^{-6}	6.4×10^{-5}
6	2.02066	7.00×10^{-9}	-1.1×10^{-6}

The problem can then be solved by using,

$$\int_0^1 w_i R dx = 0 \quad i = 1 - N \tag{3.162}$$

with $w_i = \phi_i$.

The results of this solution are presented for N values between 1 and 6 in Table 3.3. As noted before finding the coefficients for the higher order approximations is a considerable effort, and therefore, the details are omitted here. In order to evaluate the level of convergence without making a comparison to the exact solution we use two metrics. The first one is the value of the function at the mid-point (1/2) and the second one is the integral of the error in the residual,

$$\varepsilon_t = \int_0^1 R(\tilde{u})dx \tag{3.163}$$

Both of these metrics show that using $N = 4$ gives a very reasonable accuracy.

Example 3.11 Rayleigh–Ritz method

Find an approximate solution for the boundary value problem given in Example 3.10 by using Rayleigh–Ritz method,

$$\text{BVP: } 4\frac{d^2u}{dx^2} + 3u = 2x^3 \text{ in } 0 < x < 1 \tag{3.164}$$

$$\text{BCs: } \left.\frac{du}{dx}\right|_{x=0} = 1 \text{ (a)}, \quad u(1) = 2 \text{ (b)} \tag{3.165}$$

Use a two term polynomial approximation (i.e. $N = 2$).

Solution:

The Rayleigh–Ritz method works with the weak form of the boundary value problem. The weighted residual integral for the approximate for this method is,

$$\int_0^1 w_i\left(4\frac{d^2\tilde{u}}{dx^2}+3\tilde{u}-2x^3\right)dx=0 \quad i=1,2 \tag{3.166}$$

Integration by parts gives,

$$\left(w_i 4\frac{d\tilde{u}}{dx}\right)_{x=1}-\left(w_i 4\frac{d\tilde{u}}{dx}\right)_{x=0}-\int_0^1\left(4\frac{d\tilde{u}}{dx}\frac{dw_i}{dx}-3w_i\tilde{u}+2w_ix^3\right)dx=0 \tag{3.167}$$

After applying the boundary conditions in Eq. (3.143), the weak form of the boundary value problem is found as follows:

$$-4w_i(0)-\int_0^1\left(4\frac{d\tilde{u}}{dx}\frac{dw_i}{dx}-3w_i\tilde{u}+2w_ix^3\right)dx=0\ i=1,2 \tag{3.168}$$

Note that the natural boundary condition $(u_{,x})_{x=0}=1$ is now incorporated in the integral representation of the problem. We can therefore use an approximate solution that satisfies only the essential boundary conditions, as described in Section 3.5, satisfying conditions C3 – C6. Let's name this the *reduced order approximate solution*, $\tilde{u}^{(r)}$. We can also use a function that satisfies both boundary conditions, as provided by Eq. (3.151). Let's name this the *full order approximate solution*, $\tilde{u}^{(f)}$.

Here in this example, we will find the reduced order approximate solution,

$$\tilde{u}^{(r)}=\phi_0^{(r)}+C_1\phi_1^{(r)}+C_2\phi_2^{(r)} \tag{3.169}$$

It is sufficient for this equation to satisfy the essential boundary condition. This is accomplished if ϕ_0 satisfies the nonhomogeneous part of Eq. (3.165b),

$$\phi_0^{(r)}(1)=2 \tag{3.170}$$

and $\phi_j^{(r)}(j=1,2)$ satisfy the homogeneous part of (3.165b),

$$\phi_j^{(r)}(1)=0 \tag{3.171}$$

By inspection we can find, for a generic length of the solution domain L, the following function satisfies the conditions in Eq. (3.169),

$$\phi_j^{(r)}(x)=x^j-L^j \tag{3.172}$$

As $L=1$ in this problem, we have,

$$\phi_1^{(r)}(x)=x-1,\ \text{and}\ \phi_1^{(r)}(x)=x^2-1 \tag{3.173}$$

The reduced order approximate solution for $N = 2$ can be stated as follows,

$$\tilde{u}^{(r)} = 2 + C_1(x-1) + C_2\left(x^2-1\right) \tag{3.174}$$

This equation satisfies the boundary condition,

$$\left(\tilde{u}^{(r)}\right)_{x=1} = 2 \tag{3.175}$$

The derivative of (3.173) is found as follows,

$$\frac{d\tilde{u}^{(r)}}{dx} = C_1 + 2C_2x$$

The weak form of the boundary value problem (3.168) thus, becomes,

$$-4w_i(0) - \int_0^1 \left(4\frac{dw_i}{dx}(C_1+2C_2x) - 3w_i\left(2+C_1(x-1)+C_2\left(x^2-1\right)\right) + 2w_ix^3\right)dx$$
$$= 0 \text{ for } i = 1, 2 \tag{3.176}$$

where $w_1 = \phi_1$ and $w_2 = \phi_2$. By using Eq. (3.173), we can set up Eq. (3.176) as follows:

for $i = 1$,

$$-4w_1(0) - \int_0^1 \left(4\frac{dw_1}{dx}(C_1+2C_2x) - 3w_1\left(2+C_1(x-1)+C_2\left(x^2-1\right)\right) + 2w_1x^3\right)dx = 0$$

for $i = 2-4$

$$w_2(0) - \int_0^1 \left(4\frac{dw_2}{dx}(C_1+2C_2x) - 3w_2\left(2+C_1(x-1)+C_2\left(x^2-1\right)\right) + 2w_2x^3\right)dx = 0$$

These integrals result in the following set of algebraic equations:

$$\begin{aligned} 3C_1 + \frac{11}{4}C_2 &= \frac{9}{10} \\ \frac{11}{4}C_1 + \frac{56}{15}C_2 &= -\frac{1}{6} \end{aligned} \tag{3.177}$$

The coefficients are found as follows:

$$C_1 = \frac{4582}{4365}; \; C_2 = -\frac{238}{291} \tag{3.178}$$

Then the *reduced order approximate solution* by using the Rayleigh–Ritz method becomes,

$$\tilde{u}^{(r)} = 2 + \frac{4582}{4362}(x-1) - \frac{238}{291}\left(x^2-1\right) \tag{3.179}$$

TABLE 3.4 Convergence of the approximate solution $\tilde{u}^{(r)}$ for the differential equation and the boundary conditions given in Eqs. (3.142–3.143).

N	$\left(\tilde{u}^{(r)}\right)_{x=1/2}$	$\left\lvert \frac{\tilde{u}_{N+1}-\tilde{u}_N}{\tilde{u}_N} \right\rvert_{x=1/2}$	$\left(\frac{d\tilde{u}}{dx}\right)_{x=0}$
1	2.01897		1
2	2.01872	1.2×10^{-4}	1.049714
3	2.02064	9.5×10^{-4}	0.999233
4	2.02066	9.9×10^{-6}	1.001518
5	2.02066	1.2×10^{-6}	1.000004
6	2.02066	7.0×10^{-9}	0.999997

Convergence of the solution

Close inspection of the procedure outlined above will show that the following basis functions are appropriate for $\tilde{u}$ to satisfy the boundary conditions of the problem.

$$\phi_0 = 2, \ \ \phi_j(x) = x^j - 1, \tag{3.180}$$

The problem can then be solved by using Eq. (3.168) with $i = 1, N$.

The results of this solution are presented for N values between one and six in Table 3.4. We use the value of the function at the mid-point, $x = 1/2$, as a convergence metric. Both of these metrics show that using $N = 4$ gives a very reasonable accuracy. Note that the solution approaches the natural boundary condition $\left(\tilde{u}^{(r)}_{,x}\right) \to 1$ with increasing accuracy, as the number of terms in the approximate solution is increased.

Generic form of basis functions

By using the methodology outlined in Eqs. (3.93) and (3.94), we can propose the use of polynomial basis functions for problems that involve different forms of the boundary conditions. Note that the functions presented below are not unique, and as mentioned before, different polynomial functions that satisfy the boundary conditions and conditions C1–C6, may exist and could provide a more effective solutions.

The polynomial approximations are presented for Rayleigh–Ritz and Galerkin methods separately, as explained earlier in this section.

Rayleigh–Ritz method: need to satisfy the conditions C3–C5 and C6. By using the procedure outlined in Example 3.8, we can find the basis functions for different combinations of the boundary conditions for a second order differential equation,

$$\frac{d}{dx}\left[a\frac{du}{dx}\right] + cu = -q \ \text{ for } x_0 < x < x_L \tag{3.181}$$

Case-1:

If the boundary conditions for Eq. (3.181) problem are,

at $x = x_0$, $u(x_0) = u_0$ and a natural boundary condition is specified at

$$x = x_L \tag{3.182}$$

it can be shown that the following basis functions satisfy the conditions C3–C5 and C6,

$$\phi_0 = u_0, \tag{3.183a}$$

$$\phi_j = (x - x_0)x^{j-1} \tag{3.183b}$$

The approximate solution becomes,

$$\tilde{u} = u_0 + \sum_{j=1}^{N} C_j (x - x_0)x^{j-1} \tag{3.184}$$

Case-2:

If the boundary conditions for Eq. (3.181) problem are,

A natural boundary condition is specified at

$$x = x_0 \text{ and at } x = x_L,\ u(x_L) = u_L \tag{3.185}$$

it can be shown that the following basis functions satisfy the conditions C3–C5 and C6,

$$\phi_0 = u_L, \tag{3.186a}$$

$$\phi_j = (x - x_L)x^{j-1} \tag{3.186b}$$

The approximate solution becomes,

$$\tilde{u} = u_L + \sum_{j=1}^{N} C_j (x - x_L)x^{j-1} \tag{3.187}$$

Case-3:

If the boundary conditions for Eq. (3.181) problem are,

$$\text{at } x = x_0 : u(x_0) = u_0 \text{ and at } x = x_L : u(x_L) = u_L \tag{3.188}$$

it can be shown that the following basis functions satisfy the conditions C3 – C5 and C6,

$$\phi_0 = u_0\left(1 - \frac{x - x_0}{x_L - x_0}\right) + u_L\left(\frac{x - x_0}{x_L - x_0}\right), \tag{3.189a}$$

$$\phi_j = (x - x_0)(x - x_L)x^{j-1} \tag{3.189b}$$

The approximate solution becomes,

$$\tilde{u} = u_0\left(1 - \frac{x - x_0}{x_L - x_0}\right) + u_L\left(\frac{x - x_0}{x_L - x_0}\right) + \sum_{j=1}^{N} C_j (x - x_0)(x - x_L)x^{j-1} \tag{3.190}$$

Galerkin method: In this case we need to satisfy the conditions C1 – C3 and C6. By using the procedure outlined in Example 3.9, we can find the basis functions for different combinations of the boundary conditions for the second order differential equation given in (3.181).

Case-1:
If the boundary conditions for Eq. (3.181) problem are, at $x = 0$: $u(0) = u_0$ and at $x = Lu\ (x_L) = u_L$ Eq. (3.189a) can be used.

Case-2:
If the boundary conditions for Eq. (3.181) problem are,

$$\text{at } x = 0 : u(0) = u_0 \quad \text{and at} \quad x = L : a_L \frac{du}{dx}|_{x=L} = Q_L \tag{3.191}$$

it can be shown that the following basis functions satisfy the conditions C1 – C3 and C6,

$$\phi_0 = u_0 + \left(\frac{Q_L}{a_L}\right)x \tag{3.192a}$$

$$\phi_j = \left[1 - \left(\frac{j+1}{j+2}\right)\frac{x}{L}\right]x^{j+1} \tag{3.192b}$$

The approximate solution then becomes,

$$\tilde{u} = u_0 + \left(\frac{Q_L}{a_L}\right)x + \sum_{j=1}^{N} C_j\left[1 - \left(\frac{j+1}{j+2}\right)\frac{x}{L}\right]x^{j+1} \tag{3.193}$$

Case-3:
If the boundary conditions for Eq. (3.181) problem are,

$$\text{at } x = 0 : a_0 \left.\frac{du}{dx}\right|_{x=0} = Q_0 \text{ and at} x = L : u(L) = u_L \tag{3.194}$$

it can be shown that the following basis functions satisfy the conditions C1–C3 and C6,

$$\phi_0 = u_L + \frac{Q_0}{a_0}(x - L) \tag{3.195a}$$

$$\phi_j = (x - L)x^{j+1} \tag{3.196b}$$

The approximate solution then becomes,

$$\tilde{u} = u_L + \frac{Q_0}{a_0}(x - L) + \sum_{j=1}^{N} C_j(x - L)x^{j+1} \tag{3.196}$$

Case-4:
If the boundary conditions for Eq. (3.181) problem are,

$$\text{at } x = 0 : a_0 \frac{du}{dx}|_{x=0} = Q_0 \text{ and at } x = L : a_L \frac{du}{dx}|_{x=L} = Q_L \tag{3.197}$$

It can be shown that the following basis functions satisfy the conditions C1–C3 and C6,

$$\phi_0 = \frac{Q_0}{a_0}x + \frac{x^2}{2L}\left(\frac{Q_L}{a_L} - \frac{Q_0}{a_0}\right) \tag{3.198a}$$

$$\phi_j = (x - L)x^{j+2} \tag{3.198b}$$

The approximate solution then becomes,

$$\tilde{u} = \frac{Q_0}{a_0}x + \frac{x^2}{2L}\left(\frac{Q_L}{a_L} - \frac{Q_0}{a_0}\right) + \sum_{j=1}^{N} C_j(x - L)x^{j+2} \tag{3.199}$$

3.6 Problems

3.1 Find the Euler-Lagrange equations and the appropriate boundary conditions for the following functional:

$$I = \int_a^b F(x, u, u_{,x}, u_{,xx})dx$$

where $u_{,x} = du/dx$ and $u_{,xx} = d^2u/dx^2$.

3.2 Find the equilibrium equations and the admissible boundary conditions that are associated with the following functionals by using *i*) the Euler-Langrange equations and *ii*) by finding the stationary value of the functional (i.e., setting $\delta I(u) = 0$):

a) $$I(u) = \int_0^1 \left[(u_{,x})^2 + uu_{,x} + u^2\right]dx$$

b) $$I(u) = \int_0^1 \left[(u_{,x})^2 + \cos u\right]dx$$

Do both methods (*i, ii*) give the same result?

3.3 Find the equilibrium equations and the admissible boundary conditions that are associated with the following functional by using: *i*) the Euler-Langrange equations and *ii*) by finding the stationary value of the functional (i.e., setting $\delta I(u) = 0$):

$$I(u) = -\int_a^b \left[(u_{,x})^2 + u^2\right]dx + k(u(b))^2$$

3.4 The following BVP problem represents heat conduction in a 1D domain:

$$
\begin{aligned}
&-\frac{d}{dx}\left(a\frac{dT}{dx}\right)+cT=q \text{ for } 0<x<1\\
&T(0)=T_0, \text{ at } x=0\\
&\left[a\frac{dT}{dx}+\beta(T-T_\infty)\right]\Big|_{x=1}=Q_0 \text{ at } x=1,
\end{aligned}
$$

where $a=a(x)$ and $q=q(x)$ are functions of x, and β, c, T_∞ and Q_0 are constants.

a) Construct the weak form of this BVP. Show the steps of integration by parts.

b) Find the quadratic (B) and linear (l) functionals associated with this equation.

3.5 The following BVP problem represents 1D heat transfer from a fin:

$$-\frac{d}{dx}\left(kA\frac{dT}{dx}\right)+hp(T_\infty-T)+QA=0 \text{ for } 0<x<L$$

where $A=A(x)$ and $Q=Q(x)$ are functions of x, and k, h, p are constants.

a) Identify the form of the possible boundary conditions for this equation by using the weighted residual integral.

b) Find the weak form of this problem for the following boundary conditions:

$$
\begin{aligned}
&T(0)=T_0 \text{ at } x=0\\
&\left[kA\frac{dT}{dx}+\beta(T-T_\infty)\right]\Big|_{x=1}=-Q_0 \text{ at } x=L,
\end{aligned}
$$

where T_∞ and Q_0 are constants.

3.6 The following BVP problem represents the deflections (v) of a tensioned beam on an elastic foundation:

$$\frac{d^2}{dx^2}\left(b(x)\frac{d^2v}{dx^2}\right)-T\frac{d^2v}{dx^2}+kv=f(x) \text{ for } 0<x<L$$

where b ($=EI$) is the flexural rigidity f is the external traction per unit length k is the foundation stiffness and T is the tension acting on the beam.

Both b and f are functions of x, whereas the foundation stiffness k and tension T are constant.

a) Using the weighted residual form, and integration by parts, identify the essential and natural boundary conditions of this BVP.

b) Obtain the weak integral form of this BVP by considering boundary conditions given below.

c) Find the quadratic (B) and linear (l) functionals associated for this problem using these boundary conditions.

Boundary conditions:

$$\text{at } x=0: u=0 \text{ and } \frac{du}{dx}=\theta_0 \text{ and at } x=L: u=u_L \text{ and } b\frac{d^2u}{dx^2}=M_L$$

3.7 Solve the problem 3.6 by using $T=0$, and the following boundary conditions:

$$\text{at } x=0, L: w=b\frac{d^2w}{dx^2}=0$$

3.8 Solve the problem 3.6 by using $T=0$, and the following boundary conditions:

$$\text{at } x=0: u=0 \text{ and } \frac{du}{dx}=\theta_0$$

$$\text{and at } x=L: \frac{d}{dx}\left(b\frac{d^2u}{dx^2}\right)=V_L \text{ and } b\frac{d^2u}{dx^2}=M_L$$

3.9 Solve the problem 3.6 by using the following boundary conditions:

$$\text{at } x=0, L: v=b\frac{d^2v}{dx^2}=0$$

Rayleigh Ritz and Galerkin Methods

3.10 Consider the problem given in question 3.9 with $T=0, f=f_0$ and $b(x)=EI$ for constant f_0, E and I.

a) Setup the equations for N-parameter Rayleigh-Ritz approximation of this problem, using algebraic polynomial trial functions.

b) Compare the two parameter ($N=2$) and three parameter ($N=3$) Rayleigh-Ritz solutions.

3.11 Consider the problem given in question 3.9 with $k=0, f=f_0$ and $b(x)=EI$ for constant f_0, E and I.

a) Setup the equations for N-parameter Rayleigh-Ritz approximation of this problem.

b) Compare the two parameter ($N=2$) Rayleigh-Ritz solution with the exact solution by using algebraic polynomials.

$$\text{Exact Solution: } w(x)=\frac{f_0}{12EI}\left(\frac{L^3}{2}x-Lx^3+\frac{1}{2}x^4\right)$$

3.12 Consider the following boundary value problem:

$$\frac{d}{dx}\left(x^{1/2}\frac{du}{dx}\right)+xu=0 \quad \text{for } 0<x<1$$

$$\frac{du}{dx}(0)=0, \; u(1)=1.$$

a) Use N-parameter polynomial approximation functions and the Rayleigh-Ritz method to compute the coefficient matrix and the right-hand side vector.

b) Specialize your result for $N=2$ and compute the coefficients c_I and c_2.

3.13 Consider the following boundary value problem:

$$-\frac{d^2u}{dx^2} + u = \sin \pi x \text{ for } 0 < x < 1$$

$$u(0) = 0, \quad \left.\frac{du}{dx}\right|_{x=1} = 0$$

Determine a 3-parameter solution, with polynomial functions by using,

a) Rayleigh-Ritz method

b) Galerkin's method

3.14 Solve problem 3.12 for the following boundary value problem:

$$-\frac{d}{dx}\left((1+x)\frac{du}{dx}\right) = 0 \text{ for } 0 < x < 1$$

$$u(0) = 0, \ u(1) = 1.$$

3.15 Solve problem 3.13 for the following boundary value problem:

$$-\frac{d^2u}{dx^2} = \cos \pi x \text{ for } 0 < x < 1$$

$$u(0) = 0, \quad \left.\frac{du}{dx}\right|_{x=1} = 0$$

Compare with the exact solution $u(x) = (\cos \pi x - 1)/\pi^2$.

3.16 Consider the following boundary value problem:

$$-\frac{d^2u}{dx^2} + u = \sin \pi x \text{ for } 0 < x < 1$$

$$u(0) = 0, \quad \left.\frac{du}{dx}\right|_{x=1} = 0$$

a) Use N-parameter polynomial approximation functions and the Galerkin method to compute the coefficient matrix and the right-hand side vector.

b) Specialize your result for $N=3$ and compute the coefficients c_I and c_2.

3.17 The electrostatic potential ψ due to liquid-to-surface charge interactions near a submerged sphere of radius a, can be modeled by the following BVP:

$$\frac{1}{r^2}\frac{d}{dr}\left(r^2\frac{d\psi}{dr}\right) = K_0 \sinh(K_1 r) \text{ for } a < r < 10a$$

$$\frac{d\psi}{dr}(a) = \psi_o, \ \psi(10a) = 0.$$

Note that a, ψ_0, K_1 and K_2 are known constants. Compute the coefficient matrix and the right-hand side of the N-parameter Rayleigh-Ritz approximation. Use algebraic polynomials for the approximation functions. Specialize your result for $N=2$ and compute the coefficients c_I and c_2.

3.18 Solve problem 3.12 for the following boundary value problem:

$$-\frac{d}{dx}\left(\left(1+x^{3/2}\right)\frac{du}{dx}\right)=0 \text{ for } 0<x<L$$

$$\left.\frac{du}{dx}\right|_{x=0}=q_o,\ u(L)=u_L.$$

Note that q_o and u_L are known constants.

3.19 Solve problem 3.12 for the following boundary value problem:

$$-\frac{d}{dx}\left((1+x)\frac{du}{dx}\right)=0 \text{ for } 0<x<1$$

$$u(0)=0,\ u(1)=1.$$

3.20 Solve problem 3.12 for the following boundary value problem:

$$\frac{d^2u}{dx^2}+ku=f(x) \text{ for } 0<x<L$$

$$\frac{du}{dx}=\varepsilon_o,\ \text{at } x=0$$

$$u=0,\ \text{at } x=L$$

Assume $f(x)=f_o$, k and ε_o are all constant and known values.

3.21 Solve problem 3.12 for the following boundary value problem:

$$x^2\frac{d^2u}{dx^2}+2x\frac{du}{dx}=6x,\ \text{in } 1<x<2$$

$$u(1)=u(2)=0$$

Compare the results to the exact solution, $u_{ex}(x)=(6/x)+3x-9$.

3.22 Solve problem 3.12 for the following boundary value problem:

$$\frac{d^2u}{dx^2}+ku=f(x) \text{ for } 0<x<L$$

$$\frac{du}{dx}=\varepsilon_o,\quad \text{at } x=0$$

$$u=0,\quad \text{at } x=L$$

Assume, the following $f(x)=f_o$, k and ε_o are all constant and known values.

3.23 Solve problem 3.12 for the following boundary value problem:

$$x^2\frac{d^2u}{dx^2}+2x\frac{du}{dx}+x=1,\ \text{in } 1<x<2$$

$$u(1)=2 \text{ and } \left(\frac{du}{dx}+2u\right)_{x=2}=5$$

3.24 Solve problem 3.12 for the following boundary value problem:

$$x^2\frac{d^2u}{dx^2}+2xu=-4x,\ \text{in } 1<x<5$$

$$\left(\frac{du}{dx}\right)_{x=1}=2 \text{ and } u(5)=2$$

3.25 Consider the Rayleigh-Ritz solution of a BVP in the region $a < x < b$. Find two term ($N=2$) approximate functions $\tilde{u}(x)$ for each pair of boundary conditions:

a) $u(a)=u_a$ and $p(b)\left.\dfrac{du}{dx}\right|_b = Q_b$

b) $u(a)=u_a$ and $u(b)=0$

3.26 Solve problem 3.12 for the following boundary value problem:

$$\frac{d}{dx}\left(x\frac{du}{dx}\right) + u = x \text{ in } 0 < x < 1$$

$$u(0) = 0 \ \textit{and} \ u(1) = 1$$

References

[1] J.N. Reddy, An Introduction to the Finite Element Method, 2nd ed, McGraw Hill, Boston, MA, 1993.

[2] K.W. Cassel, Variational Methods, Cambridge University Press, New York, N.Y., 2013.

[3] I.H. Shames, C.L. Dym, Energy and Finite Element Methods in Structural Dynamics, McGraw Hill, New York, N.Y., 1985.

[4] C.R. Wylie, L.C. Barrett, Advanced Engineering Mathematics, McGraw Hill, New York, NY, 1995.

Chapter 4

Finite element formulation of one-dimensional boundary value problems

4.1 Introduction

In this chapter, the finite element form of a boundary value problem (BVP) presented by a second order differential equation is developed based on the weighted residual method. The general boundary value problem over the original solution domain and the concept of spatial discretization of this solution domain are presented. The boundary value problem is formulated by using the weighted residual method for a subsegment of the solution domain known as an *element* in Section 4.2. The approximate solution to the problem is introduced by using *shape functions* in Section 4.3. Derivation of the weak form is demonstrated for a single element in Section 4.4. The relations between the coeffcients of the general ordinary differential equation and the corresponding physical properties in deformation and heat transfer problems are presented in Section 4.5. Element-assembly and application of boundary conditions are presented in Sections 4.6 and 4.7, respectively and a pseudo-code and an example problem are presented in Sections 4.8 and 4.9, respectively.

4.1.1 Boundary value problem

In this chapter, we consider the solution of the Sturm–Liouville type second order differential equation,

$$\frac{d}{dx}\left(a\frac{du}{dx}\right)+pu+q=0 \text{ in } 0<x<L \tag{4.1}$$

where $a(x)$, $p(x)$ and $q(x)$ are specified functions and $u(x)$ is the dependent variable. The admissible boundary conditions for this problem are as follows:

$$\text{at } x=0: \text{either } u \text{ is specified or } -\frac{du}{dx}=\alpha_0 u+\beta_0 \tag{4.2a}$$

$$\text{at } x=L: \text{either } u \text{ is specified or } \frac{du}{dx}=\alpha_0 u+\beta_0 \tag{4.2b}$$

where α_{-} and $\beta\_$ are known coefficients on the boundaries. See Section 2.4.5.1 on the signs. The domain, $0<x<L$, is shown in Fig. 4.1. In what follows, the

Finite Element Method: Physics and Solution Methods. DOI: https://doi.org/10.1016/B978-0-12-821127-4.00009-8
Copyright © 2022 Elsevier Inc. All rights reserved.

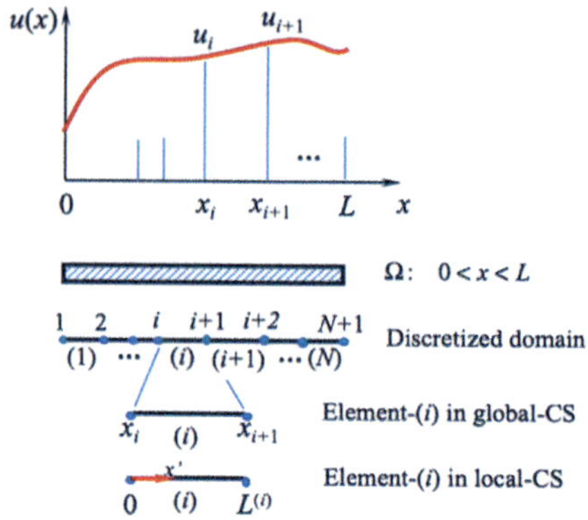

FIGURE 4.1 One-dimensional solution domain, its discretization and the definitions of the global x- and local x'-coordinate systems.

weighted residual method is implemented segment-by-segment over the solution domain.

4.1.2 Spatial Discretization

Spatial segmentation of the solution domain is known as *discretization*. In a one-dimensional problem discretization is straightforward as demonstrated in Fig. 4.1. The domain is divided into N_E interconnected segments, known as *elements*. Each element has two ends (boundaries) through which the elements are connected. Thus, there are a total of $N_E + 1$ boundary connectivity points known as *boundary-nodes*. As it will become clear in Section 4.3, there are also elements with *internal nodes* that are used to improve the order of polynomial interpolation along the element.

In a one dimensional domain, the element-to-node connectivity relationship is straightforward as shown below. This is especially true if the elements only have two nodes and the nodes are numbered consecutively starting from number one.

Element connectivity: Consider a discretization where elements are numbered consecutively as $1 \leq i \leq N_E$. It is easy to see that element-i will be connected to nodes i and $i + 1$.

Element-i will span the range $x_i < x < x_{i+1}$, and its length will be,

$$L^{(i)} = x_{i+1} - x_i$$

The element-to-node connectivity of the higher order elements, i.e. those with internal nodes, can be handled in a similar manner.

While the weighted residual method can be developed in the (global) coordinate system (x) of the boundary value problem, it is simpler to use an *element coordinate system* (x') placed on the lower numbered node of the element (Fig. 4.1). In what follows the finite element form of the weak form of the boundary value problem will be developed over the element subdomain.

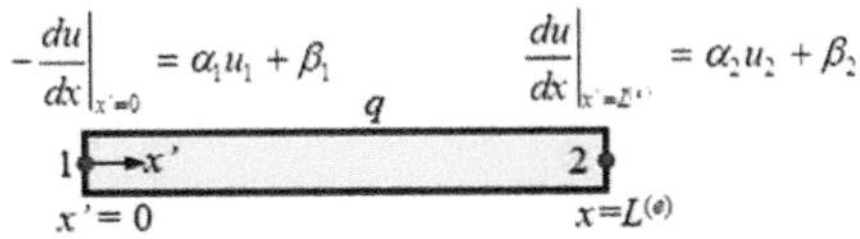

FIGURE 4.2 A one-dimensional element subjected to natural boundary conditions at its boundaries.

4.2 A second order, nonconstant coefficient ordinary differential equation over an element

Here we consider the solution of the following general boundary value problem defined over element-(e),

$$\frac{d}{dx'}\left(a\frac{du}{dx'}\right) + pu + q = 0 \text{ in } 0 < x' < L^{(e)} \tag{4.3}$$

where $a(x')$, $p(x')$ and $q(x')$ are specified functions and $u(x')$ is the dependent variable. In order to develop the finite element version of the original boundary value problem we will assume that natural boundary conditions are applied at the ends (boundaries) of this domain,

$$\text{at } x' = 0 \; : \; -\frac{du}{dx'} = \alpha_0 u + \beta_0 \tag{4.4}$$

$$\text{at } x' = L^{(e)} : \; \frac{du}{dx'} = \alpha_{L^{(e)}} u + \beta_{L^{(e)}} \tag{4.5}$$

where α_0, β_0, $\alpha_{L^{(e)}}$ and $\beta_{L^{(e)}}$ are known coefficients. We use α_1, β_1, α_2 and β_2 for notational simplicity in this chapter.

4.2.1 Deflection of a one-dimensional bar

The problem of finding the deflection $u(x')$ of the element, Fig. 4.2, is now posed as a boundary value problem. The equilibrium of this bar can be found from Eq. (2.6) as follows:

$$\frac{d}{dx'}\left(EA\frac{du'}{dx'}\right) = -f \text{ in } 0 < x' < L^{(e)} \tag{4.6}$$

where, $f(x')$ represents the distributed force per unit length acting on the bar and E and A are the elastic modulus and cross-sectional area of the bar, respectively. On the ends of the bar the following boundary conditions are specified,

$$\left(EA\frac{du'}{dx'}\right)\Bigg|_{x'=0} = -F_1' \text{ and } \left(EA\frac{du'}{dx'}\right)\Bigg|_{x'=L^{(e)}} = F_2' \tag{4.7}$$

4.2.2 Heat transfer in a one-dimensional domain

Conduction in a 1D medium that spans the length $x' = 0$ to $L^{(e)}$ is governed by Eq. (2.171),

$$AQ + \frac{d}{dx'}\left(kA\frac{dT}{dx'}\right) = 0 \text{ in } 0 < x' < L^{(e)} \tag{4.8}$$

where Q is the rate of heat generation per unit volume, k is the conductivity and A is the cross sectional area. In case the heat is transferred from a thin fin to a surrounding fluid at temperature T_∞ the energy balance can be approximated as follows:

$$AQ + \frac{d}{dx'}\left(kA\frac{dT}{dx'}\right) + hP(T_\infty - T) = 0 \text{ in } 0 < x' < L^{(e)} \tag{4.9}$$

where P is the perimeter of the fin, h is the convection coefficient. Heat flux is considered to be specified on the boundaries,

$$\left(-k\frac{dT}{dx'}\right)_0 n = f_{B1} \text{ and } \left(-k\frac{dT}{dx'}\right)_{L^{(e)}} n = f_{B2} \tag{4.10}$$

Note that n represents the unit normal of the boundary with $n = -1$ and 1 at $x = 0$ and L, respectively.

$$\begin{aligned}
&\text{Insulated Boundary} &&: f_{Bi} = f_B(t)\\
&\text{Convected Boundary} &&: f_{Bi} = h(T - T_\infty)\\
&\text{Radiated Boundary} &&: f_{Bi} = \sigma\psi\left(T^4 - T_s^4\right) = \left[\sigma\psi\left(T^3 - T_S^3\right)\right](T - T_S)
\end{aligned} \tag{4.11}$$

where $i = 1, 2$. Note that f_{Bi} for convection and radiation given in (2.4.18) are considered positive.

4.3 One-dimensional interpolation for finite element method and shape functions

We are interested in developing a polynomial approximation $\tilde{u}$ for $u(x)$ in a region $x_1' \le x' \le x_2'$ We are going to use an interpolation that passes through $M + 1$ points in this region:

$$u(x') \approx \tilde{u}(x') = a_0 + \sum_{j=1}^{M} a_j x'^j \tag{4.12}$$

where a_0 and a_j are constant coefficients. In this section we transform this polynomial approximation from one whose unknowns are a_0 and a_j, to one whose unknowns are u_j. The polynomial will be required pass through points (x'_j, u_j) where $1 \le j \le M + 1$ with $x'_1 = 0$ and $x'_M = L^{(e)}$. The values of u_j are known at $M + 1$ positions in this domain. The approach developed in the next section results in a polynomial approximation of $u(x)$ that is of order M, also known as the *Lagrange interpolation*,

$$\tilde{u}(x') = \sum_{i=1}^{M+1} N_i(x')u_i \tag{4.13}$$

where the shape functions $N_i(x')$ are defined as follows:

$$N_i(x') = \frac{(x'_1 - x')(x'_2 - x')\cdots(x'_{i-1} - x')(x'_{i+1} - x')\cdots(x'_M - x')(x'_{M+1} - x')}{(x'_1 - x'_i)(x'_2 - x'_i)\cdots(x'_{i-1} - x'_i)(x'_{i+1} - x'_i)\cdots(x'_M - x'_i)(x'_{M+1} - x'_i)} \tag{4.14}$$

General characteristics of Lagrange shape functions are as follows:

$N_i = 1$ at $x' = x'_i$, but zero at the other positions (nodes) (C1)

N_i are polynomials of the same degree (C2)

The sum of all shape functions is equal to one, i.e., $\sum_{i=1}^{M+1} N_i = 1$. (C3)

It is important to keep in mind that the Lagrange interpolation (4.13) gives a polynomial approximation that is exactly of the same polynomial order as Eq. (4.12), but the unknowns are expressed in terms of u_i instead of a_i. This will be demonstrated in the following sections.

C⁰ Continuity

Note that Lagrange shape functions use only the nodal degrees of freedom u_i. Eq. (4.13) ensures the continuity of the degrees of freedom at the nodes. Interpolation functions that only provide the continuity of the primitive variable across element boundaries are said to be C^0 continuous.

Later we will develop interpolation functions that provide higher order continuity, e.g., u_j and du_j/dx'.

4.3.1 C^0 continuous, linear shape functions

For a linear approximation let $M = 1$ in Eq. (4.12),

$$\tilde{u}(x') = a_0 + a_1 x' \tag{4.15}$$

where a_0 and a_1 are two coefficients. We next describe a method to express these two coefficients in terms of the displacements u_1 and u_2 at the boundaries of the solution region, x'_1 and x'_2, respectively (Fig. 4.3). Eq. (4.12) at the two end points of the region becomes,

$$\begin{aligned} u_1 &= a_0 + a_1 x'_1 \\ u_2 &= a_0 + a_1 x'_2 \end{aligned} \tag{4.16}$$

In matrix form this is expressed as follows:

$$\begin{Bmatrix} u_1 \\ u_2 \end{Bmatrix} = \begin{bmatrix} 1 & x'_1 \\ 1 & x'_2 \end{bmatrix} \begin{Bmatrix} a_0 \\ a_1 \end{Bmatrix} = [A]\{a\} \tag{4.17}$$

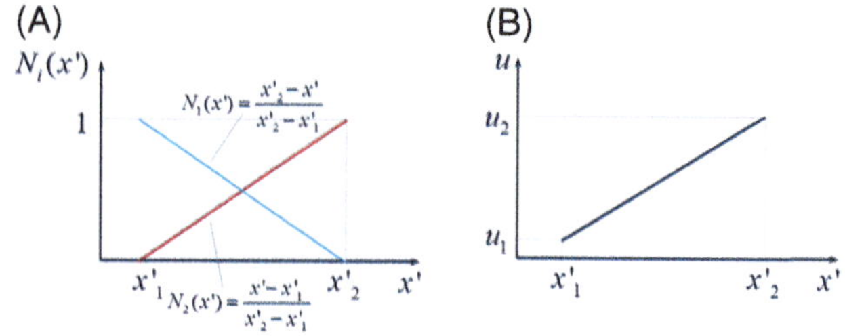

FIGURE 4.3 (A) The shape functions $N_1(x')$ and $N_2(x')$ for an element that spans the range $x'_1 \leq x' \leq x'_2$. (B) Linear interpolation of $u(x')$.

where $[A]$ is a coefficient matrix that depends on the locations of the points-1 and -2. The coefficients a_0 and a_1 can be obtained as follows:

$$\begin{Bmatrix} a_0 \\ a_1 \end{Bmatrix} = [A]^{-1} \begin{Bmatrix} u_1 \\ u_2 \end{Bmatrix} \tag{4.18}$$

with

$$[A]^{-1} = \frac{1}{x'_2 - x'_1} \begin{bmatrix} x'_2 & x'_1 \\ -1 & 1 \end{bmatrix} \tag{4.19}$$

$$a_0 = \frac{x'_2}{x'_2 - x'_1} u_1 - \frac{x'_1}{x'_2 - x'_1} u_2 \text{ and } a_1 = -\frac{u_1}{x'_2 - x'_1} + \frac{u_2}{x'_2 - x'_1} \tag{4.20}$$

Substituting Eq. (4.20) in (4.15), the interpolation of $\tilde{u}(x')$ becomes,

$$\tilde{u} = \left(\frac{x'_2}{x'_2 - x'_1} u_1 - \frac{x'_1}{x'_2 - x'_1} u_2 \right) + \left(-\frac{u_1}{x'_2 - x'_1} + \frac{u_2}{x'_2 - x'_1} \right) x' \tag{4.21}$$

This equation can be regrouped as follows:

$$\tilde{u} = \frac{x'_2 - x'}{x'_2 - x'_1} u_1 + \frac{x' - x'_1}{x'_2 - x'_1} u_2 \tag{4.22}$$

Note that this equation is the same as (4.15) where a_0 and a_1 are replaced with u_1 and u_2. Define the following functions as the *shape functions*,

$$N_1(x') = \frac{x'_2 - x'}{x'_2 - x'_1} \text{ and } N_2(x') = \frac{x' - x'_1}{x'_2 - x'_1} \tag{4.23}$$

The linear interpolation of $\tilde{u}$ defined in Eq. (4.15) can be then expressed by using the shape functions as follows:

$$\begin{aligned} \tilde{u} &= N_1(x') u_1 + N_2(x') u_2 \\ &= \sum_{i=1}^{2} N_i(x') u_i = [N]\{d\} \end{aligned} \tag{4.24}$$

where

$$\begin{array}{ll}\text{shape function matrix:} & [N] = \left[N_1(x')\; N_2(x')\right] \\ \text{degree of freedom vector:} & \{d\} = \{u_1\; u_2\}^T\end{array} \tag{4.25}$$

Note that the linear shape functions in Eq. (4.23) satisfy condition (C1), i.e., $N_1(x'_1) = 1$; $N_1(x'_2) = 0$ and $N_2(x'_1) = 0$; $N_2(x'_2) = 1$. This property enables the interpolation function $\tilde{u}$ to meet the values specified at the ends of the interpolation region as follows:

$$\tilde{u}(x'_1) = u_1 \text{ and } \tilde{u}(x'_2) = u_2 \tag{4.26}$$

4.3.2 C^0 continuous, quadratic shape functions

Consider a second order polynomial of the type given in Eq. (4.12) with $M = 2$,

$$\tilde{u} = a_0 + a_1 x' + a_2 x'^2 \tag{4.27}$$

As outlined above, evaluating this polynomial at $x' = x'_1, x'_2, x'_3$ the following $[A]$ matrix is found,

$$[A] = \begin{bmatrix} 1 & x'_1 & x'^2_1 \\ 1 & x'_2 & x'^2_2 \\ 1 & x'_3 & x'^2_3 \end{bmatrix} \tag{4.28}$$

The $[X]$ matrix in this case is,

$$[X] = \begin{bmatrix} 1 & x' & x'^2 \end{bmatrix} \tag{4.29}$$

After some algebraic manipulation the following *quadratic shape functions* can be found,

$$N_1 = \frac{(x'_2 - x')(x'_3 - x')}{(x'_2 - x'_1)(x'_3 - x'_1)}, \quad N_2 = \frac{(x'_1 - x')(x'_3 - x')}{(x'_1 - x'_2)(x'_3 - x'_2)},$$
$$\text{and } N_3 = \frac{(x'_1 - x')(x'_2 - x')}{(x'_1 - x'_3)(x'_2 - x'_3)} \tag{4.30}$$

The quadratic interpolation of $u(x)$ can then be expressed as follows:

$$\tilde{u} = \frac{(x'_2 - x)(x'_3 - x')}{(x'_2 - x'_1)(x'_3 - x'_1)} u_1 + \frac{(x'_1 - x')(x'_3 - x')}{(x'_1 - x'_2)(x'_3 - x'_2)} u_2 + \frac{(x'_1 - x')(x'_2 - x')}{(x'_1 - x'_3)(x'_2 - x'_3)} u_3. \tag{4.31}$$

This can also be expressed by using one of the following short-hand notations,

$$\begin{aligned}\tilde{u} &= N_1 u_1 + N_2 u_2 + N_3 u_3 \\ &= \sum_{i=1}^{3} N_i u_i = [N]\{d\}\end{aligned} \tag{4.32}$$

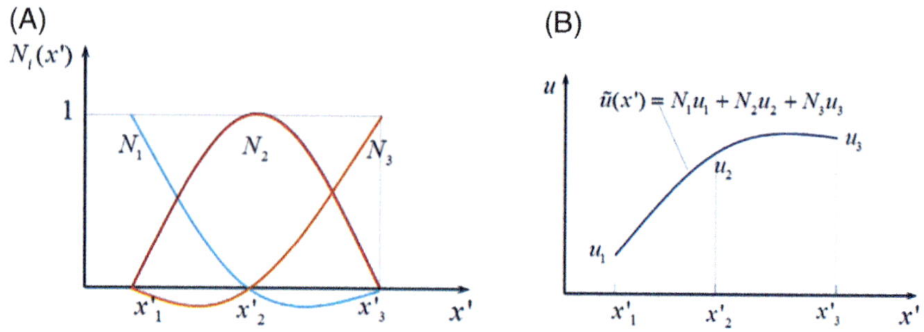

FIGURE 4.4 The quadratic shape functions N_1, N_2 and N_3, given in Eq. (4.30), are used to interpolate $\tilde{u}$ as given in Eq. (4.31).

where

shape function matrix: $[N] = [N_1(x')\ \ N_2(x')\ \ N_3(x')]$

degree of freedom vector: $\{d\} = \{u_1 \quad u_2 \quad u_3\}^T$.

Fig. 4.4 shows the shape functions N_1, N_2 and N_3 and the quadratic interpolation they enable in the range $x_1 < x' < x_3$.

4.3.3 General form of C⁰ shape functions

Consider the polynomial approximation,

$$\tilde{u} = a_0 + a_1x' + a_2x'^2 + a_3x'^3 + \ldots a_nx'^M \tag{4.33}$$

which can also be represented as follows:

$$\tilde{u} = [X]\{a\} = \sum_{i=0}^{M} a_j x'^j \tag{4.34}$$

where,

$$\begin{aligned} [X] &= [1 \quad x' \quad x'^2 \quad x'^3 \quad \cdots \quad x'^M] \\ \{a\} &= \{a_0 \quad a_1 \quad a_2 \quad a_3 \quad \ldots \quad a_M\}^T \end{aligned} \tag{4.35}$$

After we evaluate u at $(M+1)$ different x'_i locations, we get a system of $(M+1)$ algebraic equations for the unknown a_i,

$$\begin{aligned} u_1 &= u(x'_1) = a_0 + a_1x'_1 + a_2{x'^2}_1 + a_3{x'^3}_1 + \ldots a_M{x'^M}_1 \\ u_2 &= u(x'_2) = a_0 + a_1x'_2 + a_2{x'^2}_2 + a_3{x'^3}_2 + \ldots a_M{x'^M}_2 \\ &\vdots \\ u_{M+1} &= u(x'_{M+1}) = a_0 + a_1x'_{M+1} + a_2{x'^2}_{M+1} + a_3{x'^3}_{M+1} + \ldots a_M{x'^M}_{M+1} \end{aligned} \tag{4.36}$$

In matrix form, this is expressed as follows:

$$\{d\} = [A]\{a\} = \begin{bmatrix} 1 & x'_1 & x'^2_1 & \dots x'^M_1 \\ \vdots & & & \\ 1 & x'_{M+1} & x'^2_{M+1} & \dots x'^M_{M+1} \end{bmatrix} \begin{Bmatrix} a_0 \\ \vdots \\ a_M \end{Bmatrix} \tag{4.37}$$

where,

$\{d\} = \{u_1 \quad u_2 \quad u_3 \quad u_4 \quad \cdots \quad u_{M+1}\}^T$ is the d.o.f. vector.
The coefficient vector is found as follows:

$$\{a\} = [A]^{-1}\{d\} \tag{4.38}$$

The interpolation is expressed as follows:

$$\begin{aligned} \tilde{u} &= [X]\{a\} \\ &= [X]\left([A]^{-1}\{d\}\right), \text{ or} \\ &= [N]\{d\} \end{aligned} \tag{4.39}$$

where the general form of the shape function vector $[N]$ is defined as follows:

$$[N] = [X][A]^{-1} \tag{4.40}$$

4.3.4 One-dimensional, Lagrange interpolation functions

Shape functions for the domain $x' \in [0, L^{(e)}]$: In finite element method each small segment is assigned its own local coordinate system. In this case the domain of the element is defined as $x' \in [0, L^{(e)}]$. The shape functions for this domain can be easily obtained by letting,

$x_1 = 0$ and $x_2 = L^{(e)}$ for the linear shape functions, and
$x_1 = 0$, $x_2 = rL^{(e)}$ and $x_3 = L^{(e)}$ for the quadratic shape functions

in Eqs. (4.23) and (4.30), respectively. Note that r $(0 < r < 1)$ is a coefficient that ensures that node-2 is located between nodes-1 and-3. This gives,
Linear shape functions:

$$N_1(x') = 1 - \frac{x'}{L^{(e)}} \text{ and } N_2(x') = \frac{x'}{L^{(e)}} \tag{4.41}$$

Quadratic shape functions:
for $0 < r < 1$:

$$\begin{aligned} N_1(x') &= \left(1 - \frac{x'}{rL^{(e)}}\right)\left(1 - \frac{x'}{L^{(e)}}\right) \\ N_2(x') &= \frac{1}{1-r}\left(\frac{x'}{rL^{(e)}}\right)\left(1 - \frac{x'}{L^{(e)}}\right) \\ N_3(x') &= \frac{1}{r-1}\left(\frac{x'}{L^{(e)}}\right)\left(r - \frac{x'}{L^{(e)}}\right) \end{aligned} \tag{4.42}$$

for $r = 1/2$:

$$N_1(x') = \left(1 - 2\frac{x'}{L^{(e)}}\right)\left(1 - \frac{x'}{L^{(e)}}\right)$$
$$N_2(x') = 4\left(\frac{x'}{L^{(e)}}\right)\left(1 - \frac{x'}{L^{(e)}}\right)$$
$$N_3(x') = -\left(\frac{x'}{L^{(e)}}\right)\left(1 - 2\frac{x'}{L^{(e)}}\right) \tag{4.43}$$

4.4 Equilibrium equations in finite element form

The finite element formulation of this boundary value problem described by Eqs. (4.3)–(4.5) will be carried out by using the method of weighted residuals. The weighted residual integral of the boundary value problem is given as follows:

$$\int_0^{L^{(e)}} w\left[\frac{d}{dx'}\left(a\frac{du}{dx'}\right) + pu + q\right]dx' = 0 \text{ in } 0 < x' < L^{(e)} \tag{4.44}$$

where $w(x')$ is an arbitrary weight function with the properties described in Section 3.5. Integrating the first term by parts,

$$w\left(L^{(e)}\right)\left(a_{L^{(e)}}\frac{du}{dx'}|_{L^{(e)}}\right) - w(0)\left(a_0\frac{du}{dx'}|_0\right)$$
$$-\int_0^{L(e)} a\frac{du}{dx'}\frac{dw}{dx'}dx' + \int_0^{L(e)} pwudx' + \int_0^{L(e)} wqdx' = 0 \tag{4.45}$$

and using the natural boundary conditions given in Eqs. (4.4) and (4.5), the weak form of the boundary value problem is found as follows:

$$w\left(L^{(e)}\right)\left(a_{L^{(e)}}\left(\alpha_{L^{(e)}}u_{L^{(e)}} + \beta_{L^{(e)}}\right)\right) - w(0)\left(-a_0\left(\alpha_0u_0 + \beta_0\right)\right) -$$
$$\int_0^{L^{(e)}} a\frac{du}{dx'}\frac{dw}{dx'}dx' + \int_0^{L^{(e)}} pwudx' + \int_0^{L^{(e)}} wqdx' = 0 \tag{4.46}$$

Note that the weight function has the properties of the variation of u. Making the substitution $w \to \delta u$ and using the nodal notation for variables at the ends of the element, $u_1 = u(0) = u_0$ and $u_2 = u\left(L^{(e)}\right) = u_{L^{(e)}}$, the weak form is transformed as follows:

$$\delta u_2(a_2(\alpha_2u_2 + \beta_2)) - \delta u_1(-a_1(\alpha u_1 + \beta_1))$$
$$-\int_0^{L^{(e)}} a\frac{du}{dx'}\frac{d\delta u}{dx'}dx' + \int_0^{L^{(e)}} pu\,\delta u\,dx' + \int_0^{L^{(e)}} q\delta udx' = 0 \tag{4.47}$$

First, let us consider a two node, $C^{(0)}$ element shown in the Fig. 4.2. Higher order interpolations for $\tilde{u}$ can also be considered. The function $\tilde{u}$ is approximated with the linear interpolation given by Eq. (4.24) as follows:

$$\tilde{u} = [N]\{d\} \tag{4.48}$$

where $[N]$ is the shape function matrix given in Eq. (4.41),

$$[N] = [N_1 \quad N_2] \text{ with } N_1 = 1 - \frac{x'}{L^{(e)}},\ N_2 = \frac{x'}{L^{(e)}} \tag{4.49}$$

and $\{d\}$ is the degree of freedom vector.

$$\{d\} = \{u_1 \quad u_2\}^T \tag{4.50}$$

The derivative of $u(x')$ is found as follows:

$$\frac{du}{dx'} = [B]\{d\} \tag{4.51}$$

where

$$[B] = [B_1 \quad B_2] \text{ with } B_1 = -\frac{1}{L^{(e)}},\ N_2 = \frac{1}{L^{(e)}} \tag{4.52}$$

The weak form of the BVP can then be expressed as follows:

$$\{\delta d\}^T\left(\begin{Bmatrix} a_1(\alpha_1 u_1 + \beta_1) \\ a_2(\alpha_2 u_2 + \beta_2) \end{Bmatrix} - \left(\int_0^{L^{(e)}} a[B]^T[B]dx' - \int_0^{L^{(e)}} p[N]^T[N]dx'\right)\{d\} + \int_0^{L^{(e)}} q[N]^T dx'\right) = 0 \tag{4.53}$$

Eliminating $\{\delta d\}$ which is an arbitrary variation on this functional and rearranging the terms, the following equilibrium equation is obtained,

$$\left(\int_0^{L^{(e)}} a[B]^T[B]dx' - \int_0^{L^{(e)}} p[N]^T[N]dx'\right)\{d\} = \begin{Bmatrix} \alpha_1 a_1 u_1 \\ \alpha_2 a_2 u_2 \end{Bmatrix} + \begin{Bmatrix} \beta_1 a_1 \\ \beta_2 a_2 \end{Bmatrix} + \int_0^{L^{(e)}} q[N]^T dx'$$

Note that the first term on the right hand side, which involves the unknown variables $\{d\} = \{u_1 \quad u_2\}^T$ can also be expressed as follows,

$$\left(\int_0^{L^{(e)}} a[B]^T[B]dx' - \int_0^{L^{(e)}} p[N]^T[N]dx'\right)\{d\} = \begin{bmatrix} \alpha_1 a_1 & 0 \\ 0 & \alpha_2 a_2 \end{bmatrix}\{d\} + \begin{Bmatrix} \beta_1 a_1 \\ \beta_2 a_2 \end{Bmatrix} + \int_0^{L^{(e)}} q[N]^T dx' \tag{4.54}$$

Let us define the following matrices and vectors[1]

a – term stiffness matrix: $$[k_a]^{(e)} = \int_0^{L^{(e)}} a[B]^T[B]dx \tag{4.55a}$$

p – term - stiffness matrix: $$[k_p]^{(e)} = -\int_0^{L^{(e)}} p[N]^T[N]dx \tag{4.55b}$$

α – term stiffness matrix: $$[k_\alpha]^{(e)} = -\begin{bmatrix} \alpha_1 a_1 & 0 \\ 0 & \alpha_2 a_2 \end{bmatrix} \tag{4.55c}$$

Boundary – load vector: $$\{r_\beta\}^{(e)} = \begin{Bmatrix} \beta_1 a_1 \\ \beta_2 a_2 \end{Bmatrix} \tag{4.55d}$$

Consistent nodal load vector: $$\{r_q\}^{(e)} = \int_0^{L^{(e)}} q[N]^T dx' \tag{4.55e}$$

The matrices $[k_a]^{(e)}$ and $[k_p]^{(e)}$ are related to the physical effects represented by the terms that involve a and p in Eq. (4.1). The vector $\{r_q\}^{(e)}$ is the *consistent nodal load vector* and it is related to the external distributed load term represented by q in Eq. (4.1).

The vector $[k_\alpha]^{(e)}\{d\}^{(e)}$ and the vector $\{r_\beta\}^{(e)}$ represent the effects of the natural boundary conditions acting on the element boundaries. The *boundary load vector* $(\{r_\beta\}^{(e)} + [k_\alpha]^{(e)}\{d\}^{(e)})$ provides interelement continuity across element boundaries. However, on the boundaries of the actual solution domain this vector represents the load intake (if present) from outside.

In Section 4.6, we show that *if the parameter $a(x)$ is continuous in the solution domain*, the boundary load vector $(\{r_\beta\}^{(e)} + [k_\alpha]^{(e)}\{d\}^{(e)})$ will vanish for all internal nodes, except for nodes 1 and $N_E + 1$. Thus, if this condition holds, it is sufficent to use the following forms:

$$\text{at } x = 0: \quad [k_\alpha]^{(1)} = -\begin{bmatrix} \alpha_0 a_0 & 0 \\ 0 & 0 \end{bmatrix} \text{ and } \{r_\beta\}^{(1)} = \begin{Bmatrix} \beta_0 a_0 \\ 0 \end{Bmatrix} \tag{4.56a}$$

$$\text{at } x = L: \quad [k_\alpha]^{(N_E)} = -\begin{bmatrix} 0 & 0 \\ 0 & \alpha_L a_L \end{bmatrix} \text{ and } \{r_\beta\}^{(N_E)} = \begin{Bmatrix} 0 \\ \beta_L a_L \end{Bmatrix} \tag{4.56b}$$

Therefore, for a problem where $a(x)$ is continuous in $0 \le x \le L$, the element equilibrium equations for the boundary value problem given by Eqs. (4.1), (4.2a) is expressed as follows:

$$\text{Boundary element } ((e) = 1): \left([k_a]^{(1)} + [k_p]^{(1)} + [k_\alpha]^{(1)}\right)\{d\} = \{r_q\}^{(1)} + \{r_\beta\}^{(1)} \tag{4.57a}$$

1. See the following text on the interpretation of this matrix.

$$\text{Internal elements } (2 \le (e) \le N_E - 1): \left([k_a]^{(e)} + \left[k_p\right]^{(e)}\right)\{d\} = \left\{r_q\right\}^{(e)} \tag{4.57b}$$

Boundary element $((e) = N_E)$:

$$\left([k_a]^{(N_E)} + \left[k_p\right]^{(N_E)} + [k_\alpha]^{(N_E)}\right)\{d\} = \left\{r_q\right\}^{(N_E)} + \left\{r_\beta\right\}^{(N_E)} \tag{4.57c}$$

This system can also be written as follows:

$$\left[k^{(e)}\right]\{d\} = \left\{r_q^{(e)}\right\} + \left\{r_\beta^{(e)}\right\} \tag{4.58}$$

where $[k^{(e)}]$ is called the *element stiffness matrix.*

If $a(x)$ has one or multiple discontinuities in the solution domain $0 \le x \le L$, we should use the full boundary-load vector $(\{r_\beta\}^{(e)} + [k_\alpha]^{(e)}\{d\}^{(e)})$ across the discontinuities. This topic is discussed further in Section 4.6.

4.4.1 Element stiffness matrix for constant problem parameters

In case $a(x) = a$ and $p(x) = p$ are where a and p are specified constants following *stiffness matrices* can be obtained,

$$[k_a]^{(e)} = \frac{a}{L^{(e)}}\begin{bmatrix} 1 & -1 \\ -1 & 1 \end{bmatrix} \tag{4.59a}$$

$$\left[k_p\right]^{(e)} = -\frac{L^{(e)}p}{6}\begin{bmatrix} 2 & 1 \\ 1 & 2 \end{bmatrix} \tag{4.59b}$$

If natural boundary conditions are specified on the ends of the solution domain, $x = L$,

$$[k_\alpha]^{(1)} = -\begin{bmatrix} \alpha_0 a_0 & 0 \\ 0 & 0 \end{bmatrix} \text{ at } x = 0 \tag{4.60a}$$

$$[k_\alpha]^{(N)} = -\begin{bmatrix} 0 & 0 \\ 0 & \alpha_L a_L \end{bmatrix} \text{ at } x - L \tag{4.60b}$$

otherwise $[k_\alpha] = 0$. Let's call these two elements *boundary elements.*

Likewise if $q(x) = q$ is constant the following *consistent loading* vector is found,

$$\left\{r_q\right\}^{(e)} = \frac{1}{2}qL^{(e)}\begin{Bmatrix} 1 \\ 1 \end{Bmatrix} \tag{4.61}$$

In general, for elements that do not have a node on the boundary, otherwise called *internal elements*, the boundary loading vector becomes,

$$\left\{r_\beta\right\}^{(e)} = \begin{Bmatrix} a_1\beta_1 \\ a_2\beta_2 \end{Bmatrix} \tag{4.62}$$

For the boundary elements the boundary loading vector becomes,

$$\{r_\beta\}^{(1)} = \begin{Bmatrix} \beta_0 a_0 \\ \beta_2 a_2 \end{Bmatrix} \text{ at } x = 0 \tag{4.63a}$$

$$\{r_\beta\}^{(N)} = \begin{Bmatrix} \beta_N a_N \\ \beta_L a_L \end{Bmatrix} \text{ at } x = L \tag{4.63b}$$

4.4.2 Element stiffness matrix for linearly varying problem parameters *a, p,* and *q*

The finite element method allows a wide group of functions for the problem variables *a, p,* and *q*. Such variations can be approximated by developing an element type which allows a linearly varying *a, p,* and *q* along its length,

$$a(x') = a_1(1 + m_a x') \tag{4.64a}$$

$$p(x') = p_1(1 + m_p x') \tag{4.64b}$$

$$q(x') = q_1(1 + m_q x') \tag{4.64c}$$

where $m_a = \frac{a_2 - a_1}{a_1 L^{(e)}}$, $m_p = \frac{p_2 - p_1}{p_1 L^{(e)}}$, and $m_p = \frac{q_2 - q_1}{q_1 L^{(e)}}$ with a_1, a_2, p_1, p_2, q_1 and q_2 are the nodal values of the respective parameters. Using Eqs. (4.55a), (4.55b), (4.55e) along with Eq. (4.64a), the following stiffness matrices and the consistent nodal force vector can be found:

$$[k_a]^{(e)} = \frac{a_1 + a_2}{2L^{(e)}} \begin{bmatrix} 1 & -1 \\ -1 & 1 \end{bmatrix} \tag{4.65a}$$

$$[k_p]^{(e)} = -\frac{L^{(e)}}{12} \begin{bmatrix} (3p_1 + p_2) & (p_1 + p_2) \\ (p_1 + p_2) & (p_1 + 3p_2) \end{bmatrix} \tag{4.65b}$$

Likewise the *consistent loading* vector is found as follows:

$$\{r_q\}^{(e)} = \frac{L^{(e)}}{6} \begin{Bmatrix} 2q_1 + q_2 \\ q_1 + 2q_2 \end{Bmatrix} \tag{4.66}$$

The definitions of the $[k_\alpha]$ matrix and the $\{r_\beta\}^{(1)}$ and $\{r_\beta\}^{(N)}$ vectors remain as given in Eqs. (4.60a) and (4.63a), respectively.

4.5 Recovering specific physics from the general finite element form

The relationships between the functions *a*(*x*), *p*(*x*), *q*(*x*) in generic differential operator given in Eq. (4.3) and the one dimensional equation of equilibrium given in Eqs. (4.6) and (4.9), respectively are summarized in Table 4.1.

Similarly, the relationships between the general boundary conditions given in Eqs. (4.4) and (4.5) and the physical boundary conditions for the 1D elasticity

TABLE 4.1 The connection between the variables of the generic second-order differential equation and the indicated physical problems.

	$a(x)$	$p(x)$	$q(x)$
One-dimensional elastic bar	AE	-	f
One-dimensional heat transfer	kA	$-hP$	$QA+hPT_\infty$

TABLE 4.2 The connection between the variables of the boundary conditions for the generic second-order differential equation and the indicated physical problems. The natural boundary condition (NBC) is defined generically as $n\left(\frac{du}{dx}\right) = \alpha u + \beta$. The essential boundary condition is indicated by EBC.

	EBC	or	NBC	
			α	β
One-dimensional elastic bar	u-specified		0	$\frac{F_i}{AE}$
One-dimensional heat transfer	T-specified			
			Specified heat flux	
			0	$-\frac{f_{Bi}}{k}$
			Convection BC	
			$-\frac{h}{k}$	$\frac{h}{k}T_\infty$
			Radiation BC	
			$-\frac{\sigma\psi}{k}(T^3 - T_s^3)$	$\frac{\sigma\psi}{k}T_s$

problem, given in Eq. (4.7) and the 1D heat transfer problem, given in Eqs. (4.10)–(4.11), are presented in Table 4.2.

4.6 Element assembly

In order to represent the original boundary value problem in the solution domain $0 < x < L$ we note that the solution must be continuous across element boundaries. This implies continuity of the dependent variable u and the gradient of the dependent variable du/dx. Consider the node-$i+1$, which is connected to element-(i) on the left and element-($i+1$) on the right (Fig. 4.1). For this node

C1. Continuity of the dependent variable: requires,

$$u_{i+1}^{(i)} = u_{i+1}^{(i+1)} \tag{4.67}$$

The interpolation functions described in Section 4.3 enable the interelement continuity of the dependent variable.

C2. Continuity of the gradient: requires,

$$\left.\frac{du}{dx'}\right|_{i+1}^{(i)} = \left.\frac{du}{dx'}\right|_{i+1}^{(i+1)} \tag{4.68}$$

Note that the gradient of this form is related to the natural boundary condition at the element boundaries. Specifically, the first term is related to the natural boundary condition of the element-(i) at its right-node, and the second term to the natural boundary condition of the element-(i+1) at its left-node,

$$\left.\frac{du}{dx'}\right|_{i+1}^{(i)} = \alpha_{i+1}^{(i)} u_{i+1} + \beta_{i+1}^{(i)} \text{ and } \left.\frac{du}{dx'}\right|_{i+1}^{(i+1)} = -\left(\alpha_{i+1}^{(i+1)} u_{i+1} + \beta_{i+1}^{(i+1)}\right) \tag{4.69}$$

Therefore, the slope (or gradient) continuity condition can be represented as follows:

$$\left(\alpha_{i+1}^{(i)} u_{i+1} + \beta_{i+1}^{(i)}\right) + \left(\alpha_{i+1}^{(i+1)} u_{i+1} + \beta_{i+1}^{(i+1)}\right) = 0 \tag{4.70}$$

This condition can be rearranged as follows:

$$\left(\alpha_{i+1}^{(i)} + \alpha_{i+1}^{(i+1)}\right) u_{i+1} + \left(\beta_{i+1}^{(i)} + \beta_{i+1}^{(i+1)}\right) = 0 \tag{4.71}$$

and it is this condition that allows turning the element equilibrium equations to a global equilibrium equation. In fact, this condition represents the continuity of the internal force or internal heat flow for the deformable bar or heat conduction equations, respectively.

In order to implement the second continuity condition, Eq. (4.68) must be expressed for every node of the system. Consider the node-$i+1$ mentioned above. For element-(i) we have,

$$\begin{bmatrix} k_{11} & k_{12} \\ k_{21} & k_{22} \end{bmatrix}^{(i)} \begin{Bmatrix} u_i \\ u_{i+1} \end{Bmatrix} = \begin{Bmatrix} r_{q_i} \\ r_{q(i+1)} \end{Bmatrix}^{(i)} + \begin{Bmatrix} \beta_i a_i \\ \beta_{i+1} a_{i+1} \end{Bmatrix}^{(i)} + \begin{bmatrix} \alpha_i a_i & 0 \\ 0 & \alpha_{i+1} a_{i+1} \end{bmatrix}^{(i)} \begin{Bmatrix} u_i \\ u_{i+1} \end{Bmatrix} \tag{4.72}$$

and for element-($i+1$) we have,

$$\begin{bmatrix} k_{11} & k_{12} \\ k_{21} & k_{22} \end{bmatrix}^{(i+1)} \begin{Bmatrix} u_{i+1} \\ u_{i+2} \end{Bmatrix} = \begin{Bmatrix} r_{q_{i+1}} \\ r_{q(i+2)} \end{Bmatrix}^{(i+1)} + \begin{Bmatrix} \beta_{i+1} a_{i+1} \\ \beta_{i+2} a_{i+2} \end{Bmatrix}^{(i+1)} + \begin{bmatrix} \alpha_{i+1} a_{i+1} & 0 \\ 0 & \alpha_{i+2} a_{i+2} \end{bmatrix}^{(i+1)} \begin{Bmatrix} u_{i+1} \\ u_{i+2} \end{Bmatrix} \tag{4.73}$$

These two equations contribute to equilibrium at node $i+1$ as follows:

$$\begin{aligned} k_{21}^{(i)} u_i + k_{22}^{(i)} u_{i+1} &= r_{q(i+1)}^{(i)} + \beta_{i+1}^{(i)} a_{i+1}^{(i)} + \alpha_{i+1}^{(i)} a_{i+1}^{(i)} u_{i+1} \\ k_{11}^{(i+1)} u_{i+1} + k_{12}^{(i+1)} u_{i+2} &= r_{q(i+1)}^{(i+1)} + \beta_{i+1}^{(i+1)} a_{i+1}^{(i+1)} + \alpha_{i+1}^{(i+1)} a_{i+1}^{(i+1)} u_{i+1} \end{aligned} \tag{4.74}$$

Adding these two equations and after proper regrouping of the terms we find,

$$\begin{aligned} &k_{21}^{(i)}u_i + \left(k_{22}^{(i)} + k_{11}^{(i+1)}\right)u_{i+1} + k_{12}^{(i+1)}u_{i+2} = \left(r_{q(i+1)}^{(i)} + r_{q(i+1)}^{(i+1)}\right) + \\ &+\left[\left(\alpha_{i+1}^{(i)}a_{i+1}^{(i)} + \alpha_{i+1}^{(i+1)}a_{i+1}^{(i+1)}\right)u_{i+1} + \left(\beta_{i+1}^{(i)}a_{i+1}^{(i)} + \beta_{i+1}^{(i+1)}a_{i+1}^{(i+1)}\right)\right] \end{aligned} \tag{4.75}$$

Comparing the terms in the square bracket in Eq. (4.75) to Eq. (4.71) we see that the square bracket would vanish if $a(x)$ is prescribed as a continuous function, i.e. $a_{i+1}^{(i)} = a_{i+1}^{(i+1)}$, in the solution domain.

In most boundary value problems, $a(x)$ represents a material or shape related property; therefore it can be assumed to be continuous in the domain, and the square bracket would vanish upon summation as shown here. In case there is a discontinuity of $a(x)$ at a node, then its effect should be included in the assembly process as shown in Eq. (4.75). Unless otherwise stated, in the rest of this text ***we assume the material properties to be continuous in the solution domain***.

Continuity of the slope of the solution (Eq. (4.68)) leads to the procedure that results in Eq. (4.75). This condition must be imposed at all nodes in the system. This can be achieved by using the following relationship,

$$\sum_{e=1}^{N_E}\left([k_\alpha]^{(e)}\{d\}^{(e)} - \{r_q\}^{(e)}\right) = 0 \tag{4.76}$$

The procedure described in Eq. (4.76) is called the *element assembly*. This procedure will result in N_{E+1} algebraic equations for the same number of degrees of freedom, and can be represented in the following matrix form:

$$\begin{bmatrix} k_{11}^{(1)} & k_{12}^{(1)} & 0 & \cdots & & & 0 \\ k_{21}^{(1)} & \left(k_{22}^{(1)} + k_{11}^{(2)}\right) & k_{12}^{(2)} & 0 & \cdots & & 0 \\ 0 & k_{21}^{(2)} & \left(k_{22}^{(2)} + k_{11}^{(3)}\right) & k_{12}^{(3)} & 0 & \cdots & 0 \\ \vdots & 0 & \ddots & \ddots & \ddots & & \\ & \vdots & k_{21}^{(i)} & \left(k_{22}^{(i)} + k_{11}^{(i+1)}\right) & k_{12}^{(i+1)} & & 0 \\ & & 0 & & & & \\ & & & \ddots & \ddots & \ddots & \\ & & & & k_{21}^{(NE-1)} & \left(k_{22}^{(NE-1)} + k_{11}^{(NE)}\right) & k_{12}^{(NE)} \\ 0 & 0 & 0 & & 0 & k_{21}^{(NE)} & k_{22}^{(NE)} \end{bmatrix} \begin{pmatrix} u_1 \\ u_2 \\ u_3 \\ \vdots \\ u_i \\ u_{i+1} \\ u_{i+2} \\ \vdots \\ u_{NE-1} \\ u_{NE} \\ u_{NE+1} \end{pmatrix}$$

$$
= \begin{Bmatrix} r_{q_1}^{(1)} \\ r_{q_2}^{(1)} + r_{q_2}^{(2)} \\ r_{q_3}^{(2)} + r_{q_3}^{(3)} \\ \vdots \\ r_{q_i}^{(i-1)} + r_{q_i}^{(i)} \\ r_{q_{(i+1)}}^{(i)} + r_{q_{(i+1)}}^{(i+1)} \\ r_{q_{(i+2)}}^{(i+1)} + r_{q_{(i+2)}}^{(i+2)} \\ \vdots \\ r_{q_{N-1}}^{(NE-2)} + r_{q_{N-1}}^{(NE-1)} \\ r_{q_{NE}}^{(NE-1)} + r_{q_{NE}}^{(NE)} \\ r_{q_{NE+1}}^{(NE)} \end{Bmatrix} \tag{4.77}
$$

This is the *global equilibrium equation* of the system and symbolically it can be represented as follows:

$$[K]\{D\} = \{R\}.$$

where $[K]$, $\{D\}$ and $\{R\}$ are the *global stiffness matrix, global degree of freedom vector* and the *global load vector*, respectively.

4.7 Boundary conditions

The global equilibrium equation that is assembled as described in the previous section represents the finite element form of the differential equation that represents the physics of the system in the domain $0 < x < L$. Boundary conditions make the solution unique to a given specific conditions on the boundaries $x = 0$ and $x = L$. In this section we introduce how the global equilibrium equations are modified for different types of boundary conditions.

4.7.1 Natural boundary conditions

Admissible form of the natural boundary conditions for the second order system we consider here are given by Eq. (4.2a), (4.2b)

$$
\text{at } x = 0 : \; -\frac{du}{dx} = \alpha_0 u + \beta_0 \; \text{(a) and at } x = L : \; \frac{du}{dx} = \alpha_L u + \beta_L \; \text{(b)} \tag{4.78}
$$

The right-hand side terms of these equations were derived in Eqs. (4.56a), (4.56b) and are interpreted to take the following forms, depending on the boundary:

$$
\text{at } x = 0 : \begin{bmatrix} \alpha_0 a_0 & 0 \\ 0 & 0 \end{bmatrix}^{(1)} \begin{Bmatrix} u_1 \\ u_2 \end{Bmatrix} + \begin{Bmatrix} a_0 \beta_0 \\ 0 \end{Bmatrix} \tag{4.79a}
$$

$$
\text{at } x = L : \begin{bmatrix} 0 & 0 \\ 0 & \alpha_L a_L \end{bmatrix}^{(N)} \begin{Bmatrix} u_{N_E} \\ u_{N_E+1} \end{Bmatrix} + \begin{Bmatrix} 0 \\ a_L \beta_L \end{Bmatrix} \tag{4.79b}
$$

In case a natural boundary condition of the form (4.78) is specified on one of the boundaries the appropriate equation would be added to the global equilibrium equation to represent its effect on the system.

4.7.2 Essential boundary conditions

Admissible form of the essential boundary conditions for the second order system, we consider here, are given by Eqs. (4.2a) and (4.2b),

$$\text{at } x = 0 \, : \, u = u_0 \, (\text{a}) \text{ and at } x = L \, : \, u = u_L \, (\text{b}) \tag{4.80}$$

In this case, since either one or both of the $u_1 = u_0$ or $u_{N+1} = u_L$ are specified at the boundaries they will have to be eliminated from the global degree of freedom vector. For example, consider the case of $u_1 = u_0$. The global equilibrium equation is modified as follows,

$$\begin{bmatrix} \left(k_{22}^{(1)}+k_{11}^{(2)}\right) & k_{12}^{(2)} & 0 & \cdots & & 0 \\ k_{21}^{(2)} & \left(k_{22}^{(2)}+k_{11}^{(3)}\right) & k_{12}^{(3)} & 0 & \cdots & 0 \\ 0 & \ddots & \ddots & \ddots & & \\ \vdots & k_{21}^{(i)} & \left(k_{22}^{(i)}+k_{11}^{(i+1)}\right) & k_{12}^{(i+1)} & & 0 \\ & 0 & & \ddots & \ddots & \\ & & \ddots & k_{21}^{(N-1)} & \left(k_{22}^{(N_E-1)}+k_{11}^{(N_E)}\right) & k_{12}^{(N)} \\ 0 & 0 & & 0 & k_{21}^{(N_E)} & k_{22}^{(N_E)} \end{bmatrix} \begin{Bmatrix} u_2 \\ u_3 \\ \vdots \\ u_i \\ u_{i+1} \\ u_{i+2} \\ \vdots \\ u_{N_E-1} \\ u_{N_E} \\ u_{N_E+1} \end{Bmatrix}$$

$$= \begin{Bmatrix} r_{q_2}^{(1)} + r_{q_2}^{(2)} \\ r_{q_3}^{(2)} + r_{q_3}^{(3)} \\ \vdots \\ r_{q_i}^{(i-1)} + r_{q_i}^{(i)} \\ r_{q(i+1)}^{(i)} + r_{q(i+1)}^{(i+1)} \\ r_{q(i+2)}^{(i+1)} + r_{q(i+2)}^{(i+2)} \\ \vdots \\ r_{q_{N_E-1}}^{(N_E-2)} + r_{q_{N_E-1}}^{(N_E-1)} \\ r_{q_{N_E}}^{(N_E-1)} + r_{q_{N_E}}^{(N_E)} \\ r_{q_{N_E+1}}^{(N_E)} \end{Bmatrix} - \begin{Bmatrix} k_{21}^{(1)} u_0 \\ 0 \\ \\ \vdots \\ \\ 0 \end{Bmatrix} \tag{4.81}$$

Note that the size of the global stiffness matrix and the degree of freedom vector are reduced to $N_E \times N_E$ and $N_E \times 1$, respectively, and the effect of the prescribed essential boundary condition now appears as a loading term on the right-hand side of the equilibrium equation.

The treatment of the cases when $u_{N_E+1} = u_L$ is specified or both $u_1 = u_0$ or $u_{N_E+1} = u_L$ can be easily deduced from the above treatment.

TABLE 4.3 Element connectivity table is used to enter the element properties.

(e)	node-1	node-2	$L^{(e)}$	a_1	a_2	p_1	p_2	q_1	q_2
E	CT(e,1)	CT(e,2)	L(e)	a1(e)	a2(e)	p1(e)	p2(e)	q1(e)	q2(e)
1	1	2	$L^{(1)}$	a_1	a_2	p_1	p_2	q_1	q_2
2	2	3	$L^{(2)}$	a_2	a_3	p_2	p_3	q_2	q_3
...									
e	e	e + 1	...						
...									
NE	NE	$NE+1$	$L^{(NE)}$	a_{NE}	a_{NE+1}	p_{NE}	p_{NE+1}	q_{NE}	q_{NE+1}

4.8 Computer implementation

4.8.1 Main-code

A pseudo-algorithm for the solution of the one-dimensional boundary value problem presented in this chapter can be given as follows:

```
%...............................
% Main.m:
%
    clear;
    get_inputs;
    prepare_mesh;
    get_boundary_conditions;
    assemble_global_variables
    apply_BCs;
    D = inv(K)*R
    plot(x,D)
%...............................
```

4.8.2 Element connectivity table

An element connectivity table, such as the one in Table 4.3, is useful to enter the problem variables into a computer program.

4.8.3 Element assembly

Element assembly process described in Section 4.6 results in the global stiffness matrix $[K]$ and the global load vector $\{R\}$. One of the key, and typically a relatively complicated, component of a code is the assembly algorithm. The element-to-node connectivity information provided in the connectivity table is used extensively in assembly. A typical pseudo-code for the assembly of elements in a one-dimensional problem is given as follows:

```
%..............................
% Assembly algorithm for a 1D problem
%
for e =1:NE
    Compute ka matrix for element e
    Compute kp matrix for element e
    Compute rq vector for element e
    Compute ke matrix for element e

  ke = ka + kp;
    %
    i = CT(e,1);
    j = CT(e,2);
    % Assemble the K matrix
    K(i,i)   = K(i,i) + ke(1,1);
    K(i,j)   = K(i,j) + ke(1,2);
    K(j,i)   = K(j,i) + ke(2,1);
    K(j,j)   = K(j,j) + ke(2,2);
    % Assemble the R vector
    R(i)     = R(i)    + rq(1);
    R(j)     = R(j)    + rq(2);
end
%..............................
```

4.8.4 Boundary conditions

The global stiffness matrix and the load vector should be modified as described in Section 4.7, according to the boundary conditions of the problem. In this relatively simple one dimensional problem the required modifications on $[K]$ and $\{R\}$are implemented on a case-by-case basis.

4.9 Example problem

Example 4.1

a) Solve the boundary value problem given below by using four equal-length, 2-node, $C^{(0)}$ elements.

b) Write a computer program to solve the same BVP by using 2, 4, 8, 16 and 32 elements and report the change of solution value at $x = 1$ and $x = 2$ with increasing number of elements.

$$\frac{d^2u}{dx^2} + 4u = 8 + 0.5x \text{ in } 0 < x < 2 \tag{4.82}$$

$$u(0) = 5 \tag{4.83a}$$

and

$$\frac{du}{dx}\Big|_{x=2} = 10 \tag{4.83b}$$

Solution:

Comparing Eq. (4.82) to Eq. (4.3), we note that:

$$a = 1, p = 4 \text{ and } q = -(8 + 0.5x).$$

Thus we have:

$$a_0 = a_L = 1,\ p_0 = p_L = 4 \text{ and } q_0 = -8 \text{ and } q_L = -9.$$

Comparing Eq. (4.83b) to Eq. (4.2b) we note that $\alpha_L = 0$.
Considering element-4 represents the $x = L$ boundary we have find,

$$[k_\alpha]^{(4)} = 0.$$

Element equilibrium for elements $e = 1-4$: $([k_a]^{(e)} + [k_p]^{(e)})\{d\} = \{r_q\}^{(e)} + \{r_\beta\}^{(e)}$ where,

$$[k_a]^{(e)} = \frac{a_1 + a_2}{2L^{(e)}}\begin{bmatrix} 1 & -1 \\ -1 & 1 \end{bmatrix},$$

$$[k_p]^{(e)} = -\frac{L^{(e)}}{12}\begin{bmatrix} (3p_1 + p_2) & (p_1 + p_2) \\ (p_1 + p_2) & (p_1 + 3p_2) \end{bmatrix}$$

$$\{r_q\}^{(e)} = \frac{L^{(e)}}{6}\begin{Bmatrix} 2q_1 + q_2 \\ q_1 + 2q_2 \end{Bmatrix}$$

Natural boundary condition for $e = 4$:

$$\{r_\beta\}^{(4)} = \begin{Bmatrix} 0 \\ 10 \end{Bmatrix}$$

We can form the following element connectivity table.

(e) e	node-1 CT(e,1)	node-2 CT(e,2)	$L^{(e)}$ L(e)	a_1 a1(e)	a_2 a2(e)	p_1 p1(e)	p_2 p2(e)	q_1 q1(e)	q_2 q2(e)
1	1	2	0.5	1	1	4	4	−8	−8.25
2	2	3	0.5	1	1	4	4	−8.25	−8.50
3	3	4	0.5	1	1	4	4	−8.50	−8.75
4	4	5	0.5	1	1	4	4	−8.75	−9.00

Element level equilibrium equations are computed one-by-one as follows:

$$\text{Element-1: } \left(\begin{bmatrix} 2 & -2 \\ -2 & 2 \end{bmatrix} - \frac{1}{3}\begin{bmatrix} 2 & 1 \\ 1 & 2 \end{bmatrix}\right)\begin{Bmatrix} u_1 \\ u_2 \end{Bmatrix} = \begin{Bmatrix} -2.0208 \\ -2.0417 \end{Bmatrix} + \begin{Bmatrix} a_1\beta_1 \\ a_1\beta_2 \end{Bmatrix}$$

$$\text{Element-2: } \left(\begin{bmatrix} 2 & -2 \\ -2 & 2 \end{bmatrix} - \frac{1}{3}\begin{bmatrix} 2 & 1 \\ 1 & 2 \end{bmatrix}\begin{Bmatrix} u_2 \\ u_3 \end{Bmatrix} = \begin{Bmatrix} -2.0833 \\ -2.1042 \end{Bmatrix} + \begin{Bmatrix} a_2\beta_2 \\ a_3\beta_3 \end{Bmatrix}\right)$$

$$\text{Element-3: } \left(\begin{bmatrix} 2 & -2 \\ -2 & 2 \end{bmatrix} - \frac{1}{3}\begin{bmatrix} 2 & 1 \\ 1 & 2 \end{bmatrix}\right)\begin{Bmatrix} u_3 \\ u_4 \end{Bmatrix} = \begin{Bmatrix} -2.1458 \\ -2.1667 \end{Bmatrix} + \begin{Bmatrix} a_3\beta_3 \\ a_4\beta_4 \end{Bmatrix}$$

$$\text{Element-4: } \left(\begin{bmatrix} 2 & -2 \\ -2 & 2 \end{bmatrix} - \frac{1}{3}\begin{bmatrix} 2 & 1 \\ 1 & 2 \end{bmatrix}\right)\begin{Bmatrix} u_4 \\ u_5 \end{Bmatrix} = \begin{Bmatrix} -2.2083 \\ -2.2292 \end{Bmatrix} + \begin{Bmatrix} a_4\beta_4 \\ 10 \end{Bmatrix}$$

Assembly of elements (1)–(4) results in the following global equilibrium equation. Note that we directly indicate below $u_1 = 5$, i.e., the EBC given at $x = 0$.

$$\begin{bmatrix} 1.33 & -2.33 & 0 & 0 & 0 \\ -2.33 & 2.67 & -2.33 & 0 & 0 \\ 0 & -2.33 & 2.67 & -2.33 & 0 \\ 0 & 0 & -2.33 & 2.67 & -2.33 \\ 0 & 0 & 0 & -2.33 & 1.33 \end{bmatrix}\begin{Bmatrix} 5 \\ u_2 \\ u_3 \\ u_4 \\ u_5 \end{Bmatrix} = \begin{Bmatrix} -2.0208 + \beta_1 \\ -4.1250 \\ -4.2500 \\ -4.3750 \\ 7.7708 \end{Bmatrix}$$

As u_1 is known, it needs to be eliminated from the system of unknown equations. This necessitates modifying the right hand side of the global equilibrium equation. The resulting system of equations is represented as follows:

$$\begin{bmatrix} 2.67 & -2.33 & 0 & 0 \\ -2.33 & 2.67 & -2.33 & 0 \\ 0 & -2.33 & 2.67 & -2.33 \\ 0 & 0 & -2.33 & -1.33 \end{bmatrix}\begin{Bmatrix} u_2 \\ u_3 \\ u_4 \\ u_5 \end{Bmatrix} = \begin{Bmatrix} -4.1250 \\ -4.2500 \\ -4.3750 \\ 7.7708 \end{Bmatrix} - \begin{Bmatrix} -2.33 \times 5 \\ 0 \\ 0 \\ 0 \end{Bmatrix}$$

Solution this system of equations gives,

$$\begin{Bmatrix} u_2 \\ u_3 \\ u_4 \\ u_5 \end{Bmatrix} = \begin{Bmatrix} 0.3130 \\ -2.8745 \\ -1.7766 \\ 2.7190 \end{Bmatrix}$$

In order to check the convergence of this problem, it is solved by using 2, 4, 8, 16, and 32 elements. The solution is given in Fig. 4.5A. Note that by doubling the element numbers at each level of mesh refinement, we are able to keep the nodes of the previous mesh in the solution. Thus, we are able to monitor how the solution develops at the same location with mesh refinement as shown in Fig. 4.5B.

4.10 Problems

4.1 Consider the C^0 continuous, quadratic interpolations functions, given in Eq. (4.30).

(a) Verify numerically that shape functions sum to unity.

(b) What should the sum of x derivatives of the shape functions be? Verify the property numerically.

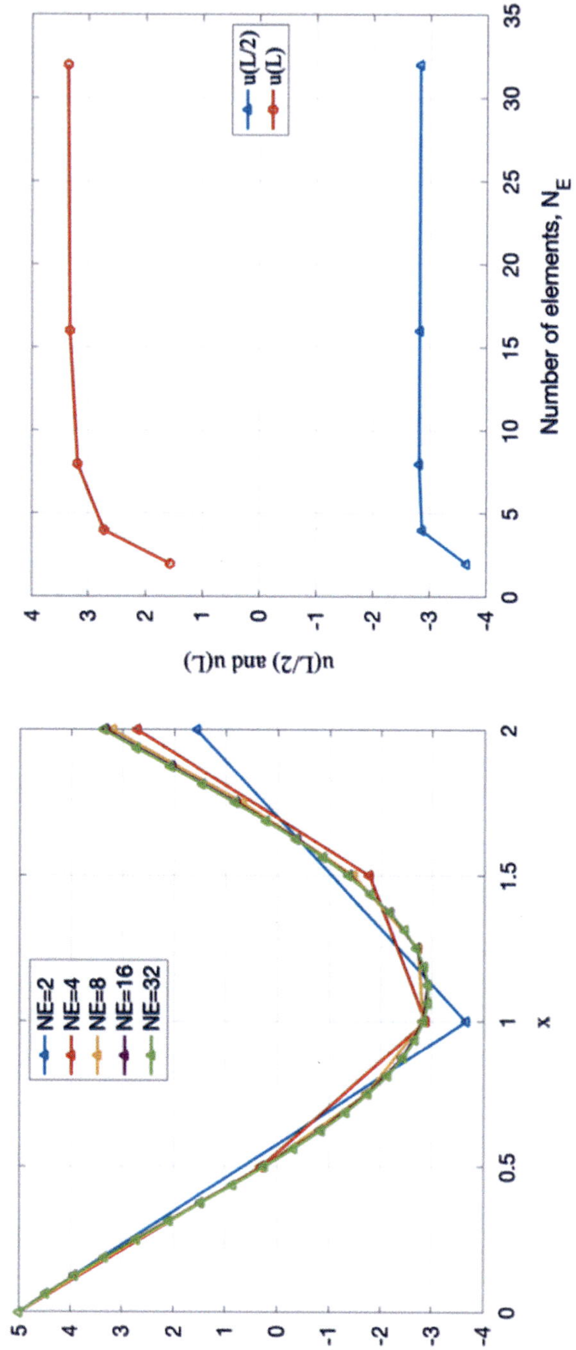

FIGURE 4.5 Solution to the differential equation by using $N_E = 2, 4, 8, 16$ and 32 elements.

4.2 Derive the finite element equations for the boundary value problem given below. Use a two node element with $C^{(0)}$ shape functions.

$$\frac{d}{dx}\left(a\frac{du}{dx}\right)+bu=-cx^{n} \text{ in } 0<x<L$$
$$\left(a\frac{du}{dx}\right)_{x=0}=Q_0 \text{ and } \left(a\frac{du}{dx}\right)_{x=L}=Q_L$$

where a, b, c, n, Q_0 and Q_L are given scalar values.

4.3 Derive the finite element equations for the boundary value problem given in Problem 4.2 by using a three-node element of length $L^{(e)}$ and $C^{(0)}$ continuous element formulation. The element has a node at each end, and a node located at $x = r.L^{(e)}$ $(0 < r < 1)$ as shown in the Fig. 4.6.

4.4 Derive the finite element equations for the boundary value problem given in Problem 4.2 by using a three-node element of length $L^{(e)}$ and $C^{(0)}$ continuous element formulation. The element has a node at each end, and a node located at $x = L^{(e)}/2$.

4.5 An elastic bar is fixed on one end and subjected to a concentrated force F_L and a distributed force q_x as shown in the Fig. 4.7. The bar has a constant cross-sectional area A, and elastic modulus E.

a) Identify the boundary value problem that describes the mechanics of this bar.

b) Find the finite element representation of this problem by using the appropriate formulation presented in this chapter.

Use $L = 1$ m, $E = 100$ GPa, $A = 1$ cm^2, $q_0 = 1$ kN/m, $F_L = 10$ kN for parts c) and d).

c) Use two elements and hand calculations to find the deflection along the bar.

d) Carryout a mesh convergence study to find the deflection along the bar.

Hint: The Matlab code provided in Appendix B.2 can be modified to solve part d).

4.6 The end temperatures of a bar are fixed as T_0 and T_L as shown in Fig. 4.8. This bar is subjected to an internal heat source Q located at middle of its length as shown in the figure. The bar is submerged in a fluid

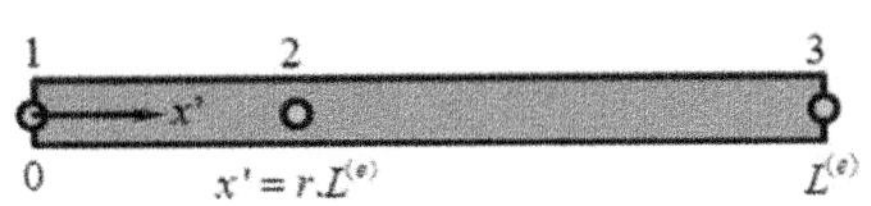

FIGURE 4.6 Problem 4.3.

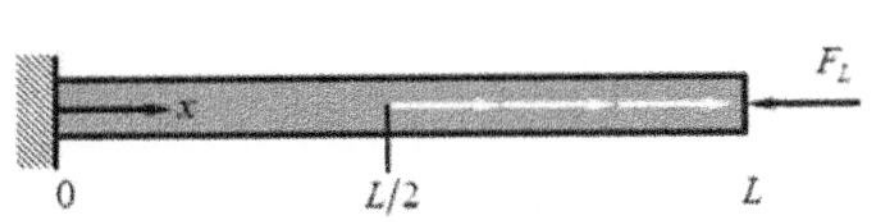

FIGURE 4.7 Problem 4.5.

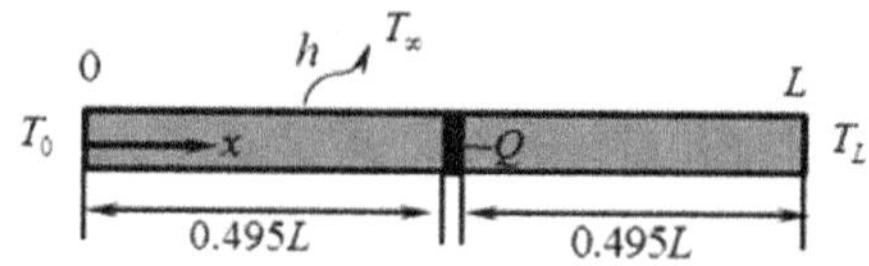

FIGURE 4.8 Problem 4.6.

with temperature T_∞ and it is thus subjected to convective heat transfer along its outer surface with convection coefficient h. The length, cross sectional area, perimeter, conductivity of the bar are given as L, A, p and k, respectively.

a) Identify the boundary value problem that describes the heat transfer in this bar.

b) Find the finite element representation of this problem by using the appropriate formulation presented in this chapter.

For parts c) and d) use the following values: $L = 1$ m, $k = 1{,}000$ W/mK, $h = 5$ W/m^{2}K, T_0, T_L and T_∞ are 20, 30 and 20 deg. C, respectively and $Q = 5$ W/m^3. The diameter of the bar is 1 cm.

c) Use three elements and hand calculations to find the temperature variation along the bar.

d) Carryout a mesh convergence study to find the temperature variation along the bar.

Hint: The Matlab code provided in Appendix B.2 can be modified to solve part d).

Use the following ordinary differential equation for problems 4.7, 4.8, and 4.9

$$\frac{d}{dx}\left(a\frac{du}{dx}\right) + bu + q = 0 \text{ in } 0 < x < L$$

where a, b, q are functions described in the problems.

4.7 Solve the ordinary differential equation given above for the following boundary conditions:

$$u(0) = u_0 \text{ and } u(L) = u_L$$

where $u_0 = 0$ and $u_L = 2$. The function $a(x)$ is described in Fig. 4.9 with $x_1 = L/2$, $L = 2$, $a_{01} = 10$ and $a_{1L} = 5$. The functions $b(x) = 0$ and $q(x) = 1$ in $0 < x < L$.

Hint: The Matlab code provided in Appendix B.2 can be modified to solve this problem.

4.8 Solve the ordinary differential equation given above for the following boundary conditions:

$$\left(a\frac{du}{dx}\right)_{x=0} = Q_0 \text{ and } \left(a\frac{du}{dx}\right)_{x=L} = Q_L$$

where $Q_0 = 1/2$ and $Q_L = 0$. The function $a(x)$ is described in Fig. 4.10

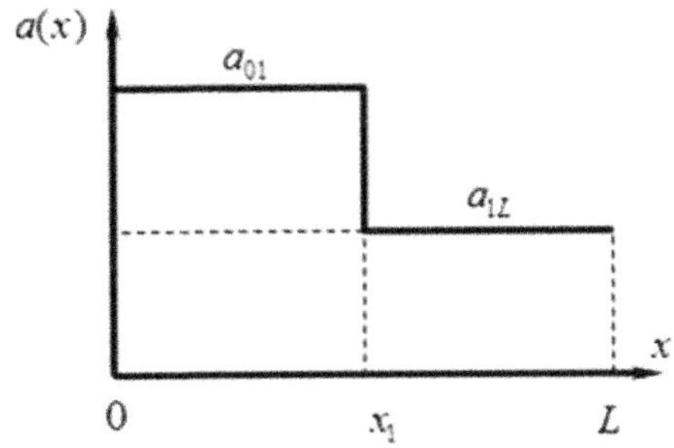

FIGURE 4.9 Problem 4.7.

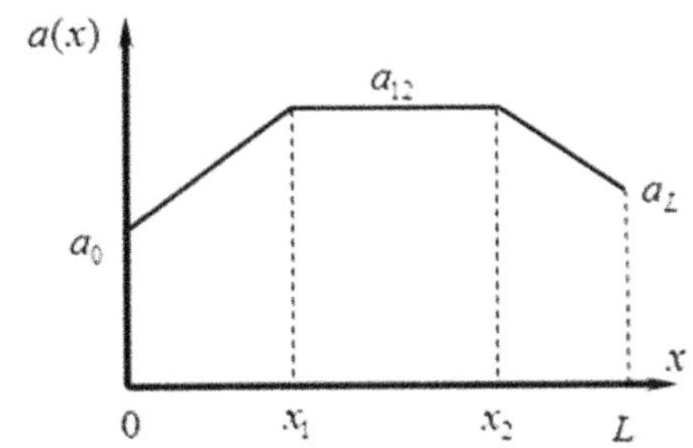

FIGURE 4.10 Problem 4.8.

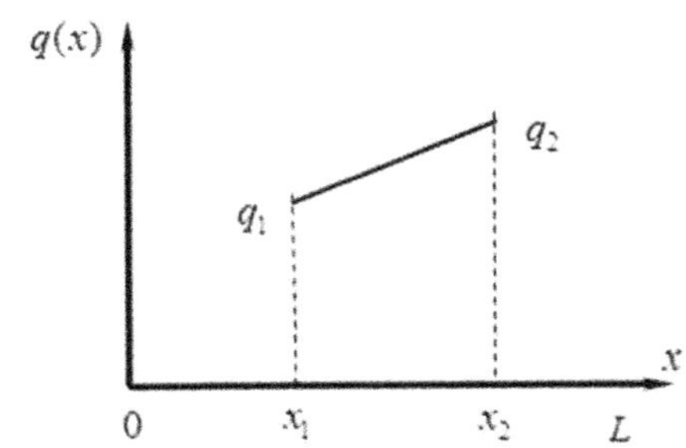

FIGURE 4.11 Problem 4.9.

with $x_1 = L/3$, $x_2 = 2L/3$, $L = 3$, $a_0 = 10$, $a_{12} = 20$ and $a_L = 15$. The functions $b(x) = 5$ and $q(x) = 1$ in $0 < x < L$.

Hint: The Matlab code provided in Appendix B.2 can be modified to solve this problem.

4.9 Solve the ordinary differential equation given above for the following boundary conditions:

$$u(0) = u_0 \text{ and } u(L) = u_L$$

where $u_0 = 2$ and $u_L = 0$. The function $q(x)$ is described in Fig. 4.11 with $x_1 = 0.25L$, $x_2 = 0.75L$, $L = 1$, $q_1 = 10$ and $q_2 = 15$. The functions $a(x) = 5$ and $b(x) = 0$ in $0 < x < L$.

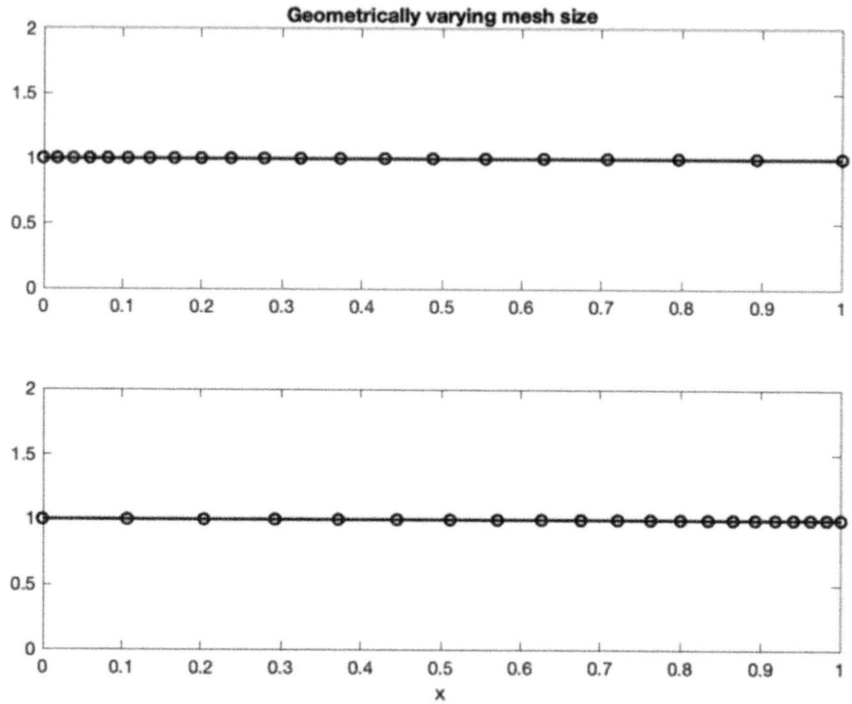

FIGURE 4.12 Problem 4.10. This figure is plotted by using $r = 1.1$, $N_E = 20$, $L = 1$.

Hint: The Matlab code provided in Appendix B.2 can be modified to solve this problem.

4.10 We need to use a fine mesh in regions of the solution domain, where the dependent variable has steep gradients. However, using a fine mesh everywhere in the solution domain may not be necessary and could lead to excessive computational effort. This dilemma can be resolved by using a gradually changing mesh size as shown in Fig. 4.12. Geometric series can be used to generate meshes as shown in this figure where we have the following relationships:

$$\frac{L^{(e+1)}}{L^{(e)}} = r \text{ for geometrically increasing element size, or}$$

$$\frac{L^{(e)}}{L^{(e+1)}} = r \text{ for geometrically decreasing element size,}$$

where r is a real number greater than 1.

Write a Matlab program, to prepare a mesh with increasing and decreasing size similar to Fig. 4.12. Use $N_E = 20$, $r = 1.2$ and $L = 1.5$m. Submit the listing of your program and a figure similar to Fig. 4.12 with your answer.

Chapter 5

Finite element analysis of planar bars and trusses

5.1 Introduction

The goal of this chapter is to develop the tools for finite element analysis of planar bars, shafts, and trusses. Longitudinal deflection of a linear, elastic bar and the twist of a linear elastic shaft will be used as examples. Admittedly the emphasis is on the deformations of an elastic bar. There is a certain amount of unavoidable overlap with this presentation and the work presented in Chapter 4. This chapter introduces the coordinate transformation equations that enable the modeling and simulation of planar truss systems. The methodology presented in this chapter is general, and it can directly be expanded to the solution of two- and three-dimensional linear elasticity problems.

Consider the one-dimensional deformable solid shown in Fig. 5.1. In most cases, the deflection of such a solid can be computed by using ***analytical methods***. However, this solution could become hard to track if the material properties, the cross-sectional area or the loading vary in a complicated way along the bar. The ***finite element method*** is a versatile numerical technique which presents a relatively straightforward solution for problems with geometric and material variabilities.

In finite element method, the solution domain is divided into small segments called ***elements***, which are connected at their ends. The ensemble of the elements is known as the ***finite element mesh***. In Fig. 5.1, the one-dimensional domain is divided into N_E elements. Each element is given an element number e, such that $e \in [1, N_E]$.

In general, the number of ***nodes*** associated with an element depends on the order of polynomial interpolation, as in Eq. (4.12), used in the solution. For a one-dimensional problem, each element will have a minimum of two nodes one at each end. If the order of the polynomial interpolation is higher than one, additional internal nodes will be part of the element (Fig. 5.2). For example,

Linear interpolation: two end nodes for a total of two nodes.

Quadratic interpolation: two end nodes and one internal node for a total of three nodes.

Finite Element Method: Physics and Solution Methods. DOI: https://doi.org/10.1016/B978-0-12-821127-4.00005-0
Copyright © 2022 Elsevier Inc. All rights reserved.

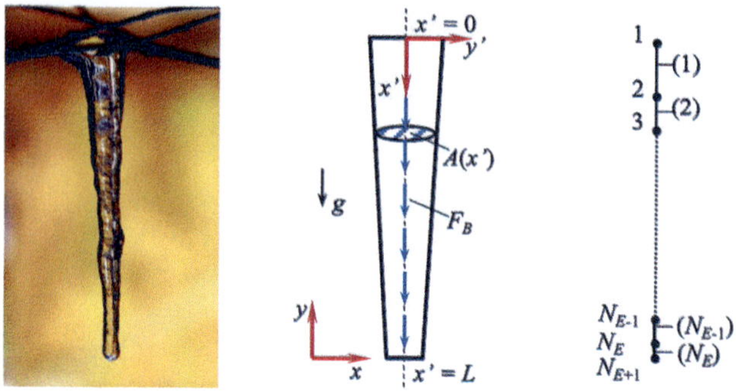

FIGURE 5.1 An icicle subjected to gravitational load can be modeled as a one-dimensional structure. (Image from: Pixbay.com, downloaded January 2018).

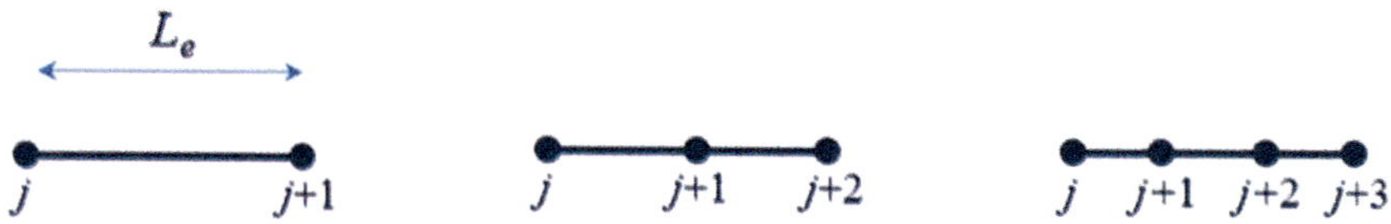

FIGURE 5.2 Elements with two, three, and four nodes. These represent linear, second-order, and third-order polynomial interpolation functions, respectively.

Third order interpolation: two end nodes and two internal nodes for a total of four nodes.

The initial examples used in this chapter involve two-node, linear interpolation elements, but the methodology can be easily expanded to higher-order interpolations.

The length of element-*e* is indicated by $L^{(e)}$. Total length of the solution domain can be easily found by using the following formula,

$$L = \sum_{e=1}^{N_E} L^{(e)} \tag{5.1}$$

It is convenient to introduce an (local) *element coordinate system* (*x'*,*y'*) to develop the method. It is also important to define the appropriate *transformation of the variables* between the element coordinate system and the *global coordinate system* in order to enable an equilibrium state that is independent of the coordinate systems.

The mechanical compliance of each element is represented by the *element stiffness matrix* formulation. The compliance of the entire structure is obtained by assembling each element stiffness matrix in to a *global stiffness matrix*. The external forces acting on the structure are either treated as natural boundary conditions, or as equivalent forces acting on the nodes of the structure. The technical term for this is *consistent nodal force*. The essential boundary

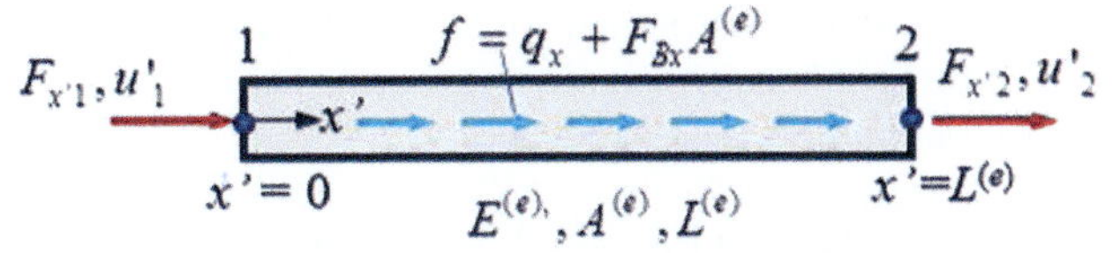

FIGURE 5.3 A finite element representing the deformation of a one-dimensional elastic medium.

conditions acting on the structure require the modification of the form or the size of the global stiffness matrix. The finite element analysis represents the compliance of the structure as linear algebraic equations for the unknown nodal displacements, known as the *global degrees of freedom*. The global equilibrium equation, obtained upon the assembly, application of the essential boundary conditions and the external forces represents the Newton's second law. In this chapter the effects of inertial forces are not considered, thus we seek to find the *steady state deformation* and *stress state* of the structure. The *secondary variables* such as the strain and the stress at each node and for each element are computed following the solution of the *global equilibrium equation* as a *postprocessing operation*.

5.2 Element equilibrium equation for a planar bar

5.2.1 Problem definition

Consider a linear elastic element subjected to external forces acting along its longitudinal axis (x'-axis), as shown in Fig. 5.3. The elastic modulus of this element is E, the cross-sectional area A, and length is $L^{(e)}$. Note that in general, the elastic modulus and the cross-sectional area could vary along the length of the element, i.e. $E = E(x')$ and $A = A(x')$.

The external forces acting on the element are the concentrated forces, $F_{x'1}$ and $F_{x'2}$ acting on the nodes-1 and -2 along x'-axis with unit $[F]$, the distributed force per unit length $q_{x'}$ with unit $[F/L]$ and the body force, $F_{Bx'}$ in axial direction with unit $[F/L^3]$.

5.2.2 Weak form of the boundary value problem

The problem of finding the deflection $u'(x')$ of the element given in Fig. 5.3, can be posed as a boundary value problem. The equilibrium of this bar can be found from Eq. (1.11) as follows:

$$\frac{d}{dx'}\left(EA\frac{du'}{dx'}\right) = -f_{x'} \text{ in } 0 < x' < L^{(e)} \tag{5.2}$$

where, $f_{x'}(x')$ represents the distributed force per unit length acting on the bar, and for this specific problem, $f_{x'}(x') = q_{x'} + F_{Bx'}A$. On the ends of the bar the following boundary conditions are specified,

$$\left(EA\frac{du'}{dx'}\right)\bigg|_{x'=0} = -F_{x'1} \text{ and } \left(EA\frac{du'}{dx'}\right)\bigg|_{x'=L^{(e)}} = F_{x'2} \tag{5.3}$$

The weak form of this boundary value problem can be found as discussed in Chapter 3. This will show that,

$$\int_0^{L^{(e)}} \left[-\frac{dw}{dx'}EA\frac{du'}{dx'} + wf_{x'}\right]dx' + \left[w\left(EA\frac{du'}{dx'}\right)\right]_0^{L^{(e)}} = 0 \tag{5.4}$$

or by using the boundary conditions,

$$\int_0^{L^{(e)}} \left[-EA\frac{dw}{dx'}\frac{du'}{dx'} + wf_{x'}\right]dx' + w(0)F_{x'1} + w\left(L^{(e)}\right)F_{x'2} = 0 \tag{5.5}$$

where w is a weight function as described in Chapter 3. A functional associated with the boundary value problem given above can be obtained by: *i*) replacing the weight function w with the variation of displacement, δu; *ii*) using $f_{x'}(x') = (q_{x'} + F_{Bx'}A)$ and *iii*) using $w(0) = \delta u'_1$ and $w(L) = \delta u'_2$:

$$I(u') = \int_0^{L^{(e)}} \left[\frac{1}{2}EA\left(\frac{du'}{dx'}\right)^2 - u'(q_{x'} + F_{Bx'}A)\right]dx' - \left(u'_1F_{x'1} + u'_2F_{x'2}\right) \tag{5.6}$$

Details of this derivation are omitted here, but can be found in the discussion provided in Chapter 3. Next, we will approach this problem from a different point of view, and find the total potential energy of this bar upon deformation.

5.2.3 Total potential energy of the system

Total potential energy π_p of the deformed bar, can be expressed as follows:

$$\pi_p = U - W \tag{5.7}$$

where, U is the *strain energy stored* in the deformed bar as follows:

$$U = \int_V u \, dV \tag{5.8}$$

with u as the *strain energy density*, and W is the *work done by the external forces* to cause the deformation u',

$$W = \int_V F_{Bx'} \, u' \, dV + \int_S q_{x'} \, u' \, dS + u'_1F_{x'1} + u'_2F_{x'2} \tag{5.9}$$

Note that when we compute the external work, the external forces are assumed to be acting at full strength during deformation.

For the element shown in Fig. 5.3, the incremental volume is $dV = Adx'$, and the incremental length of the surface $dS = dx'$. In general, the strain energy density is $u = \sigma' d\varepsilon'$. The total potential energy of the system can then be expressed as follows:

$$\pi_p = \int_0^{L^{(e)}} \int_0^{\varepsilon_{x'}} \sigma_{x'} d\varepsilon_{x'} A dx' - \left(\int_0^{L^{(e)}} u'(q_{x'} + F_{Bx'}A) dx' + u_1' F_{x'1} + u_2' F_{x'2} \right) \quad (5.10)$$

For a linear elastic material, Hooke's law provides $\sigma_{x'} = E\varepsilon_{x'}$. The normal strain is defined as $\varepsilon_{x'} = du'/dx'$ for small deformations. The total potential energy then becomes,

$$\pi_p(u', u_x') = \int_0^{L^{(e)}} \left[\frac{1}{2} EA \left(\frac{du'}{dx'} \right) - u'(q_{x'} + F_{Bx'}A) \right] dx' - \left(u_1' F_{x'1} + u_2' F_{x'2} \right) \quad (5.11)$$

Note that the total potential energy is a functional and Eq. (5.11) is identical to the form obtained from the weak form of the boundary value problem given in Eq. (5.6). Recall that the last two terms of Eq. (5.11) represent the natural boundary conditions.

5.2.4 Finite element form of the equilibrium equations of an elastic bar

The *principle of minimum total potential energy* states that equilibrium conditions are established when the variation of the total potential energy is zero. This gives,

$$\delta\pi_p = \int_0^{L^{(e)}} EA \frac{du'}{dx'} \frac{d}{dx'} (\delta u') dx' - \int_0^{L^{(e)}} \delta u' (F_{Bx'}A + q_{x'}) dx' - \delta u_1' F_{x'1} - \delta u_2' F_{x'2} = 0 \quad (5.12)$$

In order to approximate the function $u'(x')$ we use the linear approximation for $\tilde{u}(x')$ given in Eq. (4.15). Let $x'_1 = 0$ and $x'_2 = L^{(e)}$ for the linear shape functions described by Eq. (4.23), which now become,

$$N_1(x') = 1 - \frac{x'}{L^{(e)}}, \text{ and } N_2(x') = \frac{x'}{L^{(e)}} \quad (5.13)$$

The linear strain is then approximated as follows:

$$\varepsilon_{x'} \approx \frac{d\tilde{u}'}{dx'} = \frac{d}{dx'}([N]\{d'\}) = \frac{d}{dx'}([N])\{d'\} \quad (5.14)$$

where $\{d'\} = \{u_1' \ u_2'\}^T$. The strain displacement matrix $[B]$ is defined as follows:

$$[B] = \frac{d}{dx'}([N]) = \left[-\frac{1}{L^{(e)}} \ \frac{1}{L^{(e)}} \right] \tag{5.15}$$

The strain along the bar is approximated as follows:

$$\varepsilon_{x'} = [B]\{d'\} \tag{5.16}$$

and the variation of the strain then can be computed as follows:

$$\delta\varepsilon_{x'} = \frac{d}{dx'}(\delta\tilde{u}) = [B]\{\delta d'\} \tag{5.17}$$

Using this relationship in Eq. (5.12) we find,

$$\delta\pi_p = \int_0^{L^{(e)}} \{\delta d'\}^T [B]^T EA[B]\{d'\}dx' - \int_0^{L^{(e)}} \{\delta d'\}^T [N]^T (F_{Bx'}A + q_{x'})dx' - \{\delta d'\}^T \{r_F'\}^{(e)} = 0 \tag{5.18}$$

where, the concentrated forces on the nodes are expressed in the following *boundary load vector*,

$$\{r_F'\}^{(e)} = \begin{Bmatrix} F_{x'1} \\ F_{x'2} \end{Bmatrix} \tag{5.19}$$

Equation (5.18) can be further simplified as follows:

$$\delta\pi_p = \{\delta d'\}^T \left\{ \left[\int_0^{L^{(e)}} [B]^T EA[B]dx' \right] \{d'\} - \int_0^{L^{(e)}} [N]^T (F_{Bx'}A + q_{x'})dx' - \{r_F'\}^{(e)} \right\} = 0 \tag{5.20}$$

Note that the variation $\{\delta d'\}^T$ is arbitrary and can be eliminated from this equation. This results in the following equation of equilibrium:

$$\left[\int_0^{L^{(e)}} [B]^T EA[B]dx' \right] \{d'\} = \int_0^{L(e)} [N]^T (F_{Bx1}A + q_{x'})dx' + \{r_F'\}^{(e)} \tag{5.21}$$

In matrix form, this equation is expressed as follows:

$$[k_m']^{(e)}\{d'\} = \{r_B'\}^{(e)} + \{r_q'\}^{(e)} + \{r_F'\}^{(e)} \tag{5.22}$$

Element stiffness matrix: $\left[k_m{}'\right]^{(e)} = \int_0^{L^{(e)}} [B]^T EA[B]dx'$ (a)

Boundary - load vector: $\left\{r_F{}'\right\}^{(e)} = \begin{Bmatrix} F_{x'1} \\ F_{x'2} \end{Bmatrix}$ (b)

Consistent nodal load vectors: $\left\{r_B{}'\right\}^{(e)} = \int_0^{L^{(e)}} F_{Bx'}[N]^T A dx'$ (c) (5.23)

$\left\{r_q{}'\right\}^{(e)} = \int_0^{L^{(e)}} q_{x'}[N]^T dx'$ (d)

Note that the subscript $_m$ in Eq. (5.23) (a) refers to the membrane, or in-plane deformation, action in the bar. Equation (5.22) can also be expressed as follows:

$$\left[k_m{}'\right]^{(e)}\{d'\} = \{r'\}^{(e)} \tag{5.24}$$

where the *total nodal force vector* is,

$$[r']^{(e)} = \left\{r_B{}'\right\}^{(e)} + \left\{r_q{}'\right\}^{(e)} + \left\{r_F{}'\right\}^{(e)} \tag{5.25}$$

Comparing the loading vectors in Eq. (5.23) with those in Eq. (4.55), we note that the combined $(\{r_B{}'\}^{(e)} + \{r_q{}'\}^{(e)})$ vector has the function of the $\{r_q\}^{(e)}$ vector and $\{r_F{}'\}^{(e)}$ has the function of the $\{r_\beta\}^{(e)}$ vector in Eq. (4.55).

5.2.4.1 *Element stiffness matrix*

By using Eq. (5.15) in (5.23), the element stiffness matrix is computed as follows:

$$\left[k_m{}'\right]^{(e)} = \int_0^{L^{(e)}} AE \begin{bmatrix} -\dfrac{1}{L^{(e)}} \\ \dfrac{1}{L^{(e)}} \end{bmatrix} \begin{bmatrix} -\dfrac{1}{L^{(e)}} & \dfrac{1}{L^{(e)}} \end{bmatrix} dx' \tag{5.26}$$

For constant E and A values, the element stiffness matrix is found as follows:

$$\left[k_m{}'\right]^{(e)} = \left(\frac{AE}{L}\right)^{(e)} \begin{bmatrix} 1 & -1 \\ -1 & 1 \end{bmatrix} \tag{5.27}$$

5.2.4.2 *Consistent nodal forces*

Equation (5.23) defines a way to transform the effects of the *distributed forces,* $F_{Bx}{}'$ and $q_{x'}$, which act along the body of the element *to concentrated forces* acting on the nodes of the element (Fig. 5.4). Based on Eq. (5.21), it can be seen that the *consistent nodal forces* due to a general *distributed force* $f(x')$ per unit length is given by the following relationship,

$$\{r'\}^{(e)} = \int_0^{L^{(e)}} [N]^T f(x') dx' \tag{5.28}$$

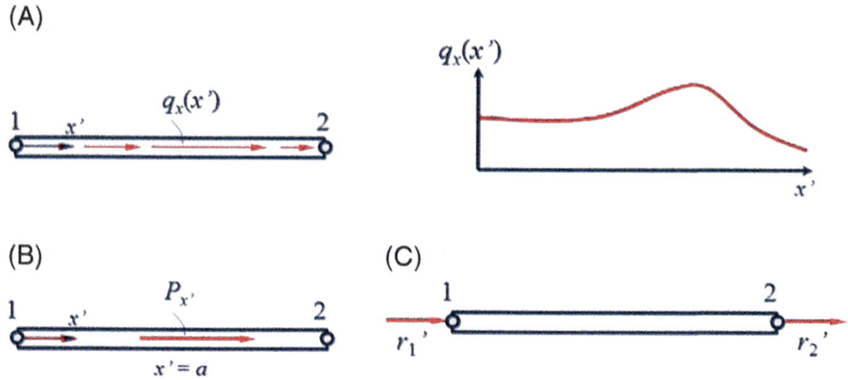

FIGURE 5.4 (A) Distributed force varying magnitude, (B) concentrated force acting at point $x' = a$, (C) consistent nodal forces that represent the effect of the distributed forces.

Distributed force $q_{x'}(x')$ with linearly varying magnitude along the bar: Consider a bar element with distributed force magnitude $q_{x'1}$ at node-1 and $q_{x'2}$ at node-2 with linear variation in between,

$$q_{x'}\left(x'\right) = \frac{q_{x'2} - q_{x'1}}{L^{(e)}} x' + q_{x'1} \tag{5.29}$$

The consistent nodal forces are computed by using Eqs. (5.13) and (5.28) as follows:

$$\left\{r_q'\right\}^{(e)} = \int_0^{L^{(e)}} \left\{ \begin{matrix} \left(\frac{q_{x'2}-q_{x'1}}{L^{(e)}}x' + q_{x'1}\right)\left(1 - \frac{x'}{L^{(e)}}\right) \\ \left(\frac{q_{x'2}-q_{x'1}}{L^{(e)}}x' + q_{x'1}\right)\left(\frac{x'}{L^{(e)}}\right) \end{matrix} \right\} dx' = \frac{L^{(e)}}{6} \left\{ \begin{matrix} (2q_{x'1} + q_{x'2}) \\ (q_{x'1} + 2q_{x'2}) \end{matrix} \right\} \tag{5.30}$$

Distributed force $q_{x'}(x') = q_{0x'}$ with constant magnitude along the bar: In this case, Eq. (5.30): reduces to the following relationship

$$\left\{r_q'\right\}^{(e)} = \frac{1}{2} q_{0x'} L^{(e)} \left\{ \begin{matrix} 1 \\ 1 \end{matrix} \right\} \tag{5.31}$$

Distributed body force $q_{x'}(x') = F_{Bx'}$ with constant magnitude along the bar:

$$\left\{r_B'\right\}^{(e)} = F_{Bx'} A \left\{ \begin{matrix} \int_0^{L^{(e)}} \left(1 - \frac{x'}{L^{(e)}}\right) dx' \\ \int_0^{L^{(e)}} \left(\frac{x'}{L^{(e)}}\right) dx' \end{matrix} \right\} = \frac{1}{2} F_{Bx'} A L^{(e)} \left\{ \begin{matrix} 1 \\ 1 \end{matrix} \right\} \tag{5.32}$$

Concentrated force $P_{x'}$ acting at $x' = a$ (Fig. 5.4): can be treated as distributed force acting over an infinitesimally small length dx' along the bar,

$$q_{x'} = \left(\frac{P_{x'}}{dx'}\right) \delta\left(x' - a\right) \tag{5.33}$$

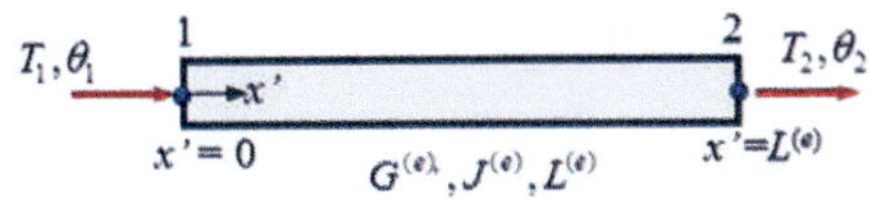

FIGURE 5.5 An element that represents the torsional resistance of a one-dimensional elastic bar.

where $\delta(x' - a)$ is the *Kroneceker* delta function that is equal to 1 when $x' = a$, and 0 otherwise. Then, by using Eq. (5.28) we can find the consistent nodal force for this loading type as follows:

$$\left\{r_q'\right\}^{(e)} = \int_0^{L^{(e)}} \left\{ \begin{array}{c} 1 - \dfrac{a}{L^{(e)}} \\ \dfrac{a}{L^{(e)}} \end{array} \right\} \left(\frac{P_{x'}}{dx'}\right) \delta\left(x' - a\right) dx' = P_{x'} \left\{ \begin{array}{c} 1 - \dfrac{a}{L^{(e)}} \\ \dfrac{a}{L^{(e)}} \end{array} \right\} \tag{5.34}$$

5.3 Finite element equations for torsion of a bar

Consider a member that is subjected pure torsion due to external torsional moments applied at its ends as shown in Fig. 5.5. The properties of this shaft are given as follows: $L^{(e)}$ is the length of the shaft; G is the shear modulus of the material $[F/L^2]$; and J is the polar moment of area of the shaft $[L^4]$.

In general, external torque T and the angle of twist $\theta_{x'}$ are related as follows:

$$\theta_{x'} = \frac{TL^{(e)}}{GJ} \tag{5.35}$$

In fact, this relationship is very similar to the elongation of a bar, presented in Section 5.2, given by,

$$u' = \frac{FL^{(e)}}{EA} \tag{5.36}$$

Without proof we give the equilibrium of an element under torsion as shown in Fig. 5.5 as follows:

$$\left[k_T'\right]^{(e)}\left\{d_\theta'\right\} = \left\{r'\right\}^{(e)} \tag{5.37}$$

where,

$$\begin{array}{lll} \text{Element stiffness matrix:} & \left[k_T'\right]^{(e)} = \dfrac{GJ}{L^{(e)}} = \begin{bmatrix} 1 & -1 \\ -1 & 1 \end{bmatrix} & \text{(a)} \\ \text{Nodal degrees of freedom:} & \left\{d_\theta'\right\} = \{\theta_1 \;\; \theta_2\}^T & \text{(b)} \\ \text{External torques acting on the nodes:} & \left\{r'\right\}^{(e)} = \{T_1 \;\; T_2\}^T & \text{(c)} \end{array} \tag{5.38}$$

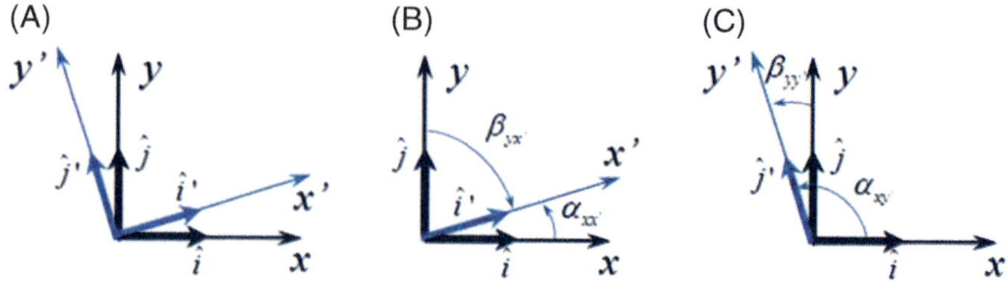

FIGURE 5.6 Base vectors $(\hat{i}, \hat{j})$ and $(\hat{i}', \hat{j}')$ of the (x, y) and (x', y') coordinate systems, respectively, and the direction cosine angles used in vector transformations.

5.4 Coordinate transformations

5.4.1 Transformation of unit vectors between orthogonal coordinate systems

Consider the two coordinate systems (x, y) and (x', y') with unit (base) vectors, $(\hat{i}, \hat{j})$ and $(\hat{i}', \hat{j}')$ respectively, as shown in Fig. 5.6. Note that the orientation of the x'-axis with respect to x- and y-axes are $\alpha_{xx'}$ and $\beta_{yx'}$, respectively. Similarly, the y'-axis is located by the angles $\alpha_{xy'}$ and $\beta_{yy'}$ with respect to the same axes, as shown. By using these direction angles, the following relationships are obtained,

$$\begin{aligned} \hat{i}' &= \left(\left|\hat{i}'\right| \cos\alpha_{xx'}\right)\hat{i} + \left(\left|\hat{i}'\right| \cos\beta_{yx'}\right)\hat{j} \\ \hat{j}' &= \left(\left|\hat{j}'\right| \cos\alpha_{xy'}\right)\hat{i} + \left(\left|\hat{j}'\right| \cos\beta_{yy'}\right)\hat{j} \end{aligned} \tag{5.39}$$

The magnitudes of $(\hat{i}', \hat{j}')$ are unity. Therefore, Eq. (5.39) becomes,

$$\begin{aligned} \hat{i}' &= \cos\,\alpha_{xx'}\hat{i} + \cos\,\beta_{yx'}\hat{j} \\ \hat{j}' &= \cos\,\alpha_{xy'}\hat{i} + \cos\,\beta_{yy'}\hat{j} \end{aligned} \tag{5.40}$$

Let's next express a generic vector $\vec{V}$ in both coordinate systems,

$$\vec{V} = u'\hat{i}' + v'\hat{j}' = u\hat{i} + v\hat{j} \tag{5.41}$$

where u' and v' are the projections of the vector $\vec{V}$ onto the (x', y') axes, and similarly u and v are the projections onto the (x, y) axes. Equation (5.41) (a) can be rewritten by using (5.40) as follows:

$$\vec{V} = u'\left(\cos\,\alpha_{xx'}\hat{i} + \cos\,\beta_{yx'}\hat{j}\right) + v'\left(\cos\,\alpha_{xy'}\hat{i} + \cos\,\beta_{yy'}\hat{j}\right) \tag{5.42}$$

After rearranging this relationship becomes,

$$\vec{V} = \left(u'\cos\,\alpha_{xx'} + v'\cos\,\alpha_{xy'}\right)\hat{i} + \left(u'\cos\,\beta_{yx'} + v'\cos\,\beta_{yy'}\right)\hat{j} \tag{5.43}$$

Comparing Eq. (5.41) (b) and (5.43), we note that,

$$\begin{aligned} u &= u' \cos \alpha_{xx'} + v' \cos \alpha_{xy'} \\ v &= u' \cos \beta_{yx'} + v' \cos \beta_{yy'} \end{aligned} \tag{5.44}$$

In matrix form Eq. (5.44) can be expressed as follows:

$$\begin{Bmatrix} u \\ v \end{Bmatrix} = \begin{bmatrix} \cos \alpha_{xx'} & \cos \alpha_{xy'} \\ \cos \beta_{yx'} & \cos \beta_{yy'} \end{bmatrix} \begin{Bmatrix} u' \\ v' \end{Bmatrix} \tag{5.45}$$

Note that the following relationships exist for this planar transformation: $\alpha_{xx'} = \alpha$, $\beta_{yx'} = \frac{\pi}{2} - \alpha$, $\alpha_{xy'} = \frac{\pi}{2} + \alpha$ and $\beta_{yy'} = \alpha$. Equation (5.45) then becomes,

$$\begin{Bmatrix} u \\ v \end{Bmatrix} = \begin{bmatrix} \cos \alpha & -\sin \alpha \\ \sin \alpha & \cos \alpha \end{bmatrix} \begin{Bmatrix} u' \\ v' \end{Bmatrix} \tag{5.46}$$

The transformation matrix in Eq. (5.46) is the inverse $[A]^{-1}$ of the *direction cosine matrix*. For this matrix it can be shown that the following property $[A]^{-1} = [A]^T$ holds. Therefore, the *forward transformation* becomes,

$$\begin{Bmatrix} u' \\ v' \end{Bmatrix} = \begin{bmatrix} \cos \alpha & \sin \alpha \\ -\sin \alpha & \cos \alpha \end{bmatrix} \begin{Bmatrix} u \\ v \end{Bmatrix} \tag{5.47}$$

The *direction cosine matrix* $[A]$ is the matrix given in Eq. (5.47), leading to following symbolic representation for vector transformation:

$$\{V'\} = [A]\{V\} \tag{5.48}$$

5.4.2 Transformation of equilibrium equations for the one-dimensional bar element

Equilibrium Eq. (5.22) for a bar element is represented in the local, x' – coordinate system, where u' is the only variable. However, when a bar element is part of a (planar) truss, it is more convenient to define the problem with respect to a global (x, y) coordinate system. Fig. 5.7 shows a bar-element that is oriented by an angle α with respect to a global coordinate system. Next, we will transform the equilibrium equations from the local coordinate system to the global coordinate system.

The degree of freedom vector of the element is $\{d'\} = \{u'_1 \quad u'_2\}^T$. This vector is transformed to the global coordinate system by using Eq. (5.47),

$$\begin{Bmatrix} u'_1 \\ u'_2 \end{Bmatrix} = \begin{bmatrix} c & s & 0 & 0 \\ 0 & 0 & c & s \end{bmatrix} \begin{Bmatrix} u_1 \\ v_1 \\ u_2 \\ v_2 \end{Bmatrix} \text{(a)} \quad \text{or} \quad \{d'\} = [T]^{(e)}\{d\} \text{ (b)} \tag{5.49}$$

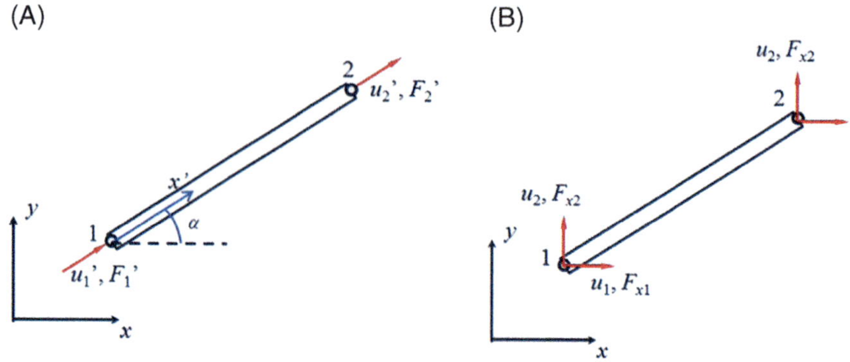

FIGURE 5.7 (A) A bar element with its displacement and loads expressed in its local coordinate system, and (B) the same element with equivalent displacements and loads in the global coordinate system.

where $s = \sin\alpha$ and $c = \cos\alpha$, $[T]^{(e)}$ is the *transformation matrix*, and $\{d\}$ is the *degree of freedom vector* in the global coordinate system,

$$\{d\} = \begin{Bmatrix} u_1 & v_1 & u_2 & v_2 \end{Bmatrix}^T \tag{5.50}$$

The superscript $^{(e)}$ is used to indicate that each element has a distinct transformation matrix. The load vector transforms similarly, according to Eq. (5.49),

$$\{r'\}^{(e)} = [T]^{(e)}\{r\}^{(e)} \tag{5.51}$$

where the load vector in the global coordinate system defined as follows:

$$\{r\}^{(e)} = \begin{Bmatrix} F_{x1} & F_{y1} & F_{x2} & F_{y2} \end{Bmatrix}^T \tag{5.52}$$

By using Eqs. (5.49) and (5.51) in the equilibrium equation derived in the local coordinate system, we can express the equilibrium equation in the global coordinate system as follows:

$$[k_m]^{(e)}\{d\} = \{r\}^{(e)} \tag{5.53}$$

where

$$[k_m]^{(e)} = [T]^{(e)T}\left[k'_m\right]^{(e)}[T]^{(e)} \tag{5.54}$$

The equilibrium equation is expressed in the local coordinate system (x') was found as follows:

$$\left[\frac{AE}{L}\right]^{(e)}\begin{bmatrix} 1 & -1 \\ -1 & 1 \end{bmatrix}\begin{Bmatrix} u_1' \\ u_2' \end{Bmatrix} = \begin{Bmatrix} F_{x'1} \\ F_{x'2} \end{Bmatrix} \tag{5.55}$$

In the global coordinate system the transformed stiffness matrix $[k_m]^{(e)}$ and loading vector $\{r\}^{(e)}$ are obtained by using Eqs. (5.54) and (5.51) as follows:

$$[k_m]^{(e)} = \left(\frac{AE}{L}\right)^{(e)} \begin{bmatrix} c^2 & cs & -c^2 & -cs \\ cs & s^2 & -cs & -s^2 \\ -c^2 & -cs & c^2 & cs \\ -cs & -s^2 & cs & s^2 \end{bmatrix}$$

$$\{r\}^{(e)} = [T]^{(e)T} \begin{Bmatrix} F_{x'1} \\ F_{x'2} \end{Bmatrix} = \begin{Bmatrix} cF_{x'1} \\ sF_{x'1} \\ cF_{x'2} \\ sF_{x'2} \end{Bmatrix} \tag{5.56}$$

The equilibrium equation is then expressed in the global coordinate system as follows:

$$\left(\frac{AE}{L}\right)^{(e)} \begin{bmatrix} c^2 & cs & -c^2 & -cs \\ cs & s^2 & -cs & -s^2 \\ -c^2 & -cs & c^2 & cs \\ -cs & -s^2 & cs & s^2 \end{bmatrix} \begin{Bmatrix} u_1 \\ v_1 \\ u_2 \\ v_2 \end{Bmatrix} = \begin{Bmatrix} cF_{x'1} \\ sF_{x'1} \\ cF_{x'2} \\ sF_{x'2} \end{Bmatrix} \tag{5.57}$$

In symbolic notation this is equivalent to Eq. (5.53).

5.5 Assembly of elements

Element assembly represents the equilibrium of the forces acting on the nodes. The boundary-load vector $\{r_F\}^{(e)}$ represents the internal forces acting on the nodes. Therefore, upon assembly $\{r_F\}^{(e)}$ should be balanced out of this equilibrium equation due to Newton's third law, or due to continuity of the slopes of the solution, as presented in Section 4.6. This results in the following system of equations:

$$\begin{gathered} \sum_{e=1}^{N_E} \{r_F\}^{(e)} = 0 \\ \sum_{e=1}^{N_E} [k_m]^{(e)}\{d\}^{(e)} = \sum_{e=1}^{N_E} \left(\{r_q\}^{(e)} + \{r_B\}^{(e)}\right) + \{P\} \end{gathered} \tag{5.58}$$

where, $\{P\}$ represents the external concentrated forces that happen to be acting on the nodes of the structure beyond those represented by $\{r_q\}^{(e)}$, $\{r_B\}^{(e)}$, and $\{r_F\}^{(e)}$. This system of equations represents the global equilibrium equation, and it is symbolically stated as follows:

$$[K]\{D\} = \{R\} \tag{5.59}$$

where, $[K]$ is the *global stiffness matrix*, $\{R\}$ is the *global force vector* and $\{D\}$ is the *global degree of freedom vector*.

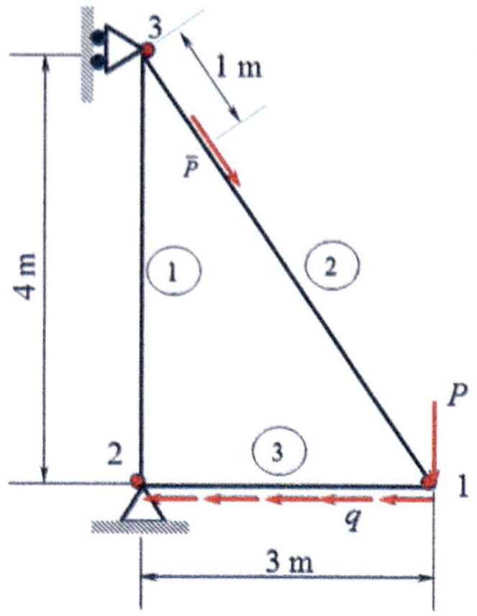

FIGURE 5.8 A three-member truss subjected to distributed forces, q, and two concentrated forces, P and $\overline{P}$ as shown.

In a general element assembly process, the key is to expand the element matrices and force vectors to the size of the global system and then perform the summation given in Eq. (5.58). This is demonstrated by an example, next.

Example 5.1 Assembly of the global stiffness matrix.

Obtain the global equilibrium equation for the three member truss shown in the Fig. 5.8. Note that the following properties are given,

$$k_1 = \left(\frac{EA}{L}\right)^{(1)},\ k_2 = \left(\frac{EA}{L}\right)^{(2)},\ k_3 = \left(\frac{EA}{L}\right)^{(3)} \tag{5.60}$$

Solution:
Element-1:

$$\begin{gathered}[k'_m]^{(1)}\{d'\}^{(1)} = \{r'\}^{(1)} \\ k_1\begin{bmatrix} 1 & -1 \\ -1 & 1 \end{bmatrix}\begin{Bmatrix} u_2' \\ u_3' \end{Bmatrix} = \begin{Bmatrix} 0 \\ 0 \end{Bmatrix}\end{gathered} \tag{5.61}$$

$\alpha = \frac{\pi}{2}$:

$$\begin{gathered}[k_m]^{(1)}\{d\}^{(1)} = \{r\}^{(1)} \\ k_1\begin{bmatrix} 0 & 0 & 0 & 0 \\ 0 & 1 & 0 & -1 \\ 0 & 0 & 0 & 0 \\ 0 & -1 & 0 & 1 \end{bmatrix}\begin{Bmatrix} u_2 \\ v_2 \\ u_3 \\ v_3 \end{Bmatrix} = \begin{Bmatrix} 0 \\ 0 \\ 0 \\ 0 \end{Bmatrix}\end{gathered} \tag{5.62}$$

This equation can be put into the global stiffness matrix size as follows:

$$k_1\begin{bmatrix} 0 & 0 & 0 & 0 & 0 & 0 \\ 0 & 0 & 0 & 0 & 0 & 0 \\ 0 & 0 & 0 & 0 & 0 & 0 \\ 0 & 0 & 0 & 1 & 0 & -1 \\ 0 & 0 & 0 & 0 & 0 & 0 \\ 0 & 0 & 0 & -1 & 0 & 1 \end{bmatrix}\begin{Bmatrix} u_1 \\ v_1 \\ u_2 \\ v_2 \\ u_3 \\ v_3 \end{Bmatrix} = \begin{Bmatrix} 0 \\ 0 \\ 0 \\ 0 \\ 0 \\ 0 \end{Bmatrix} \tag{5.63}$$

Element-2:

$$\left[k'_m\right]^{(2)}\left\{d'\right\}^{(2)} = \left\{r'\right\}^{(2)}$$

$$k_2\begin{bmatrix} 1 & -1 \\ -1 & 1 \end{bmatrix}\begin{Bmatrix} u'_1 \\ u'_3 \end{Bmatrix} = \begin{Bmatrix} F'_1 \\ F'_3 \end{Bmatrix} = -\overline{P}\begin{Bmatrix} \frac{5-4}{5} \\ \frac{4}{5} \end{Bmatrix} \tag{5.64}$$

$\alpha = \pi - \gamma = \pi - \tan^{-1}(4/3) = 180 - 53.13 = 126.87$ deg. Thus $c =$ -0.6, $s = 0.8$

$$[k_m]^{(2)}\{d\}^{(2)} = \{r\}^{(2)}$$

$$k_2\begin{bmatrix} 0.36 & -0.48 & -0.36 & 0.48 \\ -0.48 & 0.64 & 0.48 & -0.64 \\ -0.36 & 0.48 & 0.36 & -0.48 \\ 0.48 & -0.64 & -0.48 & 0.64 \end{bmatrix}\begin{Bmatrix} u_1 \\ v_1 \\ u_3 \\ v_3 \end{Bmatrix} = -\frac{\overline{P}}{5}\begin{Bmatrix} -0.6 \\ 0.8 \\ -0.6\times 4 \\ 0.8\times 4 \end{Bmatrix} = \frac{\overline{P}}{5}\begin{Bmatrix} 0.6 \\ -0.8 \\ 2.4 \\ -3.2 \end{Bmatrix} \tag{5.65}$$

In global system size this is expressed as follows:

$$k_2\begin{bmatrix} 0.36 & -0.48 & 0 & 0 & -0.36 & 0.48 \\ -0.48 & 0.64 & 0 & 0 & 0.48 & -0.64 \\ 0 & 0 & 0 & 0 & 0 & 0 \\ 0 & 0 & 0 & 0 & 0 & 0 \\ -0.36 & 0.48 & 0 & 0 & 0.36 & -0.48 \\ 0.48 & -0.64 & 0 & 0 & -0.48 & 0.64 \end{bmatrix}\begin{Bmatrix} u_1 \\ v_1 \\ u_2 \\ v_2 \\ u_3 \\ v_3 \end{Bmatrix} = \frac{\overline{P}}{5}\begin{Bmatrix} 0.6 \\ -0.8 \\ 0 \\ 0 \\ 2.4 \\ -3.2 \end{Bmatrix} \tag{5.66}$$

Element-3:

$$\left[k'_m\right]^{(3)}\left\{d'\right\}^{(3)} = \left\{r'\right\}^{(3)}$$

$$k_3\begin{bmatrix} 1 & -1 \\ -1 & 1 \end{bmatrix}\begin{Bmatrix} u'_1 \\ u'_2 \end{Bmatrix} = \frac{qL}{2}\begin{Bmatrix} 1 \\ 1 \end{Bmatrix} = \frac{3}{2}q\begin{Bmatrix} 1 \\ 1 \end{Bmatrix} \tag{5.67}$$

$\alpha = \pi$. Thus $c = -1$, $s = 0$

$$k_3 \begin{bmatrix} 1 & 0 & -1 & 0 \\ 0 & 0 & 0 & 0 \\ -1 & 0 & 1 & 0 \\ 0 & 0 & 0 & 0 \end{bmatrix} \begin{Bmatrix} u_1 \\ v_1 \\ u_2 \\ v_2 \end{Bmatrix} = \frac{3}{2} q \begin{Bmatrix} -1 \\ 0 \\ -1 \\ 0 \end{Bmatrix} \tag{5.68}$$

In global system size this is expressed as follows:

$$k_3 \begin{bmatrix} 1 & 0 & -1 & 0 & 0 & 0 \\ 0 & 0 & 0 & 0 & 0 & 0 \\ -1 & 0 & 1 & 0 & 0 & 0 \\ 0 & 0 & 0 & 0 & 0 & 0 \\ 0 & 0 & 0 & 0 & 0 & 0 \\ 0 & 0 & 0 & 0 & 0 & 0 \end{bmatrix} \begin{Bmatrix} u_1 \\ v_1 \\ u_2 \\ v_2 \\ u_3 \\ v_3 \end{Bmatrix} = \frac{3}{2} q \begin{Bmatrix} -1 \\ 0 \\ -1 \\ 0 \\ 0 \\ 0 \end{Bmatrix} \tag{5.69}$$

Equation (5.58), which represents the element assembly process, gives,

$$\left([k_m]^{(1)} + [k_m]^{(2)} + [k_m]^{(3)}\right)\{D\} = \left(\{r\}^{(1)} + \{r\}^{(2)} + \{r\}^{(3)} + \{P\}\right) \tag{5.70}$$

Note that the vector $\{P\}$ contains all external forces that happen to be acting on the nodes of the system. These forces include the external force $-P\hat{j}$ acting on node-1, as well as the reaction forces acting on nodes-2 and -3. Let's note that since node-2 is restricted in both x- and y-directions the reaction forces R_{x2} and R_{y2} will develop on this node. But, the only reaction force at node-3 will be R_{x3}, as node-3 is free to move in the y-direction.

The global equilibrium equation can then be expressed as follows:

$$\begin{bmatrix} 0.36k_2 + k_3 & -0.48k_2 & -k_3 & 0 & -0.36k_2 & 0.48k_2 \\ -0.48k_2 & 0.64k_2 & 0 & 0 & 0.48k_2 & -0.64k_2 \\ -k_3 & 0 & k_3 & 0 & 0 & 0 \\ 0 & 0 & 0 & k_1 & 0 & -k_1 \\ -0.36k_2 & 0.48k_2 & 0 & 0 & 0.36k_2 & -0.48k_2 \\ 0.48k_2 & -0.64k_2 & 0 & -k_1 & 0.48k_2 & k_1 + 0.64k_2 \end{bmatrix} \begin{Bmatrix} u_1 \\ v_1 \\ u_2 \\ v_2 \\ u_3 \\ v_3 \end{Bmatrix} = \frac{\overline{P}}{5} \begin{Bmatrix} 0.6 \\ -0.8 \\ 0 \\ 0 \\ 2.4 \\ -3.2 \end{Bmatrix} + \frac{3}{2} q \begin{Bmatrix} -1 \\ 0 \\ -1 \\ 0 \\ 0 \\ 0 \end{Bmatrix} + \begin{Bmatrix} 0 \\ -P \\ R_{x2} \\ R_{y2} \\ R_{x3} \\ 0 \end{Bmatrix} \tag{5.71}$$

This system can be expressed in compact form as follows:

$$[K]\{D\} = \{R\} \tag{5.72}$$

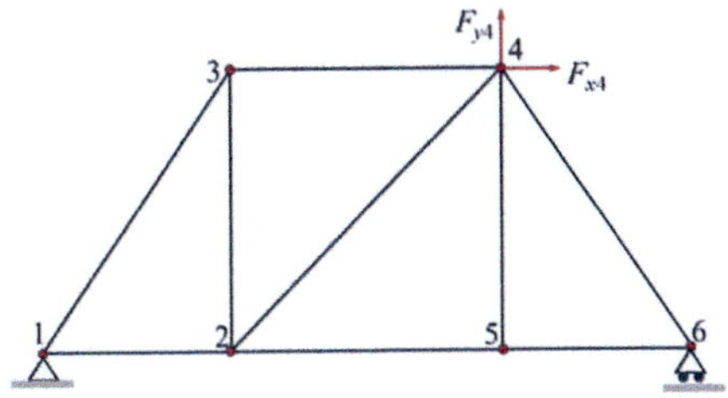

FIGURE 5.9 A nine-member truss that is loaded at node-4.

Note that the global stiffness matrix [*K*] in the form given in Eq. (5.71) is singular and cannot be inverted. This singularity can be remedied by imposing the boundary conditions.

5.6 Boundary conditions

Consider the truss shown in Fig. 5.9. This truss has nine-members which are connected at six nodes. As each node has two degrees of freedom, as described in Section 5.4.2, the global degree of freedom vector for this truss can be expressed as follows:

$$\{D\} = \begin{Bmatrix} u_1 & v_1 & u_2 & v_2 & u_3 & v_3 & u_4 & v_4 & u_5 & v_5 & u_6 & v_6 \end{Bmatrix}^T \tag{5.73}$$

The global stiffness matrix is a 12×12 matrix, and it has the following general form,

$$[K] = \begin{bmatrix} k_{11} & k_{12} & k_{13} & \cdots & k_{1(12)} \\ k_{21} & k_{22} & k_{23} & \cdots & k_{2(12)} \\ k_{31} & k_{32} & k_{33} & \cdots & k_{3(12)} \\ \vdots & \vdots & \vdots & \vdots & \vdots \\ k_{12(1)} & k_{12(2)} & k_{12(3)} & \cdots & k_{12(12)} \end{bmatrix} \tag{5.74}$$

The loading vector is given as follows:

$$\{R\} = \begin{Bmatrix} R_{x1} & R_{y1} & 0 & 0 & 0 & 0 & F_{x4} & F_{y4} & 0 & 0 & 0 & R_{y6} \end{Bmatrix}^T \tag{5.75}$$

where the reaction forces are R_{x1}, R_{y1} and R_{y6}. The global equilibrium equation is symbolically represented as follows:

$$[K]\{D\} = \{R\} \tag{5.76}$$

The stiffness matrix [*K*] given in this equation is singular and admits only rigid body displacements of the structure. In order for the structure to absorb the external effects (forces) into strain energy, the structure should be anchored.

This is accomplished by imposing the essential boundary conditions acting on the boundaries of the system on to the stiffness matrix $[K]$.

The essential boundary conditions of the truss given in this example are as follows:

$$u_1 = v_1 = v_6 = 0 \tag{5.77}$$

We will next show three different techniques to impose the essential boundary conditions. The first technique is a formal description which can be used in small scale problems.

5.6.1 Formal definition

Let's group the known and unknown degrees of freedom into two vectors as follows:

$$\begin{aligned} \{D_c\} &= \begin{Bmatrix} u_1 & v_1 & v_6 \end{Bmatrix}^T & \text{(a)} \\ \{D_x\} &= \begin{Bmatrix} u_2 & v_2 & u_3 & v_3 & u_4 & v_4 & u_5 & v_5 & u_6 \end{Bmatrix}^T & \text{(b)} \end{aligned} \tag{5.78}$$

Let's do the same for the known and unknown external forces. Noting that the reaction forces are unknown, we find,

$$\begin{aligned} \{R_x\} &= \begin{Bmatrix} R_{x1} & R_{y1} & R_{y6} \end{Bmatrix}^T & \text{(a)} \\ \{R_c\} &= \begin{Bmatrix} 0 & 0 & 0 & 0 & F_{x4} & F_{y4} & 0 & 0 & 0 \end{Bmatrix}^T & \text{(b)} \end{aligned} \tag{5.79}$$

By exchanging appropriate rows and columns Eq. (5.74) can be rearranged as follows:

$$\begin{bmatrix} [K_{xx}]_{9\times 9} & [K_{xc}]_{9\times 3} \\ [K_{cx}]_{3\times 9} & [K_{cc}]_{3\times 3} \end{bmatrix} \begin{Bmatrix} \{D_x\}_{9\times 1} \\ \{D_c\}_{3\times 1} \end{Bmatrix} = \begin{Bmatrix} \{R_c\}_{9\times 1} \\ \{R_x\}_{3\times 1} \end{Bmatrix} \tag{5.80}$$

where $[K_{ij}]$ $(i, j = x, y)$ are submatrices that result from reorganizing the stiffness matrix $[K]$ in order to preserve the original algebraic relationships between the variables. Details of this operation are described at the end of this section. Equation (5.72) can also be expressed as follows:

$$\begin{aligned} [K_{xx}]\{D_x\} + [K_{xc}]\{D_c\} &= \{R_c\} \quad \text{(a)} \\ [K_{cx}]\{D_x\} + [K_{cc}]\{D_c\} &= \{R_x\} \quad \text{(b)} \end{aligned} \tag{5.81}$$

As $\{D_c\}$ represents the known degrees of freedom (boundary conditions), Eq. (5.81) (a) can be expressed as follows:

$$[K_{xx}]\{D_x\} = \{R_c\} - [K_{xc}]\{D_c\} \tag{5.82}$$

Note that the matrix $[K_{xx}]$ is no longer a singular matrix and represents the stiffness components acting on the active degrees of freedom. Therefore, this equation is invertible. The unknown degree of freedom vector is found as follows:

$$\{D_x\} = [K_{xx}]^{-1}(\{R_c\} - [K_{xc}]\{D_c\}) \tag{5.83}$$

Once we find the displacements of the structure, the reaction forces can be found from Eq. (5.81) (b),

$$\{R_x\} = [K_{cx}]\{D_x\} + [K_{cc}]\{D_c\} \tag{5.84}$$

Repartitioning of the global stiffness matrix to the form shown in Eq. (5.80) involves row and column manipulations as shown next. This operation may not be practical for problems that have large number of degrees of freedom.

Details of matrix reordering

The general form of the equilibrium equations after element assembly gives the following matrix equation,

$$\begin{bmatrix} k_{11} & k_{12} & k_{13} & k_{14} & k_{15} & k_{16} & k_{17} & k_{18} & k_{19} & k_{1(10)} & k_{1(11)} & k_{1(12)} \\ k_{21} & k_{22} & k_{23} & k_{24} & k_{25} & k_{26} & k_{27} & k_{28} & k_{29} & k_{2(10)} & k_{2(11)} & k_{2(12)} \\ k_{31} & k_{32} & k_{33} & k_{34} & k_{35} & k_{36} & k_{37} & k_{38} & k_{39} & k_{3(10)} & k_{3(11)} & k_{3(12)} \\ k_{41} & k_{42} & k_{43} & k_{44} & k_{45} & k_{46} & k_{47} & k_{48} & k_{49} & k_{4(10)} & k_{4(11)} & k_{4(12)} \\ k_{51} & k_{52} & k_{53} & k_{54} & k_{55} & k_{56} & k_{57} & k_{58} & k_{59} & k_{5(10)} & k_{5(11)} & k_{5(12)} \\ k_{61} & k_{62} & k_{63} & k_{64} & k_{65} & k_{66} & k_{67} & k_{68} & k_{69} & k_{6(10)} & k_{6(11)} & k_{6(12)} \\ k_{71} & k_{72} & k_{73} & k_{74} & k_{75} & k_{76} & k_{77} & k_{78} & k_{79} & k_{7(10)} & k_{7(11)} & k_{7(12)} \\ k_{81} & k_{82} & k_{83} & k_{84} & k_{85} & k_{86} & k_{87} & k_{88} & k_{89} & k_{8(10)} & k_{8(11)} & k_{8(12)} \\ k_{91} & k_{92} & k_{93} & k_{94} & k_{95} & k_{96} & k_{97} & k_{98} & k_{99} & k_{9(10)} & k_{9(11)} & k_{9(12)} \\ k_{10(1)} & k_{10(2)} & k_{10(3)} & k_{10(4)} & k_{10(5)} & k_{10(6)} & k_{10(7)} & k_{10(8)} & k_{10(9)} & k_{10(10)} & k_{10(11)} & k_{10(12)} \\ k_{11(1)} & k_{11(2)} & k_{11(3)} & k_{11(4)} & k_{11(5)} & k_{11(6)} & k_{11(7)} & k_{11(8)} & k_{11(9)} & k_{11(10)} & k_{11(11)} & k_{11(12)} \\ k_{12(1)} & k_{12(2)} & k_{12(3)} & k_{12(4)} & k_{12(5)} & k_{12(6)} & k_{12(7)} & k_{12(8)} & k_{12(9)} & k_{12(10)} & k_{12(11)} & k_{12(12)} \end{bmatrix} \begin{Bmatrix} u_1 \\ v_1 \\ u_2 \\ v_2 \\ u_3 \\ v_3 \\ u_4 \\ v_4 \\ u_5 \\ v_5 \\ u_6 \\ v_6 \end{Bmatrix} = \begin{Bmatrix} R_{x1} \\ R_{y1} \\ 0 \\ 0 \\ 0 \\ 0 \\ F_{4x} \\ F_{4y} \\ 0 \\ 0 \\ 0 \\ R_{y6} \end{Bmatrix} \tag{5.85}$$

As stated above, in this form the global stiffness matrix is singular and we need to apply the boundary conditions.

In order to group the unknown variables and the known variables as described in Eq. (5.80) we would need to move the rows 1, 2, and 12 of the above equation to the bottom of the matrix. In addition, in order to keep the correct algebraic relationships between the unknown variables and the matrix entries, we would also have to move the columns 1, 2, and 12 of the above matrix equation to the right hand side of the matrix.

This follows the general rule in matrix equations which states that *if two rows of a matrix equation are interchanged, the corresponding columns also need to be interchanged*, in order to preserve the algebraic equations represented by the original system of equations intact.

This operation results in the following system of equations which are also the detailed form of Eq. (5.80),

$$\begin{bmatrix} \begin{bmatrix} k_{33} & k_{34} & k_{35} & k_{36} & k_{37} & k_{38} & k_{39} & k_{3(10)} & k_{3(11)} \\ k_{43} & k_{44} & k_{45} & k_{46} & k_{47} & k_{48} & k_{49} & k_{4(10)} & k_{4(11)} \\ k_{53} & k_{54} & k_{55} & k_{56} & k_{57} & k_{58} & k_{59} & k_{5(10)} & k_{5(11)} \\ k_{63} & k_{64} & k_{65} & k_{66} & k_{67} & k_{68} & k_{69} & k_{6(10)} & k_{6(11)} \\ k_{73} & k_{74} & k_{75} & k_{76} & k_{77} & k_{78} & k_{79} & k_{7(10)} & k_{7(11)} \\ k_{83} & k_{84} & k_{85} & k_{86} & k_{87} & k_{88} & k_{89} & k_{8(10)} & k_{8(11)} \\ k_{93} & k_{94} & k_{95} & k_{96} & k_{97} & k_{98} & k_{99} & k_{9(10)} & k_{9(11)} \\ k_{10(3)} & k_{10(4)} & k_{10(5)} & k_{10(6)} & k_{10(7)} & k_{10(8)} & k_{10(9)} & k_{10(10)} & k_{10(11)} \\ k_{11(3)} & k_{11(4)} & k_{11(5)} & k_{11(6)} & k_{11(7)} & k_{11(8)} & k_{11(9)} & k_{11(10)} & k_{11(11)} \end{bmatrix} & \begin{bmatrix} k_{31} & k_{32} & k_{3(12)} \\ k_{41} & k_{42} & k_{4(12)} \\ k_{51} & k_{52} & k_{5(12)} \\ k_{61} & k_{62} & k_{6(12)} \\ k_{71} & k_{72} & k_{7(12)} \\ k_{81} & k_{82} & k_{8(12)} \\ k_{91} & k_{92} & k_{9(12)} \\ k_{10(1)} & k_{10(2)} & k_{10(12)} \\ k_{11(1)} & k_{11(2)} & k_{11(12)} \end{bmatrix} \\ \begin{bmatrix} k_{13} & k_{14} & k_{15} & k_{16} & k_{17} & k_{18} & k_{19} & k_{1(10)} & k_{1(11)} \\ k_{23} & k_{24} & k_{25} & k_{26} & k_{27} & k_{28} & k_{29} & k_{2(10)} & k_{2(11)} \\ k_{12(3)} & k_{12(4)} & k_{12(5)} & k_{12(6)} & k_{12(7)} & k_{12(8)} & k_{12(9)} & k_{12(10)} & k_{12(11)} \end{bmatrix} & \begin{bmatrix} k_{11} & k_{12} & k_{1(12)} \\ k_{21} & k_{22} & k_{2(12)} \\ k_{12(1)} & k_{12(2)} & k_{12(12)} \end{bmatrix} \end{bmatrix} \begin{Bmatrix} \begin{Bmatrix} u_1 \\ v_1 \\ u_2 \\ v_2 \\ u_3 \\ v_3 \\ u_4 \\ v_4 \\ u_5 \\ v_5 \\ u_6 \end{Bmatrix} \\ \begin{Bmatrix} v_1 \\ u_1 \\ v_6 \end{Bmatrix} \end{Bmatrix} = \begin{Bmatrix} \begin{Bmatrix} 0 \\ 0 \\ 0 \\ 0 \\ F_{4x} \\ F_{4y} \\ 0 \\ 0 \\ 0 \end{Bmatrix} \\ \begin{Bmatrix} R_{x1} \\ R_{y1} \\ R_{y6} \end{Bmatrix} \end{Bmatrix} \tag{5.86}$$

5.6.2 Direct assembly of the active degrees of freedom

In case the essential boundary conditions are zero (fixed-boundaries), an alternative procedure is to assemble only the active degrees of freedom into a global stiffness matrix. For this example we note that there are 9 active degrees of freedom as represented in Eq. (5.78) (b). In this case, we would directly obtain the following system,

$$[K_{cc}]\{D_x\} = \{R_c\} \tag{5.87}$$

or,

$$\begin{bmatrix} k_{33} & k_{34} & k_{35} & k_{36} & k_{37} & k_{38} & k_{39} & k_{3(10)} & k_{3(11)} \\ k_{43} & k_{44} & k_{45} & k_{46} & k_{47} & k_{48} & k_{49} & k_{4(10)} & k_{4(11)} \\ k_{53} & k_{54} & k_{55} & k_{56} & k_{57} & k_{58} & k_{59} & k_{5(10)} & k_{5(11)} \\ k_{63} & k_{64} & k_{65} & k_{66} & k_{67} & k_{68} & k_{69} & k_{6(10)} & k_{6(11)} \\ k_{73} & k_{74} & k_{75} & k_{76} & k_{77} & k_{78} & k_{79} & k_{7(10)} & k_{7(11)} \\ k_{83} & k_{84} & k_{85} & k_{86} & k_{87} & k_{88} & k_{89} & k_{8(10)} & k_{8(11)} \\ k_{93} & k_{94} & k_{95} & k_{96} & k_{97} & k_{98} & k_{99} & k_{9(10)} & k_{9(11)} \\ k_{10(3)} & k_{10(4)} & k_{10(5)} & k_{10(6)} & k_{10(7)} & k_{10(8)} & k_{10(9)} & k_{10(10)} & k_{10(11)} \\ k_{11(3)} & k_{11(4)} & k_{11(5)} & k_{11(6)} & k_{11(7)} & k_{11(8)} & k_{11(9)} & k_{11(10)} & k_{11(11)} \end{bmatrix} \begin{Bmatrix} u_2 \\ v_2 \\ u_3 \\ v_3 \\ u_4 \\ v_4 \\ u_5 \\ v_5 \\ u_6 \end{Bmatrix} = \begin{Bmatrix} 0 \\ 0 \\ 0 \\ 0 \\ F_{4x} \\ F_{4y} \\ 0 \\ 0 \\ 0 \end{Bmatrix} \tag{5.88}$$

Note that in this case the reaction forces cannot be found after the solution, and we would have to do more work to get to the reaction forces. Typically that would involve using a relationship such as Eq. (5.85) where the entire global stiffness matrix has to be used.

5.6.3 Numerical implementation of the boundary conditions

5.6.3.1 *Method-1: Row and column manipulations*

One of the methods for applying the boundary conditions involves strategically eliminating the rows and the columns that are associated with the given boundary conditions, but otherwise keeping the size of the global stiffness matrix untouched.

In case the boundary conditions are zero:

In the example given in the previous section, the three boundary conditions imposed on the problem are $u_1 = v_1 = v_6 = 0$. In this case, these boundary conditions can be imposed on the global stiffness matrix as follows:

$$\begin{bmatrix} k_{11} & 0 & 0 & 0 & 0 & 0 & 0 & 0 & 0 & 0 & 0 & 0 \\ 0 & k_{22} & 0 & 0 & 0 & 0 & 0 & 0 & 0 & 0 & 0 & 0 \\ 0 & 0 & k_{33} & k_{34} & k_{35} & k_{36} & k_{37} & k_{38} & k_{39} & k_{3(10)} & k_{3(11)} & 0 \\ 0 & 0 & k_{43} & k_{44} & k_{45} & k_{46} & k_{47} & k_{48} & k_{49} & k_{4(10)} & k_{4(11)} & 0 \\ 0 & 0 & k_{53} & k_{54} & k_{55} & k_{56} & k_{57} & k_{58} & k_{59} & k_{5(10)} & k_{5(11)} & 0 \\ 0 & 0 & k_{63} & k_{64} & k_{65} & k_{66} & k_{67} & k_{68} & k_{69} & k_{6(10)} & k_{6(11)} & 0 \\ 0 & 0 & k_{73} & k_{74} & k_{75} & k_{76} & k_{77} & k_{78} & k_{79} & k_{7(10)} & k_{7(11)} & 0 \\ 0 & 0 & k_{83} & k_{84} & k_{85} & k_{86} & k_{87} & k_{88} & k_{89} & k_{8(10)} & k_{8(11)} & 0 \\ 0 & 0 & k_{93} & k_{94} & k_{95} & k_{96} & k_{97} & k_{98} & k_{99} & k_{9(10)} & k_{9(11)} & 0 \\ 0 & 0 & k_{10(3)} & k_{10(4)} & k_{10(5)} & k_{10(6)} & k_{10(7)} & k_{10(8)} & k_{10(9)} & k_{10(10)} & k_{10(11)} & 0 \\ 0 & 0 & k_{11(3)} & k_{11(4)} & k_{11(5)} & k_{11(6)} & k_{11(7)} & k_{11(8)} & k_{11(9)} & k_{11(10)} & k_{11(11)} & 0 \\ 0 & 0 & 0 & 0 & 0 & 0 & 0 & 0 & 0 & 0 & 0 & k_{12(12)} \end{bmatrix} \begin{Bmatrix} u_1 \\ v_1 \\ u_2 \\ v_2 \\ u_3 \\ v_3 \\ u_4 \\ v_4 \\ u_5 \\ v_5 \\ u_6 \\ v_6 \end{Bmatrix} = \begin{Bmatrix} 0 \\ 0 \\ 0 \\ 0 \\ 0 \\ 0 \\ F_{4x} \\ F_{4y} \\ 0 \\ 0 \\ 0 \\ 0 \end{Bmatrix} \tag{5.89}$$

Note that as the size of the matrix is kept 12×12 this stiffness matrix is no longer singular. You can convince yourself that inversion of this system would recover the imposed boundary conditions.

In case the boundary conditions are not zero:

Let us assume that an additional boundary condition is imposed on this system. The boundary conditions now become,

$$u_1 = v_1 = v_6 = 0 \text{ and } u_6 = \Delta_{x6} \tag{5.90}$$

This last condition can be imposed on Eq. (5.89) as follows:

$$\begin{bmatrix} k_{11} & 0 & 0 & 0 & 0 & 0 & 0 & 0 & 0 & 0 & 0 & 0 \\ 0 & k_{22} & 0 & 0 & 0 & 0 & 0 & 0 & 0 & 0 & 0 & 0 \\ 0 & 0 & k_{33} & k_{34} & k_{35} & k_{36} & k_{37} & k_{38} & k_{39} & k_{3(10)} & 0 & 0 \\ 0 & 0 & k_{43} & k_{44} & k_{45} & k_{46} & k_{47} & k_{48} & k_{49} & k_{4(10)} & 0 & 0 \\ 0 & 0 & k_{53} & k_{54} & k_{55} & k_{56} & k_{57} & k_{58} & k_{59} & k_{5(10)} & 0 & 0 \\ 0 & 0 & k_{63} & k_{64} & k_{65} & k_{66} & k_{67} & k_{68} & k_{69} & k_{6(10)} & 0 & 0 \\ 0 & 0 & k_{73} & k_{74} & k_{75} & k_{76} & k_{77} & k_{78} & k_{79} & k_{7(10)} & 0 & 0 \\ 0 & 0 & k_{83} & k_{84} & k_{85} & k_{86} & k_{87} & k_{88} & k_{89} & k_{8(10)} & 0 & 0 \\ 0 & 0 & k_{93} & k_{94} & k_{95} & k_{96} & k_{97} & k_{98} & k_{99} & k_{9(10)} & 0 & 0 \\ 0 & 0 & k_{10(3)} & k_{10(4)} & k_{10(5)} & k_{10(6)} & k_{10(7)} & k_{10(8)} & k_{10(9)} & k_{10(10)} & 0 & 0 \\ 0 & 0 & 0 & 0 & 0 & 0 & 0 & 0 & 0 & 0 & k_{11(11)} & 0 \\ 0 & 0 & 0 & 0 & 0 & 0 & 0 & 0 & 0 & 0 & 0 & k_{12(12)} \end{bmatrix} \begin{Bmatrix} u_1 \\ v_1 \\ u_2 \\ v_2 \\ u_3 \\ v_3 \\ u_4 \\ v_4 \\ u_5 \\ v_5 \\ u_6 \\ v_6 \end{Bmatrix} = \begin{Bmatrix} 0 \\ 0 \\ 0 \\ 0 \\ 0 \\ 0 \\ F_{4x} \\ F_{4y} \\ 0 \\ 0 \\ 0 \\ 0 \end{Bmatrix} - \begin{Bmatrix} k_{1(11)}\Delta_{x6} \\ k_{2(11)}\Delta_{x6} \\ k_{3(11)}\Delta_{x6} \\ k_{4(11)}\Delta_{x6} \\ k_{5(11)}\Delta_{x6} \\ k_{6(11)}\Delta_{x6} \\ k_{7(11)}\Delta_{x6} \\ k_{8(11)}\Delta_{x6} \\ k_{9(11)}\Delta_{x6} \\ k_{10(11)}\Delta_{x6} \\ k_{11(11)}\Delta_{x6} \\ k_{12(11)}\Delta_{x6} \end{Bmatrix} \tag{5.91}$$

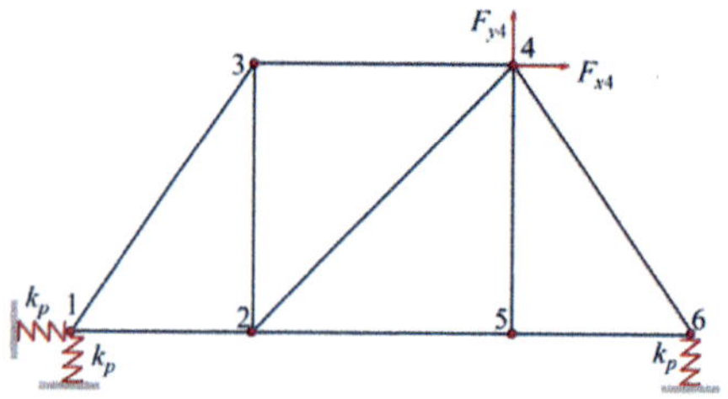

FIGURE 5.10 Implementation of the penalty method for the truss shown in Fig. 5.9 which is supported by fixed and roller supports at nodes-1 and -6, respectively. The boundary conditions we want to impose are $u_1 = v_2 = v_6 = 0$. This can be achieved by attaching very stiff springs with stiffness k_p, to restrict rigid body motion along these imposed boundary conditions.

Details description of the derivation of this method can be found in reference [1].

5.6.3.2 *Method-2: Penalty method*

In order to impose the boundary conditions described in Fig. 5.9, i.e. $u_1 = v_1 = v_6 = 0$, assume very stiff anchors are used to restrict these degrees of freedom. These anchors attach to a stationary wall and to the degree of freedom of interest (Fig. 5.10). Their compliance relationship is written in an *ad hoc* manner as follows:

$$k_p\begin{bmatrix} 1 & -1 \\ -1 & 1 \end{bmatrix}\begin{Bmatrix} u_1 \\ 0 \end{Bmatrix},\ k_p\begin{bmatrix} 1 & -1 \\ -1 & 1 \end{bmatrix}\begin{Bmatrix} v_1 \\ 0 \end{Bmatrix} \text{ and } k_p\begin{bmatrix} 1 & -1 \\ -1 & 1 \end{bmatrix}\begin{Bmatrix} v_6 \\ 0 \end{Bmatrix} \quad (5.92)$$

where k_p is the called the *penalty stiffness*. These three elements that represent the presence of a fixed boundary condition can be assembled to the global equilibrium equation as follows:

$$\begin{bmatrix}
k_{11}+k_p & k_{12} & k_{13} & k_{14} & k_{15} & k_{16} & k_{17} & k_{18} & k_{19} & k_{1(10)} & k_{1(11)} & k_{1(12)} \\
k_{21} & k_{22}+k_p & k_{23} & k_{24} & k_{25} & k_{26} & k_{27} & k_{28} & k_{29} & k_{2(10)} & k_{2(11)} & k_{2(12)} \\
k_{31} & k_{32} & k_{33} & k_{34} & k_{35} & k_{36} & k_{37} & k_{38} & k_{39} & k_{3(10)} & k_{3(11)} & k_{3(12)} \\
k_{41} & k_{42} & k_{43} & k_{44} & k_{45} & k_{46} & k_{47} & k_{48} & k_{49} & k_{4(10)} & k_{4(11)} & k_{4(12)} \\
k_{51} & k_{52} & k_{53} & k_{54} & k_{55} & k_{56} & k_{57} & k_{58} & k_{59} & k_{5(10)} & k_{5(11)} & k_{5(12)} \\
k_{61} & k_{62} & k_{63} & k_{64} & k_{65} & k_{66} & k_{67} & k_{68} & k_{69} & k_{6(10)} & k_{6(11)} & k_{6(12)} \\
k_{71} & k_{72} & k_{73} & k_{74} & k_{75} & k_{76} & k_{77} & k_{78} & k_{79} & k_{7(10)} & k_{7(11)} & k_{7(12)} \\
k_{81} & k_{82} & k_{83} & k_{84} & k_{85} & k_{86} & k_{87} & k_{88} & k_{89} & k_{8(10)} & k_{8(11)} & k_{8(12)} \\
k_{91} & k_{92} & k_{93} & k_{94} & k_{95} & k_{96} & k_{97} & k_{98} & k_{99} & k_{9(10)} & k_{9(11)} & k_{9(12)} \\
k_{10(1)} & k_{10(2)} & k_{10(3)} & k_{10(4)} & k_{10(5)} & k_{10(6)} & k_{10(7)} & k_{10(8)} & k_{10(9)} & k_{10(10)} & k_{10(11)} & k_{10(12)} \\
k_{11(1)} & k_{11(2)} & k_{11(3)} & k_{11(4)} & k_{11(5)} & k_{11(6)} & k_{11(7)} & k_{11(8)} & k_{11(9)} & k_{11(10)} & k_{11(11)} & k_{11(12)} \\
k_{12(1)} & k_{12(2)} & k_{12(3)} & k_{12(4)} & k_{12(5)} & k_{12(6)} & k_{12(7)} & k_{12(8)} & k_{12(9)} & k_{12(10)} & k_{12(11)} & k_{12(12)}+k_p
\end{bmatrix}
\begin{Bmatrix} u_1 \\ v_1 \\ u_2 \\ v_2 \\ u_3 \\ v_3 \\ u_4 \\ v_4 \\ u_5 \\ v_5 \\ u_6 \\ v_6 \end{Bmatrix}
=
\begin{Bmatrix} R_{x1} \\ R_{y1} \\ 0 \\ 0 \\ 0 \\ 0 \\ F_{4x} \\ F_{4y} \\ 0 \\ 0 \\ 0 \\ R_{y6} \end{Bmatrix} \quad (5.93)$$

As long as $k_p > 0$, this matrix is no longer singular and a solution for $\{D\}$ can be obtained by using the standard matrix inversion methods. As $k_p \to \infty$, the solution of the matrix equation approaches the exact solution and the imposed boundary conditions are recovered from the solution. This will be demonstrated in the following example.

We can take advantage of the mechanics described above to impose the boundary conditions on a matrix with minimal change. To set the $u_1, v_1, v_6 = 0$, the procedure applied above can be viewed as attaching a spring with infinite stiffness to the finite element node in the direction we wish to impose an essential boundary condition. In practice, the stiffness of the spring k_p is a number $10^2 - 10^5$ times larger than the largest value on the main diagonal of the matrix. The stiffness of the spring is called the *penalty parameter*, and imposing boundary conditions with this method is known as the *penalty method.*

Non-zero deflection components can also be applied by the penalty method. For example, if we want to fix $v_6 = d$ in the problem described above, we would need to apply a force by the amount $k_p.d$ on the right hand side of the system resulting in,

$$\begin{bmatrix} k_{11}+k_p & k_{12} & k_{13} & k_{14} & k_{15} & k_{16} & k_{17} & k_{18} & k_{19} & k_{1(10)} & k_{1(11)} & k_{1(12)} \\ k_{21} & k_{22}+k_p & k_{23} & k_{24} & k_{25} & k_{26} & k_{27} & k_{28} & k_{29} & k_{2(10)} & k_{2(11)} & k_{2(12)} \\ k_{31} & k_{32} & k_{33} & k_{34} & k_{35} & k_{36} & k_{37} & k_{38} & k_{39} & k_{3(10)} & k_{3(11)} & k_{3(12)} \\ k_{41} & k_{42} & k_{43} & k_{44} & k_{45} & k_{46} & k_{47} & k_{48} & k_{49} & k_{4(10)} & k_{4(11)} & k_{4(12)} \\ k_{51} & k_{52} & k_{53} & k_{54} & k_{55} & k_{56} & k_{57} & k_{58} & k_{59} & k_{5(10)} & k_{5(11)} & k_{5(12)} \\ k_{61} & k_{62} & k_{63} & k_{64} & k_{65} & k_{66} & k_{67} & k_{68} & k_{69} & k_{6(10)} & k_{6(11)} & k_{6(12)} \\ k_{71} & k_{72} & k_{73} & k_{74} & k_{75} & k_{76} & k_{77} & k_{78} & k_{79} & k_{7(10)} & k_{7(11)} & k_{7(12)} \\ k_{81} & k_{82} & k_{83} & k_{84} & k_{85} & k_{86} & k_{87} & k_{88} & k_{89} & k_{8(10)} & k_{8(11)} & k_{8(12)} \\ k_{91} & k_{92} & k_{93} & k_{94} & k_{95} & k_{96} & k_{97} & k_{98} & k_{99} & k_{9(10)} & k_{9(11)} & k_{9(12)} \\ k_{10(1)} & k_{10(2)} & k_{10(3)} & k_{10(4)} & k_{10(5)} & k_{10(6)} & k_{10(7)} & k_{10(8)} & k_{10(9)} & k_{10(10)} & k_{10(11)} & k_{10(12)} \\ k_{11(1)} & k_{11(2)} & k_{11(3)} & k_{11(4)} & k_{11(5)} & k_{11(6)} & k_{11(7)} & k_{11(8)} & k_{11(9)} & k_{11(10)} & k_{11(11)} & k_{11(12)} \\ k_{12(1)} & k_{12(2)} & k_{12(3)} & k_{12(4)} & k_{12(5)} & k_{12(6)} & k_{12(7)} & k_{12(8)} & k_{12(9)} & k_{12(10)} & k_{12(11)} & k_{12(12)}+k_p \end{bmatrix} \begin{Bmatrix} u_1 \\ v_1 \\ u_2 \\ v_2 \\ u_3 \\ v_3 \\ u_4 \\ v_4 \\ u_5 \\ v_5 \\ u_6 \\ v_6 \end{Bmatrix} = \begin{Bmatrix} R_{x1} \\ R_{y1} \\ 0 \\ 0 \\ 0 \\ 0 \\ F_{4x} \\ F_{4y} \\ 0 \\ 0 \\ 0 \\ R_{y6}+k_p d \end{Bmatrix} \tag{5.94}$$

Example 5.2 Implementation of boundary conditions on global stiffness matrix.

Implement the boundary conditions by the three methods described in Section 5.6, for the three-bar truss shown in Fig. 5.8.

The assembly of the element equilibrium equations presented in Example 1, resulted in a system of equations represented in the following form:

$$[K]\{D\} = \{R\} \tag{5.95}$$

$$\text{Global dof vector: } \{D\} = \{D_1\ D_2\ D_3\ D_4\ D_5\ D_6\}^T = \{u_1\ v_1\ u_2\ v_2\ u_3\ v_3\}^T \tag{5.96}$$

Global load vector: $\{R\} = \{P_{x1} \; P_{y1} \; R_{x2} \; R_{y2} \; R_{x3} \; P_{y3}\}^T$ (5.97)

Boundary conditions: $u_2 = v_2 = u_3 = 0$ or $D_3 = D_4 = D_5 = 0$ (5.98)

Active degrees of freedom: u_1, v_1, v_3 or D_1, D_2, D_6 (5.99)

In particular, the equilibrium equation for this problem was found in Eq. (5.71) as follows:

$$\begin{bmatrix} 0.36k_2+k_3 & -0.48k_2 & -k_3 & 0 & -0.36k_2 & 0.48k_2 \\ -0.48k_2 & 0.64k_2 & 0 & 0 & 0.48k_2 & -0.64k_2 \\ -k_3 & 0 & k_3 & 0 & 0 & 0 \\ 0 & 0 & 0 & k_1 & 0 & -k_1 \\ -0.36k_2 & 0.48k_2 & 0 & 0 & 0.36k_2 & -0.48k_2 \\ 0.48k_2 & -0.64k_2 & 0 & -k_1 & -48k_2 & k_1+0.64k_2 \end{bmatrix} \begin{Bmatrix} u_1 \\ v_1 \\ u_2 \\ v_2 \\ u_3 \\ v_3 \end{Bmatrix} = \begin{Bmatrix} P_{x1} \\ P_{y1} \\ R_{x2} \\ R_{y2} \\ R_{x3} \\ P_{y3} \end{Bmatrix} = \begin{Bmatrix} 0.12\bar{P} - 1.5q \\ -0.16\bar{P} - P \\ R_{x2} \\ R_{y2} \\ R_{x3} \\ -0.64\bar{P} \end{Bmatrix}$$

i) **Directly consider the active degrees of freedom:** The rows and columns related to the known degrees of freedom are eliminated from the global equilibrium equations,

$$\begin{bmatrix} k_{11} & k_{12} & k_{13} & k_{14} & k_{15} & k_{16} \\ k_{21} & k_{22} & k_{23} & k_{24} & k_{25} & k_{26} \\ k_{31} & k_{32} & k_{33} & k_{34} & k_{35} & k_{36} \\ k_{41} & k_{42} & k_{43} & k_{44} & k_{45} & k_{46} \\ k_{51} & k_{52} & k_{53} & k_{54} & k_{55} & k_{56} \\ k_{61} & k_{62} & k_{63} & k_{64} & k_{65} & k_{66} \end{bmatrix} \begin{Bmatrix} D_1 \\ D_2 \\ D_3 \\ D_4 \\ D_5 \\ D_6 \end{Bmatrix} = \begin{Bmatrix} P_{x1} \\ P_{y1} \\ R_{x2} \\ R_{y2} \\ R_{x3} \\ P_{y3} \end{Bmatrix} \tag{5.100}$$

The system of equations then reduce to,

$$\begin{bmatrix} 0.36k_2+k_3 & -0.48k_2 & 0.48k_2 \\ -0.48k_2 & 0.64k_2 & -0.64k_2 \\ 0.48k_2 & -0.64k_2 & k_1+0.64k_2 \end{bmatrix} \begin{Bmatrix} D_1 \\ D_2 \\ D_6 \end{Bmatrix} = \begin{Bmatrix} P_{x1} \\ P_{y1} \\ P_{y3} \end{Bmatrix} \tag{5.101}$$

The solution to this system is then found as follows:

$$\{D\} = \left\{ \left(\frac{P_{x1} + 0.75P_{y1}}{k_3} \right) \left(\frac{0.75P_{x1} + 0.5625P_{y1}}{k_3} + \frac{1.5625P_{y1}}{k_2} + \frac{P_{y1} + P_{y3}}{k_1} \right) \left(\frac{P_{y1} + P_{y3}}{k_1} \right) \right\}^T \tag{5.102}$$

ii) Numerical implementation of boundary conditions: Zeroing rows and columns of specified degrees of freedom: As described in Section 5.6.3.1, the boundary conditions can be imposed on the global stiffness matrix by zeroing out the rows and columns associated with the dof that are prescribed, with the exception of the diagonal entries. For $D_3 = D_4 = D_5 = \Delta$ the general form of the equilibrium equation becomes,

$$\begin{bmatrix} k_{11} & k_{12} & 0 & 0 & 0 & k_{16} \\ k_{21} & k_{22} & 0 & 0 & 0 & k_{26} \\ 0 & 0 & k_{33} & 0 & 0 & 0 \\ 0 & 0 & 0 & k_{44} & 0 & 0 \\ 0 & 0 & 0 & 0 & k_{55} & 0 \\ k_{61} & k_{62} & 0 & 0 & 0 & k_{66} \end{bmatrix} \begin{Bmatrix} D_1 \\ D_2 \\ D_3 \\ D_4 \\ D_5 \\ D_6 \end{Bmatrix} = \begin{Bmatrix} P_{x1} \\ P_{y1} \\ 0 \\ 0 \\ 0 \\ P_{y3} \end{Bmatrix} - \Delta \begin{Bmatrix} k_{13} \\ k_{23} \\ -k_{33} \\ k_{43} \\ k_{53} \\ k_{63} \end{Bmatrix} - \Delta \begin{Bmatrix} k_{14} \\ k_{24} \\ k_{34} \\ -k_{44} \\ k_{54} \\ k_{64} \end{Bmatrix} - \Delta \begin{Bmatrix} k_{15} \\ k_{25} \\ k_{35} \\ k_{45} \\ -k_{55} \\ k_{65} \end{Bmatrix} \tag{5.103}$$

with $\Delta = 0$ and by using Eq. (5.71), we get,

$$\begin{bmatrix} 0.36k_2 + k_3 & -0.48k_2 & 0 & 0 & 0 & 0.48k_2 \\ -0.48k_2 & 0.64k_2 & 0 & 0 & 0 & -0.64k_2 \\ 0 & 0 & k_3 & 0 & 0 & 0 \\ 0 & 0 & 0 & k_1 & 0 & 0 \\ 0 & 0 & 0 & 0 & 0.36k_2 & 0 \\ 0.48k_2 & -0.64k_2 & 0 & 0 & 0 & k_1 + 0.64k_2 \end{bmatrix} \begin{Bmatrix} D_1 \\ D_2 \\ D_3 \\ D_4 \\ D_5 \\ D_6 \end{Bmatrix} = \begin{Bmatrix} P_{x1} \\ P_{y1} \\ 0 \\ 0 \\ 0 \\ P_{y3} \end{Bmatrix} \tag{5.104}$$

The solution of this equation becomes,

$$\{D\} = \left\{ \left(\frac{P_{x1} + 0.75P_{y1}}{k_3} \right) \left(\frac{0.75P_{x1} + 0.5625P_{y1}}{k_3} + \frac{1.5625P_{y1}}{k_2} + \frac{P_{y1} + P_{y3}}{k_1} \right) \frac{P_{y1} + P_{y3}}{k_1} \right\} \tag{5.105}$$

iii) Numerical implementation of boundary conditions: Penalty method: The penalty method can be used to implement the boundary conditions $D_3 = D_4 = D_5 = \Delta$ by modifying the stiffness matrix as follows:

$$\begin{bmatrix} k_{11} & k_{12} & k_{13} & k_{14} & k_{15} & k_{16} \\ k_{21} & k_{22} & k_{23} & k_{24} & k_{25} & k_{26} \\ k_{31} & k_{32} & (k_{33} + k_p) & k_{34} & k_{35} & k_{36} \\ k_{41} & k_{42} & k_{43} & (k_{44} + k_p) & k_{45} & k_{46} \\ k_{51} & k_{52} & k_{53} & k_{54} & (k_{55} + k_p) & k_{56} \\ k_{61} & k_{62} & k_{63} & k_{64} & k_{65} & k_{66} \end{bmatrix} \begin{Bmatrix} D_1 \\ D_2 \\ D_3 \\ D_4 \\ D_5 \\ D_6 \end{Bmatrix} = \begin{Bmatrix} P_{x1} \\ P_{y1} \\ 0 \\ 0 \\ 0 \\ R_6 \end{Bmatrix} + \begin{Bmatrix} 0 \\ 0 \\ k_p\Delta \\ k_p\Delta \\ k_p\Delta \\ 0 \end{Bmatrix} \tag{5.106}$$

with $k_p = O(10^3 - 10^5)\times\max(k_{i,j})$. Considering $\Delta = 0$ the global equilibrium equation for this problem becomes,

$$\begin{bmatrix} 0.36k_2 + k_3 & -0.48k_2 & -k_3 & 0 & -0.36k_2 & 0.48k_2 \\ -0.48k_2 & 0.64k_2 & 0 & 0 & 0.48k_2 & -0.64k_2 \\ -k_3 & 0 & k_p + k_3 & 0 & 0 & 0 \\ 0 & 0 & 0 & k_p + k_1 & 0 & -k_1 \\ -0.36k_2 & 0.48k_2 & 0 & 0 & k_p - 0.36k_2 & -0.48k_2 \\ 0.48k_2 & -0.64k_2 & 0 & -k_1 & -0.48k_2 & k_1 + 0.64k_2 \end{bmatrix} \begin{Bmatrix} u_1 \\ v_1 \\ u_2 \\ v_2 \\ u_3 \\ v_3 \end{Bmatrix} = \begin{Bmatrix} P_{x1} \\ P_{y1} \\ 0 \\ 0 \\ 0 \\ P_{y3} \end{Bmatrix} \tag{5.107}$$

The determinant of $[K]$ now becomes,

$$\det([K]) = 0.64k_p^3 k_1 k_2 k_3 \tag{5.108}$$

Thus we see that as long as $k_p > 0$ the singularity of the system is removed. The solution of Eq. (5.107) is found as follows:

$$\begin{aligned}
u_1 &= \frac{k_p^2(P_{x1} + 0.75P_{y1}) + 0.54k_2k_3P_{y1} + k_p(k_3P_{x1} + 0.54k_2P_{y1} + 0.75k_3P_{y1})}{k_p^2k_3} \\
v_1 &= \frac{0.75P_{x1} + 0.5625P_{y1}}{k_3} + \frac{1.5625P_{y1}}{k_2} + \frac{P_{y1} + P_{y3}}{k_1} \\
&\quad + \frac{0.405k_2P_{y1}}{k_p^2} + \frac{0.75P_{x1} + 2.125P_{y1} + \frac{0.405k_2P_{y1}}{k_3} + P_{y3}}{k_p} \\
u_2 &= \frac{P_{x1} + 0.75P_{y1}}{k_p} + \frac{0.54k_2P_{y1}}{k_p^2} \\
v_2 &= \frac{P_{y1} + P_{y3}}{k_p} \\
u_3 &= \frac{0.75P_{y1}}{k_p} \\
v_3 &= \frac{P_{y1} + P_{y3}}{k_1} + \frac{P_{y1} + P_{y3}}{k_p}
\end{aligned} \tag{5.109}$$

Taking the limit as $k_p \to \infty$ we find,

$$\begin{aligned}
u_1 &= \frac{1}{k_3}(P_{x1} + 0.75P_{y1}) \\
v_1 &= \frac{P_{y3}}{k_1} + \left(\frac{1}{k_1} + \frac{1.5625}{k_2} + \frac{0.5625}{k_3}\right)P_{y1} + 0.75\frac{P_{x1}}{k_3} \\
u_2 &= 0 \\
v_2 &= 0 \\
u_3 &= 0 \\
v_3 &= \frac{1}{k_1}(P_{y1} + P_{y3})
\end{aligned} \tag{5.110}$$

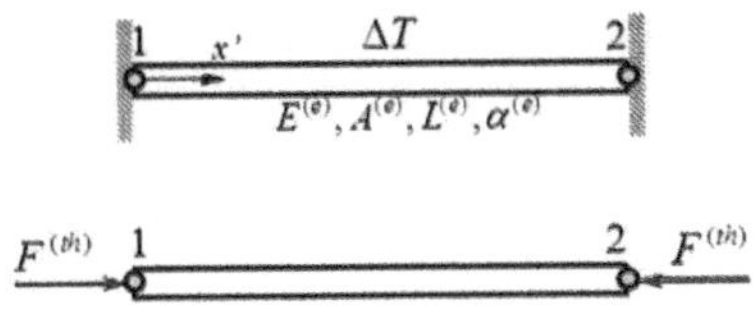

FIGURE 5.11 An element constrained to move on each end will experience thermal stresses when subjected to a temperature difference of ΔT. The nodal forces due to the thermal effects are interpreted as shown.

This result is the same as the one found in Eq. (5.105). Thus, we see that the imposed boundary conditions and the solution of the system of equations can be found by the penalty method.

5.7 Effects of initial stress or initial strain

5.7.1 Thermal stresses

The length L of a one-dimensional material subjected to temperature difference ΔT changes by,

$$\Delta L = \int_0^L \alpha \Delta T \, dx' \tag{5.111}$$

where α is the material property known as the *coefficient of thermal expansion*. Note that both α and ΔT can vary along the length of the bar. The dimensional change described in Eq. (5.111) causes a *thermal strain* which is defined as follows:

$$\varepsilon_{x'}^{th} = \frac{1}{L} \int_0^L \alpha \Delta T \, dx' \tag{5.112}$$

Thermal stress develops in the material only if the structure is statically indeterminate (e.g. fixed at both ends). If it develops, the thermal stress can be computed as follows:

$$\sigma_{x'}^{(th)} = E \varepsilon_{x'}^{Th} \tag{5.113}$$

Consider a bar of length L that is fixed on both ends, subjected to constant ΔT with constant A and α (Fig. 5.11). In this case, the thermal stress will cause the following reaction forces to develop at the ends of the bar:

$$\begin{aligned} F^{(th)} &= EA \propto \Delta T \quad \text{at } x = 0 \text{ and} \\ F^{(th)} &= -EA \propto \Delta T \quad \text{at } x = L \end{aligned} \tag{5.114}$$

If these parameters are not constant the reaction force due to thermal effects can still be found by using the general form (Eq. (5.112)) of the thermal strain. Note that, increasing temperature ($\Delta T > 0$) puts the bar in compression and decreasing temperature ($\Delta T < 0$) causes tension in the bar. The *nodal forces* due to thermal stress can then be found as follows:

$$\begin{aligned} \left\{r^{(th)}\right\} &= F^{(th)}\begin{Bmatrix} -1 \\ 1 \end{Bmatrix} \text{ if } \Delta T > 0 \text{ (compression) (a)} \\ \left\{r^{(th)}\right\} &= F^{(th)}\begin{Bmatrix} 1 \\ -1 \end{Bmatrix} \text{ if } \Delta T < 0 \text{ (tension) (b)} \end{aligned} \tag{5.115}$$

5.7.2 Initial stresses

The result presented in Eq. (5.115) can be generalized to any initial stress σ_0 as follows:

$$\begin{aligned} \left\{r^{(0)}\right\} &= \sigma_0 A\begin{Bmatrix} -1 \\ 1 \end{Bmatrix} \text{ if } \sigma_0 \text{ is compressive (a)} \\ \left\{r^{(0)}\right\} &= \sigma_0 A\begin{Bmatrix} 1 \\ -1 \end{Bmatrix} \text{ if } \sigma_0 \text{ is tensile (b)} \end{aligned} \tag{5.116}$$

5.8 Postprocessing: Computation of stresses and reaction forces

Solution of the global equilibrium equations (e.g., Eq. (5.83)) only gives the unknown degrees of freedom, which represent the deflection of the structure. The strain and stress that develops in the elements, as well as the reaction forces have to be calculated after the $\{D\}$ vector is found. This process is called *postprocessing* in finite element literature.

5.8.1 Computation of stresses in members

Even though $\{D\}$ represents the displacement degrees of freedom of the entire structure, postprocessing for strain and stress is performed on element basis. Therefore, by using the element connectivity table used during the assembly the element degree of freedom vector $\{d^{(e)}\}$ should be extracted from $\{D\}$ for each element.

Next the element solution vector is transformed from the global coordinate system to local coordinate system by using Eq. (5.49). In general, this transformation is as follows:

$$\left\{d'\right\}^{(e)} = [T]^{(e)}\{d\}^{(e)} \tag{5.117}$$

By using Eq. (5.49), the displacement components are transferred to local coordinate system as follows:

$$\begin{Bmatrix} u_1{}' \\ u_2{}' \end{Bmatrix} = \begin{bmatrix} c & s & 0 & 0 \\ 0 & 0 & c & s \end{bmatrix} \begin{Bmatrix} u_1 \\ v_1 \\ u_2 \\ v_2 \end{Bmatrix} \tag{5.118}$$

The longitudinal strain in the bar element is found by using Eq. (5.16) as follows:

$$\varepsilon_{x'}{}^{(e)} = [B]\{d'\}^{(e)} = [B][T]^{(e)}\{d\}^{(e)} \tag{5.119}$$

The longitudinal stress in the bar element is found by using the Hooke's law given in Eq. (2.59). For a one dimensional, linear elastic material, the Hooke's law becomes:

$$\sigma_{x'}{}^{(e)} = E\varepsilon_{x'}{}^{(e)} \tag{5.120}$$

By using the linear interpolation functions given by Eq. (5.13), the strain and stress variation in the element are found as follows:

$$\begin{aligned} \varepsilon_{x'}{}^{(e)} &= \frac{(u_2 - u_1)\cos\alpha + (v_2 - v_1)\sin\alpha}{L^{(e)}} \\ \sigma_{x'}{}^{(e)} &= E^{(e)}\frac{(u_2 - u_1)\cos\alpha + (v_2 - v_1)\sin\alpha}{L^{(e)}} \end{aligned} \tag{5.121}$$

Note that for the $C^{(0)}$, linear interpolation functions used in this element, both the strain and the stress are constant along the length of the element.

5.8.2 Reaction forces

Once global degrees of freedom are found, multiplying the solution vector $\{D\}$ with the original stiffness $[K]$ matrix will give the correct loading vector,

$$\{R\} = [K]\{D\} \tag{5.122}$$

Note that the newly found $\{R\}$ vector will now have the correct reaction forces acting on the boundary nodes as well as the original external forces acting on the active nodes.

Example 5.3 Strain, stress, and the reaction forces.
For the three member truss presented in Fig. 5.8, compute the strain, stress acting on the elements and the reaction forces on the boundary nodes.

Solution:
Element strain and stress: For the linear interpolation function used in this example we can use the strain–displacement matrix given in Eq. (5.15),

$$[B] = \frac{d}{dx}([N]) = \left[-\frac{1}{L^{(e)}}\ \frac{1}{L^{(e)}}\right]$$

Equation (5.119) and (5.120) are used respectively to compute the strain and stress in each element. Note that the nodal deflections are given by (any one of)

Eqs. (5.102), (5.105), or (5.110). Appropriate values should be plugged into the expressions below.

Element-1: $\alpha = \frac{\pi}{2}$, thus $c = 0$, $s = 1$; element length, $L^{(1)} = 4$ m

$$\text{Local displacement vector:} \begin{Bmatrix} u'_2 \\ u'_3 \end{Bmatrix} = \begin{bmatrix} 0 & 1 & 0 & 0 \\ 0 & 0 & 0 & 1 \end{bmatrix} \begin{Bmatrix} u_2 \\ v_2 \\ u_3 \\ v_3 \end{Bmatrix}$$

$$\text{Element strain:} \quad \varepsilon_{x'}^{(1)} = \frac{v_3 - v_2}{L_e^{(1)}}$$

$$\text{Element stress:} \quad \sigma_{x'}^{(1)} = E^{(1)} \varepsilon_{x'}^{(1)}$$

Element-2: $\alpha = \pi - \tan^{-1}(4/3)$, thus $c = -0.6$, $s = 0.8$; element length, $L^{(2)} = 5$ m

$$\text{Local displacement vector:} \begin{Bmatrix} u'_1 \\ u'_3 \end{Bmatrix} = \begin{bmatrix} -0.6 & 0.8 & 0 & 0 \\ 0 & 0 & -0.6 & 0.8 \end{bmatrix} \begin{Bmatrix} u_1 \\ v_1 \\ u_3 \\ v_3 \end{Bmatrix},$$

$$\text{Element strain:} \quad \varepsilon_{x'}^{(2)} = \frac{-0.6(u_3 - u_1) + 0.8(v_3 - v_1)}{L^{(2)}},$$

$$\text{Element stress:} \quad \sigma_{x'}^{(2)} = E^{(2)} \varepsilon_{x'}^{(2)}$$

Element-3: $\alpha = \pi$, thus $c = -1$, $s = 0$; element length, $L^{(3)} = 3$ m

$$\text{Local displacement vector:} \begin{Bmatrix} u'_1 \\ u'_2 \end{Bmatrix} = \begin{bmatrix} -1 & 0 & 0 & 0 \\ 0 & 0 & -1 & 0 \end{bmatrix} \begin{Bmatrix} u_1 \\ v_1 \\ u_2 \\ v_2 \end{Bmatrix},$$

$$\text{Element strain:} \quad \varepsilon_{x'}^{(3)} = -\frac{u_2 - u_1}{L^{(3)}},$$

$$\text{Element stress:} \quad \sigma_{x'}^{(3)} = E^{(3)} \varepsilon_{x'}^{(3)}$$

Reaction forces: Note that the original stiffness matrix, assembled in the first example, was presented in Eq. (5.71). The nodal deflections were found by (any one of) Eqs. (5.102), (5.105), or (5.110). Implementing these by using Eq. (5.122), we find the following load vector,

$$\{R\} = \begin{bmatrix} 0.36k_2 + k_3 & -0.48k_2 & -k_3 & 0 & -0.36k_2 & 0.48k_2 \\ -0.48k_2 & 0.64k_2 & 0 & 0 & 0.48k_2 & -0.64k_2 \\ -k_3 & 0 & k_3 & 0 & 0 & 0 \\ 0 & 0 & 0 & k_1 & 0 & -k_1 \\ -0.36k_2 & 0.48k_2 & 0 & 0 & 0.36k_2 & -0.48k_2 \\ 0.48k_2 & -0.64k_2 & 0 & -k_1 & -0.48k_2 & k_1 + 0.64k_2 \end{bmatrix} \begin{Bmatrix} u_1 \\ v_1 \\ u_2 \\ v_2 \\ u_3 \\ v_3 \end{Bmatrix} = \begin{Bmatrix} P_{x1} \\ P_{y1} \\ -(P_{x1} + 0.75P_{y1}) \\ -(P_{y1} + P_{y3}) \\ 0.75P_{y1} \\ P_{y3} \end{Bmatrix} \tag{5.123}$$

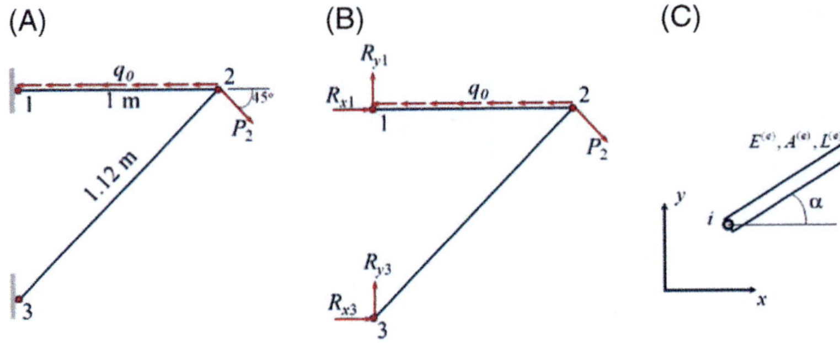

FIGURE 5.12 A) Two-member truss. B) Reaction Forces. C) Generic element properties.

Thus the reaction forces are found as follows:

$$R_{x2} = -\left(P_{x1} + 0.75P_{y1}\right),\ R_{y2} = -\left(P_{y1} + P_{y3}\right),\ R_{x3} = 0.75P_{y1} \tag{5.124}$$

Example 5.4 Consider the two member truss in Fig. 5.12 subjected to a distributed force $q_0 = 10^4$ N/m and a concentrated force $P_2 = 1$ kN as shown. The elastic modulus of member-s12 and -23 are 200 and 112 GPa, and the cross-sectional areas of the same members are 10^{-4} and 10^{-3} m^2, respectively.

a) Find the global equilibrium equation for this system.
b) Apply boundary conditions by directly eliminating the inactive dofs, and find the solution.
c) Apply boundary conditions by the penalty method and obtain the solution.
d) Analyze the results for different penalty parameters.
e) Find the reaction forces.

Solution:

Element connectivity table

(e)	i	$i+1$	α	E (Pa)	A (m^2)	L (m)	F_{xi} (N)	F_{yi}	$F_{x(i+1)}$	$F_{y(i+1)}$
1	1	2	0	2×10^{11}	10^{-4}	1	−5000	0	−5000	0
2	2	3	−153.4	112×10^{9}	10^{-3}	1.12	0	0	0	0

Element equilibrium in global coordinates:

$$[k_m]\{d\} = \{r\}$$

$$\left(\frac{AE}{L}\right)^{(e)} \begin{bmatrix} c^2 & cs & -c^2 & -cs \\ cs & s^2 & -cs & -s^2 \\ -c^2 & -cs & c^2 & cs \\ -cs & -s^2 & cs & s^2 \end{bmatrix} \begin{Bmatrix} u_i \\ v_i \\ u_{i+1} \\ v_{i+1} \end{Bmatrix} = \begin{Bmatrix} F_{xi} \\ F_{yi} \\ F_{x(i+1)} \\ F_{y(i+1)} \end{Bmatrix}$$

where $s = \sin\alpha^{(e)}$, $c = \cos\alpha^{(e)}$,$F_{xi} = cF_i'$, $F_{yi} = sF_i'$, $F_{x(i+1)} = cF_{i+1}'$, and $F_{y(i+1)} = sF_{i+1}'$

Element-1: $s = 0, c = 1$

$$2\times 10^7\begin{bmatrix}1 & 0 & -1 & 0\\ 0 & 0 & 0 & 0\\ -1 & 0 & 1 & 0\\ 0 & 0 & 0 & 0\end{bmatrix}\begin{Bmatrix}u_1\\ v_1\\ u_2\\ v_2\end{Bmatrix} = -\frac{1}{2}10^4\begin{Bmatrix}1\\ 0\\ 1\\ 0\end{Bmatrix}$$

Element-2: $c = \cos(-153.4)$, $s = \sin(-154.4)$

$$1\times 10^8\begin{bmatrix}0.8 & 0.4 & -0.8 & -0.4\\ 0.4 & 0.2 & -0.4 & -0.2\\ -0.8 & -0.4 & 0.8 & 0.4\\ -0.4 & -.02 & 0.4 & 0.2\end{bmatrix}\begin{Bmatrix}u_2\\ v_2\\ u_3\\ v_3\end{Bmatrix} = 0$$

Global vectors:

$$\{D\} = \{u_1\ v_1\ u_2\ v_2\ u_3\ v_3\}^T$$

$$\{P\} = \{0\ 0\ 707.1\ -707.1\ 0\ 0\}^T$$

$$\{\mathrm{P}_{reac}\} = \left\{R_{x1}\ R_{y1}\ 0\ 0\ R_{x3}\ R_{y3}\right\}^T$$

Element equilibrium equations in the global stiffness matrix
Element-1:

$$2\times 10^7\begin{bmatrix}1 & 0 & -1 & 0 & 0 & 0\\ 0 & 0 & 0 & 0 & 0 & 0\\ -1 & 0 & 1 & 0 & 0 & 0\\ 0 & 0 & 0 & 0 & 0 & 0\\ 0 & 0 & 0 & 0 & 0 & 0\\ 0 & 0 & 0 & 0 & 0 & 0\end{bmatrix}\begin{Bmatrix}u_1\\ v_1\\ u_2\\ v_2\\ u_3\\ v_3\end{Bmatrix} = \begin{Bmatrix}-5000\\ 0\\ -5000\\ 0\\ 0\\ 0\end{Bmatrix}$$

Element-2:

$$1\times 10^7\begin{bmatrix}0 & 0 & 0 & 0 & 0 & 0\\ 0 & 0 & 0 & 0 & 0 & 0\\ 0 & 0 & 8 & 4 & -8 & -4\\ 0 & 0 & 4 & 2 & -4 & -2\\ 0 & 0 & -8 & -4 & 8 & 4\\ 0 & 0 & -4 & -2 & 4 & 2\end{bmatrix}\begin{Bmatrix}u_1\\ v_1\\ u_2\\ v_2\\ u_3\\ v_3\end{Bmatrix} = \begin{Bmatrix}0\\ 0\\ 0\\ 0\\ 0\\ 0\end{Bmatrix}$$

Assembling the nodal forces gives the global equilibrium equation:

$$[K]\{D\} = \{R\}$$

$$1\times 10^7\begin{bmatrix}2 & 0 & -2 & 0 & 0 & 0\\ 0 & 0 & 0 & 0 & 0 & 0\\ -2 & 0 & 2+8 & 4 & -8 & -4\\ 0 & 0 & 4 & 2 & -4 & -2\\ 0 & 0 & -8 & -4 & 8 & 4\\ 0 & 0 & -4 & -2 & 4 & 2\end{bmatrix}\begin{Bmatrix}u_1\\ v_1\\ u_2\\ v_2\\ u_3\\ v_3\end{Bmatrix} = \begin{Bmatrix}-5000\\ 0\\ -5000\\ 0\\ 0\\ 0\end{Bmatrix} + \begin{Bmatrix}0\\ 0\\ 707.1\\ -707.1\\ 0\\ 0\end{Bmatrix} + \begin{Bmatrix}R_{x1}\\ R_{y1}\\ 0\\ 0\\ R_{x3}\\ R_{y3}\end{Bmatrix}$$

Apply the boundary conditions by direct assembly/elimination

- Active degrees of freedom: u_2, v_2
- Known degrees of freedom: $u_1 = v_1 = u_3 = v_3 = 0$

Eliminate the rows and the columns that correspond to the active dof.

$$10^7\begin{bmatrix} 10 & 4 \\ 4 & 2 \end{bmatrix}\begin{Bmatrix} u_2 \\ v_2 \end{Bmatrix} = -\begin{Bmatrix} 4292.9 \\ 707.1 \end{Bmatrix}$$

$$\begin{Bmatrix} u_2 \\ v_2 \end{Bmatrix} = \begin{Bmatrix} -143.9 \\ 252.5 \end{Bmatrix} \times 10^{-6} m$$

Apply the boundary conditions by using the penalty method

$$1 \times 10^7 \begin{bmatrix} 2+k_p & 0 & -2 & 0 & 0 & 0 \\ 0 & k_p & 0 & 0 & 0 & 0 \\ -2 & 0 & 2+8 & 4 & -8 & -4 \\ 0 & 0 & 4 & 2 & -4 & -2 \\ 0 & 0 & -8 & -4 & 8+k_p & 4 \\ 0 & 0 & -4 & -2 & 4 & 2+k_p \end{bmatrix}\begin{Bmatrix} u_1 \\ v_1 \\ u_2 \\ v_2 \\ u_3 \\ v_3 \end{Bmatrix} = \begin{Bmatrix} -5000 \\ 0 \\ -4292.9 \\ -707.1 \\ 0 \\ 0 \end{Bmatrix}$$

$$\begin{Bmatrix} u_1 \\ v_1 \\ u_2 \\ v_2 \\ u_3 \\ v_3 \end{Bmatrix} = \lim_{k_p \to \infty} \begin{Bmatrix} -\dfrac{7.88x10^{-4}}{k_p} \\ 0 \\ -\dfrac{1.439x10^{-4}(5.47 + k_p)}{k_p} \\ \dfrac{2.52x10^{-4}(4084 + k_p)}{k_p} \\ -\dfrac{1.41x10^{-4}}{k_p} \\ -\dfrac{7.07x10^{-4}}{k_p} \end{Bmatrix} = \begin{Bmatrix} 0 \\ 0 \\ -1.439 \times 10^{-4} \\ 2.52 \times 10^{-4} \\ 0 \\ 0 \end{Bmatrix}$$

The solution converges toward the exact solution for different penalty stiffness k_p values as follows:

k_p	$u_1(\mu m)$	$v_1(\mu m)$	$u_2(\mu m)$	$v_2(\mu m)$	$u_3(\mu m)$	$v_3(\mu m)$
10^2	-7×10^{-6}	0	151.814	264.7	-1.4×10^{-6}	-7×10^{-7}
10^4	-7×10^{-8}	0	144.014	252.637	-1.4×10^{-8}	-7×10^{-9}
10^6	-7×10^{-10}	0	143.936	252.516	-1.4×10^{-10}	-7×10^{-11}
10^8	-7×10^{-12}	0	143.935	252.515	-1.4×10^{-12}	-7×10^{-13}

Reaction forces:

$$\{P_{reac}\} = [K]\{D\} - (\{P\} + \{P_q\})$$

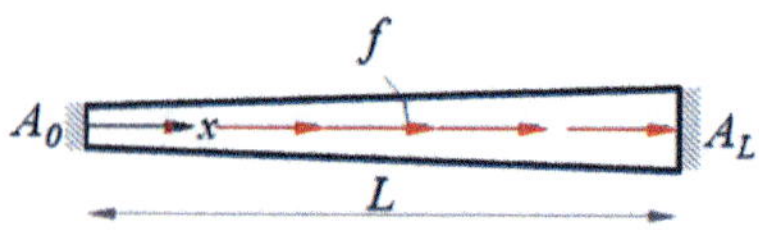

FIGURE 5.13 Analysis of deformation for $\overline{m} = -0.9$ and $\overline{f} = 10^{-4}$. a) Deflection profile computed analytically and by quadratic elements for $N = 32$. b) Comparison of strain variation along the bar for $N_E = 32$. Blue-curve represents quadratic interpolation, red-curve represents linear interpolation. Tapered bar subjected to distributed force f.

$$\begin{Bmatrix} R_{x1} \\ R_{y1} \\ 0 \\ 0 \\ R_{x3} \\ R_{y3} \end{Bmatrix} = 1\times 10^{7} \begin{bmatrix} 2 & 0 & -2 & 0 & 0 & 0 \\ 0 & 0 & 0 & 0 & 0 & 0 \\ -2 & 0 & 10 & 4 & -8 & -4 \\ 0 & 0 & 4 & 2 & -4 & -2 \\ 0 & 0 & -8 & -4 & 8 & 4 \\ 0 & 0 & -4 & -2 & 4 & 2 \end{bmatrix} \begin{Bmatrix} 0 \\ 0 \\ -143.9\times 10^{-6} \\ 252.5\times 10^{-6} \\ 0 \\ 0 \end{Bmatrix} - \begin{Bmatrix} -5000 \\ 0 \\ -4292.9 \\ -707.1 \\ 0 \\ 0 \end{Bmatrix}$$

$$\{P_{reac}\} = \{R_{x1}\ R_{y1}\ 0\ 0\ R_{x3}\ R_{y3}\}^T = \{7878.7\ \ 0\ \ 0\ \ 0\ \ 1414.2\ \ 707.1\}^T\ \text{N}$$

5.9 Error and convergence in finite element analysis

The results of an analysis carried out by using the finite element method depend on *i*) the physics of the problem, *ii*) the type of elements, and *iii*) the number of elements used in the solution. By *type of element* we mean the polynomial approximation used to generate the interpolation functions.

In this section, we demonstrate the effects of element type and number of elements by using a problem for which we have an analytical solution. This enables us to monitor the error in the solution obtained by the finite element method.

Consider the problem of finding the longitudinal deformation $u(x)$ of a tapered bar that is fixed on both ends, subjected to a distributed force f. By assuming that the loading is one-dimensional, the inertial effects are negligible and the material obeys Hooke's law Eq. (2.91) can be put into the form that was derived in Eq. (1.11).

The corresponding boundary value problem is expressed as follows:

$$\frac{d}{dx}\left(EA\frac{du}{dx}\right) = -f \quad \text{in} \quad 0 < x < L \tag{5.125}$$

$$u(0) = 0 \quad \text{and} \quad u(L) = 0 \tag{5.126}$$

The cross-sectional area of the bar is given as follows:

$$A(x) = A_0 + mx \tag{5.127}$$

with $m = (A_L - A_0)/L, A_0$ and A_L are the cross-sectional areas of the bar at its ends (Fig. 5.13).

The exact solution

For constant f and E, the solution to this problem is given as follows:

$$u = \frac{f}{Em} \frac{(\log A_0 - \log (A_0 + mx))}{(\log)A_0 - \log (A_0 + mL)} (L - x) \tag{5.128}$$

For constant cross-sectional area $A = A_0$ the exact solution becomes,

$$u = \frac{1}{2A_0} \frac{f}{E} x(L - x) \tag{5.129}$$

The finite element solution

The variation of the dependent variable $u(x)$ and the strain $\varepsilon(x)$ are approximated in the element by the following interpolations:

$$u(x') = [N]\{d\} \text{ and } \varepsilon(x') = \frac{du}{dx'} = [B]\{d\} \tag{5.130}$$

The element equilibrium equation is expressed, in its general form, as follows:

$$\left(\int_0^{L^{(e)}} EA[B]^T[B]dx' \right) \{d\} = \int_0^L [N]^T f dx' + \{P\} \tag{5.131}$$

Note that the external nodal force vector $\{P\} = 0$ in this problem, and we omit the internal nodal force vector $\{r_F\}$ to simplify notation.

Linear interpolation: The shape functions for the linear interpolation are given in Eq. (4.41),

$$N_1 = 1 - \frac{x'}{L^{(e)}}, \quad N_2 = \frac{x'}{L^{(e)}} \tag{5.132}$$

This element has two nodes located at its ends and one degree of freedom per node. Therefore, it has two degrees of freedom,

$$\{d\} = \{u_1 \ \ u_2\}^T \tag{5.133}$$

Note that strain in this element is constant and has the following value,

$$\varepsilon = \frac{1}{L^{(e)}} [-1 \ \ 1] \begin{Bmatrix} u_1 \\ u_2 \end{Bmatrix} \tag{5.134}$$

The element equilibrium equation for this element is found as follows:

$$\left(\frac{(A_1 + A_2)E}{2L^{(e)}} \right)^{(e)} \begin{bmatrix} 1 & -1 \\ -1 & 1 \end{bmatrix} \begin{Bmatrix} u_1 \\ u_2 \end{Bmatrix} = \frac{1}{2} q L^{(e)} \begin{Bmatrix} 1 \\ 1 \end{Bmatrix} \tag{5.135}$$

where A_1 and A_2 are the areas on the nodes.

Quadratic interpolation: The shape functions for the quadratic interpolation are given in Eq. (4.42). By using $r = \frac{1}{2}$, the following shape functions are obtained,

$$N_1=\left(1-2\frac{x'}{L^{(e)}}\right)\left(1-\frac{x'}{L^{(e)}}\right),\ N_2=4\frac{x'}{L^{(e)}}\left(1-\frac{x'}{L^{(e)}}\right),\ N_3=-\frac{x'}{L^{(e)}}\left(1-2\frac{x'}{L^{(e)}}\right) \tag{5.136}$$

This element has three degrees of freedom,

$$\{d\}=\{u_1\ u_2\ u_3\}^T \tag{5.137}$$

Note that strain in this element varies linearly as follows:

$$\varepsilon(x')=\frac{1}{L^{(e)}}\left[\left(-3+4\frac{x'}{L^{(e)}}\right)\ 4\left(1-2\frac{x'}{L^{(e)}}\right)\ -\left(1-4\frac{x'}{L^{(e)}}\right)\right]\begin{Bmatrix}u_1\\u_2\\u_3\end{Bmatrix} \tag{5.138}$$

The element equilibrium equation for this element is found as follows:

$$\frac{1}{6}\left(\frac{E}{L}\right)^{(e)}\begin{bmatrix}11A_1+3A_2 & -4(3A_1+A_2) & A_1+A_2\\ -4(3A_1+A_2) & 16(A_1+A_2) & -4(A_1+3A_2)\\ A_1+A_2 & -4(A_1+3A_2) & 3A_1+11A_2\end{bmatrix}\begin{Bmatrix}u_1\\u_2\\u_3\end{Bmatrix}=\frac{1}{6}qL^{(e)}\begin{Bmatrix}1\\4\\1\end{Bmatrix} \tag{5.139}$$

In case $A_1=A_3$ this equation becomes:

$$\frac{1}{3}\left(\frac{AE}{L}\right)^{(e)}\begin{bmatrix}7 & -8 & 1\\ -8 & 16 & -8\\ 1 & -8 & 7\end{bmatrix}\begin{Bmatrix}u_1\\u_2\\u_3\end{Bmatrix}=\frac{1}{6}qL^{(e)}\begin{Bmatrix}1\\4\\1\end{Bmatrix} \tag{5.140}$$

Normalization: The problem defined in Eq. (5.125) has five independent variables *E, A, m, L*, and *f*. By using the following normalized parameters,

$$\overline{x}=\frac{x}{L},\ \overline{u}=\frac{u}{L},\ \overline{m}=\frac{A_L}{A_0}-1 \text{ and } \overline{f}=\frac{fL}{EA_o}$$

the number of independent variables to describe this problem is thus reduced to $\overline{m}$ and $\overline{f}$.

The BVP given in Eq. (5.125) becomes,

$$\frac{d}{dx}\left[(1+\overline{mx})\frac{d\overline{u}}{d\overline{x}}\right]=-\overline{f}\ \text{ in } 0<x<1 \tag{5.141}$$

$$\overline{u}(0)=0 \text{ and } \overline{u}(1)=0 \tag{5.142}$$

The exact solution

The solution to this problem is given as follows:

$$\overline{u}=\frac{\overline{f}}{\overline{m}}\left(\frac{\log\ (1+\overline{mx})}{\log\ (1+\overline{m})}-\overline{x}\right) \tag{5.143}$$

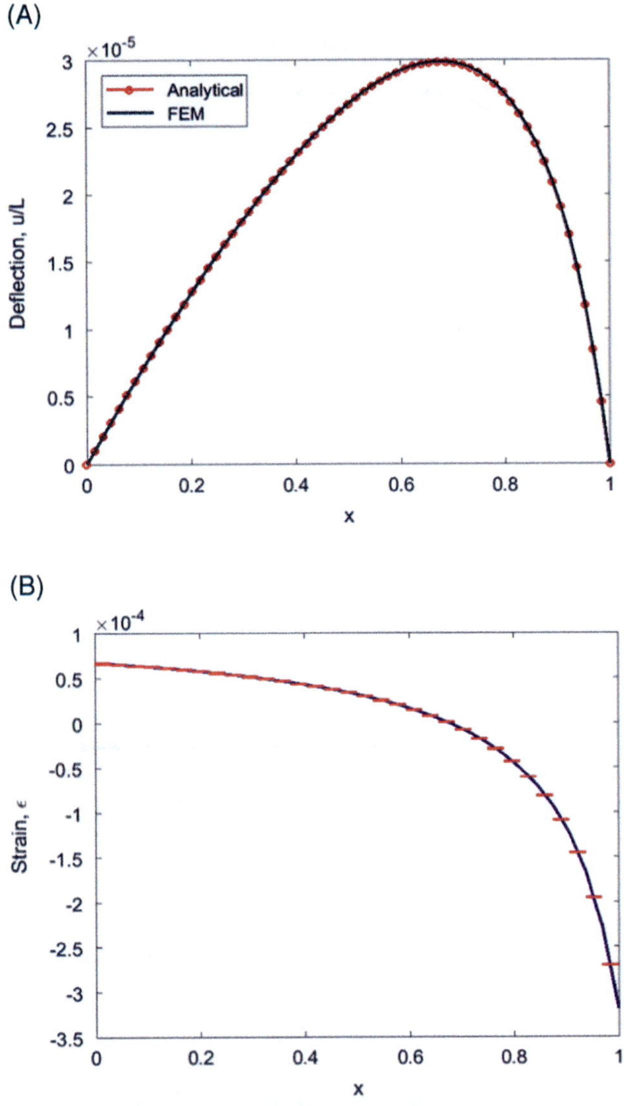

FIGURE 5.14 For the tapered bar subjected to distributed force f A) Deflection profile computed analytically and by quadratic elements for $N = 32$; B) Comparison of strain variation along the bar for $N_E = 32$. Blue-curve represents quadratic interpolation, red-curve represents linear interpolation.

For constant cross-sectional area A_0, the exact solution becomes,

$$\overline{u} = \frac{1}{2}\overline{f}(1 - \overline{x})\overline{x} \tag{5.144}$$

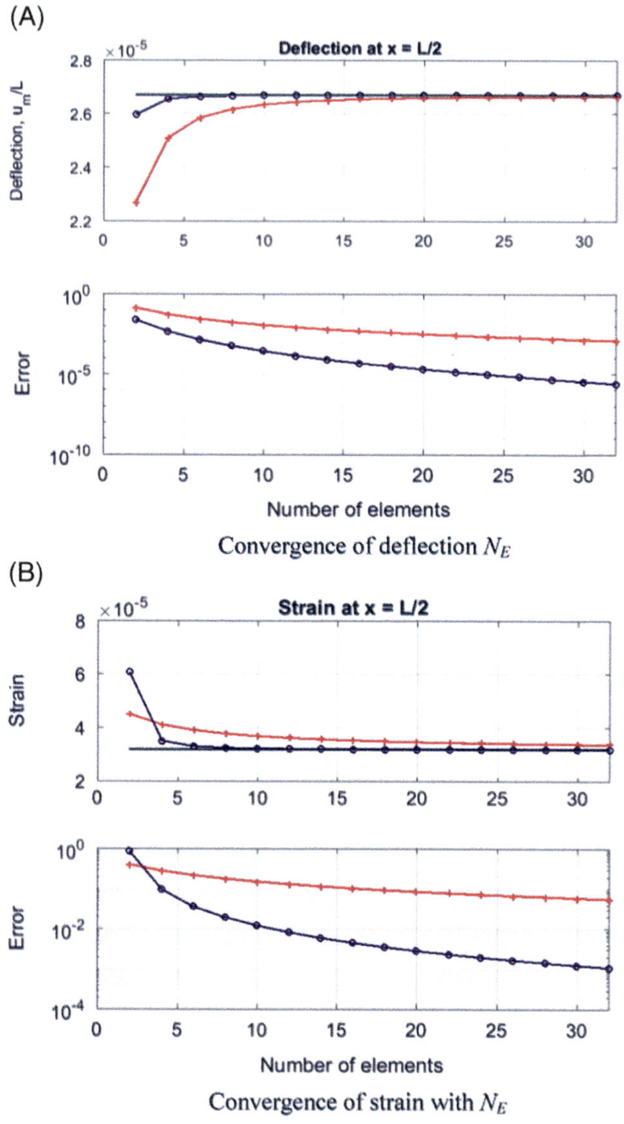

FIGURE 5.15 Analysis of deformation for $\bar{m} = -0.9$ and $\bar{f} = 10^{-4}$. Convergence of the mid-point deflection (A) and strain (B) with number of elements, N_E. Blue-curve represents quadratic interpolation, red-curve represents linear interpolation.

Nondimensional analysis can also be carried out by using the finite element method. In order to accomplish this goal no modifications to the element equations should be necessary. The following parameters are entered in to the program in order to carry out a nondimensional analysis of the same problem by the finite element method:

$$L = 1;\ E = 1;\ A_0 = 1;\ A_L \rightarrow \frac{A_L}{A_0} - 1;\ f \rightarrow \frac{fL}{EA_o} \tag{5.145}$$

Effects of order of interpolation and number of elements on convergence: Comparison of analytical and finite element solutions is presented in Fig. 5.14 for $\overline{m} = -0.9$ and $\overline{f} = 10^{-4}$.

Deflection of the bar under the given loading is shown in Fig. 5.14A. Due to the taper in the bar, the thinner end of the bar deflects more. In case the bar cross-section is uniform the solution is symmetric with respect to the center of the bar, but this solution is not shown here. For the case of 32 elements the differences between the solutions are not distinguishable in Fig. 5.14A.

Strain variation along the bar is presented for $N = 32$ for both linear and quadratic elements in Fig. 5.14B. As shown in Eq. (5.134) the strain is constant along each linear interpolation element. Interestingly, this constant strain solution approaches the exact solution as the number of elements is increased. The solution using the quadratic elements is much closer to the analytical solution despite the fact that the strain variation in the quadratic elements is linear, as shown in Eq. (5.138).

Convergence of deflection with number of elements is investigated for linear and quadratic elements in Fig. 5.15A. The error between the analytical and numerical solutions is reported by,

$$e = \left| \frac{u_{exact} - u_{FEM}}{u_{exact}} \right| \tag{5.146}$$

By using 8 elements, the solution with quadratic elements achieves an error level less than 10^{-3}, whereas at least 32 elements are required to achieve the same level of error reduction for the linear elements. Rate of convergence of deflection is much better for the quadratic elements.

Convergence of strain as function of number of elements is presented in Fig. 5.15B. The rate of convergence for strain is lower than that of the deflection, as expected. The secondary variable converges slower than the primary variable u. The strain-error is 0.1 with 32 of constant-strain linear elements where as it is smaller than 10^{-2} with the quadratic elements.

Problems

5.1 Element shown in Fig. 5.16 has a total of four nodes.

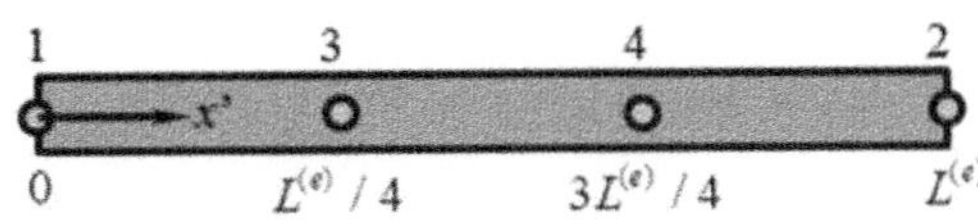

FIGURE 5.16 Problem 5.1.

a) Find the four $C^{(0)}$ shape functions to interpolate the variation of $u(x)$ along this element by using Lagrange interpolation functions. $u(x) = [N]\{d\}$with $[N] = [N_1 \ N_2 \ N_3 \ N_4]$ and $\{d\} = \{u_1 \ u_2 \ u_3 \ u_4\}^T$

b) Plot the strain in this element in the range $[0, \ L]$ for $\{d\} = \{0 \ 2.2 \times 10^{-6} \ 7.3 \times 10^{-6} - 1.1 \times 10^{-6}\}^T$ and $L = 3 \times 10^{-3}$m

c) Find the displacement and the strain in the middle of this element.

5.2 The value of a function $\varphi(x)$ is given at four positions $\{x\} = \{1 \ 3 \ 5 \ 8\}^T$ as $\{\varphi\} = \{2 \ 2 \ 2 \ 6\}^T$

a) Use Lagrange's formula to obtain an interpolating polynomial to approximate $\varphi(x)$.

b) What values of φ does the approximate formula predict at $x = 0$, at $x = 2$, at $x = 4$, and $x = 7$?

5.3 A three-node uniform bar element (with only axial deformation capability) of length $L^{(e)}$ has a node at each end, and a node located at $x = r.L^{(e)}$ $(0 < r < 1)$ as shown in Fig. 5.17. If this element is subjected to a concentrated force P_0 applied in the $+x$–direction at $x = L^{(e)}/4$ (which is not shown in the figure) determine:

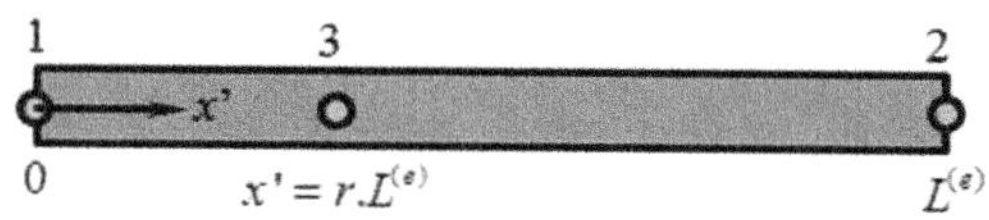

FIGURE 5.17 Problems 5.3 and 5.4.

a) The element stiffness matrix that operates on the nodal dof u_1, u_2 and u_3, and

b) The consistent nodal forces at nodes 1, 2 and 3.

Note: Use an appropriate $C^{(0)}$ continuous element formulation.

5.4 A three-node uniform bar element (with only axial deformation capability) of length $L^{(e)}$ has a node at each end, and a node located at $x = r.L^{(e)}$ $(0 < r < 1)$ as shown in Fig. 5.17. If this element is subjected to a constant distributed force q_0 applied in the $+x$–direction (which is not shown in the figure) determine the element stiffness matrix that operates on the nodal dof u_1, u_2 and u_3, and the consistent nodal forces at nodes 1, 2 and 3.

Note: Use an appropriate $C^{(0)}$ continuous element formulation.

5.5 A two-node bar element of length L has a linearly varying cross-sectional area $A(x) = b\left[\frac{h_L - h_0}{L}x + h_0\right]$ where b is the depth and h_0 and h_L are the heights of the cross-section of the element. The element is subjected to (1) concentrated forces at its ends, F_1 and F_2; (2) a distributed force q_0 of constant magnitude applied in the $+x$–direction. The elastic modulus E of the element is given and constant.

Use the *minimum total potential energy principle* to derive the finite element equations for this element. Use an appropriate $C^{(0)}$ continuous element formulation.

5.6 The bar shown in Fig. 5.18 is subjected to a distributed load q of constant magnitude between $x_1(=L/3)$ and $x_2(=2L/3)$ as shown. Find the consistent nodal force vector for the following conditions:

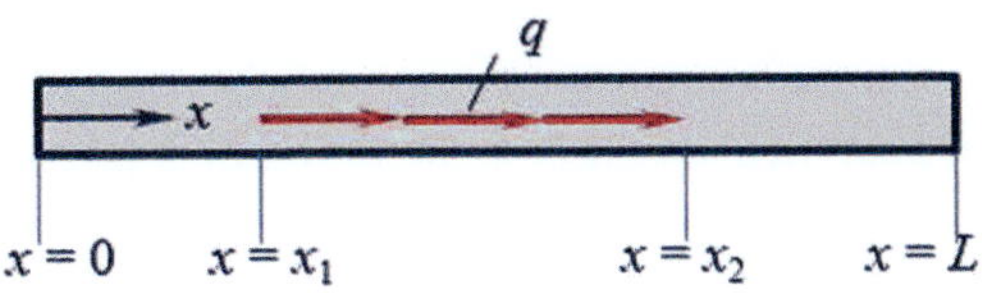

FIGURE 5.18 Problem 5.6.

a) Assume the bar is modeled by a two node element with nodes at its ends.

b) Assume the bar is modeled by a three node element where the third node is located at $x = L/2$.

5.7 Determine the axial stress distribution in the spinning bar shown in Fig. 5.19 by using the finite element method. The bar is rotating at 500 revolutions per minute. The problem can be treated as one-dimensional with the following differential equation and boundary conditions:

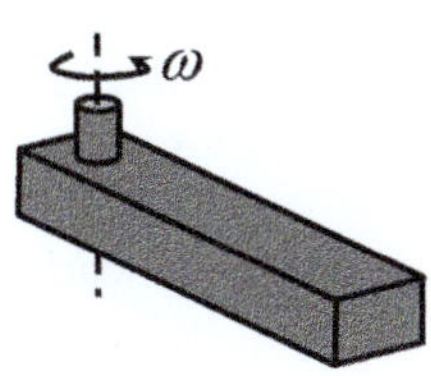

FIGURE 5.19 Problem 5.7.

$$\frac{d}{dx}\left(EA\frac{du}{dx}\right) + \left(\rho A\omega^2\right)x = 0 \text{ for } 0 < x < L,$$

BCs : $u(0) = 0$, and $EA\frac{du}{dx}|_L = 0$ where x is the coordinate along the bar, $u(x)$ is the axial displacement,

L is the length of the bar,

E is the Young's modulus,

A is the cross-sectional area

ρ is the mass density and

ω is the angular velocity in rad/s.

Use the following numerical data: $L = 0.8$m; $E = 200$ GPa; $A = 250$ mm^2; $\rho = 7850$ kg/m^3.

a) Use two $C^{(0)}$ elements of equal lengths.

b) Use one C(0) element with 3 nodes.

c) Compare the above solutions with the analytical prediction at $x = 0$, L/2 and L.

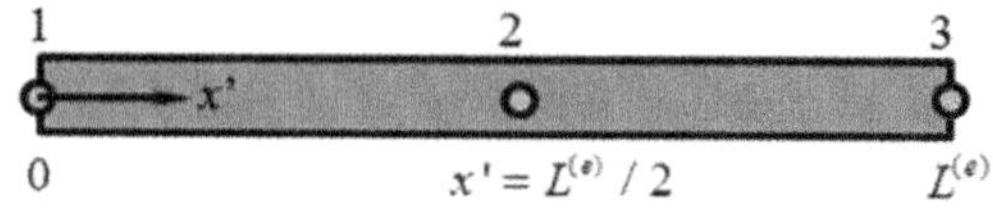

FIGURE 5.20 Problem 5.9.

An exact analytical solution to the problem is given as follows: $\sigma = \frac{1}{2}\rho\omega^2(L^2 - x^2)$. where the axial stress is $\sigma = E\varepsilon$, the axial strain is $\varepsilon = du/dx$.

5.8 Consider a two-node bar element with uniform cross-sectional area A, constant elastic modulus E and length L. This bar element is to have not only the axial displacements but also the strain at each node as nodal dof, such that $\{d\} = \{u_1 \ \ \varepsilon_1 \ \ u_2 \ \ \varepsilon_2\}^T$. Using the minimum-total potential energy principle, and assuming that the element behaves in a linear elastic manner, find the corresponding element stiffness matrix.

5.9 A bar element with uniform cross-sectional area A, constant elastic modulus E and length $L^{(e)}$, has a node at each end, and a node in the middle, as shown in Fig. 5.20.

a) Determine the interpolation function and the shape functions, related to these three nodal displacements $u_1{}'$, u'_2, and u'_3.

b) Using the minimum-total potential energy principle, and assuming that the element behaves in a linear elastic manner, find the element stiffness matrix $[k]^{(e)}$ that operates on nodal degrees of freedom $u_1{}'$, u'_2, and u'_3.

5.10 Consider the bar element with a linearly varying cross-sectional area shown in Fig. 5.21,

FIGURE 5.21 Problem 5.10.

$A(x) = A_1 + \frac{A_2 - A_1}{L}x$

Subjected to a linearly distributed external traction

$q(x) = q_1 + \frac{q_2 - q_1}{L}x$ where,

A_1, A_2: cross-sectional areas at node-1 and node-2.

q_1, q_2: magnitudes of the distributed force at node-1 and node-2.

L: the *length* of the element.

Using the minimum-total potential energy principle, and assuming that the element behaves in a linear elastic manner, find the element stiffness matrix $[k]^{(e)}$ and the external load vector $\{r\}^{(e)}$ for this element.

5.11 A two-node, bar element with axial deformation capability is considered. The equilibrium of this element in its local coordinate system is given by,

$[k']\{d'\} = \{r'\}$ with $\{d'\} = \{u_1' \quad u_2'\}^T$ and $\{r\} = \{F'_{x'1} \quad F_{x'2}\}^T$ and $[k']$ is the local stiffness matrix. When this element is oriented with respect to the (global) x-axis by an angle α, the equilibrium in (x,y) coordinates is given by,

$[k]\{d\} = \{r\}$ with $\{d\} = \{u_1\ v_1\ u_2\ v_2\}^T$, $\{r\} = \{F_{x1}\ F_{y1}\ F_{x2}\ F_{y2}\}^T$ and $[k] = [T]^T[k'][T]$

a) Write a Matlab script to find the stiffness matrix $[k]$ in the (x,y) coordinate system, for the following parameters: $\alpha = 15$ deg.; $A = 0.25\ \text{m}^2$; $E = 90$ GPa; $L = 0.15$ m; $I = 1.25 \times 10{-}5$ m4, $\{r'\} = \{100 \quad 120\}^T$N.

b) Extend the script in part-a) to have a for loop that changes the element length and orientation as follows (keep the other parameters as given in part-a)),

L	α (deg.)
0.1	20
0.25	33
1.1	120
0.39	205

For both a) and b) report the $[k]$ matrix and the $\{r\}$ vector. Submit the printout of your script

Note: The expected output is as follows:

```
For L = 0.4 and alpha = 23.4 deg.

Matrix [k] and vector {r} in local coordinate system
```

$$\begin{matrix} 0.2092e+003 & -0.1112e+003 \\ -0.1112e+003 & 0.2092e+003 \end{matrix}$$

$$\begin{matrix} 7.0453e+000 \\ -2.4647e+000 \end{matrix}$$

```
Matrix [k] and vector {r} in global coordinate system
```

$$\begin{matrix} 2.345e+003 & 2.345e+003 & -1.123e+003 & -0.321e+003 \\ 2.345e+003 & 3.345e+003 & 1.450e+003 & -2.345e+002 \\ -1.123e+003 & 1.450e+003 & 2.345e+003 & -0.923e+003 \\ -0.321e+003 & -2.345e+002 & -0.923e+003 & 3.345e+003 \end{matrix}$$

$$\begin{matrix} 0.234e+002 \\ 9.584e+002 \\ -4.374e+002 \\ -4.123e+002 \end{matrix}$$

Answer the following questions for Problems 5.12 and 5.13:

a) Prepare the element connectivity table

b) List the active degrees of freedom

c) Calculate the element equilibrium equations for each element

d) Assemble the global equilibrium equation for the active degrees of freedom, directly

e) Solve the resulting system of equations for the active degrees of freedom

f) Find the strain in each member (element).

5.12 Consider the four member truss shown in Fig. 5.22. The members of the truss have a common cross-sectional area, 10 cm^2, and are made of the same material with elastic modulus, $E = 110$ GPa. This truss is subjected to the following loading: $F_{x2} = 500$ N, $F_{y2} = 500$ N, $q_3 = 1$ kN/m.

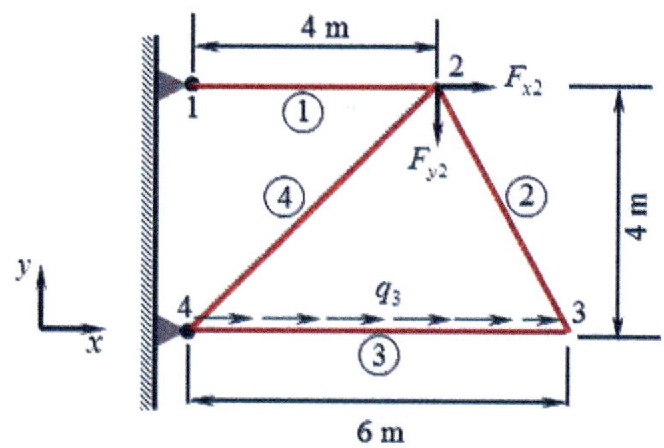

FIGURE 5.22 Problem 5.12.

5.13 Consider the five member truss shown in Fig. 5.23. The members of the truss have a common cross-sectional area, 9 cm^2, and are made of the same material with elastic modulus, $E = 20$ GPa. This truss is subjected to the following loading: $F_2 = 1$ kN and $P_4 = P_5 = 500$ N. P_4 and P_5 are loaded at $x' = L_{13}/4$ and $x' = L_{34}/4$, respectively. The lengths of the members are given as follows: $L_{13} = L_{34} = 6$ m, $L_{23} = 3$ m.

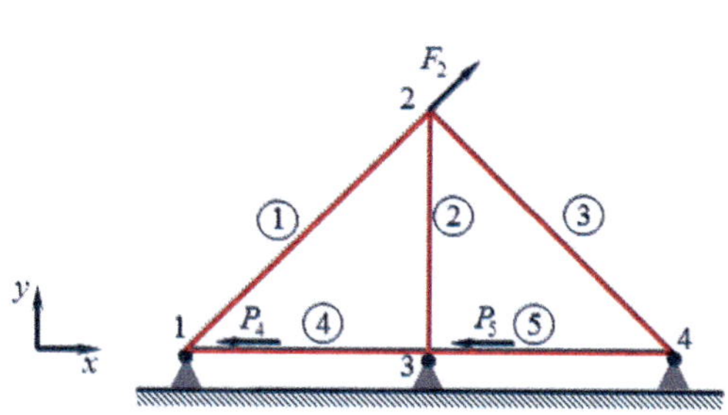

FIGURE 5.23 Problem 5.13.

5.14 Displacements of the three-member truss shown in Fig. 5.24 are confined

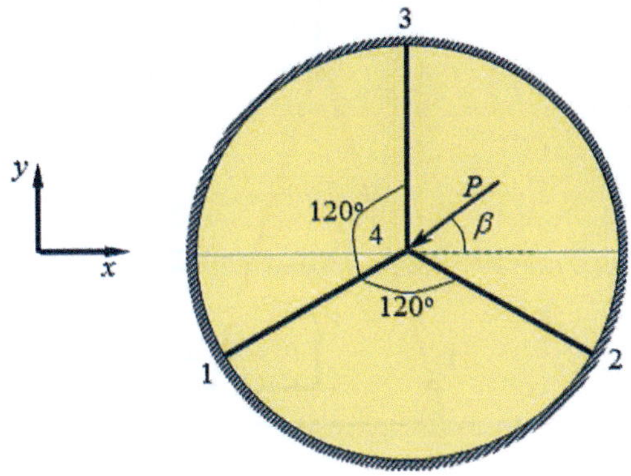

FIGURE 5.24 Problem 5.14.

to the plane of the figure, and points 1, 2 and 3 are fixed to the stationary rim. All members have the same *A*, *E*, and *L*.

a) Obtain the 2×2 stiffness matrix that operates on the horizontal and vertical degrees of freedom of the central node.

b) Obtain the corresponding global force vector.

c) Solve for the displacements and for axial stress in member (2--4), when the truss is loaded as shown.

5.15 Displacements of the two-member truss are confined to the plane of the figure (Fig. 5.25). Obtain the stiffness matrix that operates on the horizontal and vertical displacement components of the top node, as well as the corresponding force vector. Do not solve for the displacements.

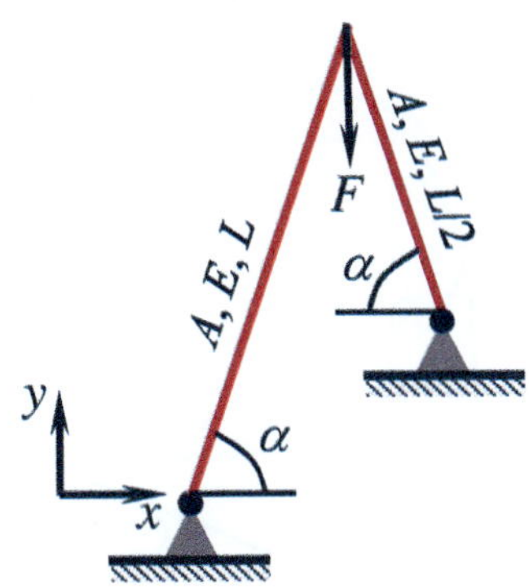

FIGURE 5.25 Problem 5.15.

5.16 Consider the two member truss shown in Fig. 5.26, with the following properties:

$E_1 = 20\times10^6$ psi, $A_1 = 1/2$ in^2,

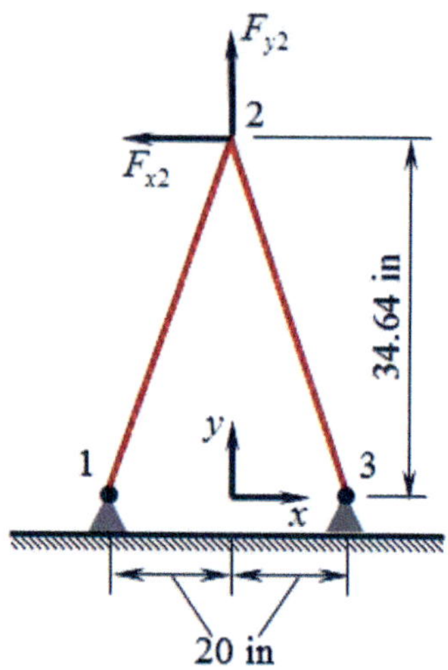

FIGURE 5.26 Problem 5.16.

$E_2 = 30\times10^6$ psi, $A_2 = 1$ in^2,
$P_1 = 200$ lbs, $P_2 = 100$ lbs.
Find the deflection of point 3 and the stresses in each truss member.

5.17 Show that if we use polynomial based, $C^{(0)}$ shape functions, the resulting element stiffness matrices for 2D two-dimensional planar elements will be symmetric.

5.18 For the structure shown in Fig. 5.27,

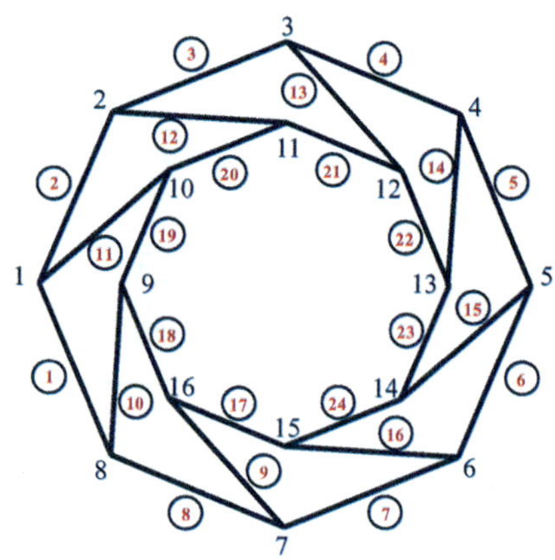

FIGURE 5.27 Problem 5.18.

a) Generate an element connectivity table that determines the following information:

Element no, node-*i*, node-*j*, *E*, *I*, *A*
Report this table by hand writing.

Elements 1-8 & 17-24: $E = 200\text{GPa}$, $I = 10^{-2}\text{m}^4$; $A = 10^{-3}\text{m}^2$.
Elements 9-16: $E = 100\text{GPa}$, $I = 5 \times 10^{-4}\text{m}^4$; $A = 10^{-5}\ \text{m}^2$.

b) Create a text file (ect.txt) that contains this element connectivity table Report by printing out this file.

c) Read this table in to Matlab by using the following command,

```
inputs_matrix = load('ect.txt);
```

Read the variables from respective columns of the `input_matrix` into individual vectors and print your results as shown below. Report your printout.

```
for e = 1:Total_El_No
    ECT(e,1) = ... ;
    ECT(e,2) = ... ;
    E(e) = ... ;
    I(e) = ... ;
    A(e) = ... ;
    fprintf('%3i %3i %12.4e %12.4e %12.4e \n', ECT(e,1),...
    ECT(e,2), E(e), I(e), A(e));
end
```

Reference

[1] R.D. Cook, D.S. Malkus, M.E. Plesha, R.J. Witt, Concepts and applications of finite element analysis, 4th ed., John Wiley & Sons, 2002.

Chapter 6

Euler–Bernoulli beam element

6.1 Introduction

In this chapter, the finite element form of the equation of equilibrium for an *Euler–Bernoulli beam* is developed. The Euler–Bernoulli beam theory is introduced in Section 2.3. In this chapter, we develop the element matrix and the consistent load vector that represents the compliance of an Euler–Bernoulli beam. The assembly, coordinate transformation, application of the boundary conditions, and the postprocessing are also described for this element. The details are similar to the development presented in Chapter 5.

6.2 C^1-Continuous interpolation function

As in Chapter 5, the element equilibrium equation is developed with respect to an element-based local coordinate system (x', y'). The independent variables are the transverse deflection component v' and the slope $\theta = dv'/dx$ of the deflected beam. Therefore, continuity of both v' and θ is required across the element boundaries.

An interpolation function that provides the continuity of the dependent variable and its derivative across element boundaries is said to be $\mathbf{C^1}$***-continuous***. In this case, a two-node element that spans the range $[0, L^{(e)}]$ would have both,

$$\left(v_1{}', \theta_1\right) \text{ and } \left(v_2{}', \theta_2\right) \tag{6.1}$$

the *displacement* and the *slope* to be specified on the boundaries. The interpolation involves a third order polynomial, as given in Eq. (4.12), which has four unknown coefficients $a_0 - a_3$. These unknown coefficients will be replaced by the displacement and slope at the nodes,

$$\begin{aligned} v' &= \sum_{j=0}^{N=3} a_j \left(x'\right)^j = a_0 + a_1 x' + a_2 \left(x'\right)^2 + a_3 \left(x'\right)^3 \\ \frac{dv'}{dx'} &= \sum_{j=0}^{N=3} a_j \frac{d}{dx} \left(x'\right)^j = a_1 + 2a_2 x' + 3a_3 \left(x'\right)^2 \end{aligned} \tag{6.2}$$

Finite Element Method: Physics and Solution Methods. DOI: https://doi.org/10.1016/B978-0-12-821127-4.00013-X
Copyright © 2022 Elsevier Inc. All rights reserved.

Evaluating this polynomial at the nodes, $x = 0$ and $x = L^{(e)}$, we find,

$$\begin{Bmatrix} v'_1 \\ \theta_1 \\ v'_2 \\ \theta_2 \end{Bmatrix} = \underbrace{\begin{bmatrix} 1 & 0 & 0 & 0 \\ 0 & 1 & 0 & 0 \\ 1 & L^{(e)} & L^{(e)2} & L^{(e)3} \\ 0 & 1 & 2L^{(e)} & 3L^{(e)2} \end{bmatrix}}_{A} \begin{Bmatrix} a_0 \\ a_1 \\ a_2 \\ a_3 \end{Bmatrix} \tag{6.3}$$

The C^1-continuous shape functions are found, as described in Section 4.3, as follows:

$$[N] = [X][A]^{-1} \tag{6.4}$$

with $[X] = [1 \quad x' \quad x'^2 \quad x'^3]$. This gives,

$$N_1^{(b)}(x') = 1 - 3\left(\frac{x'}{L^{(e)}}\right)^2 + 2\left(\frac{x'}{L^{(e)}}\right)^3 \text{(a)},\; N_2^b(x') = x'\left[1 - 2\left(\frac{x'}{L^{(e)}}\right) + \left(\frac{x'}{L^{(e)}}\right)^2\right] \text{(b)}$$

$$N_3^{(b)}(x') = 3\left(\frac{x'}{L^{(e)}}\right)^2 - 2\left(\frac{x'}{L^{(e)}}\right)^3 \text{(c)},\; N_4^b(x') = x'\left[-\left(\frac{x'}{L^{(e)}}\right) + \left(\frac{x'}{L^{(e)}}\right)^2\right] \text{(d)} \tag{6.5}$$

Note that the shape functions given in Eq. (6.5) have the following properties: at node-1 ($x' = 0$):

$$N_1^{(b)}(0) = 1 \text{ and } N_2^{(b)}(0) = N_3^{(b)}(0) = N_4^{(b)}(0) = 0$$

$$\left.\frac{dN_2^{(b)}}{dx'}\right|_{x'=0} = 1, \text{ and } \left.\frac{dN_1^{(b)}}{dx'}\right|_{x'=0} = \left.\frac{dN_3^{(b)}}{dx'}\right|_{x'=0} = \left.\frac{dN_4^{(b)}}{dx'}\right|_{x'=0} = 0 \tag{6.6}$$

at node-2 ($x' = L^{(e)}$):

$$N_3^{(b)}(L^{(e)}) = 1 \text{ and } N_1^{(b)}(L^{(e)}) = N_2^{(b)}(L^{(e)}) = N_4^{(b)}(L^{(e)}) = 0$$

$$\left.\frac{dN_4^{(b)}}{dx'}\right|_{x'=L(e)} = 1, \text{ and } \left.\frac{dN_1^{(b)}}{dx'}\right|_{x'-L^{(e)}} = \left.\frac{dN_2^{(b)}}{dx'}\right|_{x'=L^{(e)}} = \left.\frac{dN_3^{(b)}}{dx'}\right|_{x'=L^{(e)}} = 0 \tag{6.7}$$

This property ensures continuity of displacements and slopes across element boundaries. Note that the variable $v'(x')$ is now interpolated as follows:

$$v'(x') = N_1^{(b)}v'_1 + N_2^{(b)}\theta_1 + N_3^{(b)}v'^2 + N_4^{(b)}\theta_2 \tag{6.8}$$

One can easily show, by using Eqs. (6.6–6.8) that,

$$v'(0) = v'_1 \text{ and } \frac{dv'}{dx'}(0) = \theta_1$$

$$v'(L^{(e)}) = v'_2 \text{ and } \frac{dv'}{dx'}(L^{(e)}) = \theta_2 \tag{6.9}$$

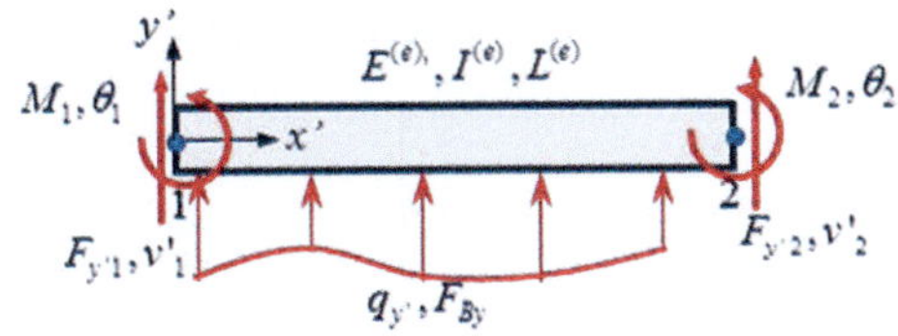

FIGURE 6.1 Definition of a beam element. The directions of the arrows show the positive sign convention for the associated variables, including $q_{y'}(x')$ and $F_{By'}(x')$.

6.3 Element equilibrium equation

6.3.1 Problem definition

Consider the beam element shown in Fig. 6.1 subjected to following external loads,

$q_{y'}(x')$: distributed force per unit length $[F/L]$
$F_{By'}(x')$: y' component of the body force $[F/L^3]$
$F_{y'1}, F_{y'2}$: concentrated forces acting on the nodes -1 and -2 $[F]$
M_1, M_2: concentrated moments acting on the nodes-1 and -2 $[FL]$

The directions of the forces and moments, shown in Fig. 6.1, are considered positive in the development of the element equations. The structural properties of this element are set by,

E: Elastic modulus $[F/L^2]$
I: Second moment of area $[L^4]$
$L^{(e)}$: Length

The degree of freedom vector $\{d'\}$ and the nodal force vector $\{P'\}$, due to the external forces and moments acting on nodes-1 and -2, are expressed in vector form,

$$\begin{aligned} \{d_b'\}^{(e)} &= \{v_1 \quad \theta_1 \quad v_2 \quad \theta_2\}^T \quad \text{(a)} \\ \{r_F'\}^{(e)} &= \{F_{y'1} \quad M_1 \quad F_{y'2} \quad M_2\}^T \quad \text{(b)} \end{aligned} \tag{6.10}$$

We expect the element equilibrium equations to be of the following matrix form,

$$[k_b']^{(e)}\{d_b'\}^{(e)} = \{r_b'\}^{(e)} \tag{6.11}$$

where $[k_b']^{(e)}$ is a 4×4 stiffness matrix, and $\{r_b'\}^{(e)}$ represents the nodal forces acting on the element including $\{r_F'\}^{(e)}$ and the consistent forces and moments due to distributed forces acting on the element. The superscript $^{(e)}$ indicates that the variables belong to a particular element. The subscript $_b$ indicates that this element is capable of deforming in bending mode.

6.3.2 Weak form of the boundary value problem

The problem of finding the transverse deflection v' of the element given in Fig. 6.1, can be posed as a boundary value problem. The equilibrium equation of a beam subjected to transverse distributed force $f_{y'}$ according to the Euler–Bernoulli beam theory is presented in Eq. (2.155) as follows:

$$\frac{d^2}{dx'^2}\left(EI\frac{d^2v'}{dx'^2}\right) = f_{y'} \text{ in } 0 < x' < L^{(e)} \tag{6.12}$$

In general, the distributed force,

$$f_{y'} = q_{y'} + F_{By'}A \tag{6.13}$$

is the sum of the y' components of the surface and body force vectors, both per unit length.

In order to find the finite element form, we consider the case where the boundaries of the element are subjected to bending moment and shear force resultants. This allows for continuity of the internal forces and moments across element boundaries. The boundary conditions are specified as follows:

$$\begin{aligned}
&\left(EI\frac{d^2v'}{dx'^2}\right)\bigg|_{x'=0} = -M_1 \text{ and } \frac{d}{dx'}\left(EI\frac{d^2v'}{dx'^2}\right)\bigg|_{x'=0} = F_{y'1} \quad \text{(a,b)}\\
&\left(EI\frac{d^2v'}{dx'^2}\right)\bigg|_{x'=L^{(e)}} = M_2 \text{ and } \frac{d}{dx'}\left(EI\frac{d^2v'}{dx'^2}\right)\bigg|_{x'=L^{(e)}} = -F_{y'2} \quad \text{(c,d)}
\end{aligned} \tag{6.14}$$

The signs of the moments M_1 and M_2, and the shear force resultants $F_{y'1}$ and $F_{y'2}$ that are acting on the element (Fig. 6.1) are determined based on the sign convention introduced in Section 2.3 (Fig. 2.19A).

The weighted residual integral for this boundary value problem is discussed in Example 3.7 of Chapter 3, and it is given as follows:

$$\int_0^{L^{(e)}} EI\frac{d^2w}{dx'^2}\frac{d^2v'}{dx'^2}dx' - \int_0^{L^{(e)}} w\, f_{y'}\, dx' + \left[w\frac{d}{dx'}\left(EI\frac{d^2v'}{dx'^2}\right) - EI\frac{dw}{dx'}\frac{d^2v'}{dx'^2}\right]_0^{L(e)} = 0 \tag{6.15}$$

By using the boundary conditions in Eq. (6.14), the weighted residual integral becomes,

$$\int_0^{L^{(e)}} \left(EI\frac{d^2w}{dx'^2}\frac{d^2v'}{dx'^2} - w\, f_{y'}\right)dx' - F_{y'1}w(0) - F_{y'2}w\left(L^{(e)}\right) - M_1\frac{dw}{dx'}\bigg|_0 - M_2\frac{dw}{dx'}\bigg|_{L^{(e)}} = 0 \tag{6.16}$$

where w is the weight function as described in Chapter 3. After replacing the weight function w with the variation of displacement, δv, the *functional form* of the boundary value problem can be found as follows:

$$I^{(e)} = \int_0^{L^{(e)}} \left(\frac{1}{2}EI\left(\frac{d^2v'}{dx'^2}\right)^2 - v'\, f_{y'}\right)dx' - F_{y1'}v'(0) - F_{y2'}v'\left(L^{(e)}\right) - M_1\frac{dv'}{dx'}\bigg|_0 - M_2\frac{dv'}{dx'}\bigg|_{L^{(e)}} \tag{6.17}$$

And, after introducing the slope of the beam,

$$\theta(x') = \frac{dv'}{dx'} \tag{6.18}$$

as a dependent a variable, and using the relationship Eq. (6.13), this functional becomes,

$$I^{(e)} = \frac{1}{2}\int_0^{L^{(e)}} EI\left(\frac{d\theta}{dx'}\right)^2 dx' - \left(\int_0^{L^{(e)}} v'(F_{By'}A + q_{y'})dx' + v_1{}'F_{y'1} + v_2{}'F_{y'2} + \theta_1 M_1 + \theta_2 M_2\right) \tag{6.19}$$

Details of this derivation are omitted here, but can be found in the discussion provided in Chapter 3. Next, we find the total potential energy of this beam upon deformation.

6.3.3 Total potential energy of a beam element

Similar to the development presented for the axial bar in Section 5.2.4, the total potential energy of the beam under the effect of external forces is expressed as follows:

$$\pi_p^{(e)} = U^{(e)} - W^{(e)} \tag{6.20}$$

$$\pi_p^{(e)} = \int_A \int_0^{L^{(e)}} \sigma' d\varepsilon' dA dx' - \left(\int_0^{L^{(e)}} v'(F_b A + q)dx' + v_1{}'F_{y1}{}' + v_2{}'F_{y2}{}' + \theta_1 M_1 + \theta_2 M_2\right) \tag{6.21}$$

where the first term represents the strain energy stored in the beam, and the second term represents the work done by the external forces. Note that for a beam we have $dA = b\,dy'$ where b is the *breadth* of the beam's cross-section.

The *bending strain*, is a longitudinal strain $\varepsilon_{x'}$, which depends on the normal distance as measured from the neutral axis of the beam (Fig. 6.2) as given in Eq. (2.138),

$$\varepsilon_{x'} = -y'\frac{d^2 v'}{dx'^2} \tag{6.22}$$

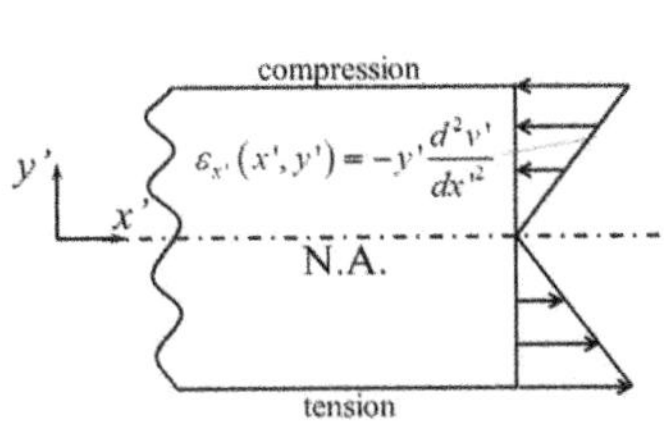

FIGURE 6.2 Bending strain definition in EB beam theory.

The strain energy component of the total potential energy then becomes,

$$U_b^{(e)} = \frac{1}{2}\int_0^{L^{(e)}} \left[\int_A E\left(-y'\frac{d^2v'}{dx'^2}\right)^2 bdy'\right]dx' = \frac{1}{2}\int_0^{L^{(e)}} E\left(\frac{d^2v'}{dx'^2}\right)^2\left(\int_{-c/2}^{c/2} y'^2 bdy'\right)dx' \tag{6.23}$$

Note that the inner integral over the area represents the definition of the *second moment of area*,

$$I = \int_{-c/2}^{c/2} y'^2 bdy' \tag{6.24}$$

The strain energy due to bending $U_b^{(e)}$ in a beam element is therefore expressed as follows:

$$U_b^{(e)} = \frac{1}{2}\int_0^{L^{(e)}} EI\left(\frac{d^2v'}{dx'^2}\right)^2 dx', \quad \text{or} \quad U_b^{(e)} = \frac{1}{2}\int_0^{L^{(e)}} EI\left(\frac{d\theta}{dx'}\right)^2 dx' \tag{6.25}$$

The total potential energy for a beam element becomes,

$$\pi_p^{(e)} = \frac{1}{2}\int_0^{L^{(e)}} EI\left(\frac{d\theta}{dx'}\right)^2 dx' - \left(\int_0^{L^{(e)}} v'(F_{By'}A + q_{y'})dx' + v_1'F_{y'1} + v_2'F_{y'2} + \theta_1 M_1 + \theta_2 M_2\right) \tag{6.26}$$

The first term of the total potential energy will lead to the element stiffness matrix, whereas the second term will lead to the consistent nodal load vector. Note that Eqs. (6.19) and (6.26) represent the same functional, as expected.

6.3.4 Finite element form of the equilibrium equations of an Euler–Bernoulli beam

The principle of minimum total potential energy states that the equilibrium conditions are found when the variation of the total potential energy functional is zero. For an Euler–Bernoulli beam element, this is expressed as follows:

$$\delta\pi_p^{(e)} = \int_0^{L^{(e)}} EI\frac{d\theta}{dx'}\frac{d\delta\theta}{dx'}dx' - \left(\int_0^{L^{(e)}} \delta v'(F_{By'}A + q_{y'})dx' + \delta v_1'F_{y'1} + \delta v_2'F_{y'2} + \delta\theta_1 M_1 + \delta\theta_2 M_2\right) \tag{6.27}$$

In order to find an expression for the curvature, $d\theta/dx'$ recall that the beam deflection v' is approximated as follows:

$$\begin{aligned} v'(x') &= N_1^{(b)}v_1' + N_2^{(b)}\theta_1 + N_3^{(b)}v_2' + N_4^{(b)}\theta_2 \quad \text{(a)} \\ &= \left[N^{(b)}\right]\left\{d_b'\right\}^{(e)} \quad \text{(b)} \end{aligned} \tag{6.28}$$

where $[N^{(b)}]$ and $\{d_b'\}^{(e)}$ are the *shape function matrix* and the *degree of freedom vector*, (Eq. (6.10)(a)) respectively,

$$\left[N^{(b)}\right] = \left[N_1^{(b)} \; N_2^{(b)} \; N_3^{(b)} \; N_4^{(b)}\right] \tag{6.29}$$

$N_i^{(b)}$ are the C^1-continuous shape functions given by Eq. (6.5). Using this approximation, the curvature vector is found as follows:

$$\frac{d\theta}{dx'} = \frac{d^2v'}{dx'^2} = \frac{d^2}{dx'^2}\left(\left[N^{(b)}\right]\left\{d_b'\right\}^{(e)}\right) = \left[B^{(b)}\right]\left\{d_b'\right\}^{(e)} \tag{6.30}$$

where $\left[B^{(b)}\right] = \left[B_1^{(b)} \; B_2^{(b)} \; B_3^{(b)} \; B_4^{(b)}\right]$ is the *curvature-displacement matrix* with,

$$\begin{aligned} B_1^{(b)} &= \frac{d^2N_1^{(b)}}{dx'^2} = \left(-\frac{6}{L^{(e)2}} + \frac{12x'}{L^{(e)3}}\right), \quad B_2^{(b)} = \frac{d^2N_2^{(b)}}{dx'^2} = \left(-\frac{4}{L^{(e)}} + \frac{6x}{L^{(e)2}}\right), \\ B_3^{(b)} &= \frac{d^2N_3^{(b)}}{dx'^2} = \left(\frac{6}{L^{(e)2}} - \frac{12x'}{L^{(e)3}}\right), \quad B_4 = \frac{d^2N_4^{(b)}}{dx'^2} = \left(-\frac{2}{L^{(e)}} + \frac{6x'}{L^{(e)2}}\right) \end{aligned} \tag{6.31}$$

The variations of the displacement and curvature are obtained as follows:

$$\begin{aligned} \delta v' &= \left[N^{(b)}\right]\left\{\delta d_b'\right\}^{(e)} \quad \text{(a)} \\ \delta\theta' &= \left[B^{(b)}\right]\left\{\delta d_b'\right\}^{(e)} \quad \text{(b)} \end{aligned} \tag{6.32}$$

By using Eqs. (6.28), (6.30), and (6.32), the minimum total potential energy expression (6.27) can be represented in finite element form as follows:

$$\begin{aligned} \delta\pi_p^{(e)} = &\int_0^{L^{(e)}} \left\{\delta d_b'\right\}^{(e)T}\left[B^{(b)}\right]^T EI\left[B^{(b)}\right]\left\{d_b'\right\}^{(e)}dx' \\ &- \int_0^{L^{(e)}} \left\{\delta d_b'\right\}^{(e)T}\left[N^{(b)}\right]^T(F_{By'}A + q_{y'})dx' - \left\{\delta d_b'\right\}^{(e)T}\left\{r_F'\right\}^{(e)} = 0 \end{aligned} \tag{6.33}$$

Eq. (6.33) can be further simplified,

$$\delta\pi_p^{(e)} = \left\{\delta d_b'\right\}^{(e)T}\left\{\left[\int_0^{L^{(e)}} \left[B^{(b)}\right]^T EI\left[B^{(b)}\right]dx'\right]\left\{d_b'\right\}^{(e)} - \int_0^{L^{(e)}} \left[N^{(b)}\right]^T(F_{By'}A + q_{y'})dx' - \left\{r_F'\right\}^{(e)}\right\} = 0 \tag{6.34}$$

Considering that the variation of the degrees of freedom $\{\delta d_b{}'\}^{(e)T}$ is arbitrary, the following equation of equilibrium can be found from Eq. (6.34):

$$\left[\int_0^{L^{(e)}} \left[B^{(b)}\right]^T EI\left[B^{(b)}\right]dx'\right]\left\{d_b{}'\right\}^{(e)} = \int_0^{L^{(e)}} \left[N^{(b)}\right]^T \left(F_{By'}A + q_{y'}\right)dx' - \left\{r_F{}'\right\}^{(e)} \tag{6.35}$$

In symbolic form the *equation of equilibrium for an Euler–Bernoulli beam element* is expressed as follows:

$$\left[k_b{}'\right]^{(e)}\left\{d_b{}'\right\}^{(e)} = \left\{r_B{}'\right\}^{(e)} + \left\{r_q{}'\right\}^{(e)} + \left\{r_F{}'\right\}^{(e)} \tag{6.36}$$

$$\begin{aligned}
&\text{Element stiffness matrix:} && \left[k_b'\right]^{(e)} = \int_0^{L(e)} \left[B^{(b)}\right]^T EI\left[B^{(b)}\right]dx' && \text{(a)}\\
&\text{Boundary - load vector:} && \left\{r_F{}'\right\}^{(e)} = \left\{F_{y'1}\ \ M_1\ \ F_{y'2}\ \ M_2\right\}^T && \text{(b)}\\
&\text{Consistent nodal load vectors:} && \left\{r_B{}'\right\}^{(e)} = \int_0^{L^{(e)}} \left[N^{(b)}\right]^T F_{By'}A\, dx' && \text{(c)}\\
& && \left\{r_q{}'\right\}^{(e)} = \int_0^{L^{(e)}} \left[N^{(b)}\right]^T q_{y'}A\, dx' && \text{(d)}
\end{aligned} \tag{6.37}$$

The equation of equilibrium (6.36) can also be expressed as follows:

$$\left[k_b{}'\right]^{(e)}\left\{d_b{}'\right\}^{(e)} = \left\{r_b{}'\right\}^{(e)} \tag{6.38}$$

where the *nodal load vector* is,

$$\left\{r_b{}'\right\}^{(e)} = \left\{r_B{}'\right\}^{(e)} + \left\{r_q{}'\right\}^{(e)} + \left\{r_F{}'\right\}^{(e)} \tag{6.39}$$

6.3.4.1 *Element stiffness matrix*

The components of the element stiffness matrix $k_{bij}{}'$ can be found as follows:

$$k_{bij}{}' = \int_0^{L^{(e)}} EI\ B_i\ B_j\ dx' \tag{6.40}$$

In case the material and the cross-sectional properties are constant along the element, the stiffness matrix can be found as follows:

$$\left[k_b{}'\right] = k_b \begin{bmatrix} 12 & 6L^{(e)} & -12 & 6L^{(e)} \\ 6L^{(e)} & 4L^{(e)2} & -6L^{(e)} & 2L^{(e)2} \\ -12 & -6L^{(e)} & 12 & -6L^{(e)} \\ 6L^{(e)} & 2L^{(e)2} & -6L^{(e)} & 4L^{(e)2} \end{bmatrix} \tag{6.41}$$

where $k_b = \dfrac{EI}{L^{(e)3}}$.

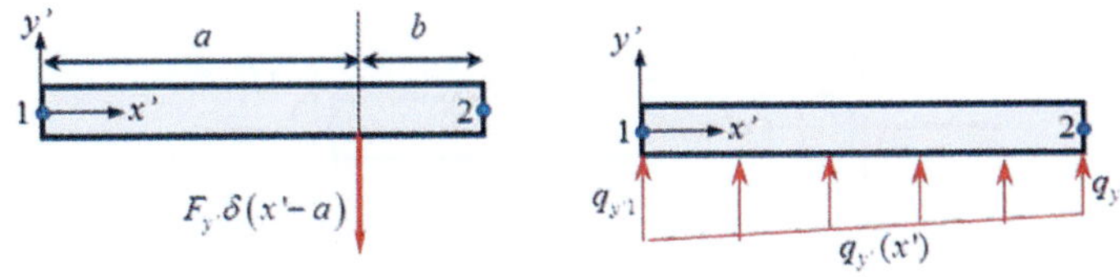

FIGURE 6.3 (A) Concentrated force, (B) linearly varying distributed force, (C) distributed force with constant magnitude, and (D) corresponding consistent nodal forces.

6.3.4.2 Consistent nodal load vector

Equations (6.37)(c,d) are used to find statically equivalent forces and moments acting on the nodes due to distributed forces acting along the element. These nodal forces (and moments) are referred to as the *consistent nodal forces* (and *moments*). Consider the concentrated and the distributed forces acting on the element shown in Fig. 6.3. By using Eqs. (6.37) and (6.29) the following consistent nodal force vectors are found.

Concentrated force: $-\ F_y\delta(x'-a)$ acting along the $-y$-axis and located at $x'=a$,

$$\{r_q'\}^{(e)} = -F_y'\left\{\frac{b^2\left(L^{(e)}+2a\right)}{L^{(e)3}}\ \frac{ab^2}{L^{(e)2}}\ \frac{a^2\left(L^{(e)}+2b\right)}{L^{(e)3}}\ -\frac{a^2b}{L^{(e)2}}\right\}^T \tag{6.42}$$

where $b=L-a$.

Linearly distributed force: $q_{y'}=q_{y'1}+\frac{q_{y'2}-q_{y'1}}{L}x'$. Note that $q_{y'}$ is positive along the $+y'$-axis.

$$\{r_q'\}^{(e)} = \left\{\frac{(7q_{y'1}+3q_{y'2})L^{(e)}}{20}\ \frac{(3q_{y'1}+2q_{y'2})L^{(e)2}}{60}\ \frac{(3q_{y'1}+7q_{y'2})L^{(e)}}{20}\ -\frac{(2q_{y'1}+3q_{y'2})L^{(e)2}}{60}\right\}^T \tag{6.43}$$

Distributed force with constant magnitude: $q_{y'}(x')=q_{y'0}$ which points along the $+y'$-axis.

$$\{r_q'\}^{(e)} = \left\{\frac{q_{y'0}L^{(e)}}{2}\ \frac{q_{y'0}L^{(e)2}}{12}\ \frac{q_{y'0}L^{(e)}}{2}\ -\frac{q_{y'0}L^{(e)2}}{12}\right\}^T \tag{6.44}$$

The signs of the forces in these vectors are determined by the sign convention defined in Fig. 6.1.

6.4 General beam element with membrane and bending capabilities

In many load bearing situations stretching and bending occurs on the same member, simultaneously (Fig. 6.4). The element stiffness matrix that represents the equilibrium of such a member is obtained by using the superposition of membrane and bending responses. Note that here we consider small deflection

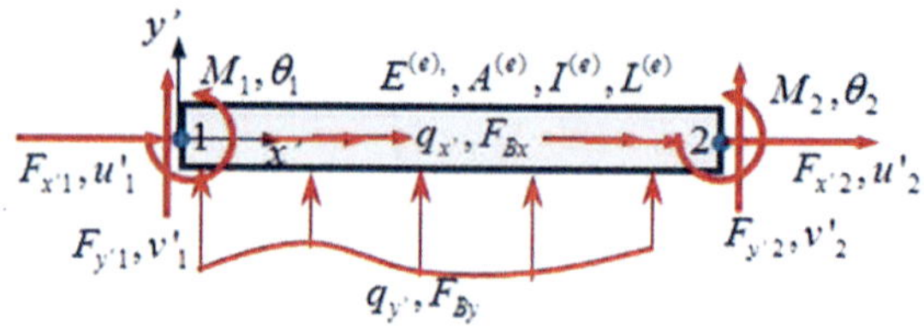

FIGURE 6.4 The EB-beam element that has bending and axial deformation capabilities.

cases where these two effects can be assumed to be uncoupled from one another. The equilibrium equation for such a *generic beam element* is represented in matrix form as follows:

$$\left[k_{mb}'\right]^{(e)}\left\{d_{mb}'\right\}^{(e)} = \left\{r_{mb}'\right\}^{(e)} \tag{6.45}$$

This relationship can be obtained by using a superposition of the membrane and beam mechanics, represented by Eqs. (5.22) and (6.38),

$$\left[k_m'\right]^{(e)}\left\{d_m'\right\}^{(e)} + \left[k_b'\right]^{(e)}\left\{d_b'\right\}^{(e)} = \left\{r_m'\right\}^{(e)} + \left\{r_b'\right\}^{(e)} \tag{6.46}$$

As the membrane and the bending actions are uncoupled (at least in this presentation) the following variables can be easily identified,

$$\left[k_{mb}'\right]^{(e)} = \begin{bmatrix} k_m & 0 & 0 & -k_m & 0 & 0 \\ 0 & 12k_b & 6k_bL^{(e)} & 0 & -12k_b & 6k_bL^{(e)} \\ 0 & 6k_bL^{(e)} & 4k_bL^{(e)2} & 0 & -6k_bL^{(e)} & 2k_bL^{(e)2} \\ -k_m & 0 & 0 & k_m & 0 & 0 \\ 0 & -12k_b & -6k_bL^{(e)} & 0 & 12k_b & -6k_bL^{(e)} \\ 0 & 6k_bL^{(e)} & 2k_bL^{(e)2} & 0 & -6k_bL^{(e)} & 4k_bL^{(e)2} \end{bmatrix} \tag{6.47}$$

with $k_m = \frac{EA}{L^{(e)}}$ and $k_b = \frac{EI}{L^{(e)3}}$

$$\left\{d_{mb}'\right\}^{(e)} = \left\{u_1' \quad v_1' \quad \theta_1' \quad u_2' \quad v_2' \quad \theta_2'\right\}^T \tag{6.48}$$

$$\left\{r_{mb}'\right\}^{(e)} = \left\{F_{x'1} \quad F_{y'1} \quad M_1 \quad F_{x'2} \quad F_{y'2} \quad M_2\right\}^T \tag{6.49}$$

6.5 Coordinate transformations

6.5.1 Vector transformation between orthogonal coordinate systems in a two-dimensional plane

In Section 5.4, coordinate transformation relationships to represent a generic planar vector $\vec{V}$ in two separate coordinate systems (x', y') and (x, y) were developed. Letting u' and v' be the projections of the vector $\vec{V}$ onto the (x', y') axes, and similarly u and v be the projections onto the (x, y), a coordinate

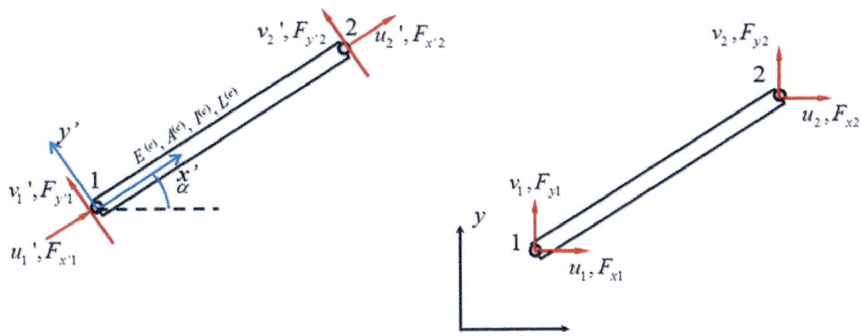

FIGURE 6.5 Degrees of freedom of the EB-beam element with axial deformation expressed in the local and global coordinates.

transformation relation was obtained in Eq. (5.47) as follows:

$$\begin{Bmatrix} u' \\ v' \end{Bmatrix} = \begin{bmatrix} \cos\alpha & \sin\alpha \\ -\sin\alpha & \cos\alpha \end{bmatrix} \begin{Bmatrix} u \\ v \end{Bmatrix} \tag{6.50}$$

or

$$\{V'\} = [A]\{V\} \tag{6.51}$$

where $[A]$ is the *direction cosine matrix* for a two-dimensional transformation between two orthogonal coordinate systems and α is the angle between the x and x' axes, as shown in Fig. 6.5.

6.5.2 Transformation of equilibrium equations for the Euler-Bernoulli beam element with axial deformation

The degree of freedom vector for the Euler-Bernoulli beam element with axial deformation capability is given in Eq. (6.48),

$$\{d_{mb}'\} = \{u_1' \quad v_1' \quad \theta_1 \quad u_2' \quad v_2' \quad \theta_2\}^T$$

This vector is defined in the local coordinate system (x',y') of the element. In order to find the representation of this vector, $\{d_{mb}\}$, in the (x,y) coordinate system (Fig. 6.5), on the same plane, the vector transformation defined in Eq. (6.50) is used. This results in the following transformation matrix for the Euler-Bernoulli beam element,

$$[T]^{(e)} = \begin{bmatrix} \cos\alpha & \sin\alpha & 0 & 0 & 0 & 0 \\ -\sin\alpha & \cos\alpha & 0 & 0 & 0 & 0 \\ 0 & 0 & 1 & 0 & 0 & 0 \\ 0 & 0 & 0 & \cos\alpha & \sin\alpha & 0 \\ 0 & 0 & 0 & -\sin\alpha & \cos\alpha & 0 \\ 0 & 0 & 0 & 0 & 0 & 1 \end{bmatrix} \tag{6.52}$$

The transformation of the degree of freedom and the load vectors are enabled accordingly as follows:

$$\begin{aligned} \left\{d_{mb}{}'\right\}^{(e)} &= [T]^{(e)}\{d_{mb}\}^{(e)} \quad \text{(a)} \\ \left\{r_{mb}{}'\right\}^{(e)} &= [T]^{(e)}\{r_{mb}\}^{(e)} \quad \text{(b)} \end{aligned} \tag{6.53}$$

In order to find the element equilibrium equation in the global coordinate system the coordinate transformations are used. Starting with the equilibrium in the local coordinate system (6.45),

$$\left[k_{mb}{}'\right]^{(e)}\left\{d_{mb}{}'\right\}^{(e)} = \left\{r_{mb}{}'\right\}^{(e)},$$

using Eq. (6.53)(a) we find,

$$\left[k_{mb}{}'\right]^{(e)}[T]^{(e)}\left\{d_{mb}{}'\right\}^{(e)} = \left\{r_{mb}{}'\right\}^{(e)}$$

multiplying both sides by $[T]^{(e)T}$,

$$[T]^{(e)T}\left[k_{mb}{}'\right]^{(e)}[T]^{(e)}\{d_{mb}\}^{(e)} = [T]^{(e)T}\left\{r_{mb}{}'\right\}^{(e)}$$

and recognizing from Eqs. (6.51) and (6.52) that for a plane transformation one can express $\{r_{mb}\}^{(e)} = [T]^{(e)T}\{r_{mb}{}'\}^{(e)}$, the equilibrium equation in the global coordinate system can be found as follows:

$$[k_{mb}]^{(e)}\{d_{mb}\}^{(e)} = \{r_{mb}\}^{(e)} \tag{6.54}$$

with

$$\begin{aligned} [k_{mb}]^{(e)} &= [T]^{(e)T}\left[k_{mb}{}'\right]^{(e)}[T]^{(e)} \\ \{r_{mb}\}^{(e)} &= [T]^{(e)T}\left\{r_{mb}{}'\right\}^{(e)} \end{aligned} \tag{6.55}$$

Note that the expression for $[k_{mb}]^{(e)}$ can be computed by using Eq. (6.55) as needed.

6.6 Assembly, boundary conditions, and reaction forces

The basics of the element assembly, application of the boundary conditions and computation of the reaction forces (and moments) have been introduced in Sections 5.5, 5.6, and 5.8 respectively. These concepts do not change and they are directly applicable for the EB-beam equations developed in this chapter.

6.7 Postprocessing and computation of stresses in members

As introduced in Section 5.8, a postprocessing operation is necessary to obtain the strain and stress variation along each element. An EB-beam element with axial deformation will develop **normal** $\sigma_{x'}$ and **shear stresses** $\tau_{x'y'}$ as described in Chapter 2. These two variables need to be computed in the element coordinate system and therefore will require access to the element's degree of freedom vector $\{d_{mb}'\}^{(e)}$.

The solution of the global equilibrium equation,

$$[K]\{D\} = \{R\} \tag{6.56}$$

gives the global degree of freedom vector $\{D\}$ which represents the displacement and slope degrees of freedom of the entire structure. The element degree of freedom vectors $\{d_{mb}\}^{(e)}$ are extracted from $\{D\}$ by using the element connectivity table. The coordinate transformation defined in Section 6.5.2 is performed to transform $\{d_{mb}\}^{(e)}$ to $\{d_{mb}'\}^{(e)}$.

The *total normal stress* $\sigma_{x'}$ has contributions due to bending σ_b and a membrane σ_m stress components,

$$\sigma_{x'} = \sigma_b + \sigma_m \tag{6.57}$$

The *bending stress* can be found as described in Section 2.3.3 as follows:

$$\sigma_b = -\frac{My'}{I} \tag{6.58}$$

where I is the second moment of area of the beam cross section, y' is the transverse coordinate axis in the thickness direction of the beam with the origin located on the neutral axis, and M is the bending moment in the beam,

$$M(x') = EI\frac{d^2v'}{dx'^2} = EI\frac{d^2}{dx'^2}\left(N_1^{(b)}v_1' + N_2^{(b)}\theta_1 + N_3^{(b)}v_2' + N_4^{(b)}\theta_2\right) \tag{6.59}$$

The *membrane stress* can be found as described in Section 5.8 as follows:

$$\sigma_m = E\varepsilon_m \tag{6.60}$$

where the membrane strain is defined in Eq. (5.14) as follows:

$$\varepsilon_m = \frac{du'}{dx'} = \frac{d}{dx'}\left(N_1^{(m)}u_1' + N_2^{(m)}u_2'\right) \tag{6.61}$$

The *average shear stress* can be found as follows,

$$\tau_{x'y'} = k_{sc}\frac{V}{A} \tag{6.62}$$

where A is the cross-sectional area of the beam, k_{sc} is the *shear correction factor* that compensates for the fact that an average shear stress is used. The shear correction coefficient k_{sc} is a tabulated value [1]. The transverse shear force resultant V, defined in Eq. (2.143), is computed as follows:

$$V = -EI\frac{d^3v'}{dx'^3} = -EI\frac{d^3}{dx'^3}\left(N_1^{(b)}v_1' + N_2^{(b)}\theta_1 + N_3^{(b)}v_2' + N_4^{(b)}\theta_2\right) \tag{6.63}$$

The stress distribution along an Euler-Bernoulli beam element with axial deformation capability can be summarized based on Eqs. (6.58–6.63) as follows:

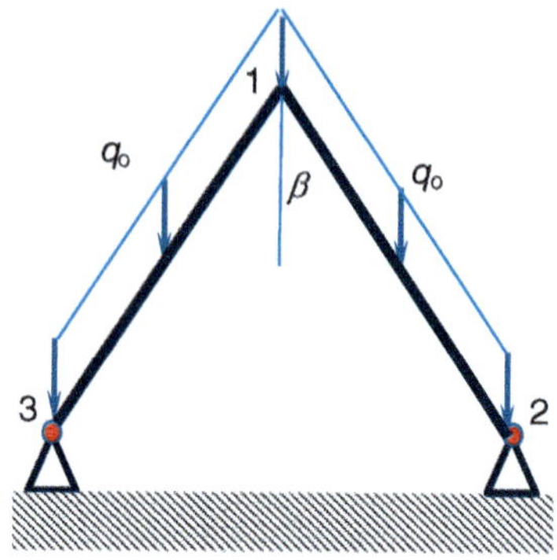

FIGURE 6.6 The frame that consists of two members is subjected to the distributed forces as shown.

TABLE 6.1 Element connectivity table.

e	*I*	*j*	α(e) (deg)	*E*(e) (Pa)	*L*(e) (m)	*A*(e) (m^2)	*I*(e) (m^4)
1	1	3	228.19	20×10^9	1.5	3×10^{-3}	4.50×10^{-7}
2	1	2	−48.19	20×10^9	1.5	3×10^{-3}	4.50×10^{-7}

Membrane stress: $$\sigma_m = E\frac{u_2' - u_1'}{L^{(e)}} \tag{6.64}$$

Bending stress: $$\sigma_b(x', y') = -Ey'\left[B_1^{(b)}v_1' + B_2^{(b)}\theta_1 + B_3^{(b)}v_2' + B_4^{(b)}\theta_2\right] \tag{6.65}$$

Shear stress: $$\tau_{x'y'} = -k_{sc}\frac{EI}{A}\left[\frac{12}{L^{(e)3}}v_1' + \frac{6}{L^{(e)2}}\theta_1 - \frac{12}{L^{(e)3}}v_2' + \frac{6}{L^{(e)2}}\theta_2\right] \tag{6.66}$$

Example 6.1

Members of the two member frame shown in Fig. 6.6 are welded to one another and the frame is pinned to the ground at joints 2 and 3, as shown. The parameters of this problem are given as follows (Table 6.1):

$L_{12} = L_{13} = 1.5\,\text{m}$, $L_{23} = 2\,\text{m}$, $E = 20\,\text{GPa}$, $h = 0.03\,\text{m}$, $b = 0.2\,\text{m}$, $q_0 = 200\,\text{N/m}$

Assume that each member can resist loading by membrane and bending actions.

a) Prepare the element connectivity table for this structure. Include all variables that will be used for computing element equilibrium equations.

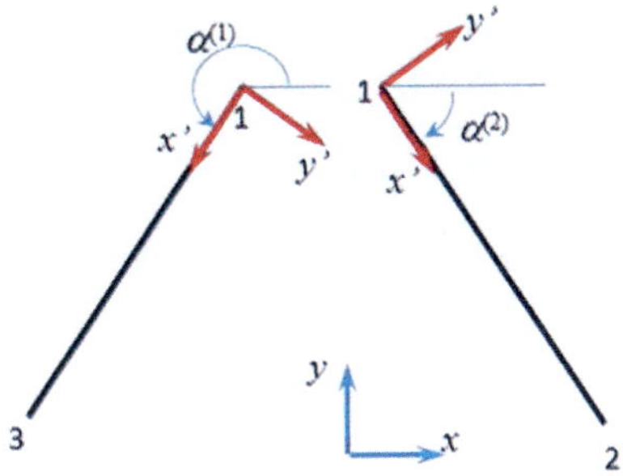

FIGURE 6.7 The local coordinate systems for elements −1 and −2.

b) Find the element equilibrium equations for each element.
c) Assemble the global equilibrium equation and find the active degrees of freedom.
d) Find the reaction forces.
e) Find the stress distribution in each member.

Solutions:
a) Geometry: $\beta = \sin^{-1}\left(\frac{2}{3}\right) = 41.81$ deg.
Element orientations: $\alpha^{(1)} = -\frac{3}{2}\pi - \beta = 228.19$ deg. and $\alpha^{(2)} = -\left(\frac{\pi}{2} - \beta\right) = -48.19$ deg.
b) Element equilibrium: in local coordinate system is given by Eqs. (6.45–6.49), and the element equilibrium equation in the global coordinate system is described in Section 6.5.
Element-1 (*nodes* 1-3): For this element we find the following parameters:

$\alpha^{(1)} = \frac{3}{2}\pi - \beta = 228.19$ deg., $c = -0.6667$, $s = -0.7454$, $k_m = \frac{EA}{L} = 8 \times 10^7$,

$$k_b = \frac{EI}{L^3} = 2{,}666.7$$

Consistent nodal forces in the local coordinate system (see Fig. 6.7 for orientation of the y' axis):

$$q_{ox'} = q_o \cos\beta \text{ and } q_{oy'} = q_o \sin\beta \tag{6.67}$$

$$\left\{r_{mb}'\right\}^{(1)} = \left\{\frac{1}{2}q_{ox'}L \quad \frac{1}{2}q_{oy'}L \quad \frac{1}{12}q_{oy'}L^2 \quad \frac{1}{2}q_{ox'}L \quad \frac{1}{2}q_{oy'}L \quad -\frac{1}{12}q_{oy'}L^2\right\}^T$$
$$= q_o\left\{\frac{L}{2}\cos\beta \quad \frac{L}{2}\sin\beta \quad \frac{L^2}{12}\sin\beta \quad \frac{L}{2}\cos\beta \quad \frac{L}{2}\sin\beta \quad -\frac{L^2}{12}\sin\beta\right\}^T \tag{6.68}$$

By using $q_o = 200$ N/m, $L = 1.5$ m, and $\beta = 41.81$ deg.:

$$\left\{r_{mb}'\right\}^{(1)} = \{111.8 \quad 100 \quad 25 \quad 111.8 \quad 100 \quad -25\}^T \tag{6.69}$$

Consistent nodal forces in the global coordinate system: $\{r_{mb}\}^{(1)} = [T]^{(1)T}\{r_{mb}'\}^{(1)}$ with $\alpha^{(1)} = 228.19$, $c = -0.6667$, $s = -0.7454$

$$\{r_{mb}\}^{(1)} = \{0 \quad -150 \quad 25 \quad 0 \quad -150 \quad -25\}^T \tag{6.70}$$

Element equilibrium equation: Note that the element stiffness matrix given below is symmetric,

$$\begin{bmatrix} 3.5573\times10^7 & 3.9736\times10^7 & 17888.5 & -3.5573\times10^7 & -3.9736\times10^7 & 17888.5 \\ & 4.4459\times10^7 & -16000 & -3.9736\times10^7 & -4.4459\times10^7 & -16000 \\ & & 24000 & -17888.5 & 16000 & 12000 \\ & & & 3.5573\times10^7 & 3.9736\times10^7 & -17888.5 \\ & & & & 4.4459\times10^7 & 16000 \\ & & & & & 24000 \end{bmatrix} \begin{Bmatrix} u_1 \\ v_1 \\ \theta_1 \\ u_3 \\ v_3 \\ \theta_3 \end{Bmatrix} = \begin{Bmatrix} 0 \\ -150 \\ 25 \\ 0 \\ -150 \\ -25 \end{Bmatrix} \tag{6.71}$$

Element-2 (*nodes* 1-2):

$$\alpha^{(2)} = -\left(\frac{\pi}{2} - \beta\right) = -48.19\,\text{deg.},\ c = 0.6667,\ s = -0.7454,$$
$$k_m = \frac{EA}{L} = 8\times10^7,\quad k_b = \frac{EI}{L^3} = 2666.7$$

Consistent nodal forces in the local coordinate system (see Fig. 6.7 for the orientation of the y' axis):

$$q_{ox'} = q_o \cos\beta \quad \text{and} \quad q_{oy'} = -q_o \sin\beta$$

$$\begin{aligned}\{r_{mb}'\}^{(2)} &= \left\{\frac{1}{2}q_{ox'}L \quad \frac{1}{2}q_{oy'}L \quad \frac{1}{12}q_{oy'}L^2 \quad \frac{1}{2}q_{ox'}L \quad \frac{1}{2}q_{oy'}L \quad -\frac{1}{12}q_{oy'}L^2\right\}^T \\ &= q_o\left\{\frac{L}{2}\cos\beta \quad -\frac{L}{2}\sin\beta \quad -\frac{L^2}{12}\sin\beta \quad \frac{L}{2}\cos\beta \quad -\frac{L}{2}\sin\beta \quad \frac{L^2}{12}\sin\beta\right\}^T\end{aligned} \tag{6.72}$$

By using $q_o = 200$ N, $L = 1.5$ m, and $\beta = 41.81$ deg.:

$$\{r_{mb}'\}^{(2)} = \{111.8 \quad -100 \quad -25 \quad 111.8 \quad -100 \quad 25\}^T \tag{6.73}$$

Consistent nodal forces in the global coordinate system: $\{r_{mb}\}^{(2)} = [T]^{(1)T}\{r_{mb}'\}^{(2)}$ with $\alpha^{(2)} = -48.19$, $c = 0.6667$, $s = -0.7454$

$$\{r_{mb}\}^{(2)} = \{0 \quad -150 \quad -25 \quad 0 \quad -150 \quad 25\}^T \tag{6.74}$$

Consistent nodal forces are summarized in Table 6.2, which is part of the element connectivity table.

TABLE 6.2 Consistent nodal loads for elements. See the text on details.

e	i	j	$r_1^{(e)}$(N)	$r_2^{(e)}$(N)	$r_3^{(e)}$(Nm)	$r_4^{(e)}$(N)	$r_5^{(e)}$(N)	$r_6^{(e)}$(Nm)
1	1	3	0	−150	25	0	−150	−25
2	1	2	0	−150	−25	0	−150	25

Element equilibrium equation: Note that the element stiffness matrix given below is symmetric,

$$\begin{bmatrix} 3.5573\times10^7 & -3.9736\times10^7 & 17888.5 & -3.5573\times10^7 & 3.9736\times10^7 & 17888.5 \\ & 4.4459\times10^7 & 16000 & 3.9736\times10^7 & -4.4459\times10^7 & 16000 \\ & & 24000 & -17888.5 & -16000 & 12000 \\ & & & 3.5573\times10^7 & -3.9736\times10^7 & -17888.5 \\ & & & & 4.4459\times10^7 & -16000 \\ & & & & & 24000 \end{bmatrix} \begin{Bmatrix} u_1 \\ v_1 \\ \theta_1 \\ u_2 \\ v_2 \\ \theta_2 \end{Bmatrix} = \begin{Bmatrix} 0 \\ -150 \\ -25 \\ 0 \\ -150 \\ 25 \end{Bmatrix} \tag{6.75}$$

c) Global equilibrium equation: Assembly of the element equilibrium equations gives,

$$\begin{bmatrix} 7.117\times10^7 & 0 & 35777.1 & -3.5573\times10^7 & 3.9736\times10^7 & 17888.5 & -3.5573\times10^7 & -3.9736\times10^7 & 17888.5 \\ 0. & 8.8917\times10^7 & 0 & 3.9736\times10^7 & -4.4459\times10^7 & 16000 & -3.9736\times10^7 & -4.4459\times10^7 & -16000 \\ 35777.1 & 0 & 48000 & -17888.5 & -16000 & 12000 & -17888.5 & 16000 & 12000 \\ -3.5573\times10^7 & 3.9736\times10^7 & -17888.5 & 3.5573\times10^7 & -3.9736\times10^7 & -17888.5 & 0 & 0 & 0 \\ 3.9736\times10^7 & -4.4459\times10^7 & -16000 & -3.9736\times10^7 & 4.4459\times10^7 & -16000 & 0 & 0 & 0 \\ 17888.5 & 16000 & 12000 & -17888.5 & -16000 & 24000 & 0 & 0 & 0 \\ -3.5573\times10^7 & -3.9736\times10^7 & -17888.5 & 0 & 0 & 0 & 3.5573\times10^7 & 3.9736\times10^7 & -17888.5 \\ -3.9736\times10^7 & -4.4459\times10^7 & 16000 & 0 & 0 & 0 & 3.9736\times10^7 & 4.4459\times10^7 & 16000 \\ 17888.5 & -16000 & 12000 & 0 & 0 & 0 & -17888.5 & 16000 & 24000 \end{bmatrix} \begin{Bmatrix} u_1 \\ v_1 \\ \theta_1 \\ u_2 \\ v_2 \\ \theta_2 \\ u_3 \\ v_3 \\ \theta_3 \end{Bmatrix}$$

$$= \begin{Bmatrix} 0 \\ -300 \\ 0 \\ 0 \\ 0 \\ 25 \\ 0 \\ 0 \\ -25 \end{Bmatrix} + \begin{Bmatrix} 0 \\ 0 \\ 0 \\ R_{x2} \\ R_{y2} \\ 0 \\ R_{x3} \\ R_{y3} \\ 0 \end{Bmatrix} \tag{6.76}$$

This equation can be expressed symbolically as,

$$[K]\{D\} = \{R_q\} + \{R_R\}$$

where $\{R_q\}$ is the global load vector due to distributed forces, and $\{R_R\}$ represents the reaction force vector. Eliminating the known degrees of freedom

$u_2 = v_2 = u_3 = v_3 = 0$ and the unknown reaction force vector the global equilibrium equation becomes,

$$\begin{bmatrix} 7.114\times10^7 & 0 & 35777.1 & 17888.5 & 17888.5 \\ 0. & 8.8917\times10^7 & 0 & 16000 & -16000 \\ 35777.1 & 0 & 48000 & 12000 & 12000 \\ 17888.5 & 16000 & 12000 & 24000 & 0 \\ 17888.5 & -16000 & 12000 & 0 & 24000 \end{bmatrix} \begin{Bmatrix} u_1 \\ v_1 \\ \theta_1 \\ \theta_2 \\ \theta_3 \end{Bmatrix} = \begin{Bmatrix} 0 \\ -300 \\ 0 \\ 25 \\ -25 \end{Bmatrix} \tag{6.77}$$

Solution gives the active degrees of freedom,

$$\begin{Bmatrix} u_1 \\ v_1 \\ \theta_1 \\ \theta_2 \\ \theta_3 \end{Bmatrix} = \begin{Bmatrix} 0 \\ -3.7497\times10^{-6} \\ 0 \\ 1\times10^{-3} \\ -1\times10^{-3} \end{Bmatrix} \tag{6.78}$$

d) Reaction forces:

The global degree of freedom vector can be obtained by using the active degrees of freedom and the given boundary conditions as follows:

$$\{D\} = \left\{0 \;\; -3.7497\times10^{-6} \;\; 0 \;\; 0 \;\; 0 \;\; 1\times10^{-3} \;\; 0 \;\; 0 \;\; -1\times10^{-3}\right\}^T \tag{6.79}$$

Multiplying the global stiffness matrix given in Eq. (6.76) with this $\{D\}$ vector we find,

$$\left\{R_q\right\} + \{R_R\} = \{0 \;\; -300 \quad 0 \;\; -167.6 \;\; 150 \;\; 25 \;\; 167.6 \;\; 150 \;\; -25\}^T \tag{6.80}$$

The reaction force vector is then identified as follows:

$$\{R_R\} = \{0 \;\; 0 \;\; 0 \;\; -167.6 \;\; 150 \;\; 0 \;\; 167.6 \;\; 150 \;\; 0\}^T \tag{6.81}$$

e) Stresses in each member:

Element degree of freedom vectors are found from Eq. (6.79) and the element connectivity table as follows:

$$\begin{aligned} \{d\}^{(1)} &= \left\{0 \;\; -3.7497\times10^{-6} \;\; 0 \;\; 0 \;\; 0 \;\; -1\times10^{-3}\right\}^T \quad \text{(a)} \\ \{d\}^{(2)} &= \left\{0 \;\; -3.7497\times10^{-6} \;\; 0 \;\; 0 \;\; 0 \;\; 1\times10^{-3}\right\}^T \quad \text{(b)} \end{aligned} \tag{6.82}$$

In order to evaluate the stress components these vectors must be transformed to the local coordinate system by using Eq. (6.53).

$$\begin{aligned} \left\{d'\right\}^{(1)} &= [T]^{(1)T}\{d\}^{(1)} = \left\{2.7946\times10^{-6} \;\; 2.4995\times10^{-6} \;\; 0 \;\; 0 \;\; 0 \;\; -1\times10^{-3}\right\}^T \quad \text{(a)} \\ \left\{d'\right\}^{(2)} &= [T]^{(2)T}\{d\}^{(2)} = \left\{2.7946\times10^{-6} \;\; 2.4995\times10^{-6} \;\; 0 \;\; 0 \;\; 0 \;\; 1\times10^{-3}\right\}^T \quad \text{(b)} \end{aligned} \tag{6.83}$$

The stress components in the members are evaluated by using Eqs. (6.64–6.66). Noting the element nodes have the following relationship $i < j$ the stress components are found by using the following equations:

$$\text{Membrane stress: } \sigma_m^{(e)} = \frac{E}{L^{(e)}}\left(u_j{}' - u_i{}'\right) \qquad \text{(a)}$$

$$\text{Bending stress: } \sigma_b^{(e)} = -Ey'\left[\left(-\frac{6}{L^{(e)2}} + \frac{12x'}{L^{(e)3}}\right)v_i{}' + \left(-\frac{4}{L^{(e)}} + \frac{6x'}{L^{(e)2}}\right)\theta_i + \left(\frac{6}{L^{(e)2}} + \frac{12x'}{L^{(e)3}}\right)v_j{}' + \left(-\frac{2}{L^{(e)}} + \frac{6x'}{L^{(e)2}}\right)\theta_j\right] \qquad \text{(b)}$$

$$\text{Shear stress: } \tau^{(e)} = -k_{sc}\frac{EI}{A}\left[\frac{12}{L^{(e)3}}v_i{}' + \frac{6}{L^{(e)2}}\theta_i - \frac{12}{L^{(e)3}}v_j{}' + \frac{6}{L^{(e)2}}\theta_j\right] \qquad \text{(c)}$$

$$(6.84)$$

Element-1:

$$\sigma_m^{(1)} = \frac{E}{L^{(1)}}\left(u_3{}' - u_1{}'\right) = -37{,}260.8\ \text{Pa}$$

$$\sigma_b^{(1)} = -Ey'\left[\left(-\frac{6}{L^{(1)2}} + \frac{12x'}{L^{(1)3}}\right)v_1{}' + \left(-\frac{4}{L^{(1)}} + \frac{6x'}{L^{(1)2}}\right)\theta_1 + \left(-\frac{6}{L^{(1)2}} + \frac{12x'}{L^{(1)3}}\right)v_3{}' + \left(-\frac{2}{L^{(1)}} + \frac{6x'}{L^{(1)2}}\right)\theta_3\right]$$
$$= (26.533 - 53.156x')y'\ \text{MPa}$$

$$\tau^{(1)} = -k_{sc}\frac{EI}{A}\left[\frac{12}{L^{(1)3}}v_1{}' + \frac{6}{L^{(1)2}}\theta_1 - \frac{12}{L^{(1)3}}v_3{}' + \frac{6}{L^{(1)2}}\theta_3\right] = 59{,}800k_{sc}\ \text{Pa}$$

$$(6.85)$$

Element-2:

$$\sigma_m^{(2)} = \frac{E}{L^{(2)}}\left(u_2{}' - u_1{}'\right) = -37{,}260.8\ \text{Pa}$$

$$\sigma_b^{(2)} = -Ey'\left[\left(-\frac{6}{L^{(2)2}} + \frac{12x'}{L^{(1)3}}\right)v_1{}' + \left(-\frac{4}{L^{(2)}} + \frac{6x'}{L^{(2)2}}\right)\theta_1 + \left(\frac{6}{L^{(2)2}} + \frac{12x'}{L^{(2)3}}\right)v_2{}' + \left(-\frac{2}{L^{(2)}} + \frac{6x'}{L^{(2)2}}\right)\theta_1\right]$$
$$= -(26.533 - 53.156x')y'\ \text{MPa}$$

$$\tau^{(2)} = -k_{sc}\frac{EI}{A}\left[\frac{12}{L^{(2)3}}v_1{}' + \frac{6}{L^{(2)2}}\theta_1 - \frac{12}{L^{(2)3}}v_2{}' + \frac{6}{L^{(2)2}}\theta_2\right] = -59{,}800k_{sc}\ \text{Pa}$$

$$(6.86)$$

We expect bending stresses on the two pinned boundaries of the frame to be zero. However, Eq. (6.85) and (6.86) show that the bending stresses are not zero on the boundaries. This prompts a mesh convergence study where increasingly more elements are used on the members−13 and −12.

TABLE 6.3 Mesh convergence analysis for the deflection at the apex and stress components at node 3 of the frame shown in Fig. 6.6.

NE	$v(L), \mu m$	$\sigma_a(0)$, Pa	$\sigma_b(0)$, Pa	$\tau(0)$, Pa
2	−3.7497	−37264.6	−833333.3	−6245.0
4	−3.7497	−46581.6	−208333.3	6254.9
8	−3.7497	−51240.1	−52083.3	12504.9
16	−3.7497	−53569.3	−13020.7	15629.9
32	−3.7497	−54733.9	−3255.2	17192.4
64	−3.7497	−55316.3	−813.8	17973.6
128	−3.7498	−55608.6	−203.5	18364.7
256	−3.7498	−55754.2	−50.9	18559.9
512	−3.7497	−55826.0	−12.7	18657.3
1024	−3.7497	−55862.4	−3.2	18706.1

A mesh convergence study is carried out by using a geometrically increasing number of elements, N_E as shown in Table 6.3. This meshing scheme ensures that the nodes of the coarser meshes are contained in the denser meshes. The total number of elements reported in Table 6.3 is the sum of the elements in both members. The results are reported with respect to a coordinate system that runs along frame where $s = 0$, L and $2L$ are correspond to points-3, 1 and 2, respectively. The bending stress for each element is calculated at $(x',y') = (0,h/2)$ for each element.

Table 6.3 shows that the primary variable, $v(L)$, the deflection component at the apex of the frame, is calculated accurately even for $N_E = 2$. However, the secondary variables, $\sigma_a(0)$, $\sigma_b(0)$ and $\tau(0)$ calculated at point-3, converge slowly. The bending stress at $s = 0$ is predicted to be −833 kPa for $N_E = 2$, this reduces to −3.2 Pa with $N_E = 1{,}024$.

The deflection and stress components along the frame, starting from point-3 ($s = 0$) and ending at point-2 ($s = 2L$), are plotted in Fig. 6.8 for $N_E = 8$ and 1,024.

Problems

6.1 Obtain the C^1 continuous shape functions for the 3-node element shown in the Fig. 6.9. The dof vector is $\{d\} = \{v_1 \;\; \theta_1 \;\; v_2 \;\; \theta_2 \;\; v_3 \;\; \theta_3\}^T$.

6.2 A two node element has three degrees of freedom per node, for a total of six degrees of freedom, $\{d\} = \{v_1 \;\; \theta_1 \;\; \kappa_1 \;\; v_2 \;\; \theta_2 \;\; \kappa_2\}^T$. Derive the C^2 shape functions, $N_i(x)$ for $i = 1$-6 to enable the following interpolation,

$$v(x) = N_1 v_1 + N_2\theta_1 + N_3\kappa_1 + N_4 v_2 + N_5\theta_2 + N_6\kappa_2$$

with $\theta = \frac{dv}{dx}$ and $\kappa = \frac{d^2v}{dx^2}$

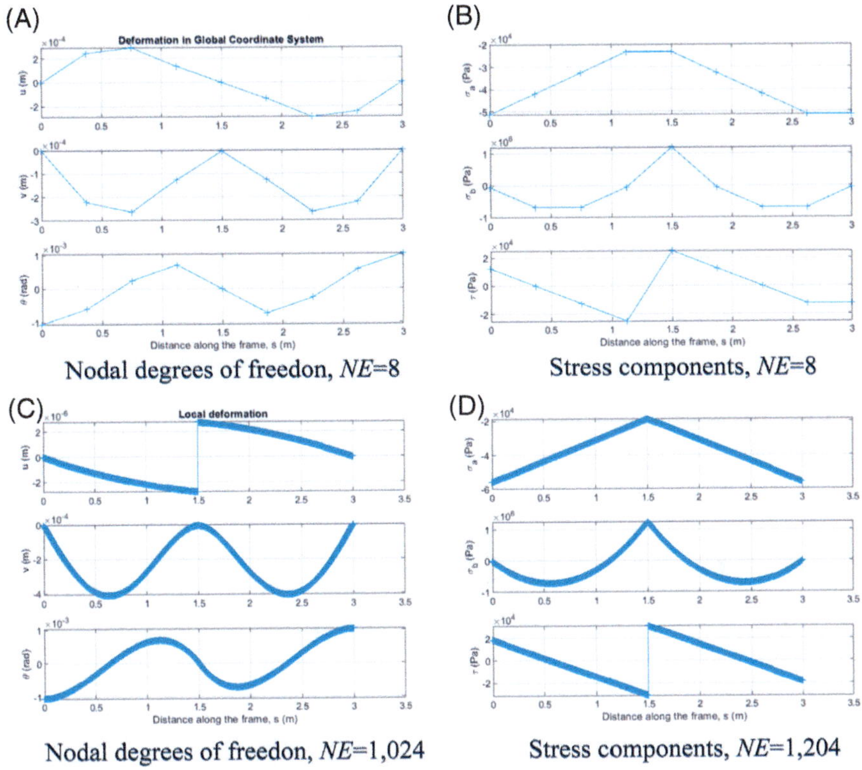

FIGURE 6.8 Degrees of freedom in global coordinates, and the element stresses for (A, B) $N_E = 8$ and (C, D) $N_E = 1024$.

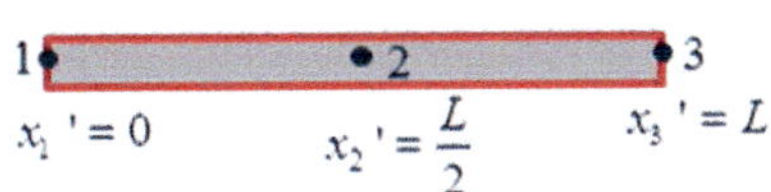

FIGURE 6.9 Problem 6.1.

6.3 Show the details of the derivation of the C^1 continuous shape functions given in Eq. (6.5).

6.4 Imagine that a curve $\phi = \phi(x)$ is to be fitted to three data values: ϕ_1 and $\phi_{,x1}$ at $x = 0$ and ϕ_2 at $x = L$ ($\phi_{,x2}$ unspecified). Determine the shape functions. Sketch them and check their behavior at $x = 0$ and $x = L$.

6.5 Consider the beam-12 supported by a cable-23 as shown in Fig. 6.10. The properties of these members are given as follows:

$$E_{12} = 70 \times 10^9 \text{ Pa}, A_{12} = 10 \text{ cm}^2, I_{12} = 10 \text{ cm}^4, L_{12} = 0.5 \text{ m},$$
$$E_{23} = 200 \times 10^9 \text{ Pa}, \text{A}_{23} = 2 \text{ cm}^2, \text{L}_{23} = 0.25 \text{ m}.$$

a) Find the magnitudes of the active degrees of freedom.

b) Find the stresses in each member at $x' = 0, L^{(e)}/2, L^{(e)}$

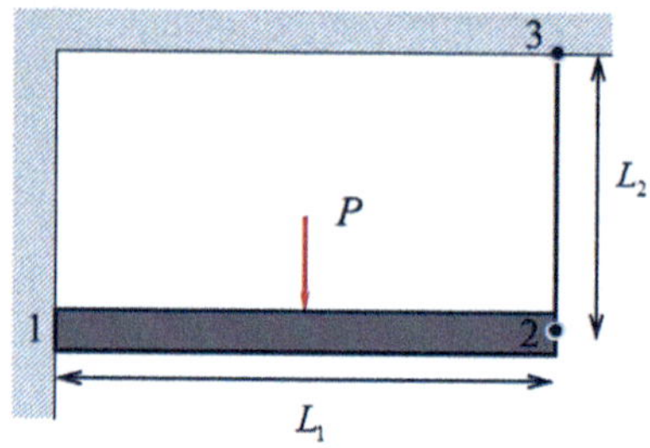

FIGURE 6.10 Problem 6.5.

Note: P is located at $L_1/2$ on the beam, with 100 N magnitude in the indicated direction.

6.6 Consider the members-12 and −23 which are welded at point-2, and supported as shown in Fig. 6.11. The member-12 is subjected to transverse loading as shown.

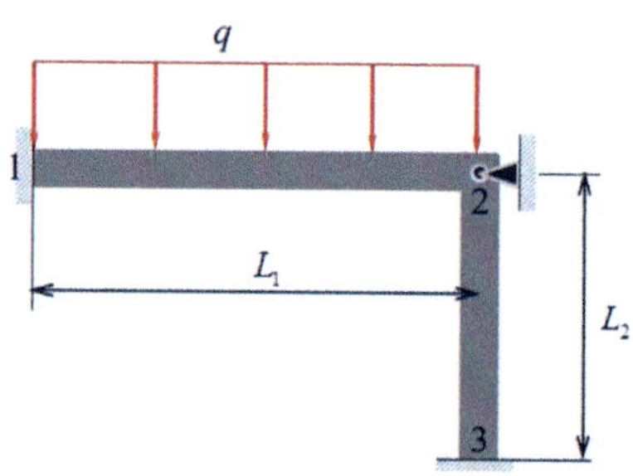

FIGURE 6.11 Problem 6.6.

$E_1 = E_2 = 14$ GPa
$b_1 = b_2 = 4$ cm; $h_1 = h_2 = 15$ cm ($I = bh^3/12$)
$A_1 = A_2$
$L_1 = 2$ m, $L_2 = 3$ m
$Q_0 = 1$ kN/m on member 12

a) Find the magnitudes of the active degrees of freedom.
b) Find the stresses in each member $x' = 0, L^{(e)}/2, L^{(e)}$

6.7 The total potential energy of a tensioned, Euler–Bernoulli beam (Fig. 6.12) is given as follows:

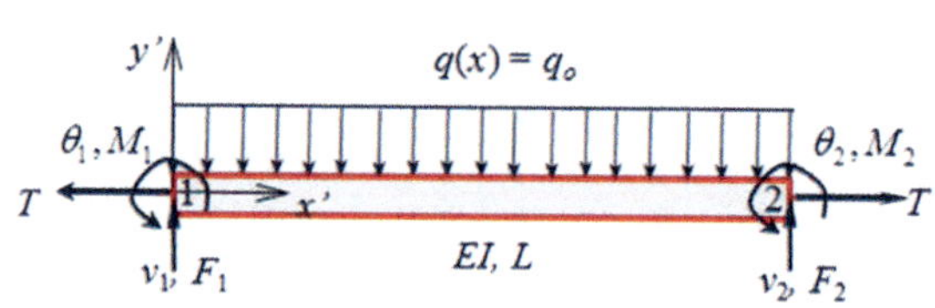

FIGURE 6.12 Problem 6.7.

$$\pi = \int_0^L \left(\frac{1}{2}EI\left(\frac{d^2 v}{dx^2}\right)^2 + \frac{1}{2}T\left(\frac{dv}{dx}\right)^2 - q_o v\right)dx + \left[v\left(F - T\frac{dv}{dx}\right)\right]_{x=0}^{x=L} - \left[\frac{dv}{dx}M\right]_{x=0}^{x=L}$$

where

- $v(x)$ is the transverse deflection
- EI: is the bending rigidity
- T: is the tension applied along the neutral axis
- q_o: is the external transverse distributed force
- F: is the nodal transverse force
- M: is the nodal moment
- L: is the length of the element

By using the C^1 continuous shape functions, determine the element equilibrium equations by using two node element with C^1 continuous shape functions. Note that $\theta = dv/dx$.

Reference

[1] I. Elishakoff, Handbook on Timoshenko-Ehrenfest Beam and Uflyand-Mindiln Plate Theories, World Scientific, 2020.

Chapter 7

Isoparametric elements for two-dimensional elastic solids

7.1 Introduction

The goal of this chapter is to develop finite element representation of two-dimensional elasticity problems. Equilibrium equations which represent the mechanics of such problems will be obtained by using the general quadrilateral elements. The methodology presented in this chapter is general, such that it can directly be expanded to the solution of one- and three-dimensional linear elasticity problems, and to using different element types.

Analytical solutions of two-dimensional elasticity problems have a rich literature and have contributed to engineering science and technology extensively over the last century [1–3]. However, analytical solutions become increasingly difficult to find if the material properties, the shape of the solution domain, or the loading vary in a complicated manner. The *finite element method* is a versatile numerical technique which presents a relatively straight forward solution for problems with geometric and material variations in the solution domain.

In finite element method, the solution domain is divided into small segments called *elements*. Elements in a two-dimensional domain are generally *quadrilateral* or *triangular* and they are connected to one another along their edges. The ensemble of the elements is known as the *finite element mesh*. In Fig. 7.1, a two-dimensional domain is divided into N_E elements. Each element is given a number e, where $e \in [1, N_E]$.

In previous chapters, we established that the number of *nodes* associated with an element depends on the *order of polynomial* used to interpolate the variables. A similar relationship exists for two-dimensional elements; however, the nature of two-dimensional polynomials, symmetry requirements, and geometry of the elements dictate certain restrictions. This will be discussed in Section 7.5.

The area of element-e is indicated by $A^{(e)}$. In general, the area A of the solution domain is only approximately equal to the sum of the element areas,

$$A \cong \sum_{e=1}^{N_E} A^{(e)} \tag{7.1}$$

Finite Element Method: Physics and Solution Methods. DOI: https://doi.org/10.1016/B978-0-12-821127-4.00002-5
Copyright © 2022 Elsevier Inc. All rights reserved.

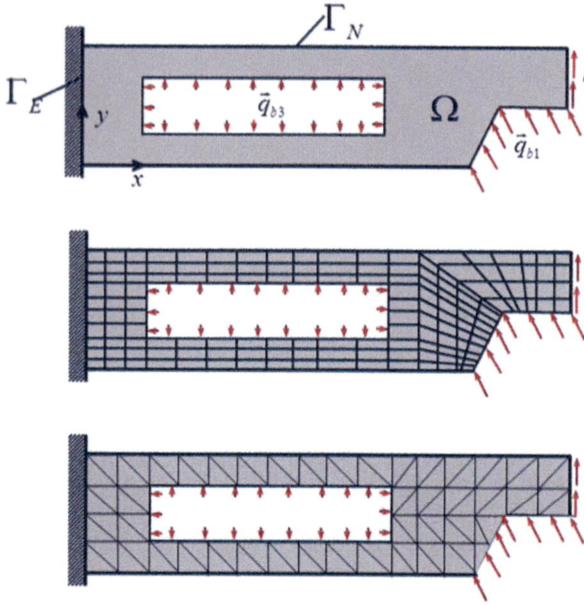

FIGURE 7.1 In finite element method, a solid domain is segmented into either quadrilateral or triangular subdomains called elements. The elements are connected across-their boundaries.

The reasons for this approximation will become clear later in this chapter.

As in the previous chapters, it is convenient to separate the *global coordinate system*(x, y) from the element coordinate system. In a general finite element mesh, the element edges are not parallel to the x- and y-axes. General quadrilateral and triangular elements are shown in Fig. 7.2. Derivation of the finite element form of the equilibrium equations require mapping of the formulation from the Cartesian coordinate system to the (local) *element coordinate system* (ξ, η). In this chapter, we will develop these equations for the so called *isoparametric formulation*.

The compliance of each element is represented by the *element stiffness matrix* formulation. The compliance of the entire structure is obtained by assembling each element stiffness matrix in to a *global stiffness matrix*. The external forces acting on the structure are either treated as natural boundary conditions, or as equivalent forces acting on the nodes of the structure. The technical term for this is *consistent nodal force*. The essential boundary conditions acting on the structure require the modification of the form or the size of the global stiffness matrix. The finite element analysis represents the compliance of the structure as linear algebraic equations for the unknown nodal displacements, known as the *global degrees of freedom*. The global equilibrium equation, obtained upon assembly, application of the essential boundary conditions and the external forces represents the Newton's second law. In this chapter the effects of inertial forces are not considered, thus we seek to find the *steady state deformation* and

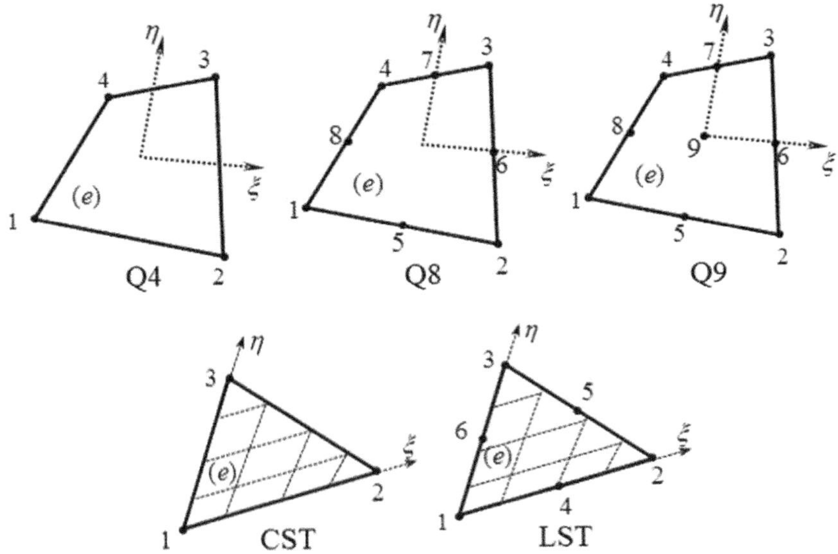

FIGURE 7.2 Quadrilateral elements with four, eight, and nine nodes. Triangular elements with three and six nodes.

stress state of the structure. The *secondary variables* such as the strain and the stress at each node and for each element are computed following the solution of the *global equilibrium equation* as a *postprocessing operation*.

7.2 Solution domain and its boundary

A two-dimensional elasticity problem is defined over a *planar domain* Ω which is an *set* of connected points where every two points in the set can be joined or connected by a polygonal path. The *boundary* Γ of a domain is a set of points such that some of the points that belong to this set belong to the domain and some do not [4]. In other words, a closed, two-dimensional, planar curve that helps enclose the solution domain Ω is the *boundary*. A planar domain could have *internal* and *external boundaries* as shown in Fig. 7.3.

The domain and the boundaries are often described with respect to a fixed global coordinate (reference) system. A Cartesian coordinate system (x, y) is depicted in Fig. 7.1. It is generally advantageous to define the boundary with respect to a tangential coordinate axis c, running along the boundary curve. In what follows we assume that both representations are available (Fig. 7.3).

7.2.1 Outward unit normal and tangent vectors along the boundary

The unit outward unit normal $\vec{n}$ and the tangent along the boundary $\vec{t}$ are two important vectors that are extensively used in defining the boundary and the boundary conditions. For a general curve that defines the boundary, these two

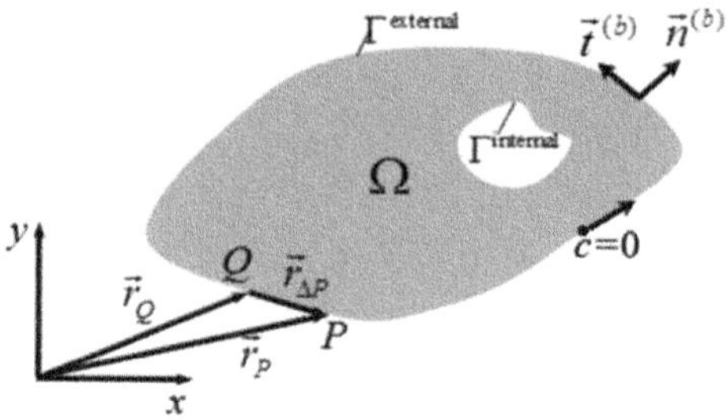

FIGURE 7.3 A two-dimensional domain Ω with internal $\Gamma^{internal}$ and external $\Gamma^{external}$ boundaries. The c-coordinate system describes the points on the boundary $\Gamma = \Gamma(c)$. The points inside the domain Ω can be referred to the global coordinate system (x, y).

vectors change from point-to-point. In this section, we introduce the general relationships to obtain these two vectors.

In the global coordinate system $\vec{n}^{(b)}$ and $\vec{t}^{(b)}$ vectors are represented as follows:

$$\vec{n}^{(b)} = n_x\hat{i} + n_y\hat{j} \tag{7.2a}$$

and

$$\vec{t}^{(b)} = t_x\hat{i} + t_y\hat{j} \tag{7.2b}$$

Let us consider the position vector $\vec{r}_P$ and the position vector $\vec{r}_Q$ of two points P and Q that are located on the boundary as shown in Fig. 7.3. If we assume that the coordinate direction c is positive in the counter clock wise direction, we find the distance $\vec{r}_{\Delta P}$ between these two points from the following relationship:

$$\vec{r}_P = \vec{r}_Q + \vec{r}_{\Delta P} \tag{7.3}$$

The components of the $\vec{r}_{\Delta P}$ vector are given as follows:

$$\vec{r}_{\Delta P} = \Delta x\,\hat{i} + \Delta y\,\hat{j} \tag{7.4}$$

In general, the tangent to the boundary at point P is defined by,

$$\vec{t}^{(b)} = \frac{d\vec{r}_P}{dc} \tag{7.5a}$$

or approximately by,

$$\vec{t}^{(b)} \approx \frac{\vec{r}_{\Delta P}}{\Delta c} \tag{7.5b}$$

Using the approximate form, the tangent vector can be expressed as follows:

$$\vec{t}^{(b)} \approx \frac{\vec{r}_{\Delta P}}{\Delta c} = \frac{\Delta x}{\Delta c}\hat{i} + \frac{\Delta y}{\Delta c}\hat{j} \tag{7.6}$$

As point P approaches point Q, i.e., $\Delta c \to 0$, along the curve Γ, we find the tangent at point P to be defined as follows:

$$\vec{t}^{(b)} = \frac{dx}{dc}\hat{i} + \frac{dy}{dc}\hat{j} \tag{7.7}$$

The unit outward normal at point P can then be found by using the orthogonality conditions, $\vec{n}^{(b)} \cdot \vec{t}^{(b)} = 0$ and $\vec{n}^{(b)} \times \vec{t}^{(b)} = 1$ as follows:

$$\vec{n}^{(b)} = -\frac{dy}{dc}\hat{i} + \frac{dx}{dc}\hat{j} \tag{7.8}$$

7.3 Equations of equilibrium for two-dimensional elastic solids

The derivation of the boundary value problem associated with loading response of a two-dimensional elastic domain is described in Section 2.2 in detail. Here, we summarize the general form of the variables and governing equations. The variables are defined in Chapter 2 and their definitions are omitted here for brevity. The boundary value problem consists of,

Equations of equilibrium:

$$\frac{\partial \sigma_{xx}}{\partial x} + \frac{\partial \tau_{xy}}{\partial y} + F_{B_x} = 0 \text{ (a)}, \quad \frac{\partial \tau_{yx}}{\partial x} + \frac{\partial \sigma_{yy}}{\partial y} + F_{B_y} = 0 \text{ (b) in } \Omega \tag{7.9}$$

Boundary conditions:

$$\vec{u} = \vec{u}^{(b)} \text{ on } \Gamma_E \text{ (a)}, \quad \vec{q} = \vec{q}^{(b)} \text{ on } \Gamma_N \text{ (b)} \tag{7.10}$$

The boundary condition vectors $\vec{u}_b$ and $\vec{q}^{(b)}$ are typically prescribed by using vector components normal and tangential to the boundary Γ.

The displacement form of the equations of equilibrium can be obtained by considering the kinematics of the deformation, strain-displacement relations and the constitutive relations summarized below.

$$\text{Deformation vector: } \vec{u} = u_x\hat{i} + u_y\hat{j} \tag{7.11}$$

where $u_x = u_x(x,y), \quad u_y = u_y(x,y)$

$$\text{Strain-displacement relationships: } \varepsilon_{xx} = \frac{\partial u_x}{\partial_x}, \quad \varepsilon_{yy} = \frac{\partial u_y}{\partial y}, \quad \gamma_{xy} = \frac{\partial u_y}{\partial x} + \frac{\partial u_x}{\partial y} \tag{7.12}$$

$$\text{Strain-compatibility condition: } \frac{\partial^2 \varepsilon_{yy}}{\partial x^2} + \frac{\partial^2 \varepsilon_{xx}}{\partial y^2} - \frac{\partial^2 \gamma_{xy}}{\partial x \partial y} = 0 \tag{7.13}$$

$$\text{Strain tensor: } [\varepsilon] = \begin{bmatrix} \varepsilon_{xx} & \gamma_{xy} \\ \gamma_{yx} & \varepsilon_{yy} \end{bmatrix} \tag{7.14}$$

$$\text{Strain-tensor in vector form: } \{\varepsilon\} = \left\{\varepsilon_{xx} \ \varepsilon_{yy} \ \gamma_{xy}\right\}^T \tag{7.15}$$

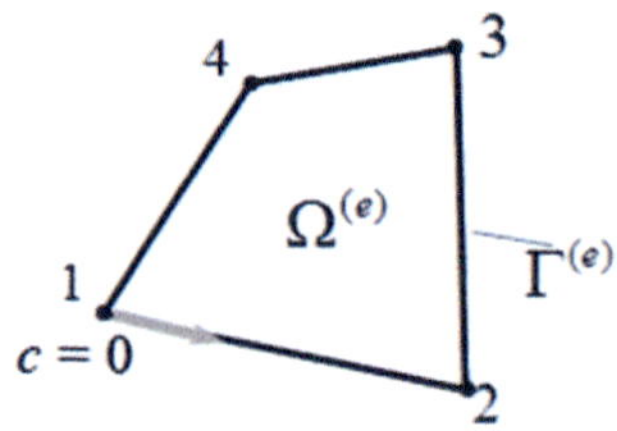

FIGURE 7.4 Element domain $\Omega^{(e)}$ and element boundary $\Gamma^{(e)}$ for a four-node quadrilateral (Q4) element. The position vector, c is used to refer to the coordinate system running along the boundary $\Gamma^{(e)}$.

Stress-tensor: $$[\sigma] = \begin{bmatrix} \sigma_{xx} & \tau_{xy} \\ \tau_{yx} & \sigma_{yy} \end{bmatrix} \tag{7.16}$$

Stress-tensor in vector form: $$\{\sigma\} = \left\{\sigma_{xx}\ \sigma_{yy}\ \tau_{xy}\right\}^T \tag{7.17}$$

Moment equilibrium: $$\tau_{xy} = \tau_{yx} \tag{7.18}$$

Traction on a surface: $$\vec{q}^{(b)} = q_x\hat{i} + q_y\hat{j} \tag{7.19}$$

Stress transformation: $$\left\{q^{(b)}\right\} = [\sigma]^T\left\{n^{(b)}\right\} \tag{7.20a}$$

$$\begin{Bmatrix} q_x \\ q_y \end{Bmatrix} = \begin{bmatrix} \sigma_{xx} & \tau_{yx} \\ \tau_{xy} & \sigma_{yy} \end{bmatrix}\begin{Bmatrix} n_x \\ n_y \end{Bmatrix}. \tag{7.20b}$$

Temperature effects and initial strains: $$\left\{\varepsilon^{(0)}\right\} = \alpha\,\Delta T\{1 \quad 1 \quad 0\}^T \tag{7.21}$$

Constitutive relation: $$\{\sigma\} = [E]\left(\{\varepsilon\} - \left\{\varepsilon^{(0)}\right\}\right) \tag{7.22}$$

Elasticity matrix:

Plane-stress: $$[E] = \frac{E}{1-\nu^2}\begin{bmatrix} 1 & \nu & 0 \\ \nu & 1 & 0 \\ 0 & 0 & (1-\nu)/2 \end{bmatrix} \tag{7.23}$$

Plane strain: $$[E] = \begin{bmatrix} \lambda + 2G & \lambda & 0 \\ \lambda & \lambda + 2G & 0 \\ 0 & 0 & G \end{bmatrix} \tag{7.24}$$

where $$\lambda = \frac{E\nu}{(1+\nu)(1-2\nu)}, \quad G = \frac{E}{2(1+\nu)} \tag{7.25}$$

7.4 General finite element form of equilibrium equations for a two-dimensional element

The general finite element form of the equations of equilibrium for a small subset of the solution domain can be obtained by using the weighted residual method. The smaller domain $\Omega^{(e)}$, known as an *element*, has a boundary $\Gamma^{(e)}$ (Fig. 7.4).

7.4.1 Variational form of the equation of equilibrium

The weighted residual integral associated with the boundary value problem described by Eq. (7.9) inside an element is expressed as follows:

$$R = \int_{\Omega^{(e)}} \left[w_{u_x} \left(\frac{\partial \sigma_{xx}}{\partial x} + \frac{\partial \tau_{xy}}{\partial y} + F_{B_x} \right) + w_{u_y} \left(\frac{\partial \tau_{yx}}{\partial x} + \frac{\partial \sigma_{yy}}{\partial y} + F_{B_y} \right) \right] d\Omega^e = 0 \tag{7.26}$$

where w_{u_x} and $w_{u_y}$are the weight functions which have the properties described in Section 3.4. Using the Green-Gauss theorem (Section 3.2), we can show that Eq. (7.26) becomes,

$$\begin{aligned} &\oint_{\Gamma^{(e)}} \left[w_{u_x} \left(\sigma_{xx} n_x + \tau_{xy} n_y \right) + w_{u_y} \left(\tau_{yx} n_x + \sigma_{yy} n_y \right) \right] d\Gamma^{(e)} \\ &- \int_{\Omega^{e0}} \left[\left(\sigma_{xx} \frac{\partial w_{u_x}}{\partial x} + \tau_{xy} \frac{\partial w_{u_x}}{\partial y} \right) + \left(\tau_{yx} \frac{\partial w_{u_y}}{\partial x} + \sigma_{yy} \frac{\partial w_{u_y}}{\partial y} \right) \right] d\Omega^{(e)} \\ &+ \int_{\Omega^{e0}} \left[w_{u_x} F_{B_x} + w_{u_y} F_{B_y} \right] d\Omega^{(e)} = 0 \end{aligned} \tag{7.27}$$

The boundary integral in Eq. (7.27) represents the traction applied on the boundary. This can be recognized by using the transformation Eq. (7.20) and the traction boundary condition in Eq. (7.10)(b). Thus we find the following relationships on the boundary:

$$q_x = \left(\sigma_{xx} n_x + \tau_{xy} n_y \right) \text{ and } q_y = \left(\tau_{yx} n_x + \sigma_{yy} n_y \right) \tag{7.28}$$

The weight functions can be replaced with the variation of the displacements as follows:

$$w_{u_x} = \delta u_x \text{ and } \mathrm{w}_{u_y} = \delta u_y,$$

as described in Section 3.4. Performing these substitutions and rearranging, we find,

$$\begin{aligned} &\int_{\Omega^{(e)}} \left[\sigma_{xx} \frac{\partial (\delta u_x)}{\partial x} + \tau_{xy} \left(\frac{\partial (\delta u_x)}{\partial y} + \frac{\partial (\delta u_y)}{\partial x} \right) + \sigma_{yy} \frac{\partial (\delta u_y)}{\partial y} \right] d\Omega^{(e)} \\ &- \oint_{\Gamma^{(e)}} \left[\delta u_x q_x + \delta u_y q_y \right] d\Gamma^{(e)} - \oint \left(\delta u_x F_{Bx} + \delta u_y F_{By} \right) d\Omega^{(e)} = 0 \end{aligned} \tag{7.29}$$

This equation can be further simplified by using the strain-displacement relationships given in Eq. (7.12),

$$\int_{\Omega^{(e)}} \left[\sigma_{xx}\,\delta\varepsilon_x + \tau_{xy}\,\delta\gamma_{xy} + \sigma_{yy}\,\delta\varepsilon_y\right]d\Omega^{(e)} - \oint_{\Gamma^{(e)}} \left[\delta u_x\, q_x + \delta u_y\, q_y\right]d\Gamma^{(e)} - \oint \left(\delta u_x\, F_{Bx} + \delta u_y\, F_{By}\right)d\Omega^{(e)} = 0 \tag{7.30}$$

In vector notation the same equation can be expressed as follows:

$$\int_{\Omega^{(e)}} \{\delta\varepsilon\}^T\{\sigma\}d\Omega^{(e)} = \oint_{\Gamma^{(e)}} \{\delta u\}^T\{q^{(b)}\}d\Gamma^{(e)} + \oint \{\delta u\}^T\{F_B\}d\Omega^{(e)} = 0 \tag{7.31}$$

where the variation of the strains is expressed by $\{\delta\varepsilon\} = \{\delta\varepsilon_{xx} \quad \delta\varepsilon_{yy} \quad \delta\gamma_{xy}\}^{\mathrm{T}}$, and the variation of the displacements is expressed by $\{\delta u\} = \{\delta u_x \quad \delta u_y\}^{\mathrm{T}}$.

7.4.2 Finite element form of the equation of equilibrium

Let us consider a generic two-dimensional element with M nodes. The shape functions associated with specific types of elements will be developed in the next section. Here, we provide a high-level derivation that can be used for triangular or rectangular elements.

The components of the deformation vector can be interpolated within the element as follows:

$$\tilde{u}_x = \sum_{i=1}^{M} N_i u_{xi} \tag{7.32a}$$

and

$$\tilde{u}_y = \sum_{i-1}^{M} N_i u_{yi} \tag{7.32b}$$

where u_{xi} and u_{yi} are the values of these variables at the nodes (*nodal variables*), N_i are the two-dimensional shape functions that enable the interpolation of the variables over the domain of the element. Note that, as before, the superscript ~ indicates the approximate nature of the interpolations used in the method. In order to avoid symbol explosion we will omit this sign in what follows. In vector notation, the displacement vector and its variation can be expressed as follows:

$$\begin{aligned} \{u\} &= \begin{Bmatrix} u_x \\ u_y \end{Bmatrix} = [N]\{d\} \\ \{\delta u\} &= \begin{Bmatrix} u_x \\ u_y \end{Bmatrix} = [N]\{\delta d\} \end{aligned} \tag{7.33}$$

where the *shape function matrix*$[N]$, the *degree of freedom vector*$\{d\}$and the variation of the degree of freedom vector $\{\delta d\}$are given as follows:

$$[N] = \begin{bmatrix} N_1 & 0 & N_2 & 0 & N_3 & \cdots & 0 \\ 0 & N_1 & 0 & N_2 & 0 & \cdots & N_M \end{bmatrix} \tag{7.34}$$

$$\{d\} = \left\{ u_{x1} \quad u_{y1} \quad \cdots \quad u_{xM} \quad u_{yM} \right\}^T \tag{7.35}$$

$$\{\delta d\} = \left\{ \delta u_{x1} \quad \delta u_{y1} \quad \cdots \quad \delta u_{xM} \quad \delta u_{yM} \right\}^T \tag{7.36}$$

respectively. Note that the element has M nodes and $2M$ degrees of freedom, therefore, the dimensions of the $[N]$ and $\{d\}$ are $2\times 2M$ and $2M\times 1$, respectively.

The strain vector in finite element form is expressed as follows:

$$\{\varepsilon\} = \left\{ \varepsilon_{xx} \quad \varepsilon_{yy} \quad \gamma_{xy} \right\}^T = [\partial]\{u\} \tag{7.37}$$

where the *derivative operator matrix* given in Eq. (2.43) is,

$$[\partial] = \begin{bmatrix} \dfrac{\partial}{\partial x} & 0 \\ 0 & \dfrac{\partial}{\partial y} \\ \dfrac{\partial}{\partial y} & \dfrac{\partial}{\partial x} \end{bmatrix} \tag{7.38}$$

Combining Eqs. (7.33)–(7.38), the strain vector is found as follows:

$$\{\varepsilon\} = [\partial][N]\{d\} = [B]\{d\} \tag{7.39}$$

where the *strain–displacement operator matrix* becomes,

$$[B] = [\partial][N] = \begin{bmatrix} \dfrac{\partial N_1}{\partial x} & 0 & \dfrac{\partial N_2}{\partial x} & 0 & \cdots & 0 \\ 0 & \dfrac{\partial N_1}{\partial y} & 0 & \dfrac{\partial N_2}{\partial y} & \cdots & \dfrac{\partial N_M}{\partial y} \\ \dfrac{\partial N_1}{\partial y} & \dfrac{\partial N_1}{\partial x} & \dfrac{\partial N_2}{\partial y} & \dfrac{\partial N_2}{\partial x} & \cdots & \dfrac{\partial N_M}{\partial x} \end{bmatrix} \tag{7.40}$$

Note that the size of the $[B]$ matrix is $3\times 2M$. The variation of the strain vector can be found by using Eq. (7.39) as follows:

$$\{\delta\varepsilon\} = [B]\{\delta d\} \tag{7.41}$$

The finite element version of the equation of static equilibrium can be obtained from Eq. (7.31) by using Eqs. (7.33) and (7.41) as follows:

$$\int\limits_{\Omega^{(e)}} \{\delta d\}^T [B]^T \{\sigma\} d\Omega^{(e)} = \int\limits_{\Omega^{(e)}} \{\delta d\}^T [N]^T \{F_B\} d\Omega^{(e)} + \oint\limits_{\Gamma^{(e)}} \{\delta d\}^T [N]^T \left\{ q^{(b)} \right\} d\Gamma^{(e)} = 0 \tag{7.42}$$

Noting that $\{\delta d\}$ is an arbitrary variation of the nodal degrees of freedom it can be eliminated from Eq. (7.42). The *finite element form of the equation of equilibrium for an element* is then found as follows:

$$\int_{\Omega^{(e)}} [B]^T\{\sigma\}d\Omega^{(e)} = \int_{\Omega^{(e)}} [N]^T\{F_B\}d\Omega^{(e)} + \oint_{\Gamma^{(e)}} [N]^T\{q^{(b)}\}d\Gamma^{(e)} \tag{7.43}$$

In case the nodes of the element are subjected external forces Eq. (7.43) can be augmented by the external nodal force vector $\{F\}^{(e)}$ as follows:

$$\int_{\Omega^{(e)}} [B]^T\{\sigma\}d\Omega^{(e)} = \int_{\Omega^{(e)}} [N]^T\{F_B\}d\Omega^{(e)} + \oint_{\Gamma^{(e)}} [N]^T\{q^{(b)}\}d\Gamma^{(e)} + \{F\}^{(e)} \tag{7.44}$$

Note that Eq. (7.44) represents the general form of equilibrium and it is valid for a material with nonlinear stress strain relationship. However, for a linearly elastic material, the constitutive Eq. (7.22) can be used to find the stress vector. Therefore, the equation of equilibrium becomes,

$$\int_{\Omega^{(e)}} [B]^T[E](\{\varepsilon\} - \{\varepsilon^{(0)}\})d\Omega^{(e)} = \int_{\Omega^{(e)}} [N]^T\{F_B\}d\Omega^{(e)} + \oint_{\Gamma^{(e)}} [N]^T\{q^{(b)}\}d\Gamma^{(e)} + \{F\}^{(e)} \tag{7.45}$$

As initial strain is a known variable it can be moved to the right hand side of this relationship.

$$\left[\int_{\Omega^{(e)}} [B]^T[E][B]d\Omega^{(e)}\right]\{d\} = \int_{\Omega^{(e)}} [N]^T\{F_B\}d\Omega^{(e)} + \oint_{\Gamma^{(e)}} [N]^T\{q^{(b)}\}d\Gamma^{(e)} + \{F\}^{(e)} + \int_{\Omega^{(e)}} [B]^T[E]\{\varepsilon^{(0)}\}d\Omega^{(e)} \tag{7.46}$$

Note that the term due to the initial strains, i.e., the last term of Eq. (7.46), has the same mathematical characteristic as the body force vector. In matrix form Eq. (7.46) is expressed as follows:

$$[k]^{(e)}\{d\} = \{r_B\}^{(e)} + \{r_q\}^{(e)} + \{r_{\varepsilon_0}\}^{(e)} + \{F\}^{(e)} \tag{7.47}$$

Element stiffness matrix: $[k]^{(e)} = \int_{\Omega^{(e)}} [B]^T[E][B]d\Omega^{(e)}$ (7.48a)

Consistent nodal force vectors

Boundary tractions: $\{r_q\}^{(e)} = \oint_{\Gamma^{(e)}} [N]^T\{q^{(b)}\}d\Gamma^{(e)}$ (7.48b)

$$\text{Body forces: } \{r_B\}^{(e)} = \int_{\Omega^{(e)}} [N]^T \{F_B\} d\Omega^{(e)} \tag{7.48c}$$

$$\text{Initial strains: } \{r_{\varepsilon 0}\}^{(e)} = \int_{\Omega^{(e)}} [B]^T [E] \{\varepsilon^{(0)}\} d\Omega^{(e)} \tag{7.48d}$$

External forces that happen to be acting on the element nodes:

$$\{F\}^{(e)} = \{F_{x1} \quad F_{y1} \quad F_{x2} \quad F_{y2} \quad \cdots \quad F_{xM} \quad F_{yM}\}^T \tag{7.48e}$$

Eq. (7.47) can also be expressed as follows:

$$[k]^{(e)}\{d\} = \{r\}^{(e)} \tag{7.49}$$

where the *total nodal force vector* is,

$$\{r\}^{(e)} = \{r_B\}^{(e)} + \{r_q\}^{(e)} + \{r_{\varepsilon_0}\}^{(e)} + \{F\}^{(e)}. \tag{7.50}$$

Note that the sizes of the $[k]^{(e)}$ matrix and the $\{r\}^{(e)}$ vector are $2M \times 2M$ and $2M \times 1$, respectively. Before these equations can be evaluated further we need to develop two-dimensional interpolation functions.

7.5 Interpolation across a two-dimensional domain

7.5.1 Two-dimensional polynomials

We next consider approximation of a two-dimensional function ϕ (x, y) defined in a Cartesian coordinate system by two-dimensional polynomials. The domain of this function is either a rectangle that spans the coordinates $-a \le x \le a$ and $-b \le y \le b$, or a right triangle that spans the coordinates $0 \le x \le a$ and $0 \le y \le b$, as shown in Fig. 7.5A. The choice of interpolation polynomials are guided by a few simple principles,

C1. The interpolation should allow for rigid body motion of the element.

C2. The interpolation should allow at least constant strain throughout the element.

C3. The interpolation of a variable (e.g., ϕ) throughout the element must be balanced, and not favor any one of the coordinate directions.

C4. The interpolation function should be a *complete polynomial* to the extent that is possible.

A family of complete, two-dimensional polynomials is represented by the summation of the terms in the Pascal's triangle (Table 7.1).

Each row of Pascal's triangle represents the order of the polynomial. In order to obtain a complete, two-dimensional polynomial of degree n, the terms on each line from 0 to n should be included in the polynomial. For example, the first three

TABLE 7.1 Pascal's triangle can be used to form two-dimensional polynomials.

N	
0	1
1	$x\ y$
2	$x^2\ xy\ y^2$
3	$x^3\ x^2y\ yx^2\ y^3$
4	$x^4\ x^3y\ x^2y^2\ yx^3\ y^4$
…	…

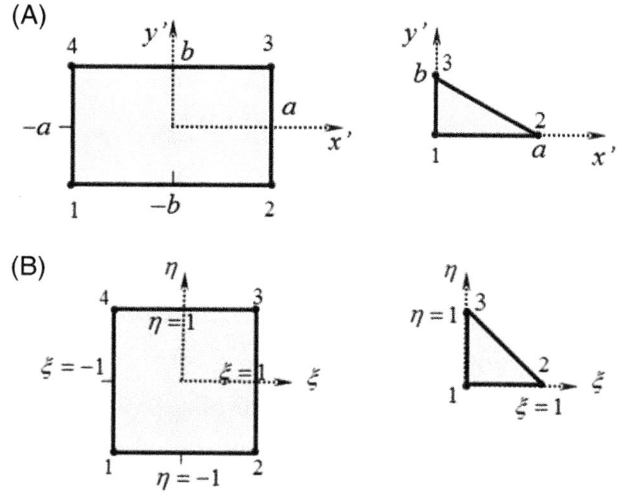

FIGURE 7.5 (A) Rectangular and triangular elements in a two-dimensional Cartesian coordinate system (x', y'). (B) The same elements in a normalized coordinate system (ξ, η).

complete polynomials are the following 3-, 6-, and 10-term equations,

$$\phi(x, y) = a_0 + a_1x + a_2y \tag{7.51}$$

$$\phi(x, y) = a_0 + a_1x + a_2y + a_3x^2 + a_4xy + a_5y^2 \tag{7.52}$$

$$\phi(x, y) = a_0 + a_1x + a_2y + a_3x^2 + a_4xy + a_5y^2 + a_6x^3 + a_7x^2y + a_8xy^2 + a_9y^3 \tag{7.53}$$

where a_j are scalar coefficients. Thus we see that the *complete two-dimensional polynomials* have the following number of terms:

Zeroth-order polynomial: single term
First-order polynomial: three terms
Second-order polynomial: six terms
Third-order polynomial: ten terms

In deciding which polynomial terms are suitable for setting the shape functions for the rectangular and triangular elements, the conditions C1–C4 stated above are considered. In addition, as shown in the previous chapters the coefficients a_j should be interchanged with the nodal unknowns ϕ_j. The first choice for the node locations are the corners of the elements.

A three node, triangular element has three nodal unknowns ϕ_1, ϕ_2, ϕ_3 and the first order, complete polynomial given in Eq. (7.51) has three coefficients: a_0, a_1, a_2. Thus, we see that the shape functions for a three node triangular element can be obtained by using Eq. (7.51). This is a linear interpolation in both directions. Incidentally, the gradient of the independent variable is always a constant for this type of element, and a three node triangular element is known as a *constant strain triangle (CST) element* for reasons that will become clear later.

A four node rectangular element has four nodal unknowns ϕ_1, ϕ_2, ϕ_3, ϕ_4, but there is no complete, two-dimensional polynomial with four terms. Nevertheless, a *symmetric*, four-term polynomial can be found from Pascal's triangle. Note that Eq. (7.52) represents a six-term complete polynomial. By neglecting the terms that involve x^2 and y^2 the following, symmetrical four-term polynomial can be obtained:

$$\phi(x, y) = a_0 + a_1 x + a_2 y + a_3 xy \tag{7.54}$$

The shape functions for a four-node rectangular element can be obtained by using Eq. (7.54). Note that the four-node quadrilateral element is also known as a *Q4-element*.

As we know from the earlier chapters, higher order interpolation functions help improve solution accuracy. In order to establish higher order interpolations, we use six-node triangles, eight-node rectangles and nine node rectangles (Fig. 7.2). These elements are known as *linear strain triangular (LST) element*, *Q8-element* and *Q9-element*, respectively.

The two-dimensional polynomial functions that satisfy the conditions given above and have number of unknown coefficients that are equal to the number of nodes of the above mentioned elements are as follows:

$$\begin{aligned}
\text{CST: } u(x, y) &= a_0 + a_1 x + a_2 y \\
\text{Q4: } u(x, y) &= a_0 + a_1 x + a_2 y + a_3 xy \\
\text{LST: } u(x, y) &= a_0 + a_1 x + a_2 y + a_3 x^2 + a_4 xy + a_5 y^2 \\
\text{Q8: } u(x, y) &= a_0 + a_1 x + a_2 y + a_3 x^2 + a_4 xy + a_5 y^2 + a_6 x^2 y + a_7 xy^2 \\
\text{Q9: } u(x, y) &= a_0 + a_1 x + a_2 y + a_3 x^2 + a_4 xy + a_5 y^2 + a_6 x^2 y + a_7 xy^2 + a_8 x^2 y^2
\end{aligned} \tag{7.55}$$

Note that of these five elements, only CST and LST represent complete, two-dimensional polynomials. For the Q4, Q8 and Q9 some kind of selection that preserves the symmetry of the polynomial has to be made. Higher order polynomials that will be useful for finite element interpolations can be obtained by keeping an eye on the symmetry, in similar fashion.

7.5.2 Two-dimensional shape functions

By interchanging the nodal degrees of freedom ϕ_i with the coefficients a_i, the polynomial interpolation functions given above can be represented in a form that is more suitable for the finite element method.

7.5.2.1 *Plane quadrilateral elements*

Consider, a four-node rectangular element with dimensions $2a\times 2b$ and centrally located origin (Fig. 7.5A). This element is defined in the following region:

$$-a \le x' \le a \text{ and } -b \le y' \le b \tag{7.56}$$

The element nodes are numbered 1 through 4 as shown in Fig. 7.5. The nodal coordinates and nodal degrees of freedom for nodes 1 – 4 are: $-a, -b)$, $(a, -b)$, (a, b), and $(-a, b)$ and ϕ_1, ϕ_2, ϕ_3, and ϕ_4, respectively. The four-term interpolation function given in Eq. (7.54) can be evaluated at the four nodes of this element to give the following matrix relationship:

$$\begin{bmatrix} 1 & -a & -b & ab \\ 1 & a & -b & -ab \\ 1 & a & b & ab \\ 1 & -a & b & -ab \end{bmatrix} \begin{Bmatrix} a_0 \\ a_1 \\ a_2 \\ a_3 \end{Bmatrix} = \begin{Bmatrix} \phi_1 \\ \phi_2 \\ \phi_3 \\ \phi_4 \end{Bmatrix} \tag{7.57}$$

Solving for the coefficients $a_0 - a_3$, substituting the result in Eq. (7.54) and rearranging the result for the nodal variables $\phi_1 - \phi_4$ the two-dimensional interpolation for a Q4-element is found as follows:

$$\phi(x, y) = N_1^{(Q4)}\phi_1 + N_2^{(Q4)}\phi_2 + N_3^{(Q4)}\phi_3 + N_4^{(Q4)}\phi_4$$

where the $Q4$ *shape functions* are the following functions:

$$\begin{aligned} N_1^{(Q4)}(x', y') &= \frac{1}{4}\left(1 - \frac{x'}{a}\right)\left(1 - \frac{y'}{b}\right), \quad N_2^{(Q4)}(x', y') = \frac{1}{4}\left(1 + \frac{x'}{a}\right)\left(1 - \frac{y'}{b}\right) \\ N_3^{(Q4)}(x', y') &= \frac{1}{4}\left(1 + \frac{x'}{a}\right)\left(1 + \frac{y'}{b}\right), \quad N_4^{(Q4)}(x', y') = \frac{1}{4}\left(1 - \frac{x'}{a}\right)\left(1 + \frac{y'}{b}\right) \end{aligned} \tag{7.58}$$

Note that while the superscript ^{Q4} is useful to distinguish the type of element, it is not typically used in the FE-literature. The reader is instead expected to recognize the elements from the context. In this section, we will keep the superscripts, but omit them in the rest of this text.

The shape functions in Eq. (7.58) have the following property,

$$\begin{aligned} &\text{at node 1:} \quad N_1^{(Q4)} = 1, \quad N_2^{(Q4)} = N_3^{(Q4)} = N_4^{(Q4)} = 0 \\ &\text{at node 2:} \quad N_2^{(Q4)} = 1, \quad N_2^{(Q4)} = N_4^{(Q4)} = N_1^{(Q4)} = 0 \\ &\text{at node 3:} \quad N_3^{(Q4)} = 1, \quad N_4^{(Q4)} = N_1^{(Q4)} = N_2^{(Q4)} = 0 \\ &\text{at node 4:} \quad N_4^{(Q4)} = 1, \quad N_1^{(Q4)} = N_2^{(Q4)} = N_3^{(Q4)} = 0 \end{aligned} \tag{7.59}$$

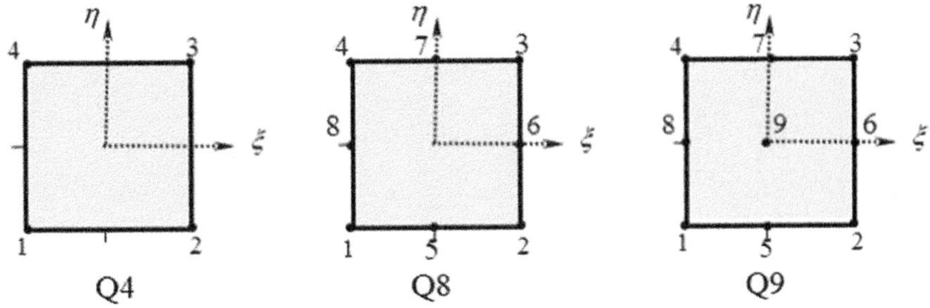

FIGURE 7.6 Common four-, eight-, and nine-node quadrilateral elements, Q4, Q8, and Q9, in generalized coordinates.

which allows for the interpolation function to meet the proper nodal values ϕ_i at the nodes.

The shape functions in Eq. (7.58) can be expressed by using the following variables,

$$\xi = \frac{x'}{a} \text{ and } \eta = \frac{y'}{b}. \tag{7.60}$$

The domain of the interpolation then becomes,

$$-1 \leq \xi \leq 1 \text{ and } -1 \leq \eta \leq 1$$

The shape functions for the Q4 elements in this nondimensional domain become,

$$\begin{aligned} N_1^{(Q4)} &= \frac{1}{4}(1-\xi)(1-\eta), \; N_1^{(Q4)} = \frac{1}{4}(1+\xi)(1-\eta) \\ N_3^{(Q4)} &= \frac{1}{4}(1+\xi)(1+\eta), \; N_4^{(Q4)} = \frac{1}{4}(1-\xi)(1+\eta) \end{aligned} \tag{7.61}$$

The nondimensionalized versions of the shape functions are used in derivation of the finite element equilibrium equations for general quadrilateral elements.

7.5.2.2 *Interpolation and shape functions for plane quadrilateral elements*

Common quadrilateral elements with four, eight or nine nodes are shown in Fig. 7.6. The locations of the nodes are chosen to preserve the symmetry of the elements and they are given in Table 7.2. In the rest of this chapter we will generally use the normalized coordinates (ξ, η) and give the shape functions in terms of these variables. One can easily obtain the shape functions in the (x', y') coordinate system by using the normalizations defined in Eq. (7.60).

TABLE 7.2 Nodal coordinates in the normalized coordinate system (ξ, η) for Q4, Q8, Q9, CST, and LST elements.

node-#	1	2	3	4	5	6	7	8	9
Q4	(−1,−1)	(1,−1)	(1,1)	(−1,1)	–	–	–	–	–
Q8	(−1,−1)	(1,−1)	(1,1)	(−1,1)	(0,−1)	(1,0)	(0,1)	(−1,0)	–
Q9	(−1,−1)	(1,−1)	(1,1)	(−1,1)	(0,−1)	(1,0)	(0,1)	(−1,0)	(0,0)
CST or T3	(0,0)	(1,0)	(0,1)	–	–	–	–	–	–
LST or T6	(0,0)	(1,0)	(0,1)	(1/2,0)	(1/2,1/2)	(0,1/2)	–	–	–

Q4-element:

The interpolation function is,

$$\phi(\xi,\eta)=\sum_{i=1}^{4}N_i^{(Q4)}\phi_i \tag{7.62}$$

The shape functions are,

$$\begin{aligned}N_1^{(Q4)}(\xi,\eta)&=\frac{1}{4}(1-\xi)(1-\eta),\ N_2^{(Q4)}(\varepsilon,\eta)=\frac{1}{4}(1+\xi)(1-\eta)\\ N_3^{(Q4)}(\xi,\eta)&=\frac{1}{4}(1+\xi)(1+\eta),\ N_4^{(Q4)}(\xi,\eta)=\frac{1}{4}(1-\xi)(1+\eta)\end{aligned} \tag{7.63}$$

Q8-element:

The interpolation function is,

$$\phi(\xi,\eta)=\sum_{i=1}^{8}N_i^{(Q8)}\phi_i \tag{7.64}$$

The shape functions are,

$$\begin{aligned}N_1^{(Q8)}&=N_1^{(Q4)}-\frac{1}{2}\left(N_8^{(Q8)}+N_5^{(Q8)}\right),\ N_2^{(Q8)}=N_2^{(Q4)}-\frac{1}{2}\left(N_5^{(Q8)}+N_6^{(Q8)}\right)\\ N_3^{(Q8)}&=N_3^{(Q4)}-\frac{1}{2}\left(N_6^{(Q8)}+N_7^{(Q8)}\right),\ N_4^{(Q8)}=N_4^{(Q4)}-\frac{1}{2}\left(N_7^{(Q8)}+N_8^{(Q8)}\right)\end{aligned}$$

with

$$\begin{aligned}N_5^{(Q8)}&=\frac{1}{2}(1-\xi^2)(1-\eta),\ N_6^{(Q8)}=\frac{1}{2}(1+\xi)(1-\eta^2)\\ N_7^{(Q8)}&=\frac{1}{2}(1-\xi^2)(1+\eta),\ N_8^{(Q8)}=\frac{1}{2}(1-\xi)(1-\eta^2)\end{aligned} \tag{7.65}$$

Q9-element:

The interpolation function is,

$$\phi(\xi,\eta)=\sum_{i=1}^{9}N_i^{(Q9)}\phi_i \tag{7.66}$$

The shape functions are,

$$\begin{aligned}N_1^{(Q9)}&=N_1^{(Q4)}-\frac{1}{2}\left(N_8^{(Q8)}+N_5^{(Q8)}\right)-\frac{1}{4}N_9^{(Q9)},\ N_2^{(Q9)}=N_2^{(Q4)}-\frac{1}{2}\left(N_5^{(Q8)}+N_6^{(Q8)}\right)-\frac{1}{4}N_9^{(Q9)}\\ N_3^{(Q9)}&=N_3^{(Q4)}-\frac{1}{2}\left(N_6^{(Q8)}+N_7^{(Q8)}\right)-\frac{1}{4}N_9^{(Q9)},\ N_4^{(Q9)}=N_4^{(Q4)}-\frac{1}{2}\left(N_7^{(Q8)}+N_8^{(Q8)}\right)-\frac{1}{4}N_9^{(Q9)}\\ N_5^{(Q9)}&=N_5^{(Q8)}-\frac{1}{4}N_9^{(Q9)},\ N_6^{(Q9)}=N_6^{(Q8)}-\frac{1}{4}N_9^{(Q9)}\\ N_7^{(Q9)}&=N_7^{(Q8)}-\frac{1}{4}N_9^{(Q9)},\ N_8^{(Q9)}=N_8^{(Q8)}-\frac{1}{4}N_9^{(Q9)}\\ N_9^{(Q9)}&=(1-\xi^2)(1-\eta^2)\end{aligned} \tag{7.67}$$

7.5.2.3 Plane triangular elements

Consider a right angled, three-node triangular element with unit edges as shown in Fig. 7.5. The nodal coordinates and nodal degrees of freedom for nodes 1–3 are: (0,0), (1,0), and (0,1) and ϕ_1, ϕ_2, and ϕ_3, respectively. The three-term interpolation function given in Eq. (7.51) can be evaluated at the three nodes of this element to give the following matrix relationship:

$$\begin{bmatrix} 1 & 0 & 0 \\ 1 & 1 & 0 \\ 1 & 0 & 1 \end{bmatrix} \begin{Bmatrix} a_0 \\ a_1 \\ a_2 \end{Bmatrix} = \begin{Bmatrix} \phi_1 \\ \phi_2 \\ \phi_3 \end{Bmatrix} \tag{7.68}$$

Solving for the coefficients $a_0 - a_2$, substituting the result in Eq. (7.51) and rearranging the result for the nodal variables $\phi_1 - \phi_3$ the finite element form of the two-dimensional interpolation for a CST-element is found as follows:

$$\phi(\xi, \eta) = N_1^{(T3)}\phi_1 + N_2^{(T3)}\phi_2 + N_3^{(T3)}\phi_3 \tag{7.69}$$

where the *CST shape functions* are,

$$N_1^{(T3)}(\xi, \eta) = 1 - \xi - \eta, \ N_2^{(T3)}(\xi, \eta) = \xi, \ N_3^{(T3)}(\xi, \eta) = \eta \tag{7.70}$$

As expected, these shape functions have the following property:

$$\begin{aligned} &\text{at node 1:} \quad N_1^{(T3)} = 1, \quad N_2^{(T3)} = N_3^{(T3)} = 0 \\ &\text{at node 2:} \quad N_2^{(T3)} = 1, \quad N_1^{(T3)} = N_3^{(T3)} = 0 \\ &\text{at node 3:} \quad N_3^{(T3)} = 1, \quad N_1^{(T3)} = N_2^{(T3)} = 0 \end{aligned} \tag{7.71}$$

7.5.2.4 Interpolation and shape functions for plane triangular elements

Triangular elements with three or six nodes are shown in Fig. 7.2. The locations of the nodes are given in Table 7.2. The shape functions are given in the normalized coordinate system (ξ, η).

CST-element:
The interpolation function is,

$$\phi(\xi, \eta) = \sum_{i=1}^{3} N_i^{(T3)}\phi_i \tag{7.72}$$

The shape functions are,

$$N_1^{(T3)}(\xi, \eta) = 1 - \xi - \eta, \ N_2^{(T3)}(\xi, \eta) = \xi, \ N_3^{(T3)}(\xi, \eta) = \eta \tag{7.73}$$

LST-element:
The interpolation function is,

$$\phi(\xi, \eta) = \sum_{i=1}^{6} N_i^{(T6)}\phi_i \tag{7.74}$$

The shape functions are,

$$\begin{aligned}
&N_1^{(T6)}(\xi,\eta) = (1-\xi-\eta)(1-2\xi-2\eta),\ N_2^{(T6)}(\xi,\eta) = \xi(2\xi-1),\\
&N_3^{(T6)}(\xi,\eta) = \eta(2\eta-1)\\
&N_4^{(T6)}(\xi,\eta) = 4\xi(1-\xi-\eta),\ N_5^{(T6)}(\xi,\eta) = 4\xi\eta,\\
&N_6^{(T6)}(\xi,\eta) = 4\eta(1-\xi-\eta)
\end{aligned} \tag{7.75}$$

7.6 Mapping between general quadrilateral and rectangular domains

Consider an M-node (M = 4, 8 or 9) rectangular element defined over the generalized coordinates (ξ,η) and an M-node *quadrilateral element* with nodal coordinates (x_i,y_i) (i = 1 − M). The *general rectangular element* defined with respect to the (ξ,η) coordinate system can be mapped to the quadrilateral element defined with respect to a Cartesian coordinate system (x,y) as follows:

$$x = \sum_{i=1}^{M} N_i(\xi,\eta)x_i \text{ and } y = \sum_{i=1}^{M} N_i(\xi,\eta)y_i \tag{7.76}$$

This relationship states that any point P located in the general rectangular element that has coordinates (ξ_P,η_P) has a corresponding (x_P,y_P) in the Cartesian coordinate system and vice versa. However, the fact that a mapping can be described does not guarantee that it is a valid mapping. The validity of mapping is discussed in some detail later in this section. The mapping is shown graphically in Fig. 7.7 for a Q4 element, for which $M = 4$.

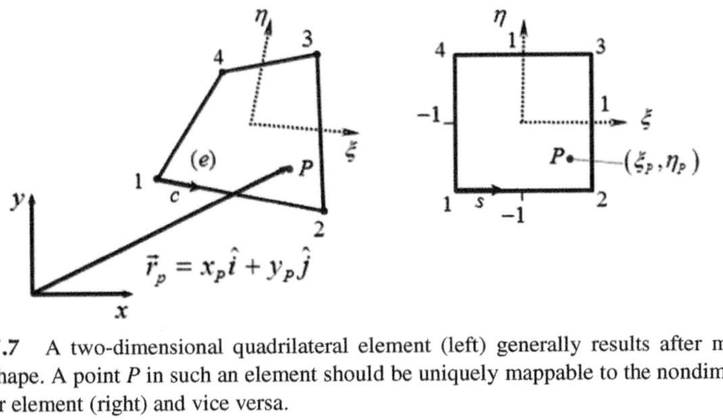

FIGURE 7.7 A two-dimensional quadrilateral element (left) generally results after meshing a complex shape. A point P in such an element should be uniquely mappable to the nondimensional, rectangular element (right) and vice versa.

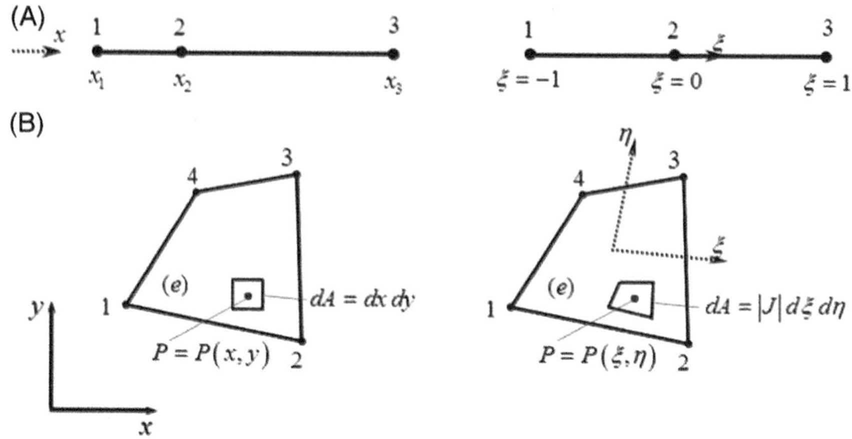

FIGURE 7.8 (A) A one-dimensional, three-node element in dimensional and generalized coordinates. (B) The differential area *dA* for a Q4 element, expressed from Cartesian and generalized coordinate perspectives.

7.6.1 Jacobian matrix and Jacobian determinant

The uniqueness of the mapping between the physical and the generalized coordinates depends on the shape of the elements and the locations of its nodes. The Jacobian is a measure that determines the scaling between the two domains, and hence it is a significant measure of the uniqueness. In what follows, we introduce the concept of Jacobian first in one-dimensional domain and then extend it to mappings in two dimensions.

7.6.1.1 Jacobian for a one-dimensional, a three-node element

Consider a one-dimensional element with three nodes, located at x_1, x_2, and x_3 and a three-node, mapped element with nodes at -1, 0, and 1, as shown in Fig. 7.8A. The shape functions for the mapped element can be obtained by using Eq. (4.43) as follows:

$$x = x(\xi, \eta) \text{ and } y = y(\xi, \eta)$$

$$N_1 = \frac{1}{2}\xi(\xi - 1), \; N_2 = 1 - \xi^2, \; N_3 = \frac{1}{2}\xi(\xi + 1) \tag{7.77}$$

The x-coordinate can then be mapped to the ξ-coordinate by using the following relationship:

$$x = N_1(\xi)x_1 + N_2(\xi)x_2 + N_3(\xi)x_3 \tag{7.78}$$

The total derivative of x can be computed as follows:

$$dx = \frac{dx}{d\xi}d\xi \tag{7.79}$$

Here $dx/d\xi$ can be regarded as a *stretch factor* between dx and $d\xi$. We define this variable as the *Jacobian*,

$$J = \frac{dx}{d\xi} = \frac{1}{2}(2\xi - 1)x_1 - 2\xi x_2 + \frac{1}{2}(2\xi + 1)x_3 \tag{7.80}$$

The mapping defined in Eq. (7.78) changes the stretch factor J as a function of position in this three-node element. Further insight can be gained by considering the element length is L and setting $x_1 = 0$, the Jacobian for this element becomes,

$$J = L\left[\frac{1}{2} + \left(1 - 2\frac{x_2}{L}\right)\xi\right] \tag{7.81}$$

We then note that only if x_2 is chosen as the mid-point ($x = L/2$) of the element, the Jacobian is a constant, $J = L/2$. This also reveals that J is positive only if the mid-node is located in the range, $L/4 < x_2 < 3L/4$. Otherwise, if the mid-node is chosen outside of this range, dx would be mapped to a negative $d\xi$ value at least in a certain region of the element! This would not be a desired feature for such elements. In most one dimensional, three-node elements x_2 is chosen at the mid-point of the element.

One can easily show that for a two-node element the J is constant with a value of $L/2$.

7.6.1.2 *Jacobian for a two-dimensional element*

Next, consider the two-dimensional mapping defined in Eq. (7.76) between the (x,y) and the (ξ,η) coordinate systems,

$$x = x(\xi, \eta) \text{ and } y = y(\xi, \eta) \tag{7.82}$$

Any one of the two-dimensional elements/interpolations defined by Eqs. (7.62), (7.64), (7.66), (7.72), or (7.74) can be considered for this mapping. The total derivatives of x and y can be evaluated as follows:

$$dx = \frac{\partial x}{\partial \xi}d\xi + \frac{\partial x}{\partial \eta}d\eta, \text{ and } dy = \frac{\partial y}{\partial \xi}d\xi + \frac{\partial y}{\partial \eta}d\eta \tag{7.83}$$

In matrix form, this relationship becomes,

$$\begin{Bmatrix} dx \\ dy \end{Bmatrix} = \begin{bmatrix} \frac{\partial x}{\partial \xi} & \frac{\partial x}{\partial \eta} \\ \frac{\partial y}{\partial \xi} & \frac{\partial y}{\partial \eta} \end{bmatrix} \begin{Bmatrix} d\xi \\ d\eta \end{Bmatrix} \tag{7.84}$$

where the Jacobian matrix for this interpolation is defined as follows:

$$[J] = \begin{bmatrix} \frac{\partial x}{\partial \xi} & \frac{\partial x}{\partial \eta} \\ \frac{\partial y}{\partial \xi} & \frac{\partial y}{\partial \eta} \end{bmatrix} \tag{7.85}$$

The *Jacobian matrix* [J] is named after German mathematician Jacobi (1804–1851). Note that, in finite element method, for a given element type, the terms of the Jacobian matrix can be easily computed by the mapping defined in Eq. (7.76) and by using the appropriate shape functions, as follows:

$$J_{11} = \frac{\partial x}{\partial \xi} = \sum_{i=1}^{M} \frac{\partial N_i}{\partial \xi} x_i \tag{7.86a}$$

$$J_{12} = \frac{\partial x}{\partial \eta} = \sum_{i=1}^{M} \frac{\partial N_i}{\partial \eta} x_i \tag{7.86b}$$

$$J_{21} = \frac{\partial y}{\partial \xi} = \sum_{i=1}^{M} \frac{\partial N_i}{\partial \xi} y_i \tag{7.86c}$$

$$J_{22} = \frac{\partial y}{\partial \eta} = \sum_{i=1}^{M} \frac{\partial N_i}{\partial \eta} y_i \tag{7.86d}$$

7.6.1.3 Jacobian along the edge of a two-dimensional element

The contours integrals such as the one in Eq. (7.48)(b) are integrated along the boundary/edge $\Gamma^{(e)}$. A distinction must be made between the coordinate axis running along the edge of the physical element c and the one running along the edge of the nondimensional (mapped) element s (Fig. 7.7). The coordinate axis s along the mapped-element becomes,

$$\text{edge-e1:}\ s = \xi,\ \eta = -1 \tag{7.87a}$$

$$\text{edge-e2:}\ s = \eta,\ \xi = 1 \tag{7.87b}$$

$$\text{edge-e3:}\ s = \xi,\ \eta = 1 \tag{7.87c}$$

$$\text{edge-e4:}\ s = \eta,\ \xi = -1 \tag{7.87d}$$

By using the mapping defined in Eq. (7.76), for a Q4-element the positions along the boundary c can be found as follows:

$$\text{edge-e1:}\ x = \frac{1}{2}(1-s)x_1 + \frac{1}{2}(1+s)x_2 \text{ and } y = \frac{1}{2}(1-s)y_1 + \frac{1}{2}(1+s)y_2 \tag{7.88a}$$

$$\text{edge-e2:}\ x = \frac{1}{2}(1-s)x_2 + \frac{1}{2}(1+s)x_3, \text{ and } y = \frac{1}{2}(1-s)y_1 + \frac{1}{2}(1+s)y_2 \tag{7.88b}$$

$$\text{edge-e3:}\ x = \frac{1}{2}(1+s)x_3 + \frac{1}{2}(1-s)x_4 \text{ and } y = \frac{1}{2}(1+s)y_3 + \frac{1}{2}(1-s)y_4 \tag{7.88c}$$

$$\text{edge-e4:}\ x = \frac{1}{2}(1+s)x_4 + \frac{1}{2}(1-s)x_1 \text{ and } y = \frac{1}{2}(1+s)y_4 + \frac{1}{2}(1-s)y_1 \tag{7.88d}$$

Thus we see that the contour can be described by a parametric relation of the form,

$$x = x(s), \quad y = y(s) \tag{7.89}$$

Note that for other element types, e.g., Q8, the relationships given in Eq. (7.88) must be modified accordingly.

The length of a small segment dc along the boundary of the element in the physical space is,

$$dc = \left(dx^2 + dy^2\right)^{1/2} \tag{7.90}$$

By using Eq. (7.89) we find,

$$dc = \left(\left(\frac{dx}{ds}\right)^2 + \left(\frac{dy}{ds}\right)^2\right)^{1/2} ds \tag{7.91}$$

The term in the brackets in Eq. (7.91) is the stretch factor that maps the differential lengths between the element defined in the physical domain and mapped-domain. In keeping with the definition in the previous sections this is called the *edge Jacobian*,

$$J = \left(\left(\frac{dx}{ds}\right)^2 + \left(\frac{dy}{ds}\right)^2\right)^{1/2} \tag{7.92}$$

7.6.2 Differential area in curvilinear coordinates

Consider the representation of a vector $\vec{r}$ in the Cartesian coordinate system,

$$\vec{r} = x\hat{i} + y\hat{j} \tag{7.93}$$

If the coordinate axes x and y are mapped to the generalized coordinates ξ and η as defined in Eq. (7.82), we can see that the vector $\vec{r}$ has a representation in the (ξ, η) coordinate system,

$$\vec{r} = x(\xi, \eta)\hat{i} + y(\xi, n)\hat{j} \tag{7.94}$$

In general, the tangent vector $\vec{a}_\xi$ in ξ-direction is defined as follows:

$$\vec{a}_\xi = \frac{\partial \vec{r}}{\partial \xi} = \vec{r}_{,\xi} = x_{,\xi}\hat{i} + y_{,\xi}\hat{j} \tag{7.95}$$

Similarly for $\vec{a}_\eta$ we get,

$$\vec{a}_\eta = \frac{\partial \vec{r}}{\partial \eta} = \vec{r}_{,\eta} = x_{,\eta}\hat{i} + y_{,\eta}\hat{j} \tag{7.96}$$

Note that $\vec{a}_\xi$ and $\vec{a}_\eta$ are not necessarily unit vectors. Also, note that the vector $\vec{a}_\xi$ maps $d\vec{a}_\xi$ to $d\xi$ and the vector $\vec{a}_\eta$ maps $d\vec{a}_\eta$ to $d\eta$, similar to the one-dimensional case shown above. Then, we can express these infinitesimally small vectors as,

$$\begin{aligned} d\vec{a}_\xi &= \left(x_{,\xi}\hat{i} + y_{,\xi}\hat{j}\right)d\xi \\ d\vec{a}_\eta &= \left(x_{,\eta}\hat{i} + y_{,\eta}\hat{j}\right)d\eta \end{aligned} \tag{7.97}$$

The differential area dA on the curvilinear coordinates is defined by the cross product of $d\vec{a}_\xi$ and $d\vec{a}_\eta$,

$$\begin{aligned} dA = d\vec{a}_\xi \times d\vec{a}_\eta &= \left(x_{,\xi}d\xi\,\hat{i} + y_{,\xi}d\xi\,\hat{j}\right) \times \left(x_{,\eta}d\eta\,\hat{i} + y_{,\eta}d\eta\,\hat{j}\right) \\ &= \left(x_{,\xi}y_{,\eta} - x_{,\eta}y_{,\xi}\right)d\xi\, d\eta\, \hat{k} \end{aligned} \tag{7.98}$$

where $\hat{k}$ is the unit normal of the area in the out-of-plane direction. After dropping $\hat{k}$ we find

$$dA = |J|d\xi\, d\eta \tag{7.99}$$

Note that $|J| = (x_{,\xi}y_{,\eta} - x_{,\eta}y_{,\xi})$ is the determinant of the Jacobian matrix $[J]$ given in Eq. (7.85).

7.6.2.1 Uniqueness of the two-dimensional mapping

As we will see in the next section, the finite element equilibrium equations are developed by using integration on the mapped rectangular element. Therefore, it is important to ensure that a unique mapping exists between the quadrilateral element in the Cartesian coordinates and the mapped rectangular element. Mapping is unique if every point in the quadrilateral element can be mapped to a point in the rectangular element and vice versa. Research has shown that, in order to ensure a unique mapping, the sign of the Jacobian determinant, $|J|$, should remain unchanged at all points of the mapped domain. This is ensured by following rules of thumb. For linear interpolation elements (Q4), no internal angle should be greater than 180 degrees. For second-order interpolation elements (Q8, Q9) no internal angle should be greater than 160 degrees, and the mid-nodes should be placed within the mid one-third of the edge [5].

The uniqueness of the mapping could come into play on two occasions. First, the geometry of every quadrilateral element in the initial mesh should ensure uniqueness of the mapping as described above. In addition, in solid mechanics problems that experience large deformation, the value of the Jacobian determinant should be monitored closely, so as not to lose uniqueness of mapping.

Example 7.1 Determinant of the Jacobian for a rectangular, Q4-element

Compute the determinant of the Jacobian for a rectangular, Q4-element.

Solution:

Nodal coordinates of a Q4-element are given as follows: (x_1,y_1), (x_2,y_2), (x_3,y_3) and (x_4,y_4). For a rectangular element the following relationships exist:

$$x_1 = x_4,\quad x_2 = x_3,\quad y_1 = y_2,\quad \text{and } y_3 = y_4 \tag{7.100}$$

The Jacobian determinant is given by Eq. (7.86). By using the Q4-interpolation functions given in Eq. (7.63) the components of the Jacobian Matrix are found as follows:

$$\begin{aligned}
J_{11} &= \frac{\partial x}{\partial \xi} = -\frac{1}{4}(1-\eta)x_1 + \frac{1}{4}(1-\eta)x_2 + \frac{1}{4}(1+\eta)x_3 - \frac{1}{4}(1+\eta)x_4 \\
J_{12} &= \frac{\partial x}{\partial \eta} = -\frac{1}{4}(1-\xi)x_1 - \frac{1}{4}(1+\xi)x_2 + \frac{1}{4}(1+\xi)x_3 + \frac{1}{4}(1-\xi)x_4 \\
J_{21} &= \frac{\partial y}{\partial \xi} = -\frac{1}{4}(1-\eta)y_1 + \frac{1}{4}(1-\eta)y_2 + \frac{1}{4}(1+\eta)y_3 - \frac{1}{4}(1+\eta)y_4 \\
J_{22} &= \frac{\partial y}{\partial \eta} = -\frac{1}{4}(1-\xi)y_1 - \frac{1}{4}(1+\xi)y_2 + \frac{1}{4}(1+\xi)y_3 + \frac{1}{4}(1-\xi)y_4
\end{aligned} \tag{7.101}$$

For the rectangular element, the determinant of the Jacobian is found by using the coordinates given in Eq. (7.100) as follows:

$$|J| = J_{11}J_{22} - J_{21}J_{12} = \frac{1}{4}(x_2 - x_1)(y_3 - y_2) = \frac{A^{(e)}}{4} \tag{7.102}$$

Thus we see that the determinant of the Jacobian of a rectangular element is constant and its value equals one-fourth of the area of the element $A^{(e)}$.

Example 7.2 Determinant of the Jacobian for a general quadrilateral, Q4-element

Compute the determinant of the Jacobian for a Q4-element whose nodal coordinates are given as follows: $(x_1,y_1) = (1,\ 1)$, $(x_2,y_2) = (2,\ 1.2)$, $(x_3,y_3) = (2.5,\ 2.5)$ and $(x_4,y_4) = (0.8,\ 1.8)$.

Solution:
By using Eq. (7.101) the following relationship is established for this element:

$$|J| = J_{11}J_{22} - J_{21}J_{12} = 0.3375 + 0.045\,\xi + 0.0825\,\eta \tag{7.103}$$

Fig. 7.9 shows the variation of $|J|$ plotted in the mapped-coordinates. Note that $|J|$ is positive throughout the element.

Example 7.3 Determinant of the Jacobian of a general Q4-element, two collinear edges

Compute the determinant of the Jacobian of a Q4-element with two collinear, consecutive edges.

Solution:
In order to attain such an element, let edges-12 and -23 of the element given in Example 7.2 be collinear. Such an element can be obtained by modifying the coordinates of node-2 as follows: $(x_2,y_2) = (1.75,\ 1.75)$. By using Eq. (7.101), the following relationship is established for this element:

$$|J| = J_{11}J_{22} - J_{21}J_{12} = 0.1875 + 0.09375(\eta - \xi) \tag{7.104}$$

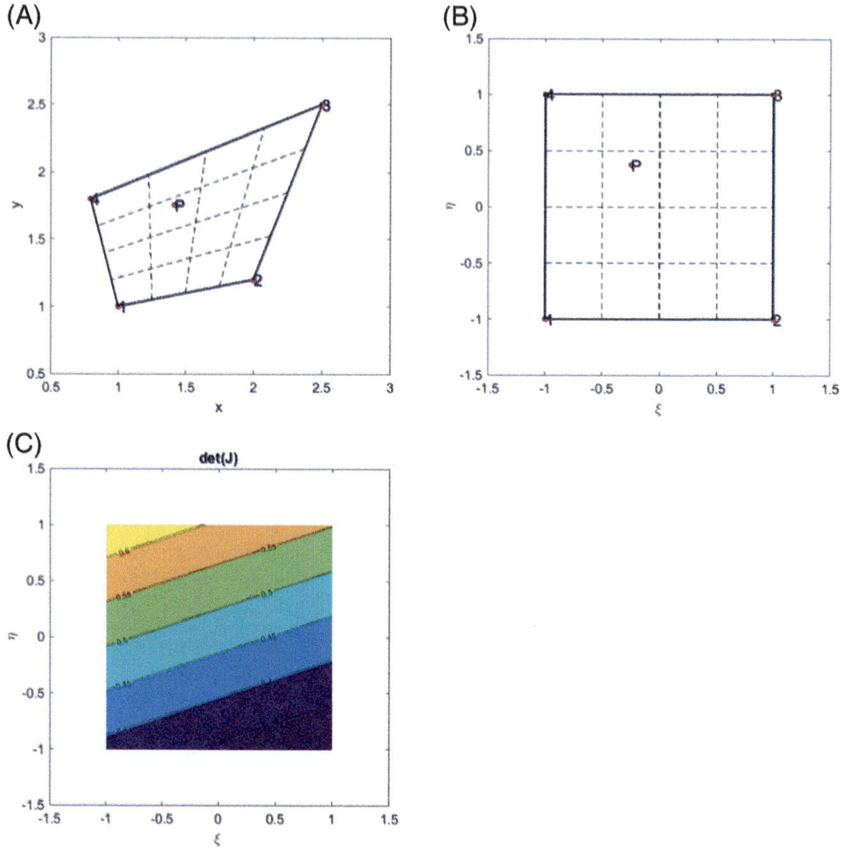

FIGURE 7.9 (A) The Q4-element described in Example 7.2 in Cartesian coordinates. (B) The mapped coordinate system, (C) Variation of the determinant of the Jacobian $|J| = 0.3375 + 0.045\,\xi + 0.0825\,\eta$ plotted in the mapped coordinate system.

Fig. 7.10 shows the variation of $|J|$ plotted in the mapped-coordinates. Note that the magnitude of $|J|$ drops toward node-2 and becomes exactly equal to zero at node-2, $(\xi, \eta) = (1, -1)$. This characteristic does not bode well for an element that has two collinear, consecutive edges because at node-2 the mapping between the Cartesian and mapped-coordinate systems is lost. Elements with such characteristics should be avoided in practice.

Example 7.4 Jacobian of a Q4-element with an internal angle great than 180°

Compute the determinant of the Jacobian of a Q4-element with an internal angle great than 180°.

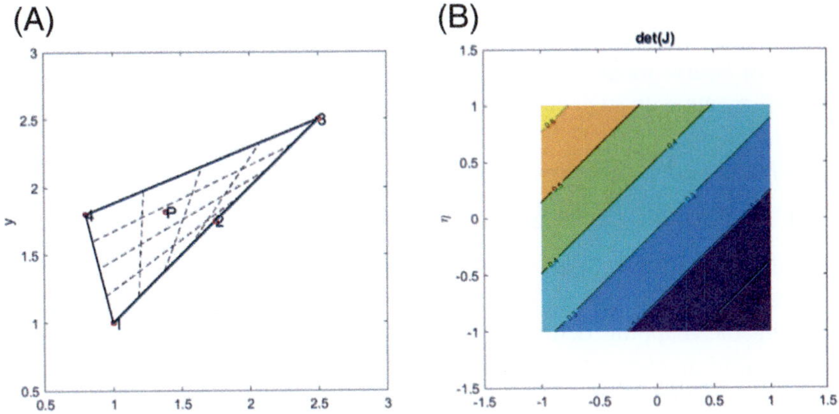

FIGURE 7.10 (A) The Q4-element described in Example 7.3 in Cartesian coordinates. (B) Variation of the determinant of the Jacobian $|J| = 0.1875 + 0.09375(n - \xi)$ plotted in the mapped coordinate system.

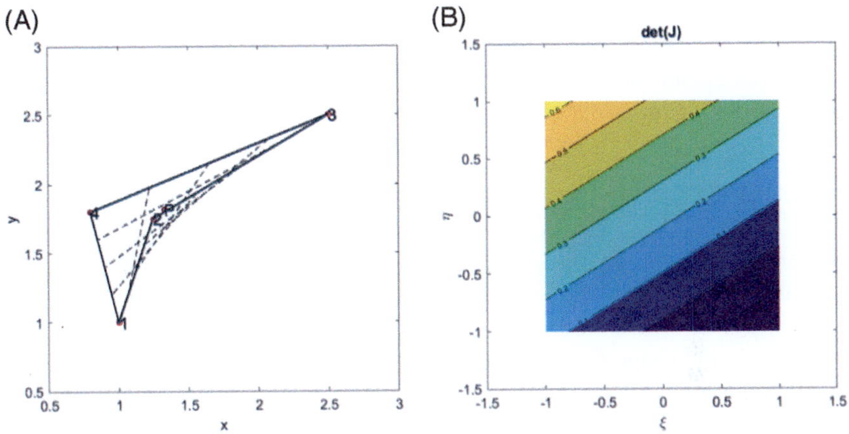

FIGURE 7.11 (A) The Q4-element described in Example 7.4 in Cartesian coordinates. (B) Variation of the determinant of the Jacobian ($|J| = 0.1875 - 0.1375\xi + 0.14375\eta$ plotted in the mapped coordinate system.

Solution:

Such an element can be obtained by modifying the coordinates of node-2 of the element given in Example 7.2 as follows: $(x_2, y_2) = (1.25, 1.75)$. By using Eq. (7.101), the following relationship is established for this element:

$$|J| = J_{11}J_{22} - J_{21}J_{12} = 0.1875 - 0.1375\xi + 0.14375\eta$$

Fig. 7.11 shows the shape of the element when the internal angle at node-2 is greater than 180° as described above. Note that for this element, some of the points in the mapped coordinate system are mapped to the outside of the actual element. Clearly this cannot be allowed. This characteristic also can be observed by noting that $|J|$ becomes negative for a substantial segment of the element. This would, of course mean, by noting Eq. (7.99), that mapping produces negative area! Elements with such characteristics should be avoided in practice.

7.7 Mapped isoparametric elements

7.7.1 Strain–displacement operator matrix, [B]

Consider the equation of equilibrium of an element as presented in Eq. (7.45). The area and contour integrals that appear in this equation can be easily evaluated by using the mapping defined in the previous section. To this end, consider a *quadrilateral element* with M-nodes with nodal coordinates (x_i,y_i), nodal displacement components (u_{xi},u_{yi}) for $i = 1 - M$.

The coordinate mapping of the isoparametric[1] elements is performed by using Eqs. (7.76),

$$x = \sum_{i=1}^{M} N_i^{(QM)} x_i \text{ and } y = \sum_{i=1}^{M} N_i^{(QM)} y_i \tag{7.105}$$

The independent variables are also interpolated with the same level of interpolation functions:

$$u_x(\xi,\eta) = \sum_{i=1}^{M} N_i^{(QM)} u_{xi} \text{ and } u_y(\xi,\eta) = \sum_{i=1}^{M} N_i^{(QM)} u_{yi} \tag{7.106}$$

where $N_i^{(QM)}$ represent the appropriate shape functions introduced for one of Q4, Q8 or Q9 elements in Section 7.5. Note that, in general, the displacement components u_x and u_y are defined in the Cartesian coordinate system (x,y), however, as the shape functions are defined in the mapped coordinate system, Eq. (7.106) states that $u_x = u_x(\xi,\eta)$ and $u_y = u_y(\xi,\eta)$. If the mapping between the two coordinate systems, defined in Eq. (7.105) is unique, knowing, for example, $u_x = u_x(\xi,\eta)$ we should be able to uniquely determine $u_x = u_x(x,y)$, and vice versa.

In order to proceed to find these relationships, let's first evaluate the partial derivatives of the displacement components by using the chain rule of differentiation as follows:

$$\frac{\partial u_x}{\partial \xi} = \frac{\partial u_x}{\partial x}\frac{\partial x}{\partial \xi} + \frac{\partial u_x}{\partial y}\frac{\partial y}{\partial \xi} \tag{7.107a}$$

$$\frac{\partial u_x}{\partial \eta} = \frac{\partial u_x}{\partial x}\frac{\partial x}{\partial \eta} + \frac{\partial u_x}{\partial y}\frac{\partial y}{\partial \eta} \tag{7.107b}$$

1. The mapping is called *isoparametric* if both the coordinate variables and the dependent variables involve the same level of interpolation. The *subparametric* and *superparametric* elements are beyond the scope of this text, and the reader is referred to other texts for their use [6].

This relationship can be expressed in vector form,

$$\begin{Bmatrix} \dfrac{\partial u_x}{\partial \xi} \\ \dfrac{\partial u_x}{\partial \eta} \end{Bmatrix} = \begin{bmatrix} \dfrac{\partial x}{\partial \xi} & \dfrac{\partial y}{\partial \xi} \\ \dfrac{\partial x}{\partial \eta} & \dfrac{\partial y}{\partial \eta} \end{bmatrix} \begin{Bmatrix} \dfrac{\partial u_x}{\partial x} \\ \dfrac{\partial u_x}{\partial y} \end{Bmatrix} \tag{7.108}$$

Note that matrix that appears in Eq. (7.108) is the *transpose of the Jacobian matrix* introduced in Eq. (7.85). Thus, we see that the partial derivatives with respect to the mapped coordinate axes are mapped to the partial derivatives with respect to the Cartesian coordinate axes through the *transpose of the Jacobian matrix* introduced in Eq. (7.85). It can easily be shown that the second displacement component has a similar relationship,

$$\begin{Bmatrix} \dfrac{\partial u_y}{\partial \xi} \\ \dfrac{\partial u_y}{\partial \eta} \end{Bmatrix} = \begin{bmatrix} \dfrac{\partial x}{\partial \xi} & \dfrac{\partial y}{\partial \xi} \\ \dfrac{\partial x}{\partial \eta} & \dfrac{\partial y}{\partial \eta} \end{bmatrix} \begin{Bmatrix} \dfrac{\partial u_y}{\partial x} \\ \dfrac{\partial u_y}{\partial y} \end{Bmatrix} \tag{7.109}$$

The transpose of the Jacobian matrix of the mapping is defined as follows:

$$[J]^T = \begin{bmatrix} J_{11} & J_{21} \\ J_{12} & J_{22} \end{bmatrix} = \begin{bmatrix} \dfrac{\partial x}{\partial \xi} & \dfrac{\partial y}{\partial \xi} \\ \dfrac{\partial x}{\partial \eta} & \dfrac{\partial y}{\partial \eta} \end{bmatrix} \tag{7.110}$$

Note that the components of the Jacobian matrix are given in Eq. (7.86). The derivatives of u_x and $u_y$with respect to the Cartesian coordinates can then be found by inverting Eqs. (7.108) and (7.109), respectively, as follows:

$$\begin{Bmatrix} \dfrac{\partial u_x}{\partial x} \\ \dfrac{\partial u_x}{\partial y} \end{Bmatrix} = [J]^{-T} \begin{Bmatrix} \dfrac{\partial u_x}{\partial \xi} \\ \dfrac{\partial u_x}{\partial \eta} \end{Bmatrix} \text{ and } \begin{Bmatrix} \dfrac{\partial u_y}{\partial x} \\ \dfrac{\partial u_y}{\partial y} \end{Bmatrix} = [J]^{-T} \begin{Bmatrix} \dfrac{\partial u_y}{\partial \xi} \\ \dfrac{\partial u_y}{\partial \eta} \end{Bmatrix} \tag{7.111}$$

The inverse of the transpose of the Jacobian matrix is given as follows:

$$[J]^{-T} = \frac{1}{|J|} \begin{bmatrix} J_{22} & -J_{21} \\ -J_{12} & J_{11} \end{bmatrix} \tag{7.112}$$

The strain components can be computed by using Eq. (7.111) as follows:

$$\varepsilon_{xx} = \frac{\partial u_x}{\partial x} = \frac{1}{|J|} \left(J_{22} \frac{\partial u_x}{\partial \xi} - J_{21} \frac{\partial u_x}{\partial \eta} \right) \tag{7.113a}$$

$$\varepsilon_{yy} = \frac{\partial u_y}{\partial y} = \frac{1}{|J|} \left(-J_{12} \frac{\partial u_y}{\partial \xi} + J_{11} \frac{\partial u_y}{\partial \eta} \right) \tag{7.113b}$$

$$\gamma_{xy} = \frac{\partial u_y}{\partial x} + \frac{\partial u_x}{\partial y} = \frac{1}{|J|} \left[J_{22} \frac{\partial u_y}{\partial \xi} - J_{21} \frac{\partial u_y}{\partial \eta} - J_{12} \frac{\partial u_x}{\partial \xi} + J_{11} \frac{\partial u_x}{\partial \eta} \right] \tag{7.113c}$$

The vector form of this equation is expressed in Eq. (7.39) where the *strain–displacement operator matrix* can be expressed as follows,

$$[B]=\begin{bmatrix}1&0&0&0\\0&0&0&1\\0&1&1&0\end{bmatrix}\begin{bmatrix}[J]^{-T}&[0]\\ [0]&[J]^{-T}\end{bmatrix}[N_\partial] \tag{7.114}$$

where,

$$[N_\partial]=\begin{bmatrix}\dfrac{\partial N_1}{\partial\xi}&0&\dfrac{\partial N_2}{\partial\xi}&0&\dfrac{\partial N_3}{\partial\xi}&0&\cdots\\ \dfrac{\partial N_1}{\partial\eta}&0&\dfrac{\partial N_2}{\partial\eta}&0&\dfrac{\partial N_3}{\partial\eta}&0&\cdots\\ 0&\dfrac{\partial N_1}{\partial\xi}&0&\dfrac{\partial N_2}{\partial\xi}&0&\dfrac{\partial N_3}{\partial\xi}&\cdots\\ 0&\dfrac{\partial N_1}{\partial\eta}&0&\dfrac{\partial N_2}{\partial\eta}&0&\dfrac{\partial N_3}{\partial\eta}&\cdots\end{bmatrix} \tag{7.115}$$

Note that the dimension of $[N_\partial]$ matrix is $4\times 2M$ and that of the $[B]$ matrix is $3\times 2M$.

7.7.2 Finite element form of the element equilibrium equations for a Q4-element

The variables of the problem are interpolated by using Eqs. (7.105) and (7.106) with $M=4$ and the shape functions given in Eq. (7.61). The finite element form of the equation of equilibrium is given in Eq. (7.45) repeated below,

$$[k]^{(e)}\{d\}=\{r_B\}^{(e)}+\{r_q\}^{(e)}+\{r_{\varepsilon_0}\}^{(e)}+\{F\}^{(e)} \tag{7.116}$$

where,

Element stiffness matrix: $$[k]^{(e)}=\int_{\Omega^{(e)}}[B]^T[E][B]d\Omega^{(e)} \tag{7.117a}$$

Consistent nodal force vectors

Body forces: $$\{r_B\}^{(e)}=\int_{\Omega^{(e)}}[N]^T\{F_B\}d\Omega^{(e)} \tag{7.117b}$$

Boundary tractions: $$\{r_q\}^{(e)}=\oint_{\Gamma^{(e)}}[N]^T\{q^{(b)}\}d\Gamma^{(e)} \tag{7.117c}$$

Initial strains: $$\{r_{\varepsilon_0}\}^{(e)}=\int_{\Omega^{(e)}}[B]^T[E]\{\varepsilon^{(0)}\}d\Omega^{(e)} \tag{7.117d}$$

External forces that happen to be acting on the element nodes:

$$\{F\}^{(e)} = \begin{Bmatrix} F_{x1} & F_{y1} & F_{x2} & F_{y2} & \cdots & F_{xM} & F_{yM} \end{Bmatrix}^T \quad (7.117e)$$

Note that the dimension of the element stiffness matrix is $2M \times 2M$ and the length of the vectors are $2M \times 1$. This equation can be now evaluated by using the shape function matrix $[N]$ defined in Eq. (7.34) and the strain–displacement matrix $[B]$ given in Eq. (7.114) with $M = 4$.

7.7.2.1 The element stiffness matrix

The element stiffness matrix is defined over the element $\Omega^{(e)}$. The integration domain for Eq. (7.117a) can be interchanged as follows, $d\Omega^{(e)} = t\, dA$. As the integration is expressed in the generalized coordinate system, we also have the following relationship, $dA = |J|\, d\xi d\eta$. Thus Eq. (7.117a) becomes,

$$[k]^{(e)} = \int_{\Omega^{(e)}} [B]^T [E][B] d\Omega^{(e)} = \int_{-1}^{1}\int_{-1}^{1} [B]^T [E][B]|J| t\, d\xi d\eta \quad (7.118)$$

Exact integration of Eq. (7.118) is not practical for a general quadrilateral element. But, it is possible for a rectangular element, as will be shown in Chapter 8. Integration of Eq. (7.118) is carried out numerically by using *Gaussian quadrature* introduced in Section 7.8.

7.7.2.2 Load vector for body forces

The body force vector given in Eq. (7.117b) involves an area integral over the solution domain, much like the integration of the element stiffness matrix. We can treat the integral similarly and show that Eq. (7.117b) becomes,

$$\{r_B\}^{(e)} = \int_{\Omega^{(e)}} [N]^T \{F_B\} d\Omega^{(e)} = \int_{-1}^{1}\int_{-1}^{1} [N]^T \{F_B\}|J| t\, d\xi\, d\eta \quad (7.119)$$

Integration of Eq. (7.119) is also carried out numerically as demonstrated in Section 7.9.

7.7.2.3 Load vector for boundary tractions

The element load vector for boundary traction is given by Eq. (7.117c),

$$\{r_q\}^{(e)} = \oint_{\Gamma^{(e)}} [N]^T \{q^{(b)}\} d\Gamma^{(e)}$$

where $[N]$ is the shape function matrix and $\{q^{(b)}\}$ is the load vector. Each edge of the element is assessed and integrated separately. In fact we can write

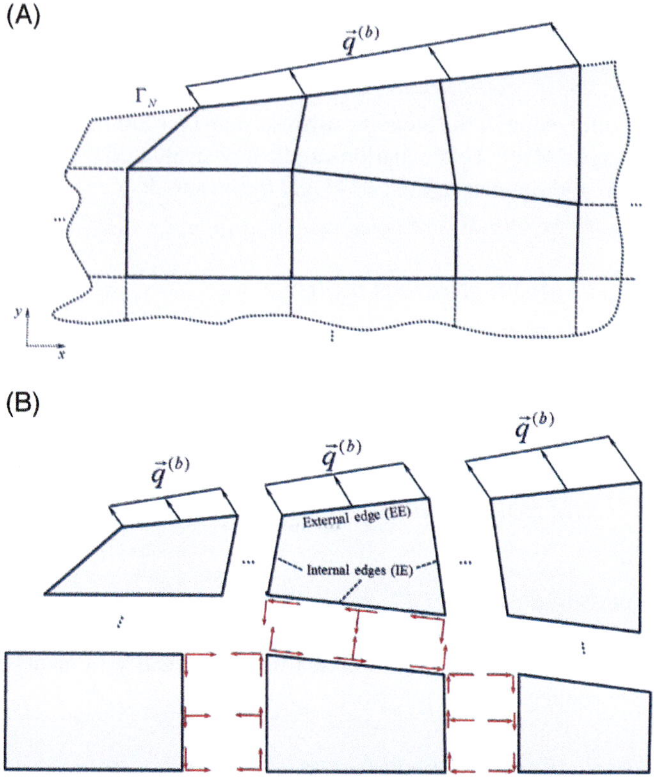

FIGURE 7.12 (A) A segment of a finite element mesh near one of the external boundaries Γ_N where external traction $\vec{q}$ is applied. (B) Exploded view of the six elements show the external and internal edges and the equilibrium that must be had between the internal edges.

Eq. (7.117c) as follows,

$$\{r_q\}^{(e)} = \sum_{ei=1}^{N_{edges}} \{r_{q-ei}\}^{(e)} \tag{7.120}$$

where the vector $\{r_{q-ei}\}^{(e)}$ represents the traction acting along the edge-*ei* of the element. Note that the number of edges, N_{edges}, is three for triangular elements and four for quadrilateral elements. By using Eq. (7.91) the traction vector acting along the edge can be represented in general as follows:

$$\{r_{q-ei}\}^{(e)} = \int [N]^T \{q_{ei}\} J\, ds \tag{7.121}$$

where J is defined in Eq. (7.92). At this point it is helpful to distinguish between the *internal* and *external edges* (Fig. 7.12) of an element. The external edge(s) of an element are shared with the boundary Γ of the solution domain. Internal edges, on the other hand are shared with the adjacent elements. Let's define the following variables to distinguish between the two:

Boundary tractions on an external edge (EE): $\{r_{qEE}\}^{(e)}$

Boundary tractions on an internal edge (IE): $\{r_{qIE}\}^{(e)}$

Of course, for an element that is surrounded by other elements all edges and hence all tractions are internal. Note that the *internal tractions acting on the elements should be balanced* according to *Newton's third law*. This observation forms the basis of element assembly as it will be introduced in Section 7.10. Analysis often does not require computation of the $\{r_{qIE}\}^{(e)}$ vector, but $\{r_{q_{EE}}\}^{(e)}$ is required to impose the natural boundary conditions of the boundary value problem.

We will next demonstrate a procedure to compute the $\{r_{qEE}\}^{(e)}$ vector for each element. To this end, we will assume that a given element has only one of its edges coincident with the external boundary, Γ_N. But, the $\{r_{qEE}\}^{(e)}$ vector for elements that share more than one edge with Γ_N can also be constructed easily by following similar steps. A Q4-element will be used in the following. Let use the edges 1 – 4 as defined in Eq. (7.87) (Fig. 7.12):

edge-1, e1: $(\xi, -1)$

edge-2, e2: $(1, \eta)$

edge-3, e3: $(\xi, 1)$

edge-4, e4: $(-1, \eta)$

In general, the external tractions are more conveniently defined with respect to the local surface normal and tangent as follows:

$$\vec{q}^{(b)}dA = \left(q_n\vec{n} + q_t\vec{t}\right)dA \tag{7.122}$$

But, the consistent nodal force vector should be defined with respect to the global coordinate system. The traction vector is expressed in the (x,y) coordinate system by using the relationships $\vec{n} = n_x\hat{i} + n_y\hat{j}$ and $\vec{t} = -n_y\hat{i} + n_x\hat{j}$, described in equations (7.7) and (7.8), as follows:

$$\{q^{(b)}\} = \begin{Bmatrix} q_x \\ q_y \end{Bmatrix} = \begin{bmatrix} n_x & -n_y \\ n_y & n_x \end{bmatrix} \begin{Bmatrix} q_n \\ q_t \end{Bmatrix} \tag{7.123}$$

For example, consider linearly varying traction components acting along edge-3 of the element as shown in Fig. 7.13. We can represent such a traction variation in finite element form as follows:

$$\begin{Bmatrix} q_x\{\xi, 1\} \\ q_y(\xi, 1) \end{Bmatrix} = \left[N^{Q4}(\xi, 1)\right]\{q\} \tag{7.124}$$

where the shape functions are evaluated on the boundary $(\xi, 1)$ and the nodal

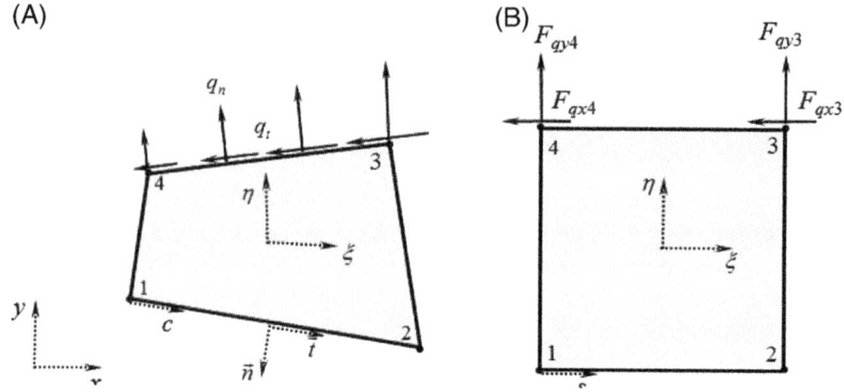

FIGURE 7.13 (A) External traction $\vec{q} = q_n\vec{n} + q_t\vec{t}$ acts along edge-e3 between nodes 3 and 4 of the element. (B) The consistent nodal force vectors that correspond to the external force in part-a) of this figure are calculated by using Eqs. (7.123), (7.127c), and they are shown on the nondimensional element.

values of the external traction (Fig. 7.13B) are expressed in the $\{q\}$vector,

$$[N^{Q4}(\xi,1)] = \begin{bmatrix} N_1^{Q4} & 0 & N_2^{Q4} & 0 & N_3^{Q4} & 0 & N_4^{Q4} & 0 \\ 0 & N_1^{Q4} & 0 & N_2^{Q4} & 0 & N_3^{Q4} & 0 & N_4^{Q4} \end{bmatrix}\Bigg|_{(\xi,1)} \tag{7.125a}$$

$$\{q\} = \{0 \quad 0 \quad 0 \quad 0 \quad q_{x3} \quad q_{y3} \quad q_{x4} \quad q_{y4}\}^T \tag{7.125b}$$

where q_{x3}, q_{y3} and q_{x4}, q_{y4} are the nodal values of traction at nodes-3 and -4 respectively. The consistent nodal force vector for edge-3 can then be expressed as follows:

$$\{r_{q-e3}\}^{(e)} = \int_{-1}^{1} [N^{(Q4)}]^T [N^{(Q4)}]\{q\}|J|d\xi \text{ with } \eta = 1 \tag{7.126}$$

Consistent nodal forces for each edge that represents the boundary of the solution domain can be computed as follows:

$$\text{edge-1: } \{r_{q-e1}\}^{(e)} = \int_{-1}^{1} [N^{(Q4)}]^T [N^{(Q4)}]\{q\}|J|d\xi \text{ with } \eta = -1 \text{ and}$$

$$\{q\} = \{q_{x1} \quad q_{y1} \quad q_{x2} \quad q_{y2} \quad 0 \quad 0 \quad 0 \quad 0\}^T \tag{7.127a}$$

$$\text{edge-2: } \{r_{q-e2}\}^{(e)} = \int_{-1}^{1} [N^{(Q4)}]^T [N^{(Q4)}]\{q\}|J|d\eta \text{ with } \xi = 1 \text{ and}$$

$$\{q\} = \{0 \quad 0 \quad q_{x2} \quad q_{y2} \quad q_{x3} \quad q_{y3} \quad 0 \quad 0\}^T \tag{7.127b}$$

$$\text{edge-3: } \{r_{q-e3}\}^{(e)} = \int_{-1}^{1} [N^{(Q4)}]^T [N^{(Q4)}]\{q\}|J|d\xi \text{ with } \eta = 1 \text{ and}$$

$$\{q\} = \{0 \quad 0 \quad 0 \quad 0 \quad q_{x3} \quad q_{y3} \quad q_{x4} \quad q_{y4}\}^T \tag{7.127c}$$

$$\text{edge-4: } \{r_{q-e4}\}^{(e)} = \int_{-1}^{1} [N^{(Q4)}]^T [N^{(Q4)}]\{q\}|J|d\eta \text{ with } \xi = -1 \text{ and}$$

$$\{q\} = \{q_{x1} \quad q_{y1} \quad 0 \quad 0 \quad 0 \quad 0 \quad q_{x4} \quad q_{y4}\}^T \tag{7.127d}$$

TABLE 7.3 Gauss quadrature points ξ_i and weight factors W_i [7].

Order, n	Quadrature point location, ξ_i	Weight factor, W_i
1	0	2
2	$\pm\frac{1}{\sqrt{3}}$	1
3	$-\sqrt{\frac{3}{5}}, 0, \sqrt{\frac{3}{5}}$	5/9, 8/9, 5/9
4	$, \pm\sqrt{\frac{3}{7}-\frac{2}{7}\sqrt{\frac{6}{5}}}, \pm\sqrt{\frac{3}{7}+\frac{2}{7}\sqrt{\frac{6}{5}}}$	$\frac{18+\sqrt{30}}{36}, \frac{18-\sqrt{30}}{36}$

7.8 Numerical integration using Gauss quadrature

Gauss quadrature is a numerical method to evaluate integrals of the type,

$$I = \int_{-1}^{1} \phi(\xi)d\xi \tag{7.128}$$

By using the Gauss quadrature this integral can be evaluated approximately as follows:

$$I \cong I_n = \sum_{i=1}^{n} W_i \phi(\xi_i) \tag{7.129}$$

where, W_i are scalar values known as *weight factors*, ξ_i are predetermined *Gauss quadrature points*, and n is the *order of quadrature*. Gaussian quadrature method is designed to evaluate polynomials exactly, as demonstrated below. An integral of the type given in Eq. (7.128) can be integrated exactly by the quadrature formula, i.e., Eq. (7.129), with degree n if the integrand $\phi(\xi)$ is a polynomial of order $2n-1$ [6]. The weights and quadrature points are tabulated values for a given order of quadrature (Table 7.3).

7.8.1 Coordinate transformation

Coordinate transformation is used to change the limits of an integral whose lower and upper limits are not -1 and 1. Consider integration of a function $f(x)$ between x_1 and x_2,

$$I = \int_{x_1}^{x_2} f(x)dx \tag{7.130}$$

Gauss quadrature method requires the lower and upper limits of the integral be -1 and 1, respectively. Thus, $x_1 \leq x \leq x_2$ range needs to be mapped to $-1 \leq \xi$

≤ 1 range. This can be achieved by using the following substitution:

$$x = \frac{1}{2}(1 - \xi)x_1 + \frac{1}{2}(1 + \xi)x_2 \tag{7.131}$$

This also gives,

$$dx = \frac{1}{2}(x_2 - x_1)d\xi = Jd\xi$$

Then the integral is transformed into the following equivalent form:

$$I = \int_{-1}^{1} f(\xi)J\,d\xi = \int_{-1}^{1} \phi(\xi)d\xi \tag{7.132}$$

7.8.2 Derivation of second-order Gauss quadrature

Next, let's next demonstrate how the parameters W_i and ξ_i are obtained for the second order ($n = 2$) Gauss quadrature. To this end, we will require Gauss quadrature of $n = 2$ to exactly integrate a third order polynomial in the interval $-1 \leq \xi \leq 1$. The third order polynomial is given as follows:

$$\phi(\xi) = a_0 + a_1\xi + a_2\xi^2 + a_3\xi^3 \tag{7.133}$$

where a_i are constant coefficients. The exact value of this integral is,

$$I = 2a_0 + \frac{2}{3}a_2 \tag{7.134}$$

By using Gauss quadrature of order $n = 2$ the numerical value of this integral becomes,

$$I_{n=2} = \sum_{i=1}^{2} W_i\,\phi(\xi_i) = W_1\,\phi(\xi_1) + W_2\,\phi(\xi_2) \tag{7.135}$$

If we assume that location of the Gauss quadrature points and the weight factors are symmetric with respect to the middle of the interval, we will find the following relationship:

$$I_{n=2} = W\,\phi(-\xi_p) + W\,\phi(\xi_p) \tag{7.136}$$

where W is the weight factor and ξ_p is the quadrature location for $n = 2$, that we are seeking. Combining Eqs. (7.133) and (7.136), the numerical value of the integral is found as follows:

$$I_{n=2} = 2W\left[a_0 + a_2\,\xi_p^2\right] \tag{7.137}$$

Letting the exact value in Eq. (7.134) be equal to the numerical value in Eq. (7.137), i.e., $I = I_{n=2}$, we find the following relationship:

$$a_0 + \frac{1}{3}a_2 = W\left(a_o + a_2\,\xi_p^2\right) \tag{7.138}$$

Since a_0 and a_2 are arbitrary constants we find that $W = 1$ and $\xi_p = \pm 1/3^{1/2}$ satisfies this equation. Gauss quadrature points ξ_i and weight factors W_i can be found for other quadrature levels in a similar way. Some of these are listed in Table 7.3.

In general, it can be shown that a polynomial of degree $2n - 1$ will be integrated exactly by n-point Gauss quadrature. Note that for a polynomial of order $2n - 1$, using Gauss quadrature order higher than n will still produce the exact value. Exact integration cannot be guaranteed if the function $\phi(\xi)$ is not a polynomial.

Example 7.1 Gaussian integration of a polynomial in one-dimension

Consider the function $y(x) = a\,x + b$ where a and b are given scalar constants. Compare the exact and numerical values of the integral of $y(x)$ in the range [3,7].

Solution:
The exact value of the integral is,

$$I = \int_3^7 (ax + b)dx = 20a + 4b \tag{7.139}$$

Using the transformation defined in Eq. (7.131), we find,

$$x = \frac{3}{2}(1 - \xi) + \frac{7}{2}(1 + \xi) \text{ and } J = 2. \tag{7.140}$$

The function $\phi(\xi)$ then becomes,

$$\phi(\xi) = 2[a(5 + 2\xi) + b] \tag{7.141}$$

The function is now in the form required in Eq. (7.132) and it can be evaluated by using Gaussian quadrature,

$$I_n = \sum_{i=1}^{n} W_i[2(a(5 + 2\xi_i) + b)] \tag{7.142}$$

We use $n = 1, 2$ and 3 to evaluate this integral numerically with the Gauss integration method. The weights and integration points are taken from Table 7.3,

$$\begin{aligned} I_{n=1} &= 2\phi(0) = 20a + 4b \\ I_{n=2} &= \phi\left(\frac{-1}{\sqrt{3}}\right) + \phi\left(\frac{1}{\sqrt{3}}\right) = 20a + 4b \\ I_{n=3} &= \frac{5}{9}\phi\left(-\sqrt{\frac{3}{5}}\right) + \frac{8}{9}\phi(0) + \frac{5}{9}\phi\left(\sqrt{\frac{3}{5}}\right) = 20a + 4b \end{aligned} \tag{7.143}$$

We see that all three cases give the same result. This is to be expected as $y(x)$ is a first order polynomial it should be evaluated exactly by a Gauss quadrature of order $n = 1$ or greater.

Example 7.2 Gaussian integration of the sine-function in one-dimension

Consider the function $y(x) = \sin \frac{\pi x}{x_2 - x_1}$. Evaluate the integral of this function between the limits $x_1 = 3$ and $x_2 = 7$.

Solution:
The exact value of the integral is,

$$I = \int_{3}^{7} \sin\left(\frac{\pi}{4}x\right)dx = -1.80063 \tag{7.144}$$

Using the transformation defined in Eq. (7.131), we find,

$$x = \frac{3}{2}(1-\xi) + \frac{7}{2}(1+\xi) = 5 + 2\xi \text{ and } J = 2 \tag{7.145}$$

The function $\phi(\xi)$ then becomes,

$$\phi(\xi) = 2\sin\left(\frac{\pi}{4}(5+2\xi)\right) \tag{7.146}$$

The function is now in the form required in Eq. (7.132) and it can be evaluated by using Gaussian quadrature,

$$I_n = \sum_{i=1}^{n} W_i\left[2\sin\left(\frac{\pi}{4}(5+2\xi_i)\right)\right] \tag{7.147}$$

We use $n = 1, 2, 3$ and 4 to evaluate the integral numerically with Gauss integration method. The weights and integration points are taken from Table 7.3,

$$\begin{aligned} I_{n=1} &= -2.82843 \\ I_{n=2} &= -1.74285 \\ I_{n=3} &= -1.80188 \\ I_{n=4} &= -1.80062 \end{aligned} \tag{7.148}$$

In this case the value of the numerical integral approaches the exact value with increasing order of Gauss quadrature.

7.8.3 Integration of two-dimensional functions by Gauss quadrature

Multidimensional Gauss rules are called Gaussian product rules. They are obtained by a successive application of the one-dimensional Gauss rules. Consider integration of the function $G(\xi,\eta)$ between the limits shown below,

$$I = \int_{-1}^{1}\int_{-1}^{1} G(\xi,\eta)d\xi\, d\eta \tag{7.149}$$

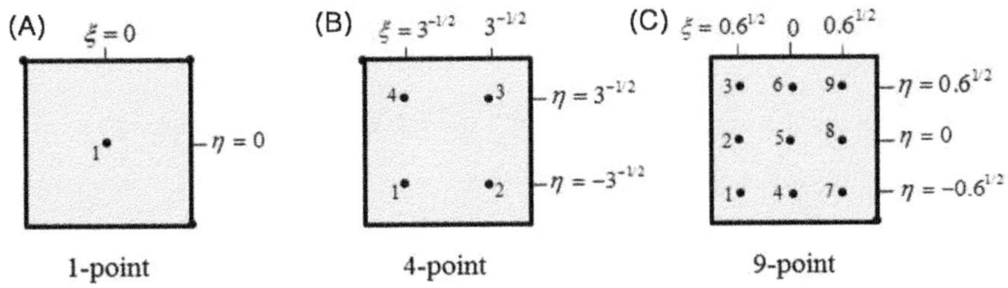

FIGURE 7.14 The locations of the Gaussian integration points for 1-, 4-, and 9-point quadrature, representing 1×1, 2×2, and 3×3 Gauss integration rules.

In general, the Gauss product rules can be applied to this function as follows:

$$I \cong I_{n\times n} = \int_{-1}^{1}\left[\sum_{i=1}^{n} W_i G(\xi_i, \eta)\right] d\eta = \sum_{i=1}^{n} W_j \left[\sum_{i=1}^{n} W_i G(\xi_i, \eta_j)\right] = \sum_{i=1}^{n}\sum_{j=1}^{n} W_i W_j G(\xi_i, \eta_j) \tag{7.150}$$

where (ξ_i, η_j) are called the Gauss integration points. Locations of these points are marked on the elements as shown in Fig. 7.14.

Two-dimensional functions can be integrated by using the Gauss integration points given in Table 7.3 as follows.

One-point Gauss integration: $I_{1\times 1} = 4G(0,0)$ (7.151)

Four-point Gauss integration:
$$I_{2\times 2} = W_1 W_1 G(\xi_1, \eta_1) + W_2 W_1 G(\xi_1, \eta_2) + W_1 W_2\, G(\xi_2, \eta_1) + W_2 W_2\, G(\xi_2, \eta_2) \tag{7.152}$$

Nine-point Gauss integration:
$$I_{3\times 3} = W_1[W_1 G_1 + W_2 G_2 + W_3 G_3] + W_2[W_1 G_4 + W_2 G_5 + W_3 G_6] + W_3[W_1 G_7 + W_2 G_8 + W_3 G_9] \tag{7.153}$$

Alternatively, we can use the following relations, where the notation G_i is used to indicate evaluation of the function $G(\xi, \eta)$ at the Gauss integration point i as shown in Fig. 7.14.

One-point Gauss integration: $I_{1\times x} = 4G_1$ (7.154)

Four-point Gauss integration: $I_{2\times 2} = G_1 + G_2 + G_3 + G_4$ (7.155)

Nine-point Gauss integration:
$$I_{3\times 3} = \frac{25}{81}(G_1 + G_3 + G_7 + G_9) + \frac{40}{81}(G_2 + G_4 + G_6 + G_8) + \frac{64}{81}G_5 \tag{7.156}$$

7.9 Numerical evaluation of the element equilibrium equations

The element equilibrium represented by Eq. (7.116) is repeated here,

$$[k]^{(e)}\{d\} = \{r_B\}^{(e)} + \left\{r_q\right\}^{(e)} + \left\{r_{\varepsilon_0}\right\}^{(e)} + \{F\}^{(e)} \tag{7.157}$$

where

Element stiffness matrix: $$[k]^{(e)} = \int_{-1}^{1}\int_{-1}^{1} [B]^T[E][B]|J|t\, d\xi d\eta \tag{7.158a}$$

Consistent nodal force vectors

Body forces: $$\{r_B\}^{(e)} = \int_{-1}^{1}\int_{-1}^{1} [N]^T\{F_B\}|J|t\, d\xi d\eta \tag{7.158b}$$

Boundary tractions: $$\left\{r_q\right\}^{(e)} = \sum_{ei=1}^{N_{edges}} \left\{r_{q-ei}\right\}^{(e)} \tag{7.158c}$$

Initial strains: $$\left\{r_{\varepsilon_0}\right\}^{(e)} = \int_{-1}^{1}\int_{-1}^{1} [B]^T[E]\{\varepsilon_0\}|J|t\, d\xi d\eta \tag{7.158d}$$

External forces that happen to be acting on the element nodes:

$$\{F\}^{(e)} = \left\{F_{x1} \quad F_{y1} \quad F_{x2} \quad F_{y2} \quad \cdots \quad F_{xM} \quad F_{yM}\right\}^T \tag{7.158e}$$

where $\{r_{q-ei}\}^{(e)}$ is one of Eqs. (7.127a–d).

In order to evaluate the *element stiffness matrix*, Eq. (7.158a), let's define the following matrix:

$$[C]^{(e)} = [B]^T[E][B]|J|t \tag{7.159}$$

Each component of the element stiffness matrix $k_{ij}^{(e)}$ can then be found as follows:

$$k_{ij}^{(e)} = \int_{-1}^{1}\int_{-1}^{1} C_{ij}^{(e)}\, d\xi\, d\eta \tag{7.160}$$

Using Gaussian quadrature the components of the element stiffness matrix can be evaluated numerically by using one of the $n \times n$ integration rules developed in Eqs. (7.154)–(7.156).

$$k_{ij}^{(e)} = \sum_{l=1}^{n}\sum_{m=1}^{n} W_l W_m C_{ij}^{(e)}(\xi_l, \eta_m) \tag{7.161}$$

In order to evaluate the *body force vector* in Eq. (7.158b) let's define the following vector:

$$\{c_B\}^{(e)} = [N]^T\{F_B\}|J|t \tag{7.162}$$

Each component of the consistent body force vector $r_{Bi}^{(e)}$ for the element can then be found as follows:

$$r_{Bi}^{(e)} = \int_{-1}^{1}\int_{-1}^{1} c_{Bi}^{(e)} d\xi d\eta \tag{7.163}$$

Similarly, this integral is integrated by using one of the $n\times n$ integration rules developed in Eqs. (7.154)–(7.156) as follows:

$$r_{Bi}^{(e)} = \sum_{l=1}^{n}\sum_{m=1}^{n} W_l W_m c_{Bi}^{(e)}(\xi_l, \eta_m) \tag{7.164}$$

Details of the *boundary traction vector* have been provided in Section 7.7.2.3. The numerical integration is carried out along the edges that are coincident with an external boundary of the domain, subjected to a natural boundary condition. If such an edge is subjected to a linearly varying traction then Eq. (7.127) represents the finite element form of the traction vector. In order to evaluate the *boundary traction vector* in Eq. (7.158c) let's define the following vector:

$$\{c_q\}^{(e)} = \left[N^{(Q4)}\right]^T\left[N^{(Q4)}\right]\{q\}|J| \tag{7.165}$$

Then Eq. (7.126) can be integrated by Gauss quadrature as follows:

$$r_{qi}^{(e)} = \sum_{l=1}^{n} W_l c_{qi}^{(e)} \tag{7.166}$$

Note that Eq. (7.117c) is a contour integral. Therefore, direction of the coordinate axis s, as expressed in Eq. (7.121), and the appropriate coordinates of the element boundaries need to be considered carefully. The vector components g_{qi} used in Eq. (7.166) are identified as follows:

$$ei = 1: c_{qi}^{(e)} = c_{qi}^{(e)}(\xi, -1) \text{ and } \{q\} = \{q_{x1}\ \ q_{y1}\ \ q_{x2}\ \ q_{y2}\ \ 0\ \ 0\ \ 0\ \ 0\}^T \tag{7.167a}$$

$$ei = 2: c_{qi}^{(e)} = c_{qi}^{(e)}(1, \eta) \text{ and } \{q\} = \{0\ \ 0\ \ q_{x2}\ \ q_{y2}\ \ q_{x3}\ \ q_{y3}\ \ 0\ \ 0\}^T \tag{7.167b}$$

$$ei = 3: c_{qi}^{(e)} = c_{qi}^{(e)}(\xi, 1) \text{ and } \{q\} = \{0\ \ 0\ \ 0\ \ 0\ \ q_{x3}\ \ q_{y3}\ \ q_{x4}\ \ q_{y4}\}^T \tag{7.167c}$$

$$ei = 4: c_{qi}^{(e)} = c_{qi}^{(e)}(-1, \eta) \text{ and } \{q\} = \{q_{x1}\ \ q_{y1}\ \ 0\ \ 0\ \ 0\ \ 0\ \ q_{x4}\ \ q_{y4}\}^T \tag{7.167d}$$

The highest order Gauss quadrature should be used for numerical integration to obtain the most accurate results, whenever possible. However, increased order of quadrature comes at a cost of increased computational expense. In two-dimensional elements, typically four-point (2×2) integration is used. This implies evaluation of the same function at four different Gauss integration points. With increasing number of elements, using high order Gauss integration adds to the overall computational cost of the solution.

For rectangular elements, the Jacobian is a constant and some element types can be integrated exactly. *Full integration* is possible for an eight- or nine-node plane element by using a 3×3 Gauss integration rule, and for a four-node plane element by 2×2 Gauss rule. In fact, use of an integration rule of less than full order is called *under-integration* or *reduced-integration*. Under-integration reduces the computational effort and reduces the over-stiffness associated with some elements [6].

7.10 Global equilibrium equations and boundary conditions

7.10.1 Assembly of global equilibrium equation

The equation of equilibrium for an element is given in Eq. (7.116) as follows:

$$[k]^{(e)}\{d\} = \{r_B\}^{(e)} + \left\{r_q\right\}^{(e)} + \left\{r_{\varepsilon_0}\right\}^{(e)} + \{F\}^{(e)} \tag{7.168}$$

The edges of a given element will be either in contact with the edges of the other elements in the mesh or they will be coincident with the external boundary of the domain, Γ_N. Here, these two types of edges are called *internal* and *external edges*, respectively. Then, as discussed in Section 7.7, the edge-load vector can be *symbolically* represented as follows:

$$[k]^{(e)}\{d\} = \{r_B\}^{(e)} + \left(\left\{r_{qIE}\right\}^{(e)} + \left\{r_{qEE}\right\}^{(e)}\right) + \left\{r_{\varepsilon_0}\right\}^{(e)} + \{F\}^{(e)} \tag{7.169}$$

where the $\{r_{qIE}\}^{(e)}$represents the traction vector acting along the internal edges and $\{r_{qEE}\}^{(e)}$ represents the traction vector acting along the external edges of the element.

The net traction on the shared internal edges of two elements should be zero, due to Newton's third law. This implies that the sum of the $\{r_{qIE}^{(e)}\}$ vectors for all elements should be zero. This summation is represented symbolically as follows:

$$\sum_{e=1}^{N_E} \left\{r_{qIE}\right\}^{(e)} = 0 \tag{7.170}$$

where N_E is the total number of elements. Note that by using Eq. (7.169) this summation can also be expressed as follows:

$$\sum_{e=1}^{N_E}[k]^{(e)}\{d\} = \sum_{e=1}^{N_E}\{r_B\}^{(e)} + \sum_{e=1}^{N_E}\left\{r_{qEE}\right\}^{(e)} + \sum_{e=1}^{N_E}\left\{r_{\varepsilon_0}\right\}^{(e)} + \sum_{e=1}^{N_E}\{F\}^{(e)} \tag{7.171}$$

Assembly of the global equations of equilibrium is defined by the summation in Eq. (7.171). This summation should be carried out by using the global node numbers. Symbolically, the global equation of equilibrium becomes,

$$[K]\{D\} = \{R\} \tag{7.172}$$

where $[K]$, $\{D\}$ and $\{R\}$ are the global stiffness matrix, degree of freedom and loading vectors, respectively. A closer look at Eq. (7.171) will show that the global loading vector has contributions due to body forces, internal forces imposed by initial strains and the tractions acting on the boundary Γ_N of the solution domain Ω,

$$\{R\} = \{R_B\} + \left\{R_q\right\} + \left\{R_{\varepsilon_0}\right\} + \{F\} \tag{7.173}$$

7.10.2 General treatment of the boundary conditions

The solution domain is subjected to essential boundary conditions (Eq. (7.10)(a)) on Γ_E, and natural boundary conditions (Eq. (7.10)(b)) on Γ_N. As described above, the natural boundary conditions are imposed through the $\{R_q\}$ vector.

There are several methods that are introduced in Sections 4.7 and 5.6 that can be used to impose the essential boundary conditions. In particular, it is helpful to remember that the degree of freedom vector can be partitioned as described in Section 5.6,

$$\{D\} = \begin{Bmatrix}\{D_x\}\\ \{D_c\}\end{Bmatrix} \tag{7.174}$$

where the subscripts $_x$ and $_c$ indicate the unknown and known (prescribed) values of the displacements, respectively. Such reordering of the degrees of freedom should be accompanied by reordering of the appropriate rows and columns of the stiffness matrix. This reordering is represented symbolically as follows:

$$\begin{bmatrix}[K_{xx}] & [K_{xc}]\\ [K_{cx}] & [K_{cc}]\end{bmatrix}\begin{Bmatrix}\{D_x\}\\ \{D_c\}\end{Bmatrix} = \begin{Bmatrix}\{R_c\}\\ \{R_x\}\end{Bmatrix} \tag{7.175}$$

Note that Eq. (7.175) represents the same equilibrium equation as (7.172), but ordered in a symbolically more convenient way. This equation can also be written as,

$$[K_{xx}]\{D_x\} + [K_{xc}]\{D_c\} = \{R_c\} \tag{7.176a}$$

$$[K_{cx}]\{D_x\} + [K_{cc}]\{D_c\} = \{R_x\} \tag{7.176b}$$

Note that the first one of these equations becomes,

$$[K_{xx}]\{D_x\} = \{R_c\} - [K_{xc}]\{D_c\} \tag{7.177}$$

considering the $\{R_c\}$ and $\{D_c\}$ represent the imposed forces (external tractions and body forces) and the displacements, respectively. The matrix $[K_{xx}]$ is the version of the global stiffness matrix on which the essential boundary conditions are imposed. This matrix is not singular and it can be inverted to solve for the unknown displacements,

$$\{D_x\} = [K_{xx}]^{-1}[\{R_c\} - [K_{xc}]\{D_c\}] \tag{7.178}$$

Note that the *reaction forces*, on the boundary where the essential boundary conditions are imposed, are contained in the vector $\{R_x\}$. This vector can be found from Eq. (7.176b),

$$\{R_x\} = [K_{cx}]\{D_x\} + [K_{cc}]\{D_c\} \tag{7.179}$$

following the solution of the $\{D_x\}$ vector in Eq. (7.178).

7.10.3 Numerical implementation of the boundary conditions

The procedure described Section 7.10.2 could be cumbersome and time consuming for problems with large number of degrees of freedom. The numerical implementations described in Section 5.6.3 can be used to apply the essential boundary conditions in two-dimensional problems.

7.11 Postprocessing of the solution

Upon finding the unknown global degrees of freedom $\{D_x\}$ the displacement vector $\{d\}$ for each element can be found by using the element connectivity table, described extensively in Chapters 4 and 5. The variation of strain in each element can then be computed by using Eq. (7.39) as follows:

$$\{\varepsilon\}^{(e)} = [B]^{(e)}\{d\} \tag{7.180}$$

where $[B]^{(e)}$ is the appropriate strain-displacement matrix for the element type. It can be shown in a straight forward manner that Eq. (7.180) represents the variation of the strain components as $\varepsilon_x = \varepsilon_x(x,y)$, $\varepsilon_y = \varepsilon_y(x,y)$ and $\gamma_{xy} = \gamma_{xy}(x,y)$ in the element (i.e., in $\Omega^{(e)}$). But, these three variables are *secondary variables* that are computed after the displacement vector. As we will see in Chapter 8, the secondary variables could have a lower order of accuracy as compared to the primary variables.

Once the strain variation in an element is obtained, the Hooke's law can be used to find the stress distribution within the element,

$$\{\sigma\}^{(e)} = [E]^{(e)}\left(\{\varepsilon\}^{(e)} - \left\{\varepsilon^{(0)}\right\}^{(e)}\right) \tag{7.181}$$

where the elasticity matrix is one of (7.23) or (7.24).

References

[1] J.R. Barber, Elasticity (Solid Mechanics and Its Applications), Springer, 2010.

[2] A.F. Bower, Applied Mechanics of Solids, CRC Press, Florida, 2010.

[3] S.P. Timoshenko, Theory of Elasticity, McGraw-Hill, New York, 1951.

[4] C.R. Wylie, L.C. Barrett, Advanced Engineering Mathematics, McGraw Hill, New York, New York, 1995.

[5] G. Strang, G. Fix, An Analysis of the Finite Element Method, Wellesley Cambridge Press, 2008.

[6] R.D. Cook, D.S. Malkus, M.E. Plesha, R.J. Witt, Concepts and Applications of Finite Element Analysis, 4th ed, John Wiley & Sons, 2002.

[7] M.A. Bhatti, Advanced Topics in Finite Element Analysis of Structures, Wiley & Sons, New York, 2006.

Chapter 8

Rectangular and triangular elements for two-dimensional elastic solids

8.1 Introduction

In this chapter, finite element formulation of two-dimensional elasticity problems is developed by using rectangular and triangular elements (Fig. 8.1). Geometrically simple rectangular and triangular elements reveal the strength and shortcomings of the interpolations used in two-dimensional problems. To this end, a brief analysis of the element characteristics is provided. The presentation used in this chapter follows the general derivation provided in Chapter 7, and it is suitable as introductory material.

In Section 8.1, we use the *minimum total potential energy principle* (Section 3.4.4) to develop the equilibrium equations for an *element* of the deformable medium. Detailed derivation of the total potential energy of an elastic solid body is presented in Section 2.2.9, and it is specialized to a two-dimensional case in Section 7.4. We specialize this to the case of a single deformable element, in the next section.

Rectangular elements are a subset of the general quadrilateral elements introduced in Chapter 7. In principle, the element coordinate system for the rectangular elements is indicated as (x',y') and the element domain is defined in the range $-a \leq x' \leq a$ and $-b \leq y' \leq b$ as shown in Fig. 8.2. As mentioned in Section 7.5, the order of two-dimensional polynomials increases with the increasing number of nodes (Fig. 8.3). We will consider rectangular elements with four, eight, or nine nodes. The triangular elements used in this chapter are chosen such that the nodes-1 and -2 are located on the x'-axis as shown in Fig. 8.2. The two-dimensional shape functions for Q4, Q8, Q9, and CST elements are presented in the (x',y') coordinate system in Section 8.2.

Equations of equilibrium for a rectangular Q4 element are derived in Section 8.3 and for a CST element in Section 8.4. The potential drawbacks and limitations of the Q4 and CST elements are demonstrated in Section 8.5. These defects can be eliminated for the most part by using higher order interpolation functions which results in increased number of nodes and degrees of freedom.

Finite Element Method: Physics and Solution Methods. DOI: https://doi.org/10.1016/B978-0-12-821127-4.00006-2
Copyright © 2022 Elsevier Inc. All rights reserved.

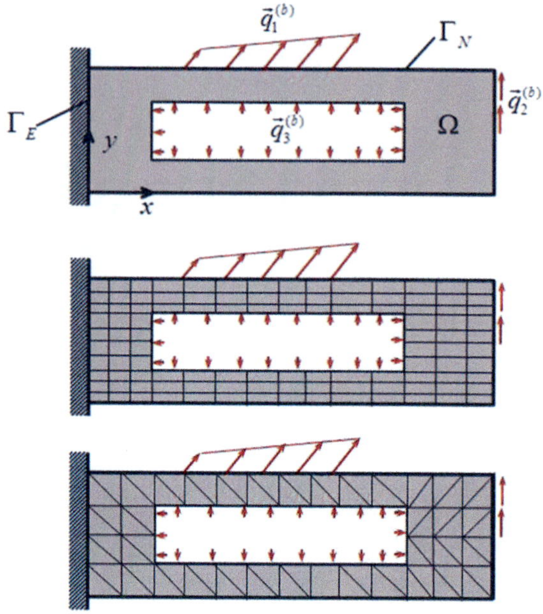

FIGURE 8.1 In finite element method, a solid domain (A) is segmented into either quadrilateral (B) or triangular (C) subdomains called elements. The elements are connected across-their boundaries.

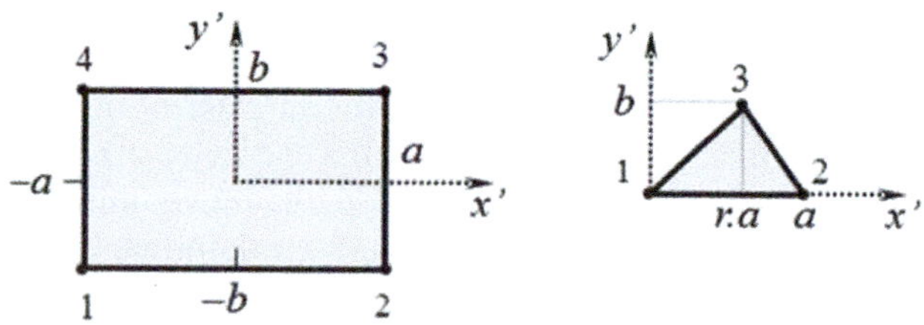

FIGURE 8.2 The four-node quadrilateral (Q4) and three-node triangular (CST) elements that are defined in the rectangular (x', y') coordinate system, used in this chapter.

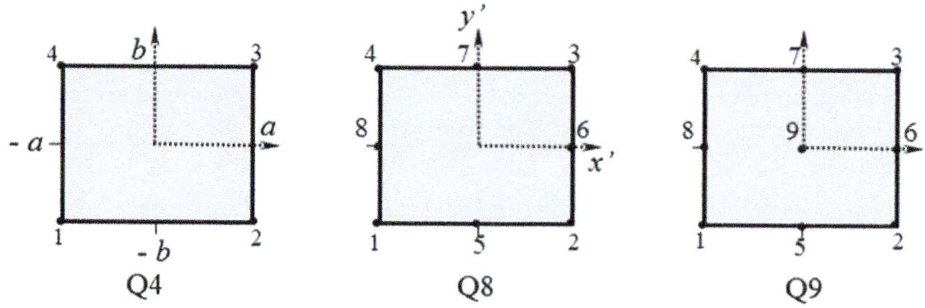

FIGURE 8.3 Quadrilateral elements with four, eight, and nine nodes, also known as Q4, Q8, and Q9 elements, respectively. The numbers of nodes represent different levels of polynomial interpolation across the element.

Reasons for the improvements introduced by the higher order elements are discussed in Section 8.6. The example given at the end of the chapter provides a top-level overview of using Q4 elements in analyzing a simple two-dimensional problem.

8.1.1 Total potential energy of an element for a two-dimensional elasticity problem

Detailed derivation of the total potential energy of an elastic solid body is presented in Section 2.2. The total potential energy for a two-dimensional element was found in Section 7.4 as follows:

$$\pi_p^{(e)} = \int_{\Omega^{(e)}} \{\varepsilon\}^T \{\sigma\} t \, d\Omega^{(e)} - \int_{\Omega^{(e)}} \{u\}^T \{F_B\} t \, d\Omega^{(e)} - \oint_{\Gamma^{(e)}} \{u\}^T \left\{q^{(b)}\right\} d\Gamma^{(e)} - \{d\}^T \{F\}^{(e)} \tag{8.1}$$

where $\Omega^{(e)}$ is the domain that describes the element, $\Gamma^{(e)}$ is the boundary and t is the thickness of the element. In addition, we have,

$$\begin{aligned}
&\text{Strain vector:} && \{\varepsilon\} = \left\{\varepsilon_{xx} \ \varepsilon_{yy} \ \gamma_{xy}\right\}^T \\
&\text{Stress vector:} && \{\sigma\} = \left\{\sigma_{xx} \ \sigma_{yy} \ \tau_{xy}\right\}^T \\
&\text{Displacement vector:} && \{u\} = \left\{u_x \ u_y\right\}^T \\
&\text{Boundary traction vector:} && \left\{q^{(b)}\right\} = \left\{q_x \ q_y\right\}^T \\
&\text{Body force vector:} && \{F_B\} = \left\{F_{Bx} \ F_{By}\right\}^T \\
&\text{Nodal force vector:} && \{F\}^{(e)} = \left\{F_{x1} \ F_{y1} \ F_{x2} \ F_{y2} \ \cdots \ F_{xM} \ F_{yM}\right\}^T \\
&\text{Nodal displacement vector:} && \{d\} = \left\{u_{x1} \ u_{y1} \ u_{x2} \ u_{y2} \ \cdots \ u_{xM} \ u_{yM}\right\}^T
\end{aligned}$$

Note that the boundary traction vector $\vec{q}^{(b)}(s)$ varies along the boundary of the element, and the body force vector $\vec{F}_B = \vec{F}_B(x, y)$ varies across the element. Also note that F_{xi} and F_{yi} represent the components of the concentrated forces that are directly applied on the element nodes and u_{xi} and u_{yi} represent the components of the nodal displacements ($i = 1$–M). The last term of Eq. (8.1) therefore represents the work due to these concentrated forces.

Equation of equilibrium can be found by using the minimum total potential energy principle, by setting, $\delta\pi_p^{(e)} = 0$. This gives,

$$\int_{\Omega^{(e)}} \{\delta\varepsilon\}^T \{\sigma\} t \, d\Omega^{(e)} = \int_{\Omega^{(e)}} \{\delta u\}^T \{F_B\} t \, d\Omega^{(e)} + \oint_{\Gamma^{(e)}} \{\delta u\}^T \left\{q^{(b)}\right\} d\Gamma^{(e)} + \{\delta d\}^T \{F\}^{(e)} \tag{8.2}$$

8.1.2 High-level derivation of the element equilibrium equations

The element equilibrium equation is obtained by using the procedure introduced in Section 7.4.2. Displacement and strain in the element $\Omega^{(e)}$ are approximated by using an interpolation of the form given in Equation (7.33),

$$\{u\} = \{u_x \ u_y\}^T = [N]\{d\} \tag{8.3a}$$

$$\{\varepsilon\} = \{\varepsilon_{xx} \ \varepsilon_{yy} \ \gamma_{xy}\}^T = [\partial]\{u\} = ([\partial][N])\{d\} = [B]\{d\} \tag{8.3b}$$

where the shape function matrix $[N]$, the strain–displacement operator matrix $[B]$ and the vector of nodal degrees of freedom (displacement) $\{d\}$ are given as follows:

$$[N] = \begin{bmatrix} N_1 & 0 & N_2 & 0 & N_3 & \cdots & 0 \\ 0 & N_1 & 0 & N_2 & 0 & \cdots & N_M \end{bmatrix} \tag{8.4a}$$

$$[B] = \begin{bmatrix} \dfrac{\partial N_1}{\partial x'} & 0 & \dfrac{\partial N_2}{\partial x'} & 0 & \cdots & 0 \\ 0 & \dfrac{\partial N_1}{\partial y'} & 0 & \dfrac{\partial N_2}{\partial y'} & \cdots & \dfrac{\partial N_M}{\partial y'} \\ \dfrac{\partial N_1}{\partial y'} & \dfrac{\partial N_1}{\partial x'} & \dfrac{\partial N_2}{\partial y'} & \dfrac{\partial N_2}{\partial x'} & \cdots & \dfrac{\partial N_M}{\partial x'} \end{bmatrix} \tag{8.4b}$$

$$\{d\} = \{u_{x1} \ u_{y1} \ \cdots \ u_{xM} \ u_{yM}\}^T \tag{8.4c}$$

Variation of the displacement and strain vectors can be expressed as follows:

$$\{\delta u\} = [N]\{\delta d\} \tag{8.5a}$$

$$\{\delta\varepsilon\} = [B]\{\delta d\} \tag{8.5b}$$

The total potential energy of an element is expressed by using Eqs. (8.2)–(8.5a),

$$\{\delta d\}^T\left(\int_{\Omega^{(e)}} [B]^T\{\sigma\}t\, d\Omega^{(e)}\right) = \{\delta d\}^T\left(\int_{\Omega^{(e)}} [N]^T\{F_B\}t\, d\Omega^{(e)} + \oint_{\Gamma^{(e)}} [N]^T\{q^{(b)}\}d\Gamma^{(e)} + \{F\}^{(e)}\right) \tag{8.6}$$

The variation of the displacement vector $\{\delta d\}$ is arbitrary and thus can be eliminated from this equation,

$$\int_{\Omega^{(e)}} [B]^T\{\sigma\}t\, d\Omega^{(e)} = \int_{\Omega^{(e)}} [N]^T\{F_B\}t\, d\Omega^{(e)} + \oint_{\Gamma^{(e)}} [N]^T\{q^{(b)}\}d\Gamma^{(e)} + \{F\}^{(e)} \tag{8.7}$$

This equation represents the equilibrium for an element with a general constitutive behavior. The term on the left hand side of this equation represents the *internal force vector* that develops in response to the *external forces* acting on the element. Note that at this point in the derivation, no assumptions about the material behavior have yet been made. If we assume that the material has a linear elastic constitutive behavior as described in Eq. (7.22),

$$\{\sigma\} = [E]\left(\{\varepsilon\} - \left\{\varepsilon^{(0)}\right\}\right) + \left\{\sigma^{(0)}\right\} \tag{8.8}$$

we find the general form of the element equilibrium equation as follows:

$$\int_{\Omega^{(e)}} [B]^T\left[[E]\left([B]\{d\} - \left\{\varepsilon^{(0)}\right\}\right) + \left\{\sigma^{(0)}\right\}\right]t\, d\Omega^{(e)}$$

$$= \int_{\Omega^{(e)}} [N]^T\{F_B\}t\, d\Omega^{(e)} + \oint_{\Gamma^{(e)}} [N]^T\left\{q^{(b)}\right\}d\Gamma^{(e)} + \{F\}^{(e)} \tag{8.9}$$

Note that in this equation the terms due to the initial strains and stresses have the same mathematical characteristic as the body force vector. In matrix form, Eq. (8.9) is expressed as follows:

$$[k]^{(e)}\{d\} = \{r_B\}^{(e)} + \left\{r_q\right\}^{(e)} + \{r_{\varepsilon 0}\}^{(e)} + \{r_{\sigma 0}\}^{(e)} + \{F\}^{(e)} \tag{8.10}$$

where

$$\text{Element stiffness matrix: } [k]^{(e)} = \int_{\Omega^{(e)}} [B]^T[E][B]t\, d\Omega^{(e)} \tag{8.11a}$$

Consistent nodal force vectors:

$$\text{Boundary tractions: } \left\{r_{qb}\right\}^{(e)} = \oint_{\Gamma^{(e)}} [N]^T\{q_b\}d\Gamma^{(e)} \tag{8.11b}$$

$$\text{Body forces: } \{r_B\}^{(e)} = \int_{\Omega^{(e)}} [N]^T[F_B]t\, d\Omega^{(e)} \tag{8.11c}$$

$$\text{Initial strains: } \{r_{\varepsilon 0}\}^{(e)} = \int_{\Omega^{(e)}} [B]^T[E]\{\varepsilon_0\}t\, d\Omega^{(e)} \tag{8.11d}$$

$$\text{Initial stresses: } \{r_{\sigma 0}\}^{(e)} = -\int_{\Omega^{(e)}} [B]^T\left\{\sigma^{(0)}\right\}t d\Omega^{(e)} \tag{8.11e}$$

$$\text{Nodal force vector: } \{F\}^{(e)} \tag{8.11f}$$

Equation (8.10) can also be expressed as follows:

$$[k]^{(e)}\{d\} = \{r\}^{(e)} \tag{8.12}$$

where the *total nodal force vector* is,

$$\{r\}^{(e)} = \{r_B\}^{(e)} + \left\{r_q\right\}^{(e)} + \{r_{\varepsilon 0}\}^{(e)} + \{r_{\sigma 0}\}^{(e)} + \{F\}^{(e)} \tag{8.13}$$

In the next sections, we evaluate the element stiffness matrix and load vector for rectangular and triangular elements.

8.2 Two-dimensional interpolation functions

Common rectangular elementals with four, eight, or nine nodes are shown in Fig. 8.3. The domain of the interpolation for these elements is as follows:

$$-a \leq x' \leq a \text{ and } -b \leq y' \leq b$$

The interpolation functions and the shape functions for general quadrilateral elements are given in Eqs. (7.62)–(7.67). These equations can be transformed to Cartesian coordinate system by using the following definitions,

$$x' = \xi a \text{ and } y' = \eta b \tag{8.14}$$

8.2.1 Interpolation and shape functions in plane quadrilateral elements

The interpolation and shape functions in rectangular elements are found from Eqs. (7.62)–(7.67) by using the definition given in Eq. (8.14).

Q4 element:
The interpolation function,

$$\phi(x', y') = \sum_{i=1}^{4} N_i^{(Q4)} \phi_i \tag{8.15}$$

The shape functions,

$$\begin{aligned} N_1^{(Q4)} &= \frac{1}{4}\left(1 - \frac{x'}{a}\right)\left(1 - \frac{y'}{b}\right), \quad N_2^{(Q4)} = \frac{1}{4}\left(1 + \frac{x'}{a}\right)\left(1 - \frac{y'}{b}\right) \\ N_3^{(Q4)} &= \frac{1}{4}\left(1 + \frac{x'}{a}\right)\left(1 + \frac{y'}{b}\right), \quad N_4^{(Q4)} = \frac{1}{4}\left(1 - \frac{x'}{a}\right)\left(1 + \frac{y'}{b}\right) \end{aligned} \tag{8.16}$$

Q8 element:
The interpolation function,

$$\phi(x', y') = \sum_{i=1}^{8} N_i^{(Q8)} \phi_i \tag{8.17}$$

The shape functions,

$$\begin{aligned} N_1^{(Q8)} &= N_1^{(Q4)} - \frac{1}{2}\left(N_8^{(Q8)} + N_5^{(Q8)}\right), \quad N_2^{(Q8)} = N_2^{(Q4)} - \frac{1}{2}\left(N_5^{(Q8)} + N_6^{(Q8)}\right) \\ N_3^{(Q8)} &= N_3^{(Q4)} - \frac{1}{2}\left(N_6^{(Q8)} + N_7^{(Q8)}\right), \quad N_4^{(Q8)} = N_4^{(Q4)} - \frac{1}{2}\left(N_7^{(Q8)} + N_8^{(Q8)}\right) \end{aligned} \tag{8.18}$$

material behavior have yet been made. If we assume that the material has a linear elastic constitutive behavior as described in Eq. (7.22),

$$\{\sigma\} = [E](\{\varepsilon\} - \{\varepsilon^{(0)}\}) + \{\sigma^{(0)}\} \tag{8.8}$$

we find the general form of the element equilibrium equation as follows:

$$\int\limits_{\Omega^{(e)}} [B]^T[[E]([B]\{d\} - \{\varepsilon^{(0)}\}) + \{\sigma^{(0)}\}]t\, d\Omega^{(e)}$$
$$= \int\limits_{\Omega^{(e)}} [N]^T\{F_B\}t\, d\Omega^{(e)} + \oint\limits_{\Gamma^{(e)}} [N]^T\{q^{(b)}\}d\Gamma^{(e)} + \{F\}^{(e)} \tag{8.9}$$

Note that in this equation the terms due to the initial strains and stresses have the same mathematical characteristic as the body force vector. In matrix form, Eq. (8.9) is expressed as follows:

$$[k]^{(e)}\{d\} = \{r_B\}^{(e)} + \{r_q\}^{(e)} + \{r_{\varepsilon 0}\}^{(e)} + \{r_{\sigma 0}\}^{(e)} + \{F\}^{(e)} \tag{8.10}$$

where

$$\text{Element stiffness matrix: } [k]^{(e)} = \int\limits_{\Omega^{(e)}} [B]^T[E][B]t\, d\Omega^{(e)} \tag{8.11a}$$

Consistent nodal force vectors:

$$\text{Boundary tractions: } \{r_{qb}\}^{(e)} = \oint\limits_{\Gamma^{(e)}} [N]^T\{q_b\}d\Gamma^{(e)} \tag{8.11b}$$

$$\text{Body forces: } \{r_B\}^{(e)} = \int\limits_{\Omega^{(e)}} [N]^T[F_B]t\, d\Omega^{(e)} \tag{8.11c}$$

$$\text{Initial strains: } \{r_{\varepsilon 0}\}^{(e)} = \int\limits_{\Omega^{(e)}} [B]^T[E]\{\varepsilon_0\}t\, d\Omega^{(e)} \tag{8.11d}$$

$$\text{Initial stresses: } \{r_{\sigma 0}\}^{(e)} = -\int\limits_{\Omega^{(e)}} [B]^T\{\sigma^{(0)}\}t d\Omega^{(e)} \tag{8.11e}$$

$$\text{Nodal force vector: } \{F\}^{(e)} \tag{8.11f}$$

Equation (8.10) can also be expressed as follows:

$$[k]^{(e)}\{d\} = \{r\}^{(e)} \tag{8.12}$$

where the *total nodal force vector* is,

$$\{r\}^{(e)} = \{r_B\}^{(e)} + \{r_q\}^{(e)} + \{r_{\varepsilon 0}\}^{(e)} + \{r_{\sigma 0}\}^{(e)} + \{F\}^{(e)} \tag{8.13}$$

In the next sections, we evaluate the element stiffness matrix and load vector for rectangular and triangular elements.

8.2 Two-dimensional interpolation functions

Common rectangular elementals with four, eight, or nine nodes are shown in Fig. 8.3. The domain of the interpolation for these elements is as follows:

$$-a \le x' \le a \text{ and } -b \le y' \le b$$

The interpolation functions and the shape functions for general quadrilateral elements are given in Eqs. (7.62)–(7.67). These equations can be transformed to Cartesian coordinate system by using the following definitions,

$$x' = \xi a \text{ and } y' = \eta b \tag{8.14}$$

8.2.1 Interpolation and shape functions in plane quadrilateral elements

The interpolation and shape functions in rectangular elements are found from Eqs. (7.62)–(7.67) by using the definition given in Eq. (8.14).

Q4 element:
The interpolation function,

$$\phi(x', y') = \sum_{i=1}^{4} N_i^{(Q4)} \phi_i \tag{8.15}$$

The shape functions,

$$\begin{aligned} N_1^{(Q4)} &= \frac{1}{4}\left(1 - \frac{x'}{a}\right)\left(1 - \frac{y'}{b}\right), \quad N_2^{(Q4)} = \frac{1}{4}\left(1 + \frac{x'}{a}\right)\left(1 - \frac{y'}{b}\right) \\ N_3^{(Q4)} &= \frac{1}{4}\left(1 + \frac{x'}{a}\right)\left(1 + \frac{y'}{b}\right), \quad N_4^{(Q4)} = \frac{1}{4}\left(1 - \frac{x'}{a}\right)\left(1 + \frac{y'}{b}\right) \end{aligned} \tag{8.16}$$

Q8 element:
The interpolation function,

$$\phi(x', y') = \sum_{i=1}^{8} N_i^{(Q8)} \phi_i \tag{8.17}$$

The shape functions,

$$\begin{aligned} N_1^{(Q8)} &= N_1^{(Q4)} - \frac{1}{2}\left(N_8^{(Q8)} + N_5^{(Q8)}\right), \quad N_2^{(Q8)} = N_2^{(Q4)} - \frac{1}{2}\left(N_5^{(Q8)} + N_6^{(Q8)}\right) \\ N_3^{(Q8)} &= N_3^{(Q4)} - \frac{1}{2}\left(N_6^{(Q8)} + N_7^{(Q8)}\right), \quad N_4^{(Q8)} = N_4^{(Q4)} - \frac{1}{2}\left(N_7^{(Q8)} + N_8^{(Q8)}\right) \end{aligned} \tag{8.18}$$

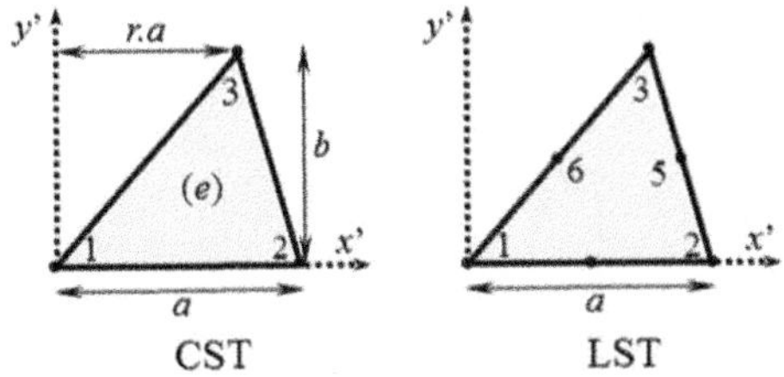

FIGURE 8.4 Triangular elements with three and six nodes, also known as constant and linear strain triangles, respectively. These elements present linear and quadratic two-dimensional interpolation capabilities across the element.

with

$$N_5^{(Q8)} = \frac{1}{2}\left(1 - \left(\frac{x'}{a}\right)^2\right)\left(1 - \frac{y'}{b}\right), \quad N_6^{(Q8)} = \frac{1}{2}\left(1 + \frac{x'}{a}\right)\left(1 - \left(\frac{y'}{b}\right)^2\right)$$
$$N_7^{(Q8)} = \frac{1}{2}\left(1 - \left(\frac{x'}{a}\right)^2\right)\left(1 + \frac{y'}{b}\right), \quad N_8^{(Q8)} = \frac{1}{2}\left(1 - \frac{x'}{a}\right)\left(1 - \left(\frac{y'}{b}\right)^2\right) \tag{8.19}$$

Q9 element:
The interpolation function,

$$\phi(x', y') = \sum_{i=1}^{9} N_i^{(Q9)}\phi_i \tag{8.20}$$

The first eight shape functions $N_i^{(Q9)}$ ($i = 1$–8) of the Q9 element are obtained by subtracting $N_9^{(Q9)}/4$ from $N_i^{(Q8)}$ of the Q8 element,

$$N_i^{(Q9)} = N_i^{(Q8)} - \frac{1}{4}N_9^{(Q9)} \text{ for } i = 1\text{–}8 \tag{8.21a}$$

$$N_9^{(Q9)} = \left(1 - \left(\frac{x'}{a}\right)^2\right)\left(1 - \left(\frac{y'}{b}\right)^2\right) \tag{8.21b}$$

8.2.2 Interpolation and shape functions in plane triangular elements

Consider the triangular element shown in Fig. 8.4. The nodes of this element are located as:

$$(x_1', y_1') = (0, 0), \ (x_2', y_2') = (a, 0) \text{ and } (x_3', y_3') = (r\ a,\ b)$$

where r is a scalar. This particular geometry is useful to demonstrate the properties of the three-node triangular element. The shape functions for this element are obtained by using the techniques described in Section 4.3. Consider,

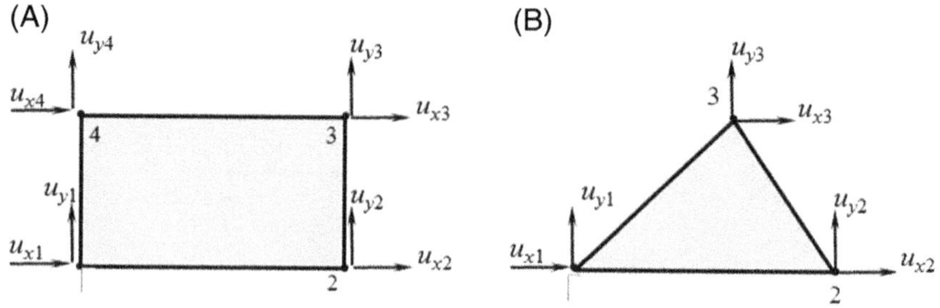

FIGURE 8.5 The nodal degrees of freedom for the four-node and the three-node elements are the projections of the displacement vector at each node along the x and y axes.

for example, interpolation of the variable $u_x(x',y')$,

$$u_x = a_0 + a_1 x' + a_2 y' \tag{8.22}$$

where $a_0 - a_2$ are scalar constants as described in Section 7.5. Upon evaluating u_x at the three nodes of the element the following set of equations is found,

$$\begin{Bmatrix} u_{x1} \\ u_{x2} \\ u_{x3} \end{Bmatrix} = \begin{bmatrix} 1 & 0 & 0 \\ 1 & a & 0 \\ 1 & ra & b \end{bmatrix} \begin{Bmatrix} a_0 \\ a_1 \\ a_2 \end{Bmatrix} \tag{8.23}$$

which leads to the following interpolation,

$$u_x = \sum_{i=1}^{3} N_i^{T3} u_{xi} \tag{8.24}$$

with shape functions,

$$N_1^{(T3)} = 1 - \frac{x'}{a} - (1-r)\frac{y'}{b}, \quad N_2^{(T3)} = \frac{x'}{a} - r\frac{y'}{b}, \text{ and } N_3^{(T3)} = \frac{y'}{b} \tag{8.25}$$

8.3 Bilinear rectangular element (Q4)

8.3.1 Element stiffness matrix

A bilinear rectangular element is a four-node element with eight degrees of freedom (Fig. 8.5A). The components of the displacement vector $\{u\}$ are interpolated across the element as follows:

$$u_x = \sum_{i-1}^{4} N_i^{Q4} u_{xi} \text{ and } u_y = \sum_{i-1}^{4} N_i^{Q4} u_{yi} \tag{8.26}$$

where u_{xi}, u_{yi} for (i = 1–4) are the displacements at the four corners of the element and N_i^{Q4} are the shape functions given in Eq. (8.16).

Eq. (8.26) can be represented in vector form as follows:

$$\{u\} = \{u_x \ u_y\}^T = [N^{(Q4)}]\{d\} \tag{8.27}$$

where the *shape function matrix* for this Q4 element is,

$$[N^{(Q4)}] = \begin{bmatrix} N_1^{(Q4)} & 0 & N_2^{(Q4)} & 0 & N_3^{(Q4)} & 0 & N_4^{(Q4)} & 0 \\ 0 & N_1^{(Q4)} & 0 & N_2^{(Q4)} & 0 & N_3^{(Q4)} & 0 & N_4^{(Q4)} \end{bmatrix} \tag{8.28}$$

and the nodal *degree of freedom vector* is,

$$\{d\}^T = \{u_{x1} \ u_{y1} \ u_{x2} \ u_{y2} \ u_{x3} \ u_{y3} \ u_{x4} \ u_{y4}\} \tag{8.29}$$

The strain vector for this problem can be computed in the finite element form as follows:

$$\{\varepsilon\} = [\partial]\{u\} = [\partial][N^{(Q4)}]\{d\} = [B^{(Q4)}]\{d\} \tag{8.30}$$

where the *strain–displacement* matrix $[B^{(Q4)}]$ is found as follows:

$$[B^{(Q4)}] = \begin{bmatrix} N_{1,x'}^{(Q4)} & 0 & N_{2,x'}^{(Q4)} & 0 & N_{3,x'}^{(Q4)} & 0 & N_{4,x'}^{(Q4)} & 0 \\ 0 & N_{1,y'}^{(Q4)} & 0 & N_{2,y'}^{(Q4)} & 0 & N_{3,y'}^{(Q4)} & 0 & N_{4,y'}^{(Q4)} \\ N_{1,y'}^{(Q4)} & N_{1,x'}^{(Q4)} & N_{2,y'}^{(Q4)} & N_{2,x'}^{(Q4)} & N_{3,y'}^{(Q4)} & N_{3,x'}^{(Q4)} & N_{4,y'}^{(Q4)} & N_{4,x'}^{(Q4)} \end{bmatrix} \tag{8.31}$$

The partial derivatives are indicated by$_{,x'}$ and$_{,y'}$. After using Eq. (8.16), these terms are computed as follows:

$$\begin{aligned} N_{1,x'}^{(Q4)} &= -\frac{1}{4a}\left(1-\frac{y'}{b}\right), \quad N_{2,x'}^{(Q4)} = \frac{1}{4a}\left(1-\frac{y'}{b}\right), \\ N_{3,x'}^{(Q4)} &= \frac{1}{4a}\left(1+\frac{y'}{b}\right), \quad N_{4,x'}^{(Q4)} = -\frac{1}{4a}\left(1-\frac{y'}{b}\right) \\ N_{1,y'}^{(Q4)} &= -\frac{1}{4b}\left(1-\frac{x'}{a}\right), \quad N_{2,y'}^{(Q4)} = -\frac{1}{4b}\left(1+\frac{x'}{a}\right), \\ N_{1,y'}^{(Q4)} &= \frac{1}{4b}\left(1+\frac{x'}{a}\right), \quad N_{4,y'}^{(Q4)} = \frac{1}{4b}\left(1-\frac{x'}{a}\right) \end{aligned} \tag{8.32}$$

The element stiffness matrix can be obtained from Eq. (8.11a) as follows:

$$\left[k^{(Q4)}\right]^{(e)} = \int_{-a}^{a}\int_{-b}^{b} \left[B^{(Q4)}\right]^T [E]\left[B^{(Q4)}\right] t\, dx'\, dy' \tag{8.33}$$

where t is the element thickness, and $d\Omega^{(e)}$ is replaced with $dx'\ dy'$. The elasticity matrix $[E]$ is expressed either by Eq. (2.75) for general–elastic behavior, or by Eq. (2.80) for plane–stress, or Eq. (2.85) plane–strain behavior.

The elasticity matrix $[E]$ for *plane–strain* formulation is,

$$[E] = \frac{E}{(1+\nu)(1-2\nu)}\begin{bmatrix} 1-\nu & \nu & 0 \\ \nu & 1-\nu & 0 \\ 0 & 0 & (1-2\nu)/2 \end{bmatrix}$$

The elasticity matrix for *plane–stress* formulation is,

$$[E] = \frac{E}{(1-\nu^2)}\begin{bmatrix} 1 & \nu & 0 \\ \nu & 1 & 0 \\ 0 & 0 & (1-\nu)/2 \end{bmatrix}$$

For general symbolic manipulation, the two-dimensional elasticity matrix for an isotropic material is introduced,

$$[E] = \begin{bmatrix} E_{11} & E_{12} & 0 \\ E_{12} & E_{11} & 0 \\ 0 & 0 & G \end{bmatrix} \tag{8.34}$$

Next let's calculate of the k_{11}-term of the element stiffness matrix. To this end, let's define the following matrix product,

$$[C] = \left[B^{(Q4)}\right]^T [E]\left[B^{(Q4)}\right] \tag{8.35}$$

The C_{11} term of this matrix can be shown to be,

$$C_{11} = \left(B_{11}^{(Q4)}\right)^2 E_{11} + \left(B_{31}^{Q4}\right)^2 G = \frac{1}{16ab^2}\left[E_{11}(b-y)^2 + G(a-x)^2\right] \tag{8.36}$$

where $G = E/2(1+\nu)$ is the shear stiffness of the material. The k_{11}-term of the element stiffness matrix can then be expressed as follows:

$$k_{11} = \int_{-a}^{1}\int_{-b}^{b} C_{11}\, t\, dx\, dy \tag{8.37}$$

which gives,

$$k_{11} = \frac{1}{3}\left(\frac{b}{a}E_{11} + \frac{a}{b}G\right) \tag{8.38}$$

For an *isotropic*, Q4 element the element stiffness matrix can be found as follows:

$$[k_{Q4}]^{(e)} = t \times \begin{bmatrix}
\frac{bE_{11}}{3a}+\frac{aG}{3b} & \frac{E_{12}+G}{4} & \frac{aG}{6b}-\frac{bE_{11}}{3a} & \frac{E_{12}-G}{4} & -\frac{bE_{11}}{6a}-\frac{aG}{6b} & -\frac{E_{12}+G}{4} & \frac{bE_{11}}{6a}-\frac{aG}{3b} & \frac{G-E_{12}}{4} \\
\frac{E_{12}+G}{4} & \frac{aE_{11}}{3b}+\frac{bG}{3a} & \frac{G-E_{12}}{4} & \frac{aE_{11}}{6b}-\frac{bG}{3a} & -\frac{E_{12}+G}{4} & -\frac{aE_{11}}{6b}+\frac{bG}{6a} & \frac{E_{12}-G}{4} & \frac{bG}{6a}-\frac{aE_{11}}{3b} \\
\frac{aG}{6b}-\frac{bE_{11}}{3a} & \frac{G-E_{12}}{4} & \frac{bE_{11}}{3a}+\frac{aG}{3b} & -\frac{E_{12}+G}{4} & \frac{bE_{11}}{6a}-\frac{aG}{3b} & \frac{E_{12}+G}{4} & -\frac{bE_{11}}{6a}-\frac{aG}{6b} & \frac{E_{12}+G}{4} \\
\frac{E_{12}-G}{4} & \frac{aE_{11}}{6b}-\frac{bG}{3a} & -\frac{E_{12}+G}{4} & \frac{aE_{11}}{3b}+\frac{bG}{3a} & \frac{G-E_{12}}{4} & \frac{bG}{6a}+\frac{aE_{11}}{3b} & \frac{E_{12}+G}{4} & -\frac{aE_{11}}{6b}-\frac{bG}{6a} \\
-\frac{bE_{11}}{6a}-\frac{aG}{6b} & -\frac{E_{12}+G}{4} & \frac{bE_{11}}{6a}-\frac{aG}{3b} & \frac{G-E_{12}}{4} & \frac{bE_{11}}{3a}+\frac{aG}{3b} & \frac{E_{12}+G}{4} & \frac{aG}{6b}-\frac{bE_{11}}{3a} & \frac{E_{12}-G}{4} \\
-\frac{E_{12}+G}{4} & -\frac{aE_{11}}{6b}-\frac{bG}{6a} & \frac{E_{12}-G}{4} & \frac{bG}{6a}-\frac{aE_{11}}{3b} & \frac{E_{12}+G}{4} & \frac{aE_{11}}{3b}+\frac{bG}{3a} & \frac{G-E_{12}}{4} & \frac{aE_{12}}{6b}-\frac{bG}{3a} \\
\frac{bE_{11}}{6a}-\frac{aG}{3b} & \frac{E_{12}-G}{4} & -\frac{bE_{11}}{6a}-\frac{aG}{6b} & \frac{E_{12}+G}{4} & \frac{aG}{6b}-\frac{bE_{11}}{3a} & \frac{G-E_{12}}{4} & \frac{bE_{11}}{3a}+\frac{aG}{3b} & -\frac{E_{12}+G}{4} \\
\frac{G-E_{12}}{4} & \frac{bG}{6a}-\frac{aE_{11}}{3b} & \frac{E_{12}+G}{4} & -\frac{aE_{11}}{6b}-\frac{bG}{6a} & \frac{E_{12}-G}{4} & \frac{aE_{11}}{6b}-\frac{bG}{3a} & -\frac{E_{12}+G}{4} & \frac{aE_{11}}{3b}+\frac{bG}{3a}
\end{bmatrix} \tag{8.39}$$

It is notable that the aspect ratio of the element (a/b) affects the element stiffness matrix. If the aspect ratio is large numerical difficulties can be encountered.

8.3.2 Consistent nodal force vector

Consistent nodal forces are work equivalent forces that are transferred to the nodes of the element to represent the effects of the distributed forces acting on the element. The total work done by consistent nodal forces moving through the nodal degrees of freedom is equal to the work done by distributed forces moving through local deformation. Consistent nodal forces are also statically equivalent to the original distributed forces. Both load systems have the same resultant force and the same moment about an arbitrarily located point.

The body force and surface traction vectors for a two-dimensional problem are given as follows:

$$\{F_B\} = \{F_{Bx} \ F_{By}\}^T \tag{8.40a}$$

$$\{q^{(b)}\} = \{q_x \ q_y\}^T \tag{8.40b}$$

and the initial strain and initial stress vectors for a two-dimensional problem are given as follows:

$$\{\varepsilon^{(0)}\} = \{\varepsilon_{xx}^{(0)} \ \varepsilon_{yy}^{(0)} \ \varepsilon_{xy}^{(0)}\}^T \tag{8.41a}$$

$$\{\sigma^{(0)}\} = \{\sigma_{xx}^{(0)} \ \sigma_{yy}^{(0)} \ \tau_{xy}^{(0)}\} \tag{8.41b}$$

The element force vector for a rectangular Q4 element can be found from Eqs. (8.11) and (8.13) as follows:

$$\left\{r^{(Q4)}\right\} = \left\{r_B{}^{(Q4)}\right\} + \left\{r_q{}^{(Q4)}\right\} + \left\{r_{\varepsilon 0}{}^{(Q4)}\right\} + \left\{r_{\sigma 0}{}^{(Q4)}\right\} + \{F\}$$

where the *consistent nodal force vectors* are,

$$\left\{r_B{}^{(Q4)}\right\} = \int_{-a}^{a}\int_{-b}^{b} \left[N^{(Q4)}\right]^T \{F_B\} t\, dx'\, dy' \tag{8.42a}$$

$$\left\{r_{\sigma 0}{}^{(Q4)}\right\} = -\int_{-a}^{a}\int_{-b}^{b} \left[B^{(Q4)}\right]^T \left\{\sigma^{(0)}\right\} t\, dx'\, dy' \tag{8.42b}$$

$$\left\{r_{\varepsilon 0}{}^{(Q4)}\right\} = \int_{-a}^{a}\int_{-b}^{b} \left[B^{(Q4)}\right]^T [E] \left\{\varepsilon^{(0)}\right\} t\, dx'\, dy' \tag{8.42c}$$

$$\left\{r_q{}^{(Q4)}\right\} = \oint_{\Gamma^{(e)}} \left[N^{(Q4)}\right]^T \left\{q^{(b)}\right\} t\, ds \tag{8.42d}$$

Note that in Eq. (8.42), we replaced $d\Omega^{(e)}$ with $t\,dx'dy'$ and $d\Gamma^{(e)}$ with ds. The *concentrated forces that happen to be acting on the nodes* are represented by,

$$\{F\} = \left\{F_{x1}\ F_{y1}\ F_{x2}\ F_{y2}\ F_{x3}\ F_{y3}\ F_{x4}\ F_{y4}\right\}^T \tag{8.43}$$

8.3.2.1 Consistent force vector due to body forces

Note that for the body force vector can be evaluated in a straight forward manner by using Eq. (8.42a). Assuming that F_{Bx} and F_{By} are constant across the element, these integrals result in the following *consistent body force vector* using Eqs. (8.28), (8.40a), and (8.42a):

$$\left\{r_B{}^{(Q4)}\right\} = abt\left\{F_{Bx}\ F_{By}\ F_{Bx}\ F_{By}\ F_{Bx}\ F_{Bx}\ F_{By}\ F_{Bx}\right\}^T \tag{8.44}$$

8.3.2.2 Consistent nodal force vector due to initial stress

The *consistent force vector due to initial stress* can be found from Eqs. (8.31), Eq. 8.41b, and (8.42b). Assuming that the initial stress is constant across the element we find this vector as follows:

$$\begin{aligned}\left\{r_{\sigma_0}{}^{(Q4)}\right\} = t\{&-\left(b\sigma_x^{(0)} + a\tau_{xy}^{(0)}\right)\ \ -\left(a\sigma_y^{(0)} + b\tau_{xy}^{(0)}\right)\ \ \left(b\sigma_x^{(0)} - a\tau_{xy}^{(0)}\right)\ \ \left(-a\sigma_y^{(0)} + b\tau_{xy}^{(0)}\right) \\ &\left(b\sigma_x^{(0)} + a\tau_{xy}^{(0)}\right)\ \ \left(a\sigma_y^{(0)} + b\tau_{xy}^{(0)}\right)\ \ \left(-b\sigma_x^{(0)} - a\tau_{xy}^{(0)}\right)\ \ \left(a\sigma_y^{(0)} - b\tau_{xy}^{(0)}\right)\}^T\end{aligned} \tag{8.45}$$

8.3.2.3 Consistent nodal force vector due to thermal strains

For two-dimensional problems, the *consistent force vector due to initial strain* depends on the elasticity matrix, hence the model that is applicable for a given

problem. In case the initial strain is due to thermal effects, the state of stress should be considered. The stress–strain relations are as follows:
Plane–strain Eq. (2.86):

$$\begin{Bmatrix} \sigma_x \\ \sigma_y \\ \tau_{xy} \end{Bmatrix} = \frac{E}{(1+\nu)(1-2\nu)} \begin{bmatrix} 1-\nu & \nu & 0 \\ \nu & 1-\nu & 0 \\ 0 & 0 & (1-2\nu)/2 \end{bmatrix} \begin{Bmatrix} \varepsilon_x \\ \varepsilon_y \\ \gamma_{xy} \end{Bmatrix} - \frac{E\alpha\Delta T}{1-2\nu} \begin{Bmatrix} 1 \\ 1 \\ 0 \end{Bmatrix}$$

Plane–stress Eq. (2.87):

$$\begin{Bmatrix} \sigma_x \\ \sigma_y \\ \tau_{xy} \end{Bmatrix} = \frac{E}{(1-\nu^2)} \begin{bmatrix} 1 & \nu & 0 \\ \nu & 1 & 0 \\ 0 & 0 & (1-\nu)/2 \end{bmatrix} \begin{Bmatrix} \varepsilon_x \\ \varepsilon_y \\ \gamma_{xy} \end{Bmatrix} - \frac{E\alpha\Delta T}{1-\nu} \begin{Bmatrix} 1 \\ 1 \\ 0 \end{Bmatrix}$$

The initial strain vectors are identified,

$$\{\varepsilon_0\} = \left\{\varepsilon^{(th)}\right\} = -\frac{E\alpha\Delta T}{1-\nu}\{1 \;\; 1 \;\; 0\}^T \text{ in } \textit{plane-stress} \tag{8.46}$$

$$\{\varepsilon_0\} = \left\{\varepsilon^{(th)}\right\} = -\frac{E\alpha\Delta T}{1-2\nu}\{1 \;\; 1 \;\; 0\}^T \text{ in } \textit{plane-strain} \tag{8.47}$$

The *consistent nodal force vector* for a Q4 element with constant material properties, subjected to temperature difference ΔT can be found by using Eq. (8.42c) as follows:

$$\left\{r_{th}^{(Q4)}\right\} = \frac{E\alpha\Delta T}{1-\nu}\{-b \;\; -a \;\; b \;\; -a \;\; b \;\; a \;\; -b \;\; a\}^T \text{ in } \textit{plane-stress} \tag{8.48}$$

$$\left\{r_{th}^{(Q4)}\right\} = \frac{E\alpha\Delta T}{1-2\nu}\{-b \;\; -a \;\; b \;\; -a \;\; b \;\; a \;\; -b \;\; a\}^T \text{ in } \textit{plane-strain} \tag{8.49}$$

8.3.2.4 Consistent nodal force vector due to boundary tractions

Recall that the solution domain is enclosed by the boundary Γ, and the *traction boundary* Γ_N is the part of the boundary where external tractions act on the deformable body. Consistent nodal forces should be calculated for elements that share an edge with the traction boundary. In this chapter, only rectangular elements are considered. Therefore, the edges of the solution domain are considered parallel to the (x', y') coordinate system. These edges are defined as follows:

edge-1, e1: $(x', -b)$

edge-2, e2: (a, y')

edge-3, e3: (x', b)

edge-4, e4: $(-a, y')$

Linearly varying traction

In general, the external traction is defined with respect to the local normal and tangential directions of the surface as indicated by Eq. (7.122),

$$\vec{q}^{(b)} = q_n \vec{n} + q_t \vec{t} = \{q_n \ q_t\}^T \tag{8.50}$$

where $\vec{n}$ and $\vec{t}$ are the *outward unit normal* and *tangent vectors* of the element boundary, $\Gamma^{(e)}$. For a two-dimensional problem, these vectors can be found as described in Section 7.2. The external traction can be transformed by using the transformation introduced in Eq. (7.123).

$$\vec{q}^{(b)} = \begin{Bmatrix} q_x \\ q_y \end{Bmatrix} = \begin{bmatrix} n_x & -n_y \\ n_y & n_x \end{bmatrix} \begin{Bmatrix} q_n \\ q_t \end{Bmatrix} \tag{8.51}$$

The consistent nodal forces due to a distributed external traction can be computed by using Eq. (8.42d),

$$\{r_q^{Q4}\} = \oint\limits_{\Gamma^{(e)}} [N^{(Q4)}]^T \{q^{(b)}\} t \, ds \tag{8.52}$$

By using Eqs. (8.28) and (8.51), we find,

$$\{r_q^{Q4}\} = \oint\limits_{\Gamma^{(e)}} \begin{bmatrix} N_1^{(Q4)} & 0 \\ 0 & N_1^{(Q4)} \\ N_2^{(Q4)} & 0 \\ 0 & N_2^{(Q4)} \\ N_3^{(Q4)} & 0 \\ 0 & N_3^{(Q4)} \\ N_4^{(Q4)} & 0 \\ 0 & N_4^{(Q4)} \end{bmatrix} \begin{Bmatrix} q_x \\ q_y \end{Bmatrix} ds \tag{8.53}$$

Details of the calculation in Eq. (8.53) depend on which element boundary the external traction is applied. For example, let's consider the linearly varying traction components q_x and q_y applied on the top edge of the rectangular domain as shown in Fig. 8.6. This boundary coincides with edge-3 of the element as shown in the figure. The linear variation of the traction components along edge-3 is calculated as follows:

$$q_x = \frac{1}{2}\left[(q_{x3} + q_{x4}) + (q_{x3} - q_{x4})\frac{x'}{a}\right] \tag{8.54a}$$

$$q_y = \frac{1}{2}\left[(q_{y3} + q_{y4}) + (q_{y3} - q_{y4})\frac{x'}{a}\right] \tag{8.54b}$$

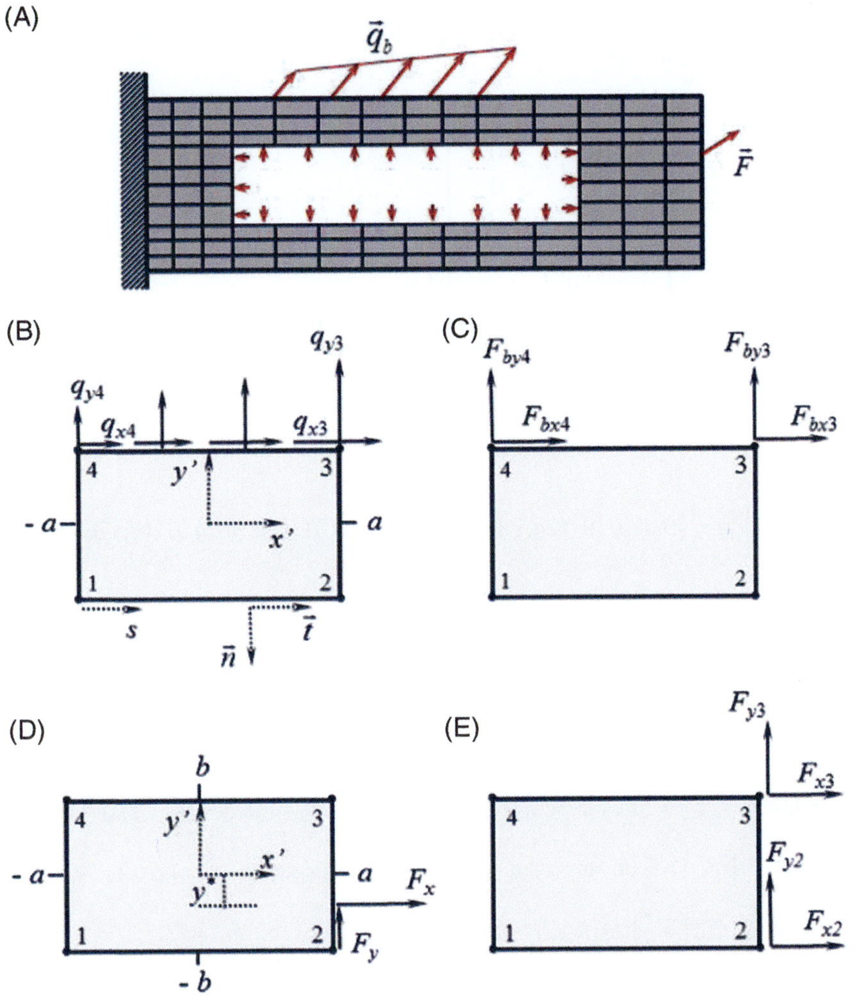

FIGURE 8.6 (A) External forces acting on the solution domain. (B) An element that lies under the distributed load on the top boundary. (C) The consistent nodal forces associated with the distributed force shown in part-b). (D) The element that coincides with the location of the concentrated force $\vec{F}$. (E) The consistent nodal force associated with the concentrated force $\vec{F}$.

The geometric location of edge-3 can be described by (x', b). Thus, we find that the shape functions, given in Eq. (8.16), become,

$$N_1^{(Q4)}(x', b) = 0, \quad N_2^{(Q4)}(x', b) = 0, \quad N_3^{(Q4)}(x', b) = \frac{1}{2}\left(1 + \frac{x'}{a}\right), \quad N_4^{(Q4)} = \frac{1}{2}\left(1 - \frac{x'}{a}\right) \tag{8.55}$$

The consistent nodal force vector $\left\{r_{qb}^{(Q4)}\right\}$ can be found by using Eqs. (8.53) and (8.54a) as follows:

$$\left\{r_q^{(Q4)}\right\} = \frac{1}{4}\int_{-a}^{a} \begin{bmatrix} 0 & 0 \\ 0 & 0 \\ 0 & 0 \\ 0 & 0 \\ 1+\frac{x'}{a} & 0 \\ 0 & 1+\frac{x'}{a} \\ 1-\frac{x'}{a} & 0 \\ 0 & 1-\frac{x'}{a} \end{bmatrix} \left\{\begin{bmatrix} (q_{x3}+q_{x4})+(q_{x3}-q_{x4})\frac{x'}{a} \\ (q_{y3}+q_{y4})+(q_{y3}-q_{y4})\frac{x'}{a} \end{bmatrix}\right\} dx' \tag{8.56}$$

$$\left\{r_{q-e3}^{(Q4)}\right\} = \frac{a}{3}\left\{0 \;\; 0 \;\; 0 \;\; 0 \;\; (2q_{x3}+q_{x4}) \;\; (2q_{y3}+q_{y4}) \;\; (q_{x3}+2q_{x4}) \;\; (q_{y3}+2q_{y4})\right\}^T \tag{8.57}$$

The consistent nodal force vectors for all of the edges, if that edge is subjected to linearly varying external traction components, can be found as follows:

$$\begin{aligned}
&\text{edge-1:}\left\{r_{q-e1}^{(Q4)}\right\} = \frac{a}{3}\left\{(2q_{x1}+q_{x2}) \;\; (2q_{y1}+q_{y2}) \;\; (q_{x1}+2q_{x2}) \;\; (q_{y1}+2q_{y2}) \;\; 0 \;\; 0 \;\; 0 \;\; 0\right\}^T \\
&\text{edge-2:}\left\{r_{q-e2}^{(Q4)}\right\} = \frac{b}{3}\left\{0 \;\; 0 \;\; (2q_{x2}+q_{x3}) \;\; (2q_{y2}+q_{y3}) \;\; (q_{x2}+2q_{x3}) \;\; (q_{y2}+2q_{y3}) \;\; 0 \;\; 0\right\}^T \\
&\text{edge-3:}\left\{r_{q-e3}^{(Q4)}\right\} = \frac{a}{3}\left\{0 \;\; 0 \;\; 0 \;\; 0 \;\; (2q_{x3}+q_{x4}) \;\; (2q_{y3}+q_{y4}) \;\; (q_{x3}+2q_{x4}) \;\; (q_{y3}+2q_{y4})\right\}^T \\
&\text{edge-4:}\left\{r_{q-e4}^{(Q4)}\right\} = \frac{b}{3}\left\{(2q_{x1}+q_{x4}) \;\; (2q_{y1}+q_{y4}) \;\; 0 \;\; 0 \;\; 0 \;\; 0 \;\; (q_{x1}+2q_{x4}) \;\; (q_{y1}+2q_{y4})\right\}^T
\end{aligned} \tag{8.58}$$

Concentrated force

Consider the concentrated force $\vec{F}$ acting on the boundary of the solution domain in Fig. 8.6A. The details of calculation of the consistent nodal forces due to a concentrated force acting between the nodes of an element is demonstrated in Section 5.2.4.2. Here let's assume that the external concentrated force is applied on edge-2 of the element between nodes 2 and 3, for example at location (a, y^*) as shown in Fig. 8.6D. In matrix notation, the force vector can be expressed as follows:

$$\{F\} = \{F_x \;\; F_y\}^T$$

where F_x and F_y are the components of the force vector along the x'- and y'-axes. The consistent nodal force vector can be found by using Eq. (8.53),

$$\left\{r_q^{Q4}\right\} = \int_{-b}^{b} \begin{bmatrix} N_1^{(Q4)}(a,y') & 0 \\ 0 & N_1^{(Q4)}(a,y') \\ N_2^{(Q4)}(a,y') & 0 \\ 0 & N_2^{(Q4)}(a,y') \\ N_3^{(Q4)}(a,y') & 0 \\ 0 & N_3^{(Q4)}(a,y') \\ N_4^{(Q4)}(a,y') & 0 \\ 0 & N_4^{(Q4)}(a,y') \end{bmatrix} \begin{Bmatrix} \dfrac{F_x}{dy}\delta(y'-y^*) \\ \dfrac{F_y}{dy}\delta(y'-y^*) \end{Bmatrix} dy' \tag{8.59}$$

where δ represents the Dirac-delta function. Note that at the load application point, the shape functions are,

$$N_1^{(Q4)}(a,y^*) = 0, \quad N_2^{(Q4)}(a,y^*) = \frac{1}{2}\left(1 - \frac{y^*}{b}\right), \quad N_3^{(Q4)}(a,y^*) = \frac{1}{2}\left(1 + \frac{y^*}{b}\right), \quad N_4^{(Q4)}(a,y^*) = 0 \tag{8.60}$$

The consistent nodal force vector is found by using Eqs. (8.59) and (8.60) as follows:

$$\left\{r_{e2}^{Q4}\right\} = \frac{1}{2}\{0 \ 0 \ F_x(1-y^*/b) \ F_y(1-y^*/b) \ F_x(1+y^*/b) \ F_y(1+y^*/b) \ 0 \ 0\}^T \ \text{on e2} \tag{8.61}$$

8.4 Constant strain triangle (CST) element

A bilinear triangular element is a three-node element with six degrees of freedom (Fig. 8.5). The components of the displacement vector $\{u\}$ are interpolated across the element as follows:

$$u_x = \sum_{i=1}^{3} N_i^{(T3)} u_{xi} \ \text{and} \ u_y = \sum_{i=1}^{3} N_i^{(T3)} u_{yi} \tag{8.62}$$

where u_{xi}, u_{yi} for ($i = 1$–3) are the displacement components at the three nodes (corners) of the element and $N_i^{(T3)}$ are the shape functions given in Eq. (8.25).

In finite element form, the displacement is interpolated as follows:

$$\{u\} = \{u_x \ u_y\}^T = \left[N^{(T3)}\right]\{d\} \tag{8.63}$$

where the shape function matrix for and the degree of freedom vector for this three-node element are defined as follows:

$$\left[N^{(T3)}\right]=\begin{bmatrix} N_1^{(T3)} & 0 & N_2^{(T3)} & 0 & N_3^{(T3)} & 0 \\ 0 & N_1^{(T3)} & 0 & N_2^{(T3)} & 0 & N_3^{(T3)} \end{bmatrix} \tag{8.64}$$

$$\{d\}=\left\{u_{x1} \quad u_{y1} \quad u_{x2} \quad u_{y2} \quad u_{x3} \quad u_{y3}\right\}^T \tag{8.65}$$

The strain vector can be expressed by using the following general relationship,

$$\{\varepsilon\}=[\partial]\,\{u\}=[\partial]\left[N^{(T3)}\right]\{d\}=\left[B^{(T3)}\right]\{d\} \tag{8.66}$$

where the strain–displacement matrix$[B^{(T3)}]$ for this linear triangle element is defined as follows:

$$\left[B^{(T3)}\right]=[\partial]\left[N^{(T3)}\right]=\begin{bmatrix} N_{1,x'}^{(T3)} & 0 & N_{2,x'}^{(T3)} & 0 & N_{3,x'}^{(T3)} & 0 \\ 0 & N_{1,y'}^{(T3)} & 0 & N_{2,y'}^{(T3)} & 0 & N_{3,y'}^{(T3)} \\ N_{1,y'}^{(T3)} & N_{1,x'}^{(T3)} & N_{2,y'}^{(T3)} & N_{2,x'}^{(T3)} & N_{3,y'}^{(T3)} & N_{3,x'}^{(T3)} \end{bmatrix} \tag{8.67}$$

The partial derivatives of the shape functions are found by using Eq. (8.25) as follows:

$$N_{1,x'}^{(T3)}=-\frac{1}{a},\ N_{2,x'}^{(T3)}=\frac{1}{a},\ N_{3,x'}^{(T3)}=0,\ N_{1,y'}^{(T3)}=-\frac{1-r}{b},\ N_{2,y'}^{(T3)}=-\frac{r}{b},\ N_{2,y'}^{(T3)}=\frac{1}{b} \tag{8.68}$$

The strain vector can then be found as follows:

$$\begin{Bmatrix} \varepsilon_{xx} \\ \varepsilon_{yy} \\ \gamma_{xy} \end{Bmatrix}=\begin{bmatrix} -1/a & 0 & 1/a & 0 & 0 & 0 \\ 0 & -(1-r)/b & 0 & -r/b & 0 & 1/b \\ -(1-r)/b & -1/a & -r/b & 1/a & b & 0 \end{bmatrix}\begin{Bmatrix} u_{x1} \\ u_{y1} \\ u_{x2} \\ u_{y2} \\ u_{x3} \\ u_{y3} \end{Bmatrix} \tag{8.69}$$

This equation shows that all three strain components ε_x, ε_y, γ_{xy} are constant across this element and depend on the nodal displacements and the geometric positions of the nodes. Therefore, the three-node triangular element described by the shape functions given in Eq. (8.25) is known as the *constant strain triangle*.

The stiffness matrix for this element can found by using Eq. (8.11a)

$$[k_{T3}]^{(e)}=\int_A \left[B^{(T3)}\right]^T [E]\left[B^{(T3)}\right] t\, dA \tag{8.70}$$

As all of the components of the $[B^{(T3)}]$ matrix are constant, the stiffness matrix can be found in a straightforward manner as follows:

$$[k_{T3}]^{(e)} = \left[B^{(T3)}\right]^T [E]\left[B^{(T3)}\right] t \int_A dA = \left[B^{(T3)}\right]^T [E]\left[B^{(T3)}\right] t\, A \tag{8.71}$$

The elasticity matrix $[E]$ depends on the type of problem considered and it is defined in Eq. (8.34). The stiffness matrix for the CST element can be found as follows:

$$[k_{T3}]^{(e)} = \frac{1}{2} abt \times$$
$$\begin{bmatrix} \frac{E_{11}}{a^2}+\frac{G(r-1)^2}{b^2} & \frac{(E_{12}+G)(1-r)}{ab} & -\frac{E_{11}}{a^2}+\frac{G(1-r)r}{b^2} & \frac{E_{12}r+G(r-1)r}{ab} & \frac{G(r-1)}{b^2} & -\frac{E_{12}}{ab} \\ \frac{(E_{12}+G)(1-r)}{ab} & \frac{E_{11}(r-1)^2}{b^2}+\frac{G}{a^2} & \frac{E_{12}(r-1)+Gr}{ab} & \frac{E_{11}(1-r)r}{b^2}-\frac{G}{a^2} & -\frac{G}{ab} & \frac{E_{11}(r-1)}{b^2} \\ \frac{G(1-r)r}{b^2}-\frac{E_{11}}{a^2} & \frac{E_{12}(r-1)+Gr}{ab} & \frac{E_{11}}{a^2}+\frac{Gr^2}{b^2} & -\frac{(E_{12}+G)r}{ab} & -\frac{Gr}{b^2} & \frac{E_{12}}{ab} \\ \frac{E_{12}r+G(r-1)}{ab} & \frac{E_{11}(1-r)r}{b^2}-\frac{G}{a^2} & -\frac{(E_{12}+G)r}{ab} & \frac{E_{11}r^2}{b^2}+\frac{G}{a^2} & \frac{G}{ab} & -\frac{E_{11}r}{b^2} \\ \frac{G(r-1)}{b^2} & -\frac{G}{ab} & -\frac{E_{12}}{b^2} & \frac{G}{ab} & \frac{G}{b^2} & 0 \\ -\frac{E_{12}}{ab} & \frac{E_{11}(r-1)}{b^2} & \frac{E_{12}}{ab} & -\frac{E_{11}r}{b^2} & 0 & \frac{E_{11}}{b^2} \end{bmatrix} \tag{8.72}$$

It should be noted that that the aspect ratio (a/b) of this element also plays a role in the stiffness of this matrix, similar to the case of the Q4 element.

8.5 Element defects

8.5.1 Constant strain triangle element

The constant strain triangle (CST) is one of the earliest elements developed [1]. Use of this element in solid mechanics problems has certain drawbacks. This element does not represent bending properly and is prone to predict spurious shear strains. Prediction of shear strains for a problem in which we would not expect them (e.g., beam bending) causes a mesh that consists of these linear triangular elements to behave too stiffly. Therefore, use of the CST element in solid mechanics problems should be handled with care.

Consider a beam subjected to pure bending that is modeled by a three-node linear interpolation (CST) element as shown in Fig. 8.7. A beam subjected only to an external moment M_0 at its ends is said to be in *pure bending*. According to the Euler–Bernoulli beam theory, a beam subjected to pure bending will develop bending stress σ_b in the axial direction of magnitude,

$$\sigma_x = \sigma_b = \frac{M_0 y}{I} \tag{8.73}$$

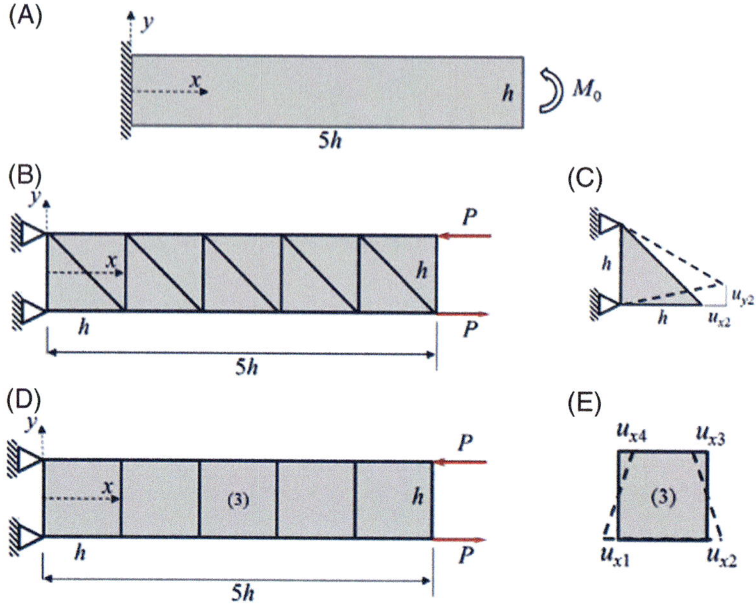

FIGURE 8.7 (A) A cantilevered beam subjected to bending moment M_0 at its free edge. (B) Finite element model of pure bending of a beam by CST elements. (C) Deflection of the element attached to the boundary. (D) Finite element model of pure bending of a beam by a Q4 element. (E) Deflection of an element (e.g., (3)). Figure modified from reference [1] with permission.

where y is the transverse coordinate axis located on the neutral axis of the beam. The *bending strain* is proportional to distance from the neutral axis, i.e. $\varepsilon_x \propto -y$. A *transverse strain* of magnitude $\varepsilon_y = -\nu\varepsilon_x$ develops in the material, according to the exact solution of the problem [2]. No *shear strain* should develop in the beam, i.e. $\gamma_{xy} = 0$, under pure bending condition.

In the model shown in Fig. 8.7B, the bending moment is approximated by a force couple. The left most element of the model connects the beam to the boundary (Fig. 8.7C). The positions of the nodes 1–3 are (0, 0), (h, 0), and (0, h). A positive bending moment such as the one shown in the figure would cause node-2 to be displaced by (u_{x2},u_{y2}). Nodes-1 and -3 are fixed. The displacement vector for this element becomes,

$$\{d\} = \{0 \quad 0 \quad u_{x2} \quad u_{y2} \quad 0 \quad 0\}^T \tag{8.74}$$

By using Eq. (8.69), it can be shown that the strain in this element is:

$$\{\varepsilon\} = \left\{\frac{u_{x2}}{h} \quad 0 \quad \frac{u_{y2}}{h}\right\}^T \tag{8.75}$$

The beam should not have any transverse shear strain as it is subjected to pure-bending. However, Eq. (8.75) shows that the CST element would predict a nonzero shear strain. This *spurious* (or *parasitic*) *shear strain* is unavoidable in this element type. The spurious shear strain absorbs energy, and as a result it makes the CST element behave stiffer than it should be. We have shown in Eq. (8.69) that this element takes its name from the fact it can only predict constant strain through its volume. Therefore, unless the exact solution of the problem involves a constant strain, the element will have issues in solving the two-dimensional elasticity problems properly.

8.5.2 Bilinear rectangle (Q4)

Similar to the CST, the Q4 element cannot exhibit pure bending properly, and predicts spurious shear behavior.

Consider the finite element representation of the beam that consists of five Q4 elements as shown in Fig. 8.7D. The moment M_0 is again approximated by using a couple force of magnitude P. The deflected shape of, for example, element-3, is shown in Fig. 8.7E. Note that all of the edges remain straight after deformation, as the element only has linear interpolation capability. Simple thought experiment will demonstrate that the deflection vector for this element will be as follows:

$$\{d\} = \begin{Bmatrix} -u^* & 0 & u^* & 0 & -u^* & 0 & u^* & 0 \end{Bmatrix}^T \tag{8.76}$$

Note that nodal displacements have the following relationship $|u_{x1}| = u_{x2} = |u_{x3}| = u_{x4} = u^*$.

The strain distribution in the element can be computed by using Eq. (8.30) as follows:

$$\{\varepsilon\} = -\frac{u^*}{h^2}\begin{Bmatrix} y' & 0 & x' \end{Bmatrix}^T \tag{8.77}$$

where (x', y') are the local coordinates in the element. This shows that the bending strain ε_x is predicted to vary linearly in the transverse direction. While this is the correct form of the bending strain, the Q4 element fails to predict a transverse normal strain ε_y and it predicts a spurious shear strain γ_{xy}. It can be shown that the correct form of the strain components for this element should be $\varepsilon_x = -u^* y'/h^2$, $\varepsilon_y = -\nu\varepsilon_x$ and $\gamma_{xy} = 0$ [1].

Similar to the case of the CST element the spurious shear strain predicted by the Q4 element in a pure bending problem absorbs some of the external work as strain energy. As a result the Q4 element subjected to pure bending conditions behaves stiffer than it should be.

8.6 Higher order elements

8.6.1 Quadratic triangle (linear strain triangle)

This is a six-node triangular element which uses a two-dimensional quadratic interpolation. In addition to the three corners, the element has three nodes on the sides (Fig. 8.4). This element has a total of 12 nodal degrees of freedom,

(u_{xi}, u_{yi}) $(i = 1, \ldots 6)$. The displacement components u_x and u_y are interpolated by using a six-coefficient polynomials, as the element has six nodes. The following polynomials can be obtained from Pascal's triangle and close inspection of Table 7.1 shows that this is a complete, second order, two-dimensional polynomial.

$$
\begin{aligned}
u_x &= a_0 + a_1 x + a_2 y + a_3 x^2 + a_4 xy + a_5 y^2 \\
u_y &= b_0 + b_1 x + b_2 y + b_3 x^2 + b_4 xy + b_5 y^2
\end{aligned}
\tag{8.78}
$$

where a_i and b_i are the constant coefficients. The shape functions $N_i^{(T6)}(x, y)$, for $i = 1$–6, can be obtained by using the procedure described in Section 4.3. Eq. (8.78) can then be expressed as follows:

$$
u_x = \sum_{i=1}^{6} N_i^{(T6)} u_{xi} \text{ and } u_y = \sum_{i=1}^{6} N_i^{(T6)} u_{yi}
\tag{8.79}
$$

where u_{xi} and u_{yi} are the displacement components at the nodes.

Note that by using the interpolation functions given in Eq. (8.78), we can get a general idea about the how the strain varies across the element. The strain components can be found as follows:

$$
\begin{aligned}
\varepsilon_x &= a_1 + 2a_3 x + a_4 y \\
\varepsilon_y &= b_2 + b_4 x + 2b_5 y \\
\gamma_{xy} &= (a_2 + b_1) + (a_4 + 2b_3)x + (b_4 + 2a_5)y
\end{aligned}
\tag{8.80}
$$

Eq. (8.80) shows that strain varies linearly across this element. This behavior gives rise to the term *linear strain triangle* (LST). LST element is capable of modeling linear variations in strain, and thus is a good element for using in simulations that involve members in bending. Numerical experiments show that LST elements do not display spurious shear behavior when used in simulations of pure bending [1].

8.6.2 Q8 quadratic rectangle

This is an eight-node rectangular element which uses a two-dimensional quadratic interpolation. In addition to the four corners, the element has four nodes on the edges of the element (Fig. 8.3B). This element has a total of 16 nodal degrees of freedom, (u_{xi}, u_{yi}) for $i = 1$–8. The displacement components u_x and u_y are interpolated by using an eight-coefficient polynomials, as the element has eight nodes. The following polynomials are obtained from Pascal's triangle by using polynomial symmetry arguments mentioned in Section 7.5:

$$
\begin{aligned}
u_x &= a_0 + a_1 x + a_2 y + a_3 x^2 + a_4 xy + a_5 y^2 + a_6 x^2 y + a_7 xy^2 \\
u_y &= b_0 + b_1 x + b_2 y + b_3 x^2 + b_4 xy + b_5 y^2 + b_6 x^2 y + b_7 xy^2
\end{aligned}
\tag{8.81}
$$

The shape functions $N_i^{(Q8)}(x, y)$ for $i = 1$–8 are given in Section 8.2. Eq. (8.81) can then be expressed as follows:

$$u_x = \sum_{i=1}^{8} N_i^{(Q8)} u_{xi} \text{ and } u_y = \sum_{i=1}^{8} N_i^{(Q8)} u_{yi} \tag{8.82}$$

where u_{xi} and u_{yi} are the displacement components at the nodes. Strain components can be obtained by using Eqs. (8.18) and (8.82). However, here we are only interested in the effect of the polynomial terms on the predicted strains. Therefore, we simply use Eq. (8.81) to find the strain values,

$$\begin{aligned}
\varepsilon_x &= a_1 + 2a_3x + a_4y + 2a_6xy + a_7y^2 \\
\varepsilon_y &= b_2 + b_4x + 2b_5y + b_6x^2 + 2b_7xy \\
\gamma_{xy} &= (a_2 + b_1) + (a_4 + 2b_3)x + (b_4 + 2a_5)y + a_6x^2 + 2(a_7 + b_6)xy + b_7y^2
\end{aligned} \tag{8.83}$$

Note that the strains predicted by the Q8 element are quadratic functions and thus should be able to model pure bending situations properly. Numerical experiments show that Q8 elements do not display spurious shear behavior when used in simulations of pure bending [1].

8.6.3 Q9 quadratic rectangle

This is a nine-node rectangular element which uses a two-dimensional quadratic interpolation. In addition to the four corners, the element has four nodes on the edges of the element and an additional node in the center of the element (Fig. 8.3C). This element has a total of 18 nodal degrees of freedom, (u_{xi}, u_{yi}) for $i = 1$–9. The displacement components u_x and u_y are interpolated by using nine-coefficient polynomials. The following polynomials are obtained from Pascal's triangle by using polynomial symmetry arguments mentioned in Section 7.5. The same polynomials can also be obtained by multiplying the quadratic one-dimensional polynomials in x- and y-directions. This element is also known as the *Lagrange element*.

$$\begin{aligned}
u_x &= a_0 + a_1x + a_2y + a_3x^2 + a_4xy + a_5y^2 + a_6x^2y + a_7xy^2 + a_8x^2y^2 \\
u_y &= b_0 + b_1x + b_2y + b_3x^2 + b_4xy + b_5y^2 + b_6x^2y + b_7xy^2 + b_8x^2y^2
\end{aligned} \tag{8.84}$$

The shape functions $N_i^{(Q9)}(x, y)$ for $i = 1$–9 are given in Section 8.2. Eq. (8.84) can then be expressed as follows:

$$u_x = \sum_{i=1}^{9} N_i^{(Q9)} u_{xi} \text{ and } u_y = \sum_{i=1}^{9} N_i^{(Q9)} u_{yi} \tag{8.85}$$

where u_{xi} and u_{yi} are the displacement components at the nodes. The ninth node can be added to the center ($x = 0$; $y = 0$) of the Q8 element described above.

This of course increases the order of approximation. Neverthless, the following improvements to the displacement interpolation and strain take place,

$$\begin{aligned} &u \rightarrow a_{17}x^2y^2, \quad v \rightarrow a_{18}x^2y^2 \\ &\varepsilon_x \rightarrow 2a_{17}xy^2, \quad \varepsilon_y \rightarrow 2a_{18}yx^2, \quad \gamma_{xy} \rightarrow 2\left(a_{17}xy^2 + a_{18}yx^2\right) \end{aligned} \tag{8.86}$$

Strain components can be obtained by using Eqs. (8.21a) and (8.85). Here, we simply use Eq. (8.84) to find the strain values,

$$\begin{aligned} \varepsilon_x &= a_1 + 2a_3x + a_4y + 2a_6xy + a_7y^2 + 2a_8xy^2 \\ \varepsilon_y &= b_2 + b_4x + 2b_5y + b_6x^2 + 2b_7xy + 2b_8yx^2 \\ \gamma_{xy} &= (a_2 + b_1) + (a_4 + 2b_3)x + (b_4 + 2a_5)y + a_6x^2 + 2(a_7 + b_6)xy + b_7y^2 + 2\left(a_8x^2y + b_8xy^2\right) \end{aligned} \tag{8.87}$$

Note that the strains predicted by the Q9 element are also quadratic functions and thus should be able to model pure bending situations properly. Numerical experiments show that Q9 elements do not display spurious shear behavior when used in simulations of pure bending [1].

8.6.4 Q6 quadratic rectangle

The four-node plane, rectangular element (Q4) and its quadrilateral counterpart develop spurious shear strain. As a result, these elements are known to behave in an excessively stiff manner when they are used in modeling domains dominated by bending. To remedy this issue, bending modes can be added to the element displacement fields. For a rectangular element ($2a\times2b$) this entails adding the following quadratic terms:

$$u_x = \sum_{i=1}^{4} N_i^{(Q4)}u_{xi} + a_1\left(1 - \frac{x^2}{a^2}\right) + a_2\left(1 - \frac{y^2}{b^2}\right) \tag{8.88a}$$

$$u_x = \sum_{i=1}^{4} N_i^{(Q4)}u_{yi} + b_1\left(1 - \frac{x^2}{a^2}\right) + b_2\left(1 - \frac{y^2}{b^2}\right) \tag{8.88b}$$

where $N_i^{(Q4)}$ are the shape functions given in Eq. (8.16), u_{xi} and u_{yi} are the nodal degrees of freedom and a_1, a_2, b_1, and b_2 are the *nodeless degrees of freedom.* The nodeless degrees of freedom can be eliminated from the list of unknown degrees of freedom by using the *static condensation* process.

For the more general four-node quadrilateral (Q4) element described in Chapter 7, the Q6 element can be obtained by adding the following quadratic terms:

$$u_x = \sum_{i=1}^{4} N_i^{(Q4)}u_{xi} + a_1\left(1 - \xi^2\right) + a_2\left(1 - \eta^2\right) \tag{8.89a}$$

$$u_y = \sum_{i=1}^{4} N_i^{(Q4)}u_{yi} + b_1\left(1 - \xi^2\right) + b_2\left(1 - \eta^2\right) \tag{8.89b}$$

where $N_i^{(Q4)}$ are the shape functions given in Eq. (7.61). The quadratic terms added for this element can be defined as,

$$N_1^{(a)} = 1 - \xi^2 \text{ and } N_2^{(a)} = 1 - \eta^2 \tag{8.90}$$

The displacement modes associated with the nodeless degrees of freedom are incompatible as they allow overlaps or gaps to develop between the elements at locations other than the nodes. Despite this, the element can be useful to prevent spurious shear strain [1].

The derivation of the element stiffness matrix for the Q6 element is presented next. The degree of freedom vector can be initially expressed as follows:

$$\{d'\} = \{\{d\}^T \quad \{a\}^T\} \tag{8.91}$$

where $\{d\}$ is the nodal displacement vector of the Q4 element given for example in Eq. (8.29), and $\{a\}$ is a temporary vector that represents the nodeless degrees of freedom, $\{a\} = \{a_1 \quad a_2 \quad b_1 \quad b_2\}^T$. With this definition the displacement interpolation given in Eq. (8.89) can be expressed in vector form as follows:

$$\begin{Bmatrix} u_x \\ u_y \end{Bmatrix} = [[N^{(Q4)}][N^{(a)}]]\begin{Bmatrix} \{d\} \\ \{a\} \end{Bmatrix} \tag{8.92}$$

where

$$[N^{(Q4)}] = \begin{bmatrix} N_1^{(Q4)} & 0 & N_2^{(Q4)} & 0 & N_3^{(Q4)} & 0 & N_4^{(Q4)} & 0 \\ 0 & N_1^{(Q4)} & 0 & N_2^{(Q4)} & 0 & N_3^{(Q4)} & 0 & N_4^{(Q4)} \end{bmatrix} \tag{8.93a}$$

$$[N^{(a)}] = \begin{bmatrix} N_1^{(a)} & 0 & N_2^{(a)} & 0 \\ 0 & N_1^{(a)} & 0 & N_2^{(a)} \end{bmatrix} \tag{8.93b}$$

The strain–displacement matrix for this element is symbolically defined as,

$$[B^{(Q6)}] = [[B^{(Q4)}] \quad [B^{(a)}]] \tag{8.94}$$

where $[B^{(Q4)}]$ is the strain–displacement matrix that operates on the nodal degrees of freedom, $\{d\}$ and $[B^{(a)}]$ is the part that operates on the nodeless degrees of freedom. At a high level, the element stiffness matrix can be obtained by using Eq. (7.118),

$$[k]^{(e)} = \int_{\Omega^{(e)}} [B^{(Q6)}]^T [E][B^{(Q6)}] t d\Omega^{(e)} = \int_{-1}^{1}\int_{-1}^{1} [B^{(Q6)}]^T [E][B^{(Q6)}] t|J| d\xi d\eta \tag{8.95}$$

This process results in the following equation of equilibrium,

$$\begin{bmatrix} [k_{dd}] & [k_{da}] \\ [k_{ad}] & [k_{aa}] \end{bmatrix}\begin{Bmatrix} \{d\} \\ \{a\} \end{Bmatrix} = \begin{Bmatrix} \{r\} \\ \{0\} \end{Bmatrix} \tag{8.96}$$

where,

$$[k_{dd}] = \int_{-1}^{1}\int_{-1}^{1} \left[B^{(Q4)}\right]^T [E]\left[B^{(Q4)}\right]|J|t\, d\xi\, d\eta \tag{8.97a}$$

$$[k_{aa}] = \int_{-1}^{1}\int_{-1}^{1} \left[B^{(a)}\right]^T [E]\left[B^{(a)}\right]|J|t d\xi\, d\eta \tag{8.97b}$$

$$[k_{ad}] = \int_{-1}^{1}\int_{-1}^{1} \left[B^{(a)}\right]^T [E]\left[B^{(Q4)}\right]|J|t\, d\xi\, d\eta \tag{8.97c}$$

$$[k_{da}] = [k_{ad}]^T \tag{8.97d}$$

and $\{r\}$ is the usual nodal load vector as defined in Eq. (7.116). There are no external forces associated with the nodeless degrees of freedom. These can be eliminated from the system of equations by using the static condensation process. To this end, let's note that Eq. (8.96) can be expressed as follows:

$$[k_{dd}]\{d\} + [k_{da}]\{a\} = \{r\} \tag{8.98a}$$

$$[k_{ad}]\{d\} + [k_{aa}]\{a\} = \{0\} \tag{8.98b}$$

we can find the nodeless degree of freedom vector from Eq. (8.98b) as follows:

$$\{a\} = -[k_{aa}]^{-1}([k_{ad}]\{d\}) \tag{8.99}$$

The $\{a\}$ vector can be eliminated from the system by substituting Eq. (8.99) in to Eq. (8.98a),

$$\left([k_{dd}] - [k_{da}][k_{aa}]^{-1}[k_{ad}]\right)\{d\} = \{r\} \tag{8.100}$$

Note that $[k_{dd}] = [k^{(Q4)}]$. With this substitution the element stiffness matrix for the Q6 element is found to be equal to,

$$\left[k^{(Q6)}\right] = \left[k^{(Q4)}\right] - [k_{da}][k_{aa}]^{-1}[k_{ad}] \tag{8.101}$$

Note that this *condensed element* operates on the $\{d\}$ vector of the Q4 element and the nodeless degrees of freedom have been eliminated from the system of equations. With this modification we see that Q6 element is a four node element with quadratic interpolation characteristics.

We note that the Q6 elements work very well in bending dominated problems, but fail to represent constant state of stress. This drawback can be eliminated if the Jacobian of the element is evaluated at its center, i.e., $\xi = \eta = 0$ [1,3].

8.7 Assembly, boundary conditions, solution, and postprocessing

The concepts related to element assembly, imposing boundary conditions, solution, and postprocessing of the solution do not change for rectangular elements. The reader is referred to Sections 7.10 and 7.11.

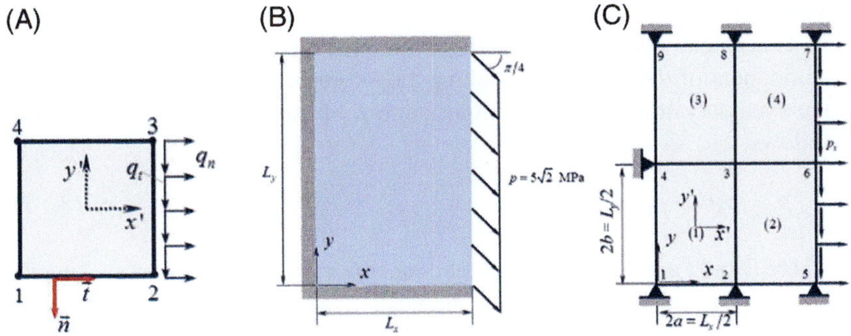

FIGURE 8.8 (A) A Q4 element subjected to external tractions q_n and q_t along its edge 23. (B) The two-dimensional solid is constrained along three edges and subjected to external tractions p_x and p_y as shown.

Example 8.1

a) Consider the rectangular Q4 element (Fig. 8.8A) of size $2a\times 2b$ whose equilibrium equation is given as follows,

$$\begin{bmatrix} k_{11} & k_{12} & k_{13} & k_{14} & k_{15} & k_{16} & k_{17} & k_{18} \\ & k_{22} & k_{23} & k_{24} & k_{25} & k_{26} & k_{27} & k_{28} \\ & & k_{33} & k_{34} & k_{35} & k_{36} & k_{37} & k_{38} \\ & & & k_{44} & k_{45} & k_{46} & k_{47} & k_{48} \\ & & & & k_{55} & k_{56} & k_{57} & k_{58} \\ & sym & & & & k_{66} & k_{67} & k_{68} \\ & & & & & & k_{77} & k_{78} \\ & & & & & & & k_{88} \end{bmatrix}^{(e)} \begin{Bmatrix} u_{x1} \\ u_{y1} \\ u_{x2} \\ u_{y2} \\ u_{x3} \\ u_{y3} \\ u_{x4} \\ u_{y4} \end{Bmatrix}^{(e)} = \begin{Bmatrix} r_{x1} \\ r_{y1} \\ r_{x2} \\ r_{y2} \\ r_{x3} \\ r_{y3} \\ r_{x4} \\ r_{y4} \end{Bmatrix}^{(e)} \tag{8.102}$$

The components $k_{ij}^{(e)}$ and the element stiffness matrix $[k]^{(e)}$ are given in Eq. (8.39). Find the components of the consistent loading vector $\{r\}^{(e)}$ if this element is subjected to the distributed forces q_n and q_t along the edge-2 as shown in Fig. 8.8A.

b) Consider the rectangular, elastic domain subjected to the distributed load p applied along its left edge as shown in Fig. 8.8B. The remaining edges of the rectangle are restrained. Assemble the global stiffness matrix and load vector for this domain by using only the active degrees of freedom, if this domain is discretized to 4 rectangular elements (Fig. 8.8C).

c) If the four elements are identical in size, find the displacement, strain, and stress distributions in this two-dimensional elastic solid. Assume the solid to be in plane stress with $t = 2$ mm, $L_x = 10$ cm, $L_y = 16$ cm, $E = 5$ GPa, $\nu = 0.3$, $p = 5\sqrt{2}$ MPa.

Solution:

a) Consistent nodal force vector: Note that in Fig. 8.8A the tangential component of the traction acts along the $-\vec{t}$ direction. Thus, in vector form the boundary traction acting along edge-2 of the element is expressed as follows:

$$\vec{p} = p_n\vec{n} - p_t\vec{t} = \begin{Bmatrix} p_n \\ -p_t \end{Bmatrix} \tag{8.103}$$

where $\vec{n}$ and $\vec{t}$ are the unit normal and tangent vectors, respectively, along the boundary of the element. Along the edge-2, we see that $\vec{n} = \hat{i}$ and $\vec{t} = \hat{j}$. The traction vector given in Eq. (8.103) can be transformed to the (x', y') coordinate system by using Eq. (7.123), and noting that $n_x = 1$ and $n_y = 0$,

$$\begin{Bmatrix} q_x \\ q_y \end{Bmatrix} = \begin{bmatrix} 1 & 0 \\ 0 & 1 \end{bmatrix} \begin{Bmatrix} q_n \\ -q_t \end{Bmatrix} \tag{8.104}$$

The consistent nodal force vector can then be found by using Eq. (8.58)(b),

$$\left\{ r_{rb}^{(Q4)} \right\} = \left\{ 0 \quad 0 \quad q_n b^{(e)} \quad -q_t b^{(e)} \quad q_n b^{(e)} \quad -q_t b^{(e)} \quad 0 \quad 0 \right\}^T \tag{8.105}$$

b) The two-dimensional domain is divided into four elements as shown in Fig. 8.8C. The nodes and the elements of the domain are numbered as shown in the figure. The active nodes are nodes-3 and -6, and therefore the active degree of freedom vector is $\{D\} = \left\{ u_{x3} \quad u_{y3} \quad u_{x6} \quad u_{y6} \right\}$

Equilibrium equations for each element can be found as follows:

$$\text{Element-1:} \begin{bmatrix} k_{55}^{(1)} & k_{56}^{(1)} \\ k_{65}^{(1)} & k_{66}^{(1)} \end{bmatrix} \begin{Bmatrix} u_{x3} \\ u_{y3} \end{Bmatrix} = \begin{Bmatrix} 0 \\ 0 \end{Bmatrix} \text{ in global coordinate system,} \tag{8.106}$$

$$\text{Element-4:} \begin{bmatrix} k_{33}^{(4)} & k_{34}^{(4)} \\ k_{43}^{(4)} & k_{44}^{(4)} \end{bmatrix} \begin{Bmatrix} u_{x3} \\ u_{y3} \end{Bmatrix} = \begin{Bmatrix} 0 \\ 0 \end{Bmatrix} \text{ in global coordinate system,} \tag{8.107}$$

$$\text{Element-2:} \begin{bmatrix} k_{55}^{(2)} & k_{56}^{(2)} & k_{57}^{(2)} & k_{58}^{(2)} \\ k_{65}^{(2)} & k_{66}^{(2)} & k_{67}^{(2)} & k_{68}^{(2)} \\ k_{75}^{(2)} & k_{76}^{(2)} & k_{77}^{(2)} & k_{78}^{(2)} \\ k_{85}^{(2)} & k_{86}^{(2)} & k_{87}^{(2)} & k_{78}^{(2)} \end{bmatrix} \begin{Bmatrix} u_{x6} \\ u_{y6} \\ u_{x3} \\ u_{y3} \end{Bmatrix} = b^{(2)} \begin{Bmatrix} p_x \\ -p_y \\ 0 \\ 0 \end{Bmatrix} \begin{matrix} \text{in element} \\ \text{coordinate system,} \end{matrix} \tag{8.108}$$

The nodes in Eq. (8.108) are ordered according to the element node numbering scheme. By rearranging the appropriate rows and columns the element stiffness matrix can be expressed in the global coordinate system as follows:

$$\text{Element-2: } \begin{bmatrix} k_{77}^{(2)} & k_{78}^{(2)} & k_{57}^{(2)} & k_{67}^{(2)} \\ k_{78}^{(2)} & k_{88}^{(2)} & k_{58}^{(2)} & k_{68}^{(2)} \\ k_{57}^{(2)} & k_{58}^{(2)} & k_{55}^{(2)} & k_{56}^{(2)} \\ k_{67}^{(2)} & k_{68}^{(2)} & k_{56}^{(2)} & k_{66}^{(2)} \end{bmatrix} \begin{Bmatrix} u_{x3} \\ u_{y3} \\ u_{x6} \\ u_{y6} \end{Bmatrix} = b^{(2)} \begin{Bmatrix} 0 \\ 0 \\ p_x \\ -p_y \end{Bmatrix} \text{ in global coordinate system} \tag{8.109}$$

$$\text{Element-3: } \begin{bmatrix} k_{11}^{(3)} & k_{12}^{(3)} & k_{13}^{(3)} & k_{14}^{(3)} \\ k_{21}^{(3)} & k_{22}^{(3)} & k_{23}^{(3)} & k_{24}^{(3)} \\ k_{31}^{(3)} & k_{32}^{(3)} & k_{33}^{(3)} & k_{34}^{(3)} \\ k_{41}^{(3)} & k_{42}^{(3)} & k_{43}^{(3)} & k_{44}^{(3)} \end{bmatrix} \begin{Bmatrix} u_{x3} \\ u_{y3} \\ u_{x6} \\ u_{y6} \end{Bmatrix} = b^{(3)} \begin{Bmatrix} 0 \\ 0 \\ p_x \\ -p_y \end{Bmatrix} \text{ in global coordinate system} \tag{8.110}$$

Global equilibrium equation is found by summing the element stiffness matrices Eqs. (8.106), (8.107), (8.109), and (8.110),

$$\sum_{e=1}^{4} \left([k]^{(e)}\{d\} - \{r_q\}^{(e)} \right) = 0$$

$$\begin{bmatrix} \left(k_{55}^{(1)} + k_{77}^{(2)} + k_{11}^{(3)} + k_{33}^{(4)}\right) & \left(k_{56}^{(1)} + k_{78}^{(2)} + k_{12}^{(3)} + k_{34}^{(4)}\right) & \left(k_{57}^{(2)} + k_{13}^{(3)}\right) & \left(k_{67}^{(2)} + k_{14}^{(3)}\right) \\ \left(k_{65}^{(1)} + k_{78}^{(2)} + k_{21}^{(3)} + k_{43}^{(4)}\right) & \left(k_{66}^{(1)} + k_{88}^{(2)} + k_{22}^{(3)} + k_{44}^{(4)}\right) & \left(k_{58}^{(2)} + k_{23}^{(3)}\right) & \left(k_{68}^{(2)} + k_{24}^{(3)}\right) \\ \left(k_{57}^{(2)} + k_{31}^{(3)}\right) & \left(k_{58}^{(2)} + k_{32}^{(3)}\right) & \left(k_{55}^{(2)} + k_{33}^{(3)}\right) & \left(k_{56}^{(2)} + k_{34}^{(3)}\right) \\ \left(k_{67}^{(2)} + k_{41}^{(3)}\right) & \left(k_{68}^{(2)} + k_{41}^{(3)}\right) & \left(k_{56}^{(2)} + k_{43}^{(3)}\right) & \left(k_{66}^{(2)} + k_{44}^{(3)}\right) \end{bmatrix} \begin{Bmatrix} u_{x3} \\ u_{y3} \\ u_{x6} \\ u_{y6} \end{Bmatrix}$$

$$= \begin{Bmatrix} 0 \\ 0 \\ \left(b^{(2)} + b^{(3)}\right)p_x \\ -\left(b^{(2)} + b^{(3)}\right)p_y \end{Bmatrix} \tag{8.111}$$

c) The element connectivity table for this assembly is given in Table 8.1. The traction components are $p_x = 5$ and $p_y = -5$ MPa. The material stiffness matrix for the plane stress condition is given in Eq. (8.35) where $E_{11} = E_{22} = E/(1 - \nu^2)$, $E_{12} = \nu E_{11}$ and $G = E/2(1 + \nu)$. The element stiffness matrices are identical, as all four elements are identical in all aspects. By using Eq. (8.39) and the properties listed in the element connectivity table,

TABLE 8.1 Element connectivity table.

(e)	Node-1, 2, 3, 4	$a^{(e)}$, $b^{(e)}$, $t^{(e)}$ (mm)	E (Pa), ν	$\left(q_{x1}^{(e)}, q_{y1}^{(e)}\right), \left(q_{x2}^{(e)}, q_{y2}^{(e)}\right), \left(q_{x3}^{(e)}, q_{y3}^{(e)}\right), \left(q_{x4}^{(e)}, q_{y4}^{(e)}\right)$
1	1, 2, 3, 4	25, 40, 2	5×10^9, 0.3	(0, 0), (0, 0), (0, 0), (0, 0)
2	2, 5, 6, 3	25, 40, 2	5×10^9, 0.3	(0, 0), (p_x, p_y), (p_x, p_y), (0, 0)
3	3, 6, 7, 8	25, 40, 2	5×10^9, 0.3	(0, 0), (p_x, p_y), (p_x, p_y), (0, 0)
4	4, 3, 8, 9	25, 40, 2	5×10^9, 0.3	(0, 0), (0, 0), (0, 0), (0, 0)

the element stiffness matrices for $e = 1$–4 can be found as follows:

$$[k]^{(e)} = 2\times 10^6 \begin{bmatrix} 3.3310 & 0.8929 & -2.7301 & -0.0687 & -1.6655 & -0.8929 & 1.0646 & 0.0687 \\ 0.8929 & 2.1703 & 0.0687 & -0.4533 & -0.8929 & -1.0852 & -0.0687 & -0.6319 \\ -2.7301 & 0.0687 & 3.3310 & -0.8929 & 1.0646 & -0.0687 & -1.6655 & 0.8929 \\ -0.0687 & -0.4533 & -0.8929 & 2.1703 & 0.0687 & -0.6319 & 0.8929 & -1.0852 \\ -1.6655 & -0.8929 & 1.0646 & 0.0687 & 3.3310 & 0.8929 & -2.7301 & -0.0687 \\ -0.8929 & -1.0852 & -0.0687 & -0.6319 & 0.8929 & 2.1703 & 0.0687 & -0.4533 \\ 1.0646 & -0.0687 & -1.6655 & 0.8929 & -2.7301 & 0.0687 & 3.3310 & -0.8929 \\ 0.0687 & -0.6319 & 0.8929 & -1.0852 & -0.0687 & -0.4533 & -0.8929 & 2.1703 \end{bmatrix} \tag{8.112}$$

The load vectors for elements 1–4 can be found by using Eq. (8.105) as follows,

$$\{r\}^{(1)} = \{r\}^{(3)} = 0, \ \{r\}^{(2)} = \{r\}^{(4)} = \left\{0 \quad 0 \quad 400 \quad -400 \quad 400 \quad -400 \quad 0 \quad 0\right\}^T \mathrm{N} \tag{8.113}$$

The global equilibrium equation can be assembled by using Eqs. (8.111), (8.112), and (8.114). The displacement vector can be found by solving the global equilibrium equation.

$$\{D\}_{2\times2} = \left\{37.047 \quad -9.838 \quad 90.405 \quad -94.207\right\}^T \mu\mathrm{m} \tag{8.114}$$

For each element ($e = 1\ldots4$), variation of displacement, strain, and stress components across the solution domain can be calculated by using Eq. (8.27), Eqs. (8.30)–(8.32), and Eq. 8.34 and the Hooke's law, respectively. The resulting contours are combined for the entire solution domain and plotted in Fig. 8.9A–8.11A for the case of the 2×2 mesh. Based on these results, the solution does not seem to be accurate.

A mesh convergence study is presented in Table 8.2. The mesh is varied in such a way that a finer mesh includes the nodes of the older mesh. For this geometry this is accomplished by using a meshing scheme with $2^n \times 2^{n+1}$ elements in the x and y-directions, respectively. Nodal displacement, strain and stress components are reported for the mid-point of the domain (0.5, 8) mm and the mid-point of edge-2, (10, 8) mm.

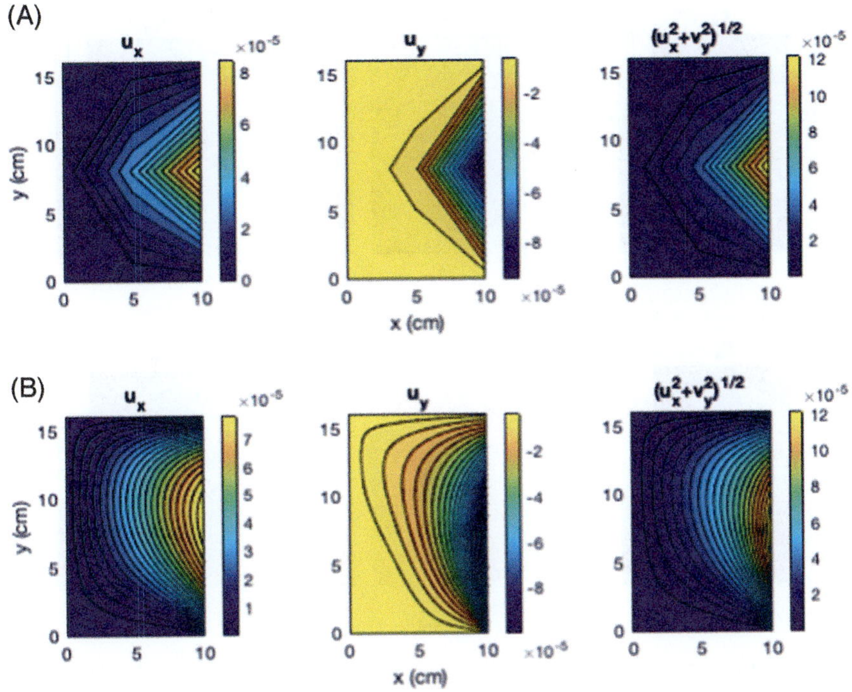

FIGURE 8.9 Displacement variation in the solution domain for (A) 2×2 and (B) 64×128 elements.

Note that displacements are the primary degrees of freedom at the nodes. They are directly obtained by inverting the equation of equilibrium. Strain and stress values are secondary degrees of freedom. The elements surrounding a given node make a contribution to the nodal values of the stress and strain. In this example, to compute the strain and stress values, the nodal values from the surrounding elements are averaged. For example, strain value ε_{xx} at node-3 is computed as follows,

$$\varepsilon_{xx}\Big|_{node\text{-}3} = \frac{1}{4}\left(\varepsilon_{xx}\Big|_{node\text{-}3}^{(1)} + \varepsilon_{xx}\Big|_{node\text{-}3}^{(2)} + \varepsilon_{xx}\Big|_{node\text{-}3}^{(3)} + \varepsilon_{xx}\Big|_{node\text{-}3}^{(4)}\right) \tag{8.115}$$

Table 8.2 shows that the all variables converge monotonically toward a constant value. For the 64 × 128 mesh, we find the solutions as follows:

$$\{D\}_{64\times128} = \{38.461 \quad -17.821 \quad 81.827 \quad -97.417\}^T \mu\text{m} \tag{8.116}$$

Along edge-2, we expect the internal stress values σ_x and τ_{xy} to be equal to the applied boundary tractions. The normal σ_{xx} and shear τ_{xy} stresses on the boundary location (10, 8) mm approach the applied traction $p_x = 5$ MPa and $p_y = -5$ MPa, respectively, with increasing number of elements. The variations

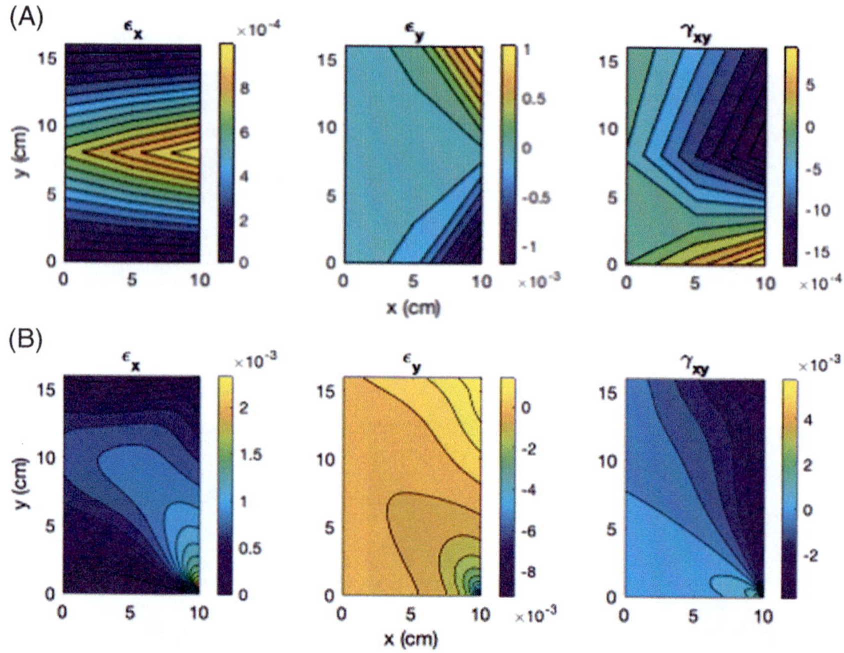

FIGURE 8.10 Strain variation in the solution domain for (A) 2×2 and (B) 64×128 elements.

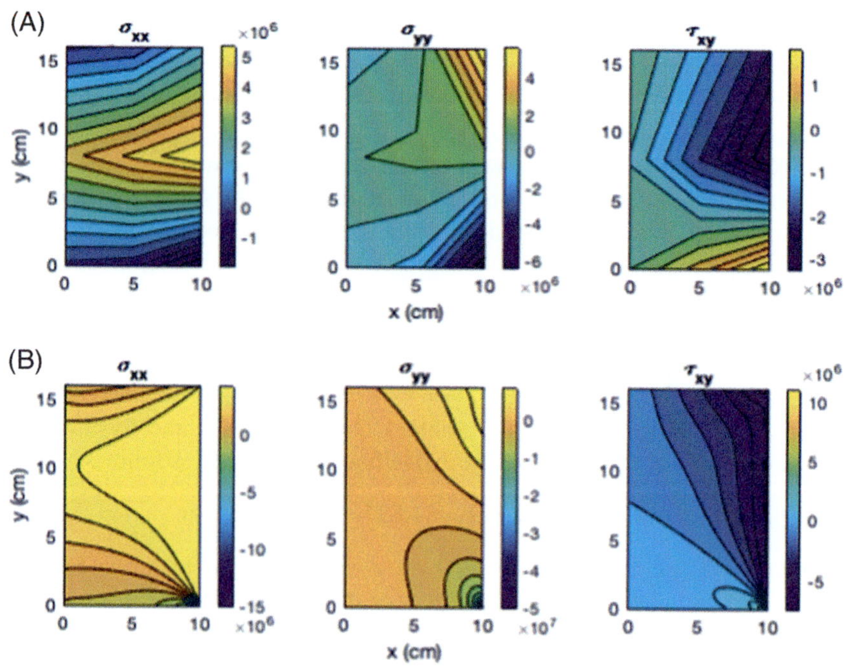

FIGURE 8.11 Stress variation in the solution domain for (A) 2×2 and (B) 64×128 elements.

TABLE 8.2 Convergence study for the displacement, strain and stress components at a) the center $(x, y) = (5, 8)$ mm and b) $(x, y) = (10, 8)$ mm at middle of the right edge of the solution domain.

(5,8) mm	u_x μm	u_y μm	ε_{xx}	ε_{yy}	ε_{xy}	σ_{xx} MPa	σ_{yy} MPa	τ_{xy} MPa
2 x 2	37.047	−9.838	9.041×10^{-4}	-1.009×10^{-13}	-9.421×10^{-4}	4.967	1.490	−1.812
2 x 4	38.281	−12.492	8.056×10^{-4}	-1.851×10^{-4}	-6.640×10^{-4}	4.121	0.311	−1.277
4 x 8	38.027	−16.239	8.445×10^{-4}	-1.514×10^{-4}	-5.600×10^{-4}	4.391	0.560	−1.077
8 x 16	38.280	−17.359	8.506×10^{-4}	-1.399×10^{-4}	-5.369×10^{-4}	4.443	0.633	−1.033
16 x 32	38.397	−17.691	8.520×10^{-4}	-1.360×10^{-4}	-5.339×10^{-4}	4.457	0.657	−1.027
32 x 64	38.444	−17.791	8.525×10^{-4}	-1.347×10^{-4}	-5.337×10^{-4}	4.462	0.665	−1.026
64 x 128	38.461	−17.821	8.527×10^{-4}	-1.342×10^{-4}	-5.339×10^{-4}	4.464	0.668	−1.027
(10, 8) mm	u_x μm	u_y μm	ε_{xx}	ε_{yy}	ε_{xy}	σ_{xx} MPa	σ_{yy} MPa	σ_{xy} MPa
2 x 2	90.405	−94.207	1.067×10^{-3}	9.822×10^{-14}	-1.687×10^{-3}	5.864	1.759	−3.245
2 x 4	80.557	−90.812	8.455×10^{-4}	2.189×10^{-4}	-1.479×10^{-3}	5.007	2.596	−2.844
4 x 8	81.311	−96.009	8.548×10^{-4}	3.133×10^{-4}	-1.987×10^{-3}	5.213	3.130	−3.820
8 x 16	81.643	−97.083	8.391×10^{-4}	3.226×10^{-4}	-2.288×10^{-3}	5.142	3.156	−4.399
16 x 32	81.764	−97.330	8.270×10^{-4}	3.230×10^{-4}	-2.442×10^{-3}	5.076	3.138	−4.697
32 x 64	81.810	−97.397	8.202×10^{-4}	3.232×10^{-4}	-2.521×10^{-3}	5.039	3.128	−4.848
64 x 128	81.827	−97.417	8.166×10^{-4}	3.233×10^{-4}	-2.560×10^{-3}	5.020	3.122	−4.924

of the variables for the solution obtained with 64×128 mesh are plotted in Figs. 8.9B–8.11B.

References

[1] R.D. Cook, D.S. Malkus, M.E. Plesha, R.J. Witt, Concepts and Applications of Finite Element Analysis, 4th ed., John Wiley & Sons, 2002.

[2] S.P. Timoshenko, Theory of Elasticity, McGraw-Hill, New York, 1951.

[3] M.A. Bhatti, Advanced Topics in Finite Element Analysis of Structures, Wiley & Sons, New York, 2006.

Chapter 9

Finite element analysis of one-dimensional heat transfer problems

9.1 Introduction

Conduction can be assumed to be one dimensional in a large slab of solid that is exposed to external effects only at its two ends (Fig. 2.20.B). The energy balance in such a solid is given by Eq. (2.171). In case the solid has a slender shape (Fig. 2.20.C), such as that of a fin, cooling an electronic device, or an icicle, hanging from the roof line, convection from the periphery of the solid should also be considered. In this case, the energy balance is represented by Eq. (2.174).

The goal of this chapter is to develop the finite element form of the one-dimensional heat transfer equations, at steady state. The presentation given here is similar to the one given in Chapter 4 for a general second-order ordinary differential equation. The boundary value problem (BVP) and its admissible boundary conditions that are specific to heat transfer in a one-dimensional domain are presented in Section 9.2. The finite element form of this boundary value problem is developed for a general M-node element in Section 9.3. This section involves detailed derivations of the special treatment required for the natural boundary conditions. Element equilibrium equation for a two-node, C^0 element is also given in this section. The similarities between the finite element forms of the second-order ordinary differential equation introduced in Chapter 4 and the equations developed in Section 9.3 are presented in detail in Section 9.4. Element assembly is redescribed by using the variables that are relevant for the heat transfer problem in Section 9.5. Treatments of natural and essential boundary conditions are described in Section 9.6. The example given at the end of the chapter provides details of implementation of the concepts discussed in the chapter.

9.2 One-dimensional heat transfer

The one-dimensional boundary value problem is stated in the solution domain, Ω, defined in $0 < x < L$ and on its boundary, Γ, located at $x = 0$ and $x = L$.

Finite Element Method: Physics and Solution Methods. DOI: https://doi.org/10.1016/B978-0-12-821127-4.00011-6
Copyright © 2022 Elsevier Inc. All rights reserved.

Conduction in the one-dimensional solid is governed by Eq. (2.171),

$$QA + \frac{\partial}{\partial x}\left(kA\frac{\partial T}{\partial x}\right) = \rho cA\frac{\partial T}{\partial t} \text{ in } 0 < x < L \tag{9.1}$$

where Q is the rate of internal heat generation per unit volume (W/m^3); k is the thermal conductivity in x-direction (W/mK); ρ is the mass density (kg/m^3); c is the specific heat (J/kg K); $T(x, t)$ is the temperature as a function of space coordinate, x, and time, t.

In case the heat is transferred from a thin fin to a surrounding fluid at temperature T_∞ the energy balance is governed by Eq. (2.174),

$$AQ + \frac{\partial}{\partial x}\left(kA\frac{\partial T}{\partial x}\right) + hP(T_\infty - T) = \rho cA\frac{\partial t}{\partial T} \text{ in } 0 < x < -L \tag{9.2}$$

where A is the cross-sectional area of the fin; P is the perimeter of the fin (m); h is the convection coefficient (W/m^{2}K).

9.2.1 Boundary conditions for one-dimensional heat transfer

Essential boundary condition: $T = T_b$, temperature specified on the boundary (9.3)

$$\text{Natural boundary conditions: } \left(-k\frac{\partial T}{\partial x}\right)n = f_{B0} \text{ at } x = 0 \quad \text{(a)}$$
$$\left(-k\frac{\partial T}{\partial x}\right)n = f_{BL} \text{ at } x = L \quad \text{(b)} \tag{9.4}$$

where T_b is the specified temperature value on the boundary and n represents the unit outward normal of the boundary such that, $n = -1$ at $x = 0$ and $n = 1$ at $x = L$.

The heat flux on the boundary f_B can be one or superposition of the following effects,

$$\begin{aligned}
&\text{Specified heat flux on the boundary:} && f_B = f_B(t)\\
&\text{Convection on the boundary:} && f_B = h(T - T_\infty)\\
&\text{Radiation on the boundary:} && f_B = \sigma\psi\left(T^4 - T_2^4\right)
\end{aligned} \tag{9.5}$$

where h (W/m^{2}K) is the *convection coefficient*, T_∞ is the far distance *temperature* of the fluid in which the domain is submerged, σ is the Stefan–Boltzmann constant (Wm^{2}K^4) and ψ is the *view factor* between the two surfaces. Note that, in general, each term in Eq. (9.5) can potentially depend on time for the transient version of the problem. Refer to Section 2.4 for more extensive background on these relationships.

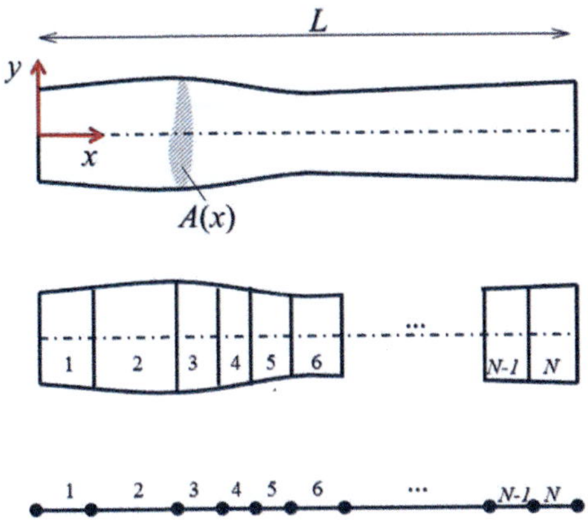

FIGURE 9.1 A one-dimensional solution domain discretized in to N elements.

9.3 Finite element formulation of the one-dimensional, steady state, heat transfer problem

The discretization of the one-dimensional domain, described in Section 5.1, applies here as well. The solution domain is divided into small segments called ***elements***, which are connected at their ends. The ensemble of the elements is known as the ***finite element mesh***. In Fig. 9.1 the one-dimensional domain is divided into N elements. Each element is given a number e, where $e \in [1, N]$. The size (length) of element-e is indicated by $L^{(e)}$. The total length of the solution domain can be easily found by using Eq. (5.1).

A (local) ***element coordinate system***, x', is used in order to separate the *element* and the *global, coordinate systems*, x. Also note that as temperature is a scalar variable there is no need for coordinate transformations.

At steady state, the temperature is only a function of space, the transient effects are neglected, and Eqs. (9.1) and (9.2) become,

$$\frac{d}{dx}\left(kA\frac{dT}{dx}\right) + QA = 0 \tag{9.6}$$

$$\frac{d}{dx}\left(kA\frac{dT}{dx}\right) + hp(T_\infty - T) + AQ = 0 \tag{9.7}$$

The boundary conditions remain the same as in Eq. (9.4) except f_B no longer varies in time. In order to derive the element-level energy balance equations for one-dimensional heat transfer, the *internal elements* and the *elements that share a node with one of the physical boundaries* ($x = 0$ or L) of the solution domain have to be treated slightly differently. The internal elements are connected to one another at the nodes. The internal heat flow $f_i A_i$ between the elements is balanced. Therefore, the internal elements are considered to be subjected to internal heat flow at their connecting nodes, which will disappear upon assembly. The elements that are exposed to a physical boundary experience the physical

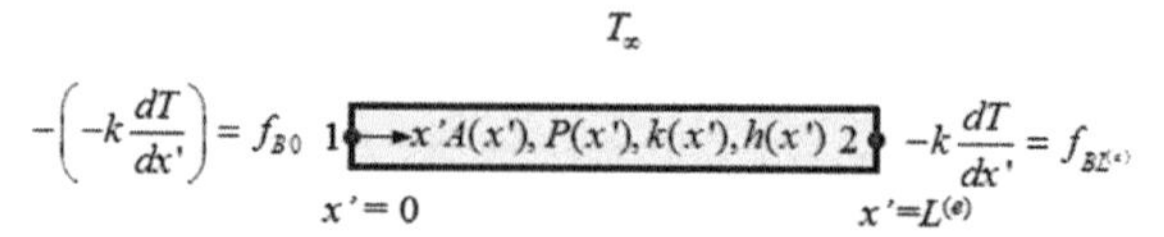

FIGURE 9.2 One-dimensional domain representative of an element subjected to natural boundary conditions.

boundary condition (Eq. (9.4)) on that boundary as well as the internal heat flow on their inner boundary. These effects are treated separately in the next two sections.

9.3.1 Element equilibrium equations for a generic one-dimensional element

The derivation will be demonstrated for the mathematically more general form of the one-dimensional heat transfer problem given by Eq. (9.7) for the domain shown in Fig. 9.2. The problem is stated as a BVP spanning the length of an element $0 < x' < L^{(e)}$, with the boundaries $x' = 0$ and $x' = L^{(e)}$. The cross-sectional area and the perimeter of the element are$A(x')$ and $P(x')$, conduction and convection coefficients are $k(x')$ and $h(x')$, and the rate of internal heat generation is $Q(x')$.

At steady state, we use Eq. (9.7),

$$\frac{d}{dx'}\left(kA\frac{dT}{dx'}\right) + hP(T_\infty - T) + AQ = 0 \,\text{in}\, 0 < x' < L^{(e)} \tag{9.8}$$

The finite element form for Eq. (9.6) can be obtained by setting $hP = 0$.

9.3.1.1 *Internal elements*

In order to ensure that the internal heat flow $f_i A_i$ is continuous across the element boundaries the problem is formulated such that the ends 0 and $L^{(e)}$ of the element are subjected to internal heat flux values of f_{i0} and $f_{iL^{(e)}}$, respectively,

$$\begin{aligned} -\left(-k\frac{dT}{dx'}\right)_{x'=0} &= f_{i0} \quad \text{(a)} \\ \left(-k\frac{dT}{dx'}\right)_{x'=L^{(e)}} &= f_{iL^{(e)}} \quad \text{(b)} \end{aligned} \tag{9.9}$$

Note that the values of the unit normal n are directly implemented in Eq. (9.9). The *weighted residual integral* for Eq. (9.8) is expressed as follows:

$$\int_0^{L(e)} w\left[\frac{d}{dx'}\left(kA\frac{dT}{dx'}\right) + hP(T_\infty - T) + AQ\right]dx' = 0 \tag{9.10}$$

where w is a weight function with the properties of the variation δT on temperature, as described in Chapter 3. Integration of the first term of (9.10) by parts results in the following relationship,

$$\left[\delta T\left(kA\frac{dT}{dx'}\right)\right]_0^{L^{(e)}} - \int_0^{L(\varepsilon)} kA\frac{d(\delta T)}{dx'}\frac{dT}{dx'}dx' + \int_0^{L(\varepsilon)} \delta T(AQ + hPT_\infty)dx' - \int_0^{L(\varepsilon)} hP\,\delta T\,T dx' = 0 \tag{9.11}$$

Using the boundary conditions, (9.9), the *weak form* of Eq. (9.8) is found as follows:

$$-\left[\delta T(0)(f_{i0}A_0) + \delta T\left(L^{(e)}\right)(f_{iL^{(e)}}A_{L^{(e)}})\right]$$
$$- \int_0^{L(e)} kA\frac{d\delta T}{dx'}\frac{dT}{dx'}dx' + \int_0^{L(\theta)} \delta T(AQ + hPT_\infty)dx' - \int_0^{L(\theta)} hP\,\delta T\,T dx' = 0 \tag{9.12}$$

where A_0 and $A_{L^{(e)}}$ are the cross-sectional areas of the element at the element boundaries. We will use the nodal notation for the variables at the ends of the element. At this point we are not specifying the order of interpolation, therefore we are assuming that the element has M nodes. The variable at the ends of the element are therefore defined as follows:

$$T_1 = T(0), \quad T_M = T\left(L^{(e)}\right)$$
$$A_1 = A_0, \quad A_M = A_{L^{(e)}}$$
$$f_{i1} = f_{i0}, \quad f_{iM} = f_{iL^{(e)}}$$

In order to obtain the finite element form of the heat transfer equation, the temperature and its gradient are interpolated as follows:

$$\begin{aligned} &T \simeq \tilde{T} = [N]\{d_T\} && \text{(a)} \\ &\frac{dT}{dx'} \simeq \frac{d\tilde{T}}{dx'} = [B]\{d_T\} && \text{(b)} \end{aligned} \tag{9.13}$$

The shape function matrix $[N]$ and the temperature-gradient matrix $[B]$ are defined in the usual manner as described before, and as stated above the order of interpolation is not specified at this stage. In what follows we will drop the approximation sign from $\tilde{T}$ and simply use T. The nodal temperature vector is,

$$\{d_T\} = \{T_1 \cdots T_M\}^T \tag{9.14}$$

Using Eq. (9.13) in (9.12), the following relationship is obtained,

$$\{\delta d_T\}^T\left(\int_0^{L(e)} kA[B]^T[B]\{d_T\}dx' + \int_0^{L(e)} hP[N]^T[N]\{d_T\}dx' - \int_0^{L(e)} [N]^T(AQ + hPT_\infty)dx' - \{r_i\}^{(e)}\right) = 0 \tag{9.15}$$

where

$$\{r_i\}^{(e)} = \{-f_{i1}A_1 \cdots - f_{iM}A_M\}^T \tag{9.16}$$

is the *nodal heat flow* vector. As $\{\delta d_T\}$ represents an arbitrary variation of the nodal temperatures, it can be dropped from Eq. (9.15). This results in the finite element form of the energy balance equation as follows:

$$\begin{aligned}&\left(\left[\int_0^{L(e)} kA[B]^T[B]dx'\right] + \left[\int_0^{L(e)} hP[N]^T[N]dx'\right]\right)\{d_T\}\\&= \int_0^{E(e)} [N]^T(AQ + hPT_\infty)dx' + \{r_i\}^{(e)}\end{aligned} \tag{9.17}$$

Let's define the following matrices and vectors,

Conductivity matrix: $[k_k]^{(e)} = \int_0^{L(e)} kA[B]^T[B]dx'$ (a)

Surface convection matrix: $[k_h]^{(e)} = \int_0^{L(e)} hP[N]^T[N]dx'$ (b)

Internal heat generation and surface convection vector:

$$\{r_q\}^{(e)} = \int_0^{L(e)} [N]^T(AQ + hPT_\infty)dx' \quad \text{(c)} \tag{9.18}$$

Nodal heat flow vector: $\{r_i\}^{(e)} = \{-f_{i1}A_1 \cdots - f_{iM}A_M\}^T$ (d)

The finite element form of the conduction heat transfer equation then becomes,

$$\left([k_k]^{(e)} + [k_h]^{(e)}\right)\{d_T\} = \{r_q\}^{(e)} + \{r_i\}^{(e)} \tag{9.19}$$

Eq. (9.19) is used for *internal elements* of a finite element mesh, where the nodal heat flow vector (9.16) vanishes at the interelement boundaries, as the heat flow must be continuous across element boundaries.

9.3.1.2 Elements on the external boundaries subjected to convection

In case the element is exposed to one of the physical boundaries ($x = 0$, or L) of the solution domain, the nature of the heat flux f_B (Eq. (9.5)) on that boundary must be considered. To this end, the boundary term of Eq. (9.11) is modified.

In case convection is specified at the $x = 0$ as $f_{B0} = h(T(0) - T_\infty)$, the term evaluated on the boundary in Eq. (9.11) becomes,

$$\left[\delta T\left(kA\frac{dT}{dx'}\right)\right]_0^{L^{(e)}} = -(\delta T_1 f_{B0}A_0 + \delta T_M f_{iM}A_M) = -(\delta T_1[hA_0(T_1 - T_\infty)] + \delta T_M f_{iM}A_M) \tag{9.20}$$

Note that Eq. (9.20) can be put into vector form as follows:

$$\left[\delta T\left(kA\frac{dT}{dx'}\right)\right]_0^{[e(e)} = -\{\delta d_T\}^T\left(\begin{bmatrix} hA_0 & \cdots & 0 \\ \vdots & \cdots & \vdots \\ 0 & \cdots & 0 \end{bmatrix}\{d_T\} - \begin{Bmatrix} hA_0T_\infty \\ \vdots \\ -f_{iM}A_M \end{Bmatrix}\right)$$
$$= -\{\delta d_T\}^T\left([k_{h-b0}]^{(e=1)}\{d_T\} - \{f_{h-b0}\}^{(e=1)}\right)$$

where the *boundary convection matrix* and the *boundary heat flow vector* at $x = 0$ boundary are defined as follows:

$$[k_{h-b0}]^{(e=1)} = \begin{bmatrix} hA_0 & \cdots & 0 \\ \vdots & \cdots & \vdots \\ 0 & \cdots & 0 \end{bmatrix} \quad \text{and} \quad \{r_{h-b0}\}^{(e=1)} = \begin{Bmatrix} hA_0T_\infty \\ \vdots \\ -f_{iM}A_M \end{Bmatrix} \quad \text{(a, b)} \tag{9.21}$$

respectively.

Similarly, in case convection is specified at $x = L$ as $f_{BL} = h(T(L) - T_\infty)$, the boundary term of (9.11) can be shown to be,

$$\left[\delta T\left(kA\frac{dT}{dx'}\right)\right]\Big|_0^{L_0^{(e)}} = -(\delta T_1 f_{i1}A_1 + \delta T_M[hA_L(T_M - T_\infty)]+) = -\{\delta d_T\}^T\left([k_{h-bL}]^{(e=N)}\{d_T\} - \{f_{h-bL}\}^{(e=N)}\right) \tag{9.22}$$

where the *boundary convection matrix* and the *boundary heat flow vector* at $x = L$ boundary are defined as follows:

$$[k_{h-bL}]^{(e=N)} = \begin{bmatrix} 0 & \cdots & 0 \\ \vdots & \cdots & \vdots \\ 0 & \cdots & hA_L \end{bmatrix} \quad \text{and} \quad \{r_{h-bL}\}^{(e=N)} = \begin{Bmatrix} -f_{i1}A_1 \\ \vdots \\ hA_LT_\infty \end{Bmatrix} \tag{9.23}$$

respectively.

9.3.1.3 Elements on the external boundaries subjected to prescribed heat flux

In case a constant heat flux f_{B0} is specified at $x = 0$ or a constant heat flux f_{BL} is specified at $x = L$ the *nodal heat flow vectors* become,

$$\{r_{b0}\}^{(e=1)} = \begin{Bmatrix} -f_{B0}A_0 \\ \vdots \\ -f_{iM}A_M \end{Bmatrix}, \qquad \{r_{bL}\}^{(e=N)} = \begin{Bmatrix} -f_{i1}A_1 \\ \vdots \\ -f_{BL}A_L \end{Bmatrix} \quad \text{(a, b)} \tag{9.24}$$

9.3.1.4 The finite element form of the one-dimensional heat transfer equations

Summarizing Eqs. (9.19–9.24) the finite element form of the one-dimensional conduction heat transfer equation is found as follows:

$$\begin{aligned}
&\text{Boundary element } (e=1): \quad ([k_k]^{(1)} + [k_h]^{(1)} + [k_{h-b0}]^{(1)})\{d_T\} = \{r_q\}^{(1)} + \{r_{h-b0}\}^{(1)} \\
&\text{Internal elements } (2 \le e \le N-1): \quad ([k_k]^{(e)} + [k_h]^{(e)})\{d_T\} = \{r_q\}^{(e)} + \{r_i\}^{(e)} \\
&\text{Boundary element } (e=N): ([k_k]^{(N)} + [k_h]^{(N)} + [k_{h-bL}]^{(N)})\{d_T\} = \{r_q\}^{(N)} + \{r_{h-bL}\}^{(N)}
\end{aligned} \tag{9.25}$$

Note that the boundary convection matrix and the boundary heat flux vectors depend on the prescribed boundary as described below.

Natural boundary condition:

$x = 0$:

Convection boundary condition: $[k_{h-b0}]^{(1)}$, $\{r_{h-b0}\}^{(1)}$ are given by Eq. (9.21).

Constant heat flux: $[k_{b0}]^{(1)} = 0$, $\{r_{b0}\}^{(1)}$ is given by Eq. (9.24a).

$x = L$:

Convection boundary condition: $[k_{h-bL}]^{(N)}$, $\{r_{h-b0}\}^{(N)}$ are given by Eq. (9.23).

Constant heat flux: $[k_{bL}]^{(N)} = 0$, $\{r_{bL}\}^{(N)}$ is given by Eq. (9.24b).

Essential boundary condition:

$$x = 0\text{: } T(0) = T_{b0}\text{:} \quad [k_{h-b0}]^{(1)} = 0, \quad \{r_{h-b0}\}^{(1)} = \{-f_{b0}A_0 \quad -f_{i2}A_2\}^T \tag{9.26}$$

$$x = L\text{: } T(L) = T_{bL}\text{:} \quad [k_{h-bL}]^{(N)} = 0, \quad \{r_{h-bL}\}^{(N)} = \{-f_{iN}A_N \quad -f_{bL}A_L\}^T \tag{9.27}$$

where T_{b0} and T_{bL} are the prescribed temperature values on the $x = 0$ and $x = L$ boundaries, respectively. Note that *when essential boundary conditions are prescribed, the boundary flux values f_{b0} and f_{bL} can only be found after the solution is obtained*. As such, this is similar to finding the reaction forces at a fixed boundary of a deformable medium.

9.3.2 Finite element form with linear interpolation

We next present the finite element form of the one-dimensional heat transfer problem for a two-node element ($M = 2$). Temperature variation along this element is interpolated as follows:

$$T(x') = \sum_{i=1}^{2} N_i T_i = [N]\{d_T\} \tag{9.28}$$

where the shape function matrix is $[N] = [N_1, N_2]$, the nodal temperature vector is $\{d_T\} = \{T_1, T_2\}^T$ with the nodal temperatures T_1 and T_2. The C^0 continuous, linear interpolation functions are given in Section 4.3,

$$N_1 = 1 - \frac{x'}{L^{(e)}}, \quad N_2 = \frac{x'}{L^{(e)}}, \tag{9.29}$$

The general form of the equation of equilibrium is given by Eq. (9.25). The *conductivity* and *surface convection matrices* are defined for a generic M-node element in Eqs. (9.18a, b) and the *boundary convection matrices* are defined in Eqs. (9.21a) and (9.23a). We are going to assume that all the parameters (k, h, A, P, T_∞) are constant along the element. Thus, by using the relevant equations just mentioned, the matrices that form the element stiffness matrix can be calculated as follows. Note that the details of these calculations are left as an exercise.

Conductivity: $$[k_k]^{(e)} = \int_0^{L^{(e)}} kA[B]^T[B]dx' = \left(\frac{kA}{L}\right)^{(e)} \begin{bmatrix} 1 & -1 \\ -1 & 1 \end{bmatrix} \quad \text{(a)}$$

Surface convection: $$[k_b]^{(e)} = \int_0^{E^{(e)}} hP[T]^T[N]dx' = \left(\frac{hPL}{3}\right)^{(e)} \begin{bmatrix} 1 & \frac{1}{2} \\ \frac{1}{2} & 1 \end{bmatrix} \quad \text{(b)}$$

Boundary convection at $x = 0$: $$[k_{h-b0}]^{(e)} = \begin{bmatrix} hA_0 & 0 \\ 0 & 0 \end{bmatrix} \quad \text{(c)}$$

Boundary convection at $x = L$: $$[k_{h-bL}]^{(e)} = \begin{bmatrix} 0 & 0 \\ 0 & hA_L \end{bmatrix} \quad \text{(d)}$$

(9.30)

The internal heat generation and surface convection vectors are defined in Eq. (9.18c), the nodal heat flow vector is defined in Eq. (9.18d) and boundary heat flow vectors are defined in Eqs. (9.21b) and (9.23b) for a generic M-node element. In case of the two-node element, we find the following vectors:

Internal heat generation and surface convection:

$$\{r_q\}^{(e)} = \int_0^{(L^e)} [N]^T (AQ + hPT_\infty) dx' = \frac{1}{2}\left(QAL^{(e)} + hPLT_\infty\right)\begin{Bmatrix}1\\1\end{Bmatrix} \quad \text{(a)}$$

Boundary convection at $x = 0 (e = 1)$: $$\{r_{h-b0}\}^{(e)} = \begin{Bmatrix} hA_0 T_\infty \\ -f_{i2}A_2 \end{Bmatrix} \quad \text{(b)}$$

Boundary convection at $x = L (e = N)$: $$\{r_{h-bL}\}^{(e)} = \begin{Bmatrix} -f_{i(N-1)}A_{(N-1)} \\ hA_L T_\infty \end{Bmatrix} \quad \text{(c)}$$

Boundary constant heat flux at $x = 0 (e = 1)$: $$\{r_{b0}\}^{(e)} = \begin{Bmatrix} -f_{B0}A_1 \\ -f_{i2}A_2 \end{Bmatrix} \quad \text{(d)}$$

Boundary constant heat flux at $x = L (e = N)$: $$\{r_{bL}\}^{(e)} = \begin{Bmatrix} -f_{i(N-1)}A_{(N-1)} \\ -f_{BL}A_L \end{Bmatrix} \quad \text{(e)}$$

Nodal heat flow: $(1 < e < N)$: $$\{r_i\}^{(e)} = \begin{Bmatrix} -f_{i1}A_1 \\ -f_{i2}A_2 \end{Bmatrix} \quad \text{(f)}$$

(9.31)

9.4 Element equilibrium equations: general ordinary differential equation

In Chapter 4, the finite element solution of a general boundary value problem is developed. This problem is defined over the domain $0 < x < L$ with natural boundary conditions at $x = 0$ and $x = L$ as follows:

$$\frac{d}{dx}\left(a\frac{dT}{dx}\right) + pT + q = 0 \text{ in } 0 < x < L \tag{9.32}$$

$$-\frac{dT}{dx} = \alpha T + \beta \text{ at } x = 0 \quad \text{(a)}$$

$$\frac{dT}{dx} = \alpha T + \beta \text{ at } x = L \quad \text{(b)}$$

(9.33)

where $a(x)$, $p(x)$, $q(x)$, α, and β are given parameters of the problem. Table 4.1 can be used to identify the functions a, p, and q for the one-dimensional heat transfer problem, and Table 4.2 can be used to identify α and β for the various possible heat flux conditions.

The details of the derivation of the weak form of Eq. (9.32) for C^0 continuous, linear interpolation functions are given in Section 4.3. Naturally, this derivation is the same as the one used in the previous section, and results in the following matrix equation,

$$\left([k_a]^{(e)} + [k_p]^{(e)} + [k_\alpha]^{(e)}\right)\{d\} = \{r_q\}^{(e)} + \{r_\beta\}^{(e)} \tag{9.34}$$

where $\{d\}$ is the nodal degree of freedom vector. The $[k_a]^{(e)}$, $[k_p]^{(e)}$ and $[k_\alpha]^{(e)}$ matrices are defined as follows:

$$[k_a] = \int_0^{L(e)} a\,[B]^T[B]dx' \qquad \text{(a)}$$

$$[k_p] = -\int_0^{L(e)} p\,[N]^T\,[N]dx' \qquad \text{(b)}$$

$$[k_\alpha] = -\begin{bmatrix} (\alpha a)_{x=0} & \cdots & 0 \\ \vdots & \cdots & \vdots \\ 0 & \cdots & 0 \end{bmatrix} \quad \text{or} \quad [k_\alpha] = -\begin{bmatrix} 0 & \cdots & 0 \\ \vdots & \cdots & \vdots \\ 0 & \cdots & (\alpha a)_{x=L} \end{bmatrix} \qquad \text{(c)}$$

(9.35)

And the vectors $\{r_\beta\}^{(e)}$ and $\{r_q\}^{(e)}$ are defined as follows:

$$\{r_q\} = \int_0^{L(e)} [N]^T\,q dx' \qquad \text{(a)}$$

$$\{r_\beta\} = \{(a\beta)_{x=0} \quad \cdots \quad 0 \quad 0\}^T \quad \text{or } \{r_\beta\} = \{0 \cdots (a\beta)_{x=L}\}^T \qquad \text{(b)}$$

(9.36)

Details of these derivations are given in Section 4.3. Note that $[k_\alpha]$ and $\{r_\beta\}$ depend on the boundary conditions.

For the one-dimensional, steady state heat transfer problem given by Eq. (9.6) inspection shows the following parallels can be drawn between the general differential equation and the heat transfer equation,

$$a = kA,\ p = 0,\ q = QA \qquad (9.37)$$

Similarly, for the one-dimensional, steady state heat transfer from a fin described by Eq. (9.3.2) the following equivalencies can be surmised:

$$a = kA,\ p = -hP,\ q = QA + hPT_\infty \qquad (9.38)$$

The formulation given above assumes that both nodes of the element are subjected to natural boundary conditions given in the form of Eqs. (9.33). The element equilibrium equations depend on whether it is internal-element, or one of its nodes is in contact with a physical boundary of the problem. By using the physical boundary conditions in Eqs. (9.2.4) and (9.2.5) the variables α and β are identified as follows.

Natural boundary conditions:

Internal element: heat flux between the adjacent elements is in balance, no further boundary effects need to be considered,

$$\alpha = 0, \quad \beta = 0 \qquad (9.39)$$

Specified flux on a boundary: f_{B0} or f_{BL}

$$\alpha = 0, \quad \beta = \frac{f_{B\_}}{k} \qquad (9.40)$$

TABLE 9.1 Element stiffness matrices and the load vectors for one-dimensional heat transfer as determined from the finite element form of the general second-order ordinary differential equation.

	a(x)	p(x)	q(x)
		One-dimensional heat transfer	
	kA	0	QA
$[k_a]$	$\int_0^{L^{(e)}} kA[B]^T[B]dx'$		
$[k_p]$		[0]	
$\{r_q\}$			$\int_0^{L(e)} QA[N]^T dx'$
		One-dimensional heat transfer from a fin	
	kA	$-hp$	$QA + hpT_\infty$
$[k_a]$	$\int_0^{L^{(e)}} kA[B]^T[B]dx'$		
$[k_p]$		$-\int_0^{L^{(e)}} (-hP)[N]^T[N]dx'$	
$\{r_q\}$			$\int_0^{L^{(e)}} (QA + hPT_\infty)[N]^T dx'$

Convective heat flux on a boundary: $f_{B_-} = h(T - T_\infty)$

$$\alpha = -\frac{h}{k}, \quad \beta = \frac{hT_\infty}{k} \tag{9.41}$$

Essential boundary conditions:

$$T \text{ is specified therefore } \alpha = 0, \quad \beta = 0 \tag{9.42}$$

Table 9.1 gives the form of the stiffness matrices and loading vectors and Table 9.2 gives the form of the boundary conditions for the one-dimensional heat transfer problems.

9.5 Element assembly

Detailed explanation of the element assembly process is given in Section 4.6. There we demonstrate that in order to represent the original boundary value problem in the global solution domain Ω defined in $0 < x < L$ the temperature and heat flow must be continuous across element boundaries, that is, at the nodes. In fact element assembly represents the equilibrium of the internal heat flow on the nodes. The nodal heat flow can be found from the element equilibrium Eq. (9.25) as follows:

$$\{r_i\}^{(e)} = \left([k_k]^{(e)} + [k_h]^{(e)}\right)\{d_T\} - \{r_q\}^{(e)} \tag{9.43}$$

TABLE 9.2 The $[k_\alpha]$ matrix and the load vectors for the one-dimensional heat transfer problem as deduced form the finite element form of a general second-order ordinary differential equation. Note that if the element shares a node with a boundary ($x = 0$ or L) the equations need to be modified accordingly.

	α	ß
EBC temperature specified	0	0
$[k_\alpha]$	[0]	
$\{r_\beta\}$		{0}
Specified flux at x = 0	0	$-\frac{f_{B0}}{k}$
$[k_\alpha]$	[0]	
$\{r_\beta\}$		$\begin{Bmatrix} -f_{B0}A_0 \\ 0 \end{Bmatrix}$
Specified flux at x = L	0	$-\frac{f_{BL}}{k}$
$[k_\alpha]$		[0]
$\{r_\beta\}$		$\begin{Bmatrix} 0 \\ -f_{BL}A_L \end{Bmatrix}$
Convection at x = 0	$-\frac{h}{k}$	$\frac{hT_\infty}{k}$
$[k_\alpha]$	$-\begin{bmatrix} -hA_0 & 0 \\ 0 & 0 \end{bmatrix}$	
$\{r_\beta\}$		$\begin{Bmatrix} hA_0T_\infty \\ 0 \end{Bmatrix}$
Convection at x = L	$-\frac{h}{k}$	$\frac{hT_\infty}{k}$
$[k_\alpha]$	$-\begin{bmatrix} 0 & 0 \\ 0 & -hA_L \end{bmatrix}$	
$\{r_\beta\}$		$\begin{Bmatrix} 0 \\ hA_LT_\infty \end{Bmatrix}$

As shown in Section 4.6, the sum of heat flow at the *internal nodes* must be zero. This can be represented by the following summation over the elements:

$$\sum_{N}^{e=1} \{r_i\}^{(e)} = 0 \tag{9.44}$$

Note that at the boundaries of the solution domain Ω, that is, nodes−1 and −(N+1), there will be a net heat flow. Therefore, a more strict interpretation of this equation would be,

$$\sum_{N}^{e=1} \{r_i\}^{(e)} = \{R_B\} \tag{9.45}$$

where $\{R_B\} = \{q_{B0}0\cdots 0q_{BL}\}^T$ is the global heat flow vector that includes the effects on the boundaries. But, if the elements are adjusted appropriately to include the boundary effects, as shown in this chapter in Eqs. (9.25–9.27) then Eqs. (9.44) and (9.45) can be used without problems. By using Eq. (9.43) this equation becomes,

$$\sum_{N}^{e=1}\left[\left([k_k]^{(e)}+[k_h]^{(e)}\right)\{d_T\}-\{r_q\}^{(e)}\right]=0 \tag{9.46}$$

Thus we can see that the element summation as described in Eq. (9.46) represents the balance of heat flow at each internal node. This summation results in the assembly of the global equilibrium equation. Let's consider a linear solution domain discretized into N-elements and $N+1$ nodes as shown in Fig. 9.1. Upon assembly the global equilibrium equation will be in the following form:

$$\begin{bmatrix} k_{11}^{(1)} & k_{12}^{(1)} & 0 & \cdots & & & 0 \\ k_{21}^{(1)} & \left(k_{22}^{(1)}+k_{11}^{(2)}\right) & k_{12}^{(2)} & 0 & \cdots & & 0 \\ 0 & k_{21}^{(1)} & \left(k_{22}^{(1)}+k_{11}^{(1)}\right) & k_{13}^{(2)} & 0 & \cdots & 0 \\ \vdots & 0 & & & & & \\ & \vdots & \ddots & \ddots & \ddots & & \\ & & k_{21}^{(i)} & \left(k_{22}^{(i)}+k_{11}^{(i+1)}\right) & k_{12}^{(i+1)} & & 0 \\ & & 0 & & & & \\ & & & \ddots & \ddots & \ddots & \\ & & & & k_{21}^{(N-1)} & \left(k_{22}^{(N-1)}+k_{11}^{(N)}\right) & k_{12}^{(N)} \\ & & & & & k_{21}^{(N)} & k_{22}^{(N)} \end{bmatrix} \begin{Bmatrix} T_1 \\ T_2 \\ T_3 \\ \vdots \\ T_i \\ T_{i+1} \\ T_{i+2} \\ \vdots \\ T_{N-1} \\ T_N \\ T_{N+1} \end{Bmatrix}$$

$$= \begin{pmatrix} r_{q_1}^{(1)} \\ r_{q_2}^{(1)}+r_{q_2}^{(2)} \\ r_{q_3}^{(2)}+r_{q_3}^{(3)} \\ \vdots \\ r_{q_i}^{(i-1)}+r_{q_i}^{(i)} \\ r_{q_{i+1}}^{(i)}+r_{q_{i+1}}^{(i+1)} \\ r_{q_{i+2}}^{(i+1)}+r_{q_{i+2}}^{(i+2)} \\ \vdots \\ r_{N-1}^{(N-2)}+r_{q_{N-1}}^{(N-1)} \\ r_{q_N}^{(N-1)}+r_{q_N}^{(N)} \\ r_{q_{N+1}}^{(N)} \end{pmatrix} \tag{9.47}$$

This equation can be represented in a more compact, symbolic form as follows:

$$[K]\{D_T\} = \{R\} \tag{9.48}$$

where $[K]$ is the *global stiffness matrix*, $\{R\}$ is the *global heat load vector*, and $\{D_T\}$ is the *global degree of freedom vector* which contains the *nodal temperatures* for the entire solution domain.

9.6 Boundary conditions

The global equilibrium equation represents the finite element form of the heat equation for the entire domain, Ω, that spans the range $0 < x < L$. Boundary conditions make the solution unique to a given set of specific conditions on the boundaries $x = 0$ and $x = L$. As introduced in Chapter 3, at a given boundary this boundary value problem can admit *either* an essential *or* a natural boundary condition.

9.6.1 Natural boundary conditions

Admissible form of the natural boundary conditions for the heat equation are given in Eq. (9.4),

$$\text{at } x = 0 \quad : -\left(-k\frac{dT}{dx}\right) = f_{B0} \quad \text{(a) and at } x = L: \quad \left(-k\frac{dT}{dx}\right) = f_{BL} \quad \text{(b)} \tag{9.49}$$

The terms on the *right-hand side* these equations were given in Eqs. (9.30) and (9.31). and are interpreted to take the following forms.

$$\begin{aligned}
&\text{if } f_{B0} = h(T - T_\infty) \text{ at } x = 0: && \begin{bmatrix} hA_0 & 0 \\ 0 & 0 \end{bmatrix}^{(1)} \begin{Bmatrix} T_1 \\ T_2 \end{Bmatrix} + \begin{Bmatrix} hA_0T_\infty \\ 0 \end{Bmatrix} && \text{(a)} \\
&\text{if } f_{B0} \text{ is constant at } x = 0: && \begin{Bmatrix} f_{B0}A_0 \\ 0 \end{Bmatrix} && \text{(b)} \\
&\text{if } f_{BL} = h(T - T_\infty) \text{ at } x = L: && \begin{bmatrix} 0 & 0 \\ 0 & hA_L \end{bmatrix}^{(N)} \begin{Bmatrix} T_N \\ T_{N+1} \end{Bmatrix} + \begin{Bmatrix} 0 \\ hA_LT_\infty \end{Bmatrix} && \text{(c)} \\
&\text{if } f_{BL} \text{ is constant at } x = L: && \begin{Bmatrix} 0 \\ f_{BL}A_L \end{Bmatrix} && \text{(d)}
\end{aligned} \tag{9.50}$$

In case a natural boundary condition of the form (9.49) is specified on one of the boundaries the appropriate equation would be added to the global equilibrium equation to represent its effect on the system.

9.6.2 Essential boundary conditions

Admissible essential boundary conditions on the boundaries are,

$$\text{at}\, x = 0 : T = T_{b0} \quad \text{(a)}, \qquad \text{and at}\, x = L : T = T_{bL} \quad \text{(b)} \tag{9.51}$$

Once the problem is discretized these boundary conditions become,

$$\text{at}\, x = 0 : T_1 = T_{b0} \quad \text{(a)} \quad \text{and at}\, x = L : T_{N+1} = T_{bL} \quad \text{(b)} \tag{9.52}$$

In this case, as the temperature value is specified on a given boundary, it is no longer an unknown and it has to be eliminated from the system of equations. For example, consider the case of $T_1 = T_{b0}$. The global equilibrium equation is modified as follows,

$$\begin{bmatrix} (k_{22}^{(1)} + k_{11}^{(2)}) & k_{12}^{(2)} & 0 & \cdots & & 0 \\ k_{21}^{(2)} & (k_{22}^{(2)} + k_{11}^{(3)}) & k_{12}^{(3)} & 0 & \cdots & 0 \\ \vdots & \ddots & \ddots & \ddots & & \\ & k_{21}^{(i)} & (k_{22}^{(i)} + k_{11}^{(i+1)}) & k_{12}^{(i+1)} & & 0 \\ & 0 & & & & \\ & & \ddots & \ddots & \ddots & \\ & & & k_{21}^{(N-1)} & (k_{22}^{(N-1)} + k_{11}^{(N)}) & k_{12}^{(N)} \\ 0 & 0 & & 0 & k_{21}^{(N)} & k_{21}^{(N)} \end{bmatrix} \begin{Bmatrix} T_2 \\ T_3 \\ \vdots \\ T_i \\ T_{i+1} \\ T_{i+2} \\ \vdots \\ T_{N-1} \\ T_N \\ T_{N+1} \end{Bmatrix}$$

$$= \begin{Bmatrix} r_{q2}^{(1)} + r_{q2}^{(2)} \\ r_{q3}^{(2)} + r_{q3}^{(3)} \\ \vdots \\ r_{qi}^{(i-1)} + r_{qi}^{(i)} \\ r_{q(i+1)}^{(i)} + r_{q(i+1)}^{(i+1)} \\ r_{q(i+2)}^{(i+1)} + r_{q(i+2)}^{(i+2)} \\ \vdots \\ r_{qN-1}^{(N-2)} + r_{qN-1}^{(N-1)} \\ r_{qN}^{(N-1)} + r_{qN}^{(N)} \\ r_{qN+1}^{(N)} \end{Bmatrix} - \begin{Bmatrix} k_{21}^{(1)} T_{b0} \\ 0 \\ \vdots \\ 0 \\ 0 \\ 0 \\ \vdots \\ 0 \\ 0 \\ 0 \end{Bmatrix} \tag{9.53}$$

Note that the size of the global stiffness matrix and the degree of freedom vector are now reduced to $N \times N$ and $N \times 1$, respectively, and the effect of the prescribed essential boundary condition now appears as a loading term on the right-hand side of the equilibrium equation.

The treatment of the cases when $T_{N+1} = T_{bL}$ is specified or the cases when both $T_1 = T_{b0}$ and $T_{N+1} = T_{bL}$ are specified is similar to process demonstrated in

Eq. (9.53). Example 9.1 demonstrates the details of application of both essential and natural boundary conditions in a one-dimensional heat transfer problem.

9.7 Computer implementation

The computer implementation of the assembly and imposition of boundary conditions for a second-order ordinary differential equation is explained in Section 4.8.3. The reader is referred to that section for the details.

Note that the essential boundary conditions can be imposed by using any one of the methods described in Section 5.6. These include: elimination of the prescribed essential boundary condition from the system of equations as described in Section 9.6.2: zeroing of rows and columns of the prescribed degrees of freedom as described in Section 5.6.3.1; and the penalty method as described in Section 5.6.3.2.

Example 9.1 Consider the fin shown in Fig. 9.3 subjected to a constant temperature T_b at $x = 0$ and convection at $x = L$. Note that this fin interacts with the surrounding air through convection along its outer surface. By using three C^0 Lagrange elements of equal length, compute the temperature values at $x = 0$, $L/3$, $2L/3$ and L. Assume that k, h, A, P, L, T_∞, T_b are constant and given as $k = 1000\,\text{W/m·C}, L = 0.15\,\text{m}, A = 0.1\,\text{m}^2, h = 1000\,\text{W/C·m}^2, P = 0.02\,\text{m}, T_b = 100°\text{C}, T_\infty = 20°\text{C}$.

Solution:

BVP: Represented by Eq. (9.8).

BCs: $$T(0) = T_b \text{ and } \left(-k\frac{\partial T}{\partial x}\right)_{x=L} = h(T(L) - T_\infty)$$

Element Eqs. (9.30) or (9.34): $([k_a]+[k_p]+[k_\alpha])\{d\}=\{r_q\}+\{r_\beta\}$

Interpolation: $$T(x') = [N_1 N_2]\begin{Bmatrix} T_1 \\ T_2 \end{Bmatrix}^{(e)}$$

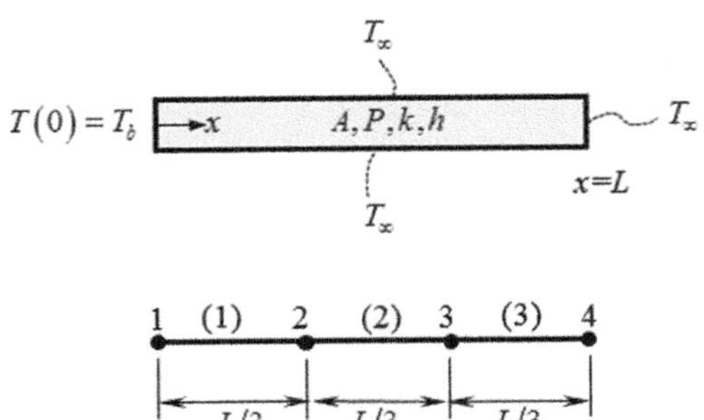

FIGURE 9.3 The solution domain and its finite element representation with three elements.

Conductivity matrix: $[k_k]^{(e)} = \left(\frac{kA}{L}\right)^{(e)} \begin{bmatrix} 1 & -1 \\ -1 & 1 \end{bmatrix}$

Surface convection matrix: $[k_h]^{(e)} = \left(\frac{hPL}{3}\right)^{(e)} \begin{bmatrix} 1 & 1/2 \\ 1/2 & 1 \end{bmatrix}$

Boundary convection matrix

(only for El-3): $[k_{h-bL}]^{(3)} = -\begin{bmatrix} 0 & 0 \\ 0 & -hA \end{bmatrix}$

Internal heat generation vector: $\{r_q\}^{(e)} = \frac{1}{2}(hPL)^{(e)} T_\infty \begin{Bmatrix} 1 \\ 1 \end{Bmatrix}$

Boundary convection vector (only for El-3): $\{r_{h-bL}\}^{(3)} = \begin{Bmatrix} 0 \\ hAT_\infty \end{Bmatrix}$

For $e = 1, 2$:

$$\begin{bmatrix} \frac{kA}{L^{(e)}} + \frac{hPL^{(e)}}{3} & -\frac{kA}{L^{(e)}} + \frac{hPL^{(e)}}{6} \\ -\frac{kA}{L^{(e)}} + \frac{hPL^{(e)}}{6} & \frac{kA}{L^{(e)}} + \frac{hPL^{(e)}}{3} \end{bmatrix} \begin{Bmatrix} T_i \\ T_{i+1} \end{Bmatrix} = \begin{bmatrix} k_{11} k_{12} \\ k_{21} k_{22} \end{bmatrix} \begin{Bmatrix} T_i \\ T_{i+1} \end{Bmatrix}$$

$$= \frac{1}{2}(hPLT_\infty)^{(e)} \begin{Bmatrix} 1 \\ 1 \end{Bmatrix} \quad \text{note that } i = e.$$

For $e = 3$:

$$\begin{bmatrix} \frac{kA}{L^{(e)}} + \frac{hPL^{(e)}}{3} & -\frac{kA}{L^{(e)}} + \frac{hPL^{(e)}}{6} \\ -\frac{kA}{L^{(e)}} + \frac{hPL^{(e)}}{6} & \frac{kA}{L^{(e)}} + \frac{hPL^{(e)}}{3} + hA \end{bmatrix} \begin{Bmatrix} T_3 \\ T_4 \end{Bmatrix}$$

$$= \begin{bmatrix} k_{11} & k_{12} \\ k_{21} & k_{22} + k_\alpha \end{bmatrix} \begin{Bmatrix} T_3 \\ T_4 \end{Bmatrix} = \frac{1}{2}(hPLT_\infty)^{(e)} \begin{Bmatrix} 1 \\ 1 \end{Bmatrix} + \begin{Bmatrix} 0 \\ hAT_\infty \end{Bmatrix}$$

Global energy balance is found by assembling elements 1–3:

$$\begin{bmatrix} k_{11} & k_{12} & 0 & 0 \\ k_{21} & (k_{11} + k_{22}) & k_{12} & 0 \\ 0 & k_{21} & (k_{11} + k_{22}) & k_{12} \\ 0 & 0 & k_{21} & (k_{22} + k_\alpha) \end{bmatrix} \begin{Bmatrix} T_1 \\ T_2 \\ T_3 \\ T_4 \end{Bmatrix} = hT_\infty \begin{Bmatrix} \frac{1}{2} PL^{(e)} \\ PL^{(e)} \\ PL^{(e)} \\ \frac{1}{2} PL^{(e)} + A \end{Bmatrix}$$

where $k_{11} = k_{22} = \frac{kA}{L^{(e)}} + \frac{hPL^{(e)}}{3}$, $k_{21} = k_{12} = -\frac{kA}{L^{(e)}} + \frac{hPL^{(e)}}{6}$, $k_\alpha = hA$

Note that $T_l = T_b$ is specified. We can use any of the methods described in Section 5.6 to impose the boundary condition. Using the procedure outlined in Section 5.6.1 ((5.6.8))

$$\begin{bmatrix} [k_{11}] & \begin{bmatrix} k_{12} & 0 & 0 \end{bmatrix} \\ \begin{bmatrix} k_{21} \\ 0 \\ 0 \end{bmatrix} & \begin{bmatrix} (k_{11} + k_{22}) & k_{12} & 0 \\ k_{21} & (k_{11} + k_{22}) & k_{12} \\ 0 & k_{21} & (k_{22} + k_\alpha) \end{bmatrix} \end{bmatrix} \begin{Bmatrix} T_b \\ T_2 \\ T_3 \\ T_4 \end{Bmatrix} = hT_\infty \begin{Bmatrix} \frac{1}{2} PL^{(e)} \\ PL^{(e)} \\ PL^{(e)} \\ \frac{1}{2} PL^{(e)} + A \end{Bmatrix}$$

We identify the following submatrices. Note that $\{D_c\} = T_b$ in this case.

$$[K_{11}]\{D_c\} + [K_{12}]\{D_x\} = \{R_x\} \qquad \text{(a)}$$
$$[K_{21}]\{D_c\} + [K_{22}]\{D_x\} = \{R_c\} \qquad \text{(b)}$$

From (b) we get

$$[K_{22}]\{D_x\} = \{R_c\} - [K_{21}]\{D_c\}$$

or

$$\begin{bmatrix} (k_{11}+k_{22}) & k_{12} & 0 \\ k_{21} & (k_{11}+k_{12}) & k_{12} \\ 0 & k_{21} & (k_{22}+k_{\alpha}) \end{bmatrix} \begin{Bmatrix} T_2 \\ T_3 \\ T_4 \end{Bmatrix} = hT_{\infty} \begin{Bmatrix} PL^{(e)} \\ PL^{(e)} \\ \frac{1}{2}PL + A \end{Bmatrix} - T_b \begin{Bmatrix} k_{21} \\ 0 \\ 0 \end{Bmatrix}$$

Solution of this system will give the unknown temperatures at nodes-2, -3, and -4.

For the following numerical values, $k = 1000\,\text{W/m}\cdot\text{C}$, $L = 0.05\,\text{m}$, $A = 0.1\,\text{m}^2$, $h = 1000\,\text{W/C}\cdot\text{m}^2$, $P = 0.02\,\text{m}$, $T_b = 100°C$, $T_{\infty} = 20°\text{C}$ we find the solution as follows:

$$\{d_x\} = \begin{Bmatrix} T_2 \\ T_3 \\ T_4 \end{Bmatrix} = \begin{Bmatrix} 98.7197 \\ 97.4437 \\ 96.1720 \end{Bmatrix} °\text{C}.$$

The analytical solution for this problem is found as follows:

$$T(x) = T_{\infty} + \frac{e^{x\lambda}(T_{\infty} - T_b)(h - k\lambda)}{\left(e^{2L\lambda}(h + k\lambda) - (h - k\lambda)\right)} - \frac{e^{2L\lambda - x\lambda}(T_{\infty} - T_b)(h + k\lambda)}{\left(e^{2L\lambda}(h + k\lambda) - (h - k\lambda)\right)}$$

with $\lambda = \left(\frac{hP}{kA}\right)^{1/2}$. Evaluating the temperatures at $x = L/3$, $2L/3$ and L we find 98.7197, 97.4437, and 96.1720 degrees. Thus, we see that even the three-element solution to the problem gives the exact solution within three digit accuracy.

Problems

9.1 Find the finite element equilibrium form of Eq. (9.8) for a two-node element with linearly variable cross-sectional area. Solve the problem for the following two cases and assume all other parameters of the problem to be constant.

a) Rectangular cross-sectional area $A(x') = bh(x')$, where h is the height and b is the depth of the cross section.

b) Circular cross-sectional area.

9.2 Find the finite element equilibrium form of Eq. (9.8) for the three-node element shown in Fig. 9.4. Note that node-2 is located at$x' = L^{(e)}/2$.

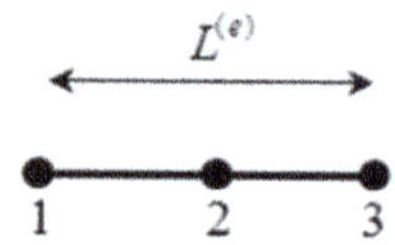

FIGURE 9.4 Problem 9.2.

9.3 Find the finite element equilibrium form of Eq. (9.8) for the four-node element shown in Fig. 9.5. Note that nodes-2 and -3 are located at $x' = L^{(e)}/3$ and $x' = 2L^{(e)}/3$, respectively.

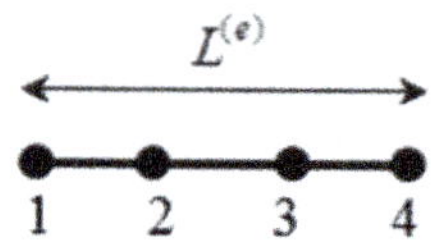

FIGURE 9.5 Problem 9.3.

9.4 Consider a one-dimensional thermoelasticity problem where the thermal expansion of a heated bar is considered (Fig. 9.6). The problem can be formulated by using the equation of equilibrium of an elastic bar and the one-dimensional energy balance in equation.

a) Derive the boundary value problem that represents the steady state equilibrium of an elastic bar subjected to a thermal strain. Assume the bar to be at a uniform initial temperature $T^{(0)}$ and also temperature to vary linearly along the bar. The constitutive behavior of such a bar is given as $\sigma = E(\varepsilon - \varepsilon^{(th)})$, where $\varepsilon^{(th)} = \alpha\Delta T$.

b) Assume that the thermal properties (e.g., α) are not coupled to bar's displacement and express the boundary value problem that represents the steady state energy balance in the bar. Assume the convective heat transfer from the outer surface of the bar is significant.

c) Find the finite element form of both of these problems for a C^0, two-node element.

d) Combine your results in to a single element type as follows:

$$[k_{uT}]^{(e)}\{d_{uT}\} = \{r_{uT}\}^{(e)}$$

where $[k_{uT}]^{(e)}$ is the 4×4element stiffness matrix that represents the coupled stiffness,

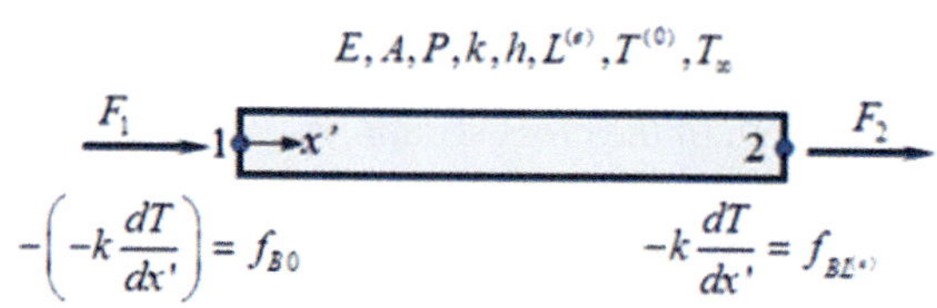

FIGURE 9.6 Problem 9.4.

$\{d_{uT}\} = \{u_1 \quad T_1 \quad u_2 \quad T_2\}^T$ is the nodal degree of freedom vector,
$\{r_{uT}\}^{(e)}$ is the 4×1 mechanical and thermal loading vector.

Note: All the independent parameters of the bar $E, A, P, k, \alpha, h, L^{(e)}$, $T^{(0)}$, T_∞ are constant. The definition of these parameters is the same as those used throughout the text.

9.5 Derive the steady state heat transfer equation for a one-dimensional domain by considering the effects of both convective and radiative heat transfer along its surface (Fig. 9.7).

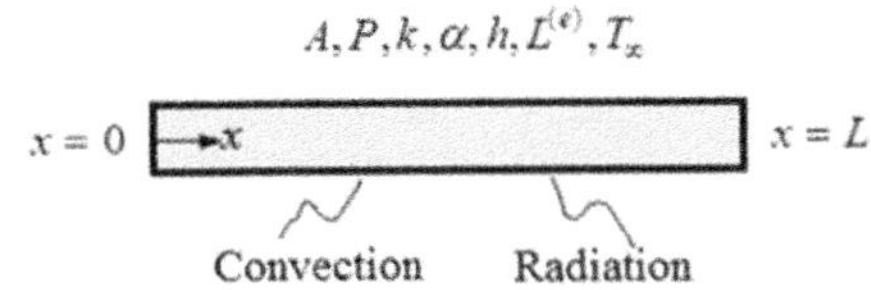

FIGURE 9.7 Problem 9.5.

9.6 Solve Example 9.1 by using the penalty method.

Chapter 10

Heat transfer problems in two-dimensions

10.1 Introduction

Certain geometric and heating conditions allow a three-dimensional problem to be modeled in a two-dimensional domain. The example shown in Fig. 10.1 is one of these cases. The long columns in the ancient Greek city of Ephesus are exposed to solar radiation and convective cooling. If we are interested in the temperature distribution in a general horizontal cross section of these columns, we can use a two-dimensional model as shown in the figure. The regions of the column that are sufficiently far away from its top and bottom surfaces would be expected to experience the same or similar heating and cooling effects.

In this chapter, the finite element formulation of heat transfer in a planar two-dimensional domain $\Omega(x, y)$ is presented. The heat equation and its admissible boundary conditions, presented in Section 2.4, are reintroduced in Section 10.3. The weak form of the two-dimensional heat equation is obtained by using the variational method, in Section 10.4. The finite element approximation for an element with generic interpolation level is derived by using the isoparametric elements, in Section 10.5, and specialized to Q4-element. Natural boundary conditions have a significant effect on the element equilibrium equation: effects of constant flux, convection and radiation are developed in Section 10.6. A summary of the different type of element equilibrium equations thus obtained is provided in Section 10.7. Gauss-quadrature formulas are used to integrate the equilibrium equations as shown in Section 10.8. Element assembly and essential boundary conditions are presented in Sections 10.9 and 10.10, respectively. The formulation shown in chapter can be easily expanded to three-dimensional heat equation.

10.2 Solution domain and its boundary

Boundary conditions play a significant role in the solution of the heat transfer problem. The volume in which heat transfer takes place is called the *solution domain*. In a two-dimensional problem, the solution domain is expressed symbolically as $\Omega = \Omega(x, y)$ and it has a uniform thickness t in the third (z-)

Finite Element Method: Physics and Solution Methods. DOI: https://doi.org/10.1016/B978-0-12-821127-4.00003-7
Copyright © 2022 Elsevier Inc. All rights reserved.

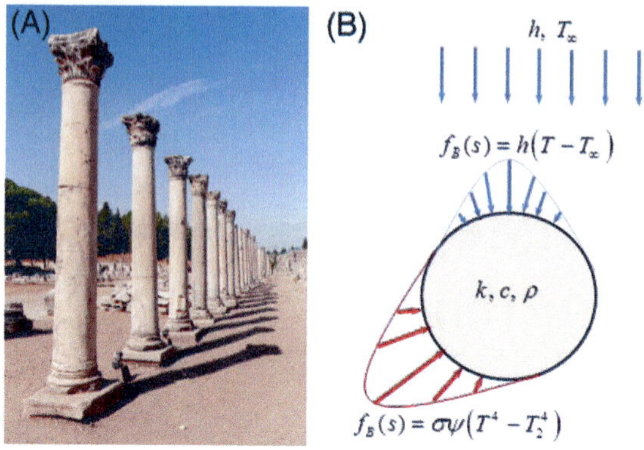

FIGURE 10.1 (A) On its outer boundary, these columns are exposed to solar radiation and convective cooling by the wind coming from the valley. It would be reasonable to assume that the temperature distribution of the mid-height cross section of this column can be modeled by using the two-dimensional heat transfer equation. (B) Schematic depiction of the solar radiation and convective heat flux on this two-dimensional model of heating of the columns. (Image (A) from canstockphoto.com used with permission).

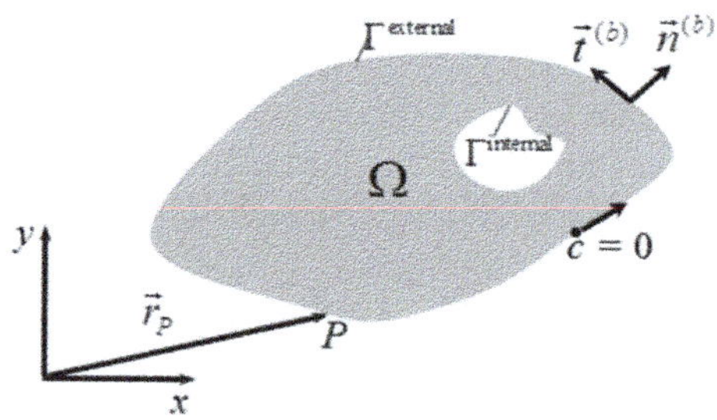

FIGURE 10.2 Two-dimensional domain, Ω (x, y), its boundary Γ (s) and the unit tangent $\vec{t}^{(b)}$ and outward unit normal $\vec{n}^{(b)}$ vectors of the boundary.

direction. A given solution domain has one external boundary and could also have multiple internal boundaries (Fig. 10.2). In a two-dimensional problem, the boundary is a two-dimensional closed curve, typically defined parametrically as $\Gamma(c)$ where c is the coordinate direction running along the boundary, measured relative to an arbitrary datum. The unit tangent $\vec{t}^{(b)}$ and the unit normal $\vec{n}^{(b)}$ vectors along the boundary are expressed in a Cartesian coordinate frame as follows:

$$\underline{t}^{(\mathrm{b})} = t_z\hat{i} + t_y\hat{j} \qquad (a), \vec{n}^{(b)} = n_x\hat{i} + n_y\hat{j} \qquad (\mathrm{b}) \tag{10.1}$$

Consider a point P on the boundary with the position vector,

$$\vec{r}_P = x\hat{i} + y\hat{j} \tag{10.2}$$

expressed in a Cartesian system. As shown in Section 7.2, the tangent $\vec{t}$and normal $\vec{n}$ vectors to the boundary can be found as follows:

$$\begin{aligned} \vec{t}^{(b)} &= \frac{dx}{dc}\hat{i} + \frac{dy}{dc}\hat{j} \qquad &\text{(a)} \\ \vec{n}^{(b)} &= -\frac{dy}{dc}\hat{i} + \frac{dx}{dc}\hat{j} \qquad &\text{(b)} \end{aligned} \tag{10.3}$$

This equation can be approximated numerically to find the tangent and normal vectors on the boundary represented by discrete points.

10.3 The heat equation and its boundary conditions

Heat transfer in a medium in which *thermal conductivity* k (W/mK) is isotropic and heat is generated at a rate of Q (Watts/m^3) is governed by the following Eq. (2.184):

$$Q + \nabla \cdot (k\nabla T) = c\rho \frac{\partial T}{\partial t} \tag{10.4}$$

where c is the heat capacity (J/kgK) and ρ is the mass density (kg/m^3) of the medium. The *del operator* is defined in the two-dimensional Cartesian coordinate system as follows:

$$\nabla(\,) \equiv \frac{\partial}{\partial x}(\,)\hat{i} + \frac{\partial}{\partial y}(\,)\hat{j} \tag{10.5}$$

10.3.1 Boundary conditions for heat transfer in two-dimensional domain

In general, the solution domain could be exposed to prescribed temperature and heat flux conditions on its boundary Γ. The segment of the boundary on which the temperature is prescribed is called the *temperature boundary* Γ_T and the segment over which heat flux is prescribed is called the *flux boundary* Γ_f. Note that these two segments have the following properties: $\Gamma_T \cap \Gamma_f = 0$ and $\Gamma = \Gamma_T \cup \Gamma_f$. The *prescribed temperature* is an essential boundary condition and *prescribed heat flux condition* is a natural boundary condition. These conditions are given as follows (Eqs. (2.186) and (2.187)):

$$\text{Essential Boundary Condition:} \quad T(x,t) = T_b(t) \quad \text{on} \quad \Gamma_T \tag{10.6}$$

$$\text{Natural Boundary Condition:} \; -k\nabla T\vec{n}^{(b)} = f_B(t) \quad \text{on} \quad \Gamma_f \tag{10.7}$$

The term on the left-hand side of Eq. (10.7) represents the heat flux *in the solid domain* that is normal to the boundary and it can also be expressed as follows:

$$-k\left(\frac{\partial T}{\partial x}n_x + \frac{\partial T}{\partial y}n_y\right) = f_B \tag{10.8}$$

The heat flux on the boundary f_B can be one or superposition of the following effects as also presented in Sections 2.4 and 9.2:

$$\begin{array}{ll} \text{Specified heat flux on the boundary:} & f_B = f_B(t) \\ \text{Convection on the boundary:} & f_s = h(T - T_\infty) \\ \text{Radiation on the boundary:} & f_B = \sigma\psi\left(T^4 - T_2^4\right). \end{array} \tag{10.9}$$

where h (W/m^2 K) is the *convection coefficient*, T_∞ is the far distance temperature of the fluid in which the domain is submerged, σ is the Stefan–Boltzmann constant, and ψ is the view factor between the two surfaces in radiation heat transfer. Note that, in general, each term in Eq. (10.9) can potentially depend on time if transient effects are taken in to account. If the boundary is insulated we will have$f_B = 0$. Refer to Section 2.4 for more detailed background on these boundary conditions.

10.4 The weak form of heat transfer equation in two dimensions

The weighted residual form of Eq. (10.4) is expressed as follows:

$$\int_\Omega w\left[Q + \nabla\cdot(k\nabla T) - c\rho\frac{\partial T}{\partial t}\right]d\Omega = 0 \tag{10.10}$$

It is useful to recall that the heat flux vector at any point in the domain point is governed by Fourier's law of conduction,

$$\vec{f} = -k\nabla T \tag{10.11}$$

and, the dot product of the *del* operator with the heat flux vector,

$$\nabla\cdot\vec{f} = -\nabla.(k\nabla T) \tag{10.12}$$

represents the *divergence* of heat flux, a measure of the quantity of the heat flux emanating from any point of the vector field.

The divergence theorem (Gauss theorem) in the plane states that *the area integral of the divergence of any continuously differentiable vector is the closed contour integral of the outward normal component of the vector.* Thus if we consider the heat flux, the divergence theorem can be stated as follows:

$$\begin{aligned} &\int_\Omega \nabla\cdot\vec{f}d\Omega = \oint_\Gamma \vec{f}\cdot\vec{n}^{(b)}d\Gamma && \text{(a)} \\ &\int_\Omega \nabla\cdot(-k\nabla T)d\Omega = -\oint_\Gamma k\nabla T\cdot\vec{n}^{(b)}d\Gamma && \text{(b)} \end{aligned} \tag{10.13}$$

The Green–Gauss theorem states that for a continuously differentiable scalar function $w(x,y)$ the following relationship holds,

$$\int_{\Omega} \nabla \cdot (w\vec{f}) d\Omega = \oint_{\Gamma} w\vec{f} \cdot \vec{n}^{(b)} d\Gamma \tag{10.14}$$

By using chain differentiation, the following relationship can be established:

$$\nabla \cdot (w\vec{f}) = (\nabla w) \cdot \vec{f} + w\nabla \cdot \vec{f} \tag{10.15}$$

Therefore, the Green–Gauss theorem in Eq. (10.14) can be expressed by using Eq. (10.15) as follows:

$$\int_{\Omega} w\nabla \cdot \vec{f} d\Omega = \oint_{\Gamma} w\vec{f} \cdot \vec{n}^{(b)} d\Gamma - \int_{\Omega} \nabla w \cdot \vec{f} d\Omega \tag{10.16}$$

By using this result, the weighted residual form of the heat Eq. (10.10) can be expressed as follows:

$$\begin{aligned} \int w\left(\nabla \cdot (k\nabla T) + Q - c\rho\frac{\partial T}{\partial t}\right) d\Omega &= \oint_{\Gamma} w(k\nabla T)\vec{n}^{(b)} d\Gamma \\ -\int_{0} \left(k\nabla w \cdot \nabla T - wQ + c\rho w\frac{\partial T}{\partial t}\right) d\Omega &= 0 \end{aligned} \tag{10.17}$$

Replacing the weight function w with the variation of temperature δT the following relationship is established,

$$\oint_{\Gamma} \delta T(k\nabla T) \cdot \vec{n}^{(b)} d\Gamma - \int_{\Omega} \left(k\nabla\delta T \cdot \nabla T - \delta T Q + c\rho\, \delta T\frac{\partial T}{\partial t}\right) d\Omega = 0 \tag{10.18}$$

As the boundary can be subjected to either an essential or a natural boundary condition we can partition the first term of Eq. (10.18) as follows:

$$\int_{\Gamma_T} \delta T(k\nabla T) \cdot \vec{n}^{(b)} d\Gamma + \int_{\Gamma f} \delta T(k\nabla T) \cdot \vec{n}^{(b)} d\Gamma - \int_{\Omega} \left(k\nabla\delta T \cdot \nabla T - \delta T Q + c\rho\delta T\frac{\partial T}{\partial t}\right) d\Omega \tag{10.19}$$

The variation of specified temperature on Γ_T should be zero as discussed in Chapter 3. Therefore, the first term of Eq. (10.19) can be neglected. On the boundary where the heat flux is prescribed Γ_f we note that the term $k\nabla T.\vec{n}^{(b)}$ in the boundary integral, represents the heat flux $-f_B$ (Eq. (10.7)). Implementing this substitution and after using the commutative property of the variation operator (Section 3.3.4), Eq. (10.19) becomes,

$$\delta\left\{-\int_{\Gamma_f} T f_B \, d\Gamma - \int_{\Omega}\left(\frac{1}{2}k\nabla T\cdot\nabla T - TQ + c\rho T\frac{\partial T}{\partial t}\right)d\Omega\right\} = 0 \quad (10.20)$$

Finally, the following functional for the heat transfer problem can be established:

$$\Pi = \int_{\Omega}\left(\frac{1}{2}k\nabla T\cdot\nabla T - TQ + c\rho T\frac{\partial T}{\partial t}\right)d\Omega + \int_{\Gamma_f} T f_B \, d\Gamma \quad (10.21)$$

10.5 The finite element form of the two-dimensional heat transfer problem

The functional Π associated with transient heat transfer in a two-dimensional domain is given in Eq. (10.21). In order to find the finite element equations for a two-dimensional heat transfer we express Eq. (10.21) over the element domain $\Omega^{(t)}$. We also assume that all boundaries (edges) of the element are subjected to the natural boundary conditions. Thus we find,

$$\Pi^{(e)} = \int_{\Omega^{(e)}}\left[\frac{1}{2}k\left(\left(\frac{\partial T}{\partial x}\right)^2 + \left(\frac{\partial T}{\partial y}\right)^2\right) - TQ + c\rho T\frac{\partial T}{\partial t}\right]d\Omega^{(t)} + \oint_{\Gamma^{(e)}} T f_B d\Gamma^{(e)} \quad (10.22)$$

Note that the temperature variation across the element is also a function of time. Thus, we have $T = T(x, y, t)$. The spatial variation of temperature across the element can be interpolated by using any one of the two-dimensional interpolations introduced in Section 7.5, as follows:

$$T \simeq \tilde{T} = \sum_{i=1}^{M} N_i T = [N]\{d_T\} \quad (10.23)$$

where N_i are the shape functions, T_i are the nodal temperatures, M is the number of nodes of the element, $[N]$ is the shape function matrix, and $\{d_T\}$ is the nodal temperature vector. In what follows, we will drop the approximation sign on the temperature and use the symbol T for the sake of simplicity. Note that in this representation, that is, Eq. (10.23), the nodal temperatures represent the transient nature of the solution, that is, $T_i = T_i(t)$. The time rate of change of temperature can then be found as follows:

$$\frac{\partial T}{\partial t} = [N]\frac{\partial}{\partial t}\{d_T\} = [N]\{\dot{d}_T\} \quad (10.24)$$

where the time derivative is expressed by the following notation, $\partial(T_i)/\partial t = \dot{T}_i$.

The temperature gradient vector ∇T can be expressed in finite element form as follows:

$$\nabla T = \{T_\partial\} = \left\{ \frac{\partial T}{\partial x} \quad \frac{\partial T}{\partial y} \right\}^T = [B]\{d_T\} \tag{10.25}$$

where $[B]$ is the numerical gradient operator matrix that will be introduced below. Next we define a *heat transfer coefficient matrix*,

$$[K] = k[I] \tag{10.26}$$

where $[I]$ is a 2×2 identity matrix. The heat transfer functional in Eq. (10.22) can then be expressed in the finite element form, as follows:

$$\Pi^{(e)} \int_{\Omega^{(e)}} \frac{1}{2}\{T_\partial\}^T[K]\{T_\partial\}d\Omega^{(e)} - \int_{\Omega^{(e)}} \{d_T\}^T[N]^T Q\, d\Omega^{(c)} + \oint_{\Gamma_f^{(e)}} \{d_T\}^T[N]^T f_B\, d\Gamma^{(e)}$$
$$+ \int_{\Omega^{(e)}} c\rho\{d_T\}^T[N]^T[N]\{\dot{d}_T\}d\Omega^{(e)}$$

By using Eq. (10.25) and taking the variational derivative Eq. (10.26) we find,

$$\delta\Pi^{(e)} = \{\delta d_T\}^T \left\{ \left[\int_{\Omega^{(e)}} [B]^T[K][B]d\Omega^{(e)} \right]\{d_T\} - \int_{\Omega^{(e)}} [N]^T Q\, d\Omega^{(e)} \right.$$
$$\left. + \oint_{\Gamma_f^{(e)}} [N]^T f_B\, d\Gamma^{(e)} + \left[\int_{\Omega^{(e)}} c\rho[N]^T[N]d\Omega^{(e)} \right]\{\dot{d}_T\} \right\} = 0 \tag{10.27}$$

Eliminating the $\{\delta d_T\}^T$ vector, which represents arbitrary variations of nodal temperatures, the finite element form of the two-dimensional heat equation can be found as follows:

$$\left[\int_{\Omega^{(e)}} [B]^T[\kappa][B]d\Omega^{(e)} \right]\{d_T\} + \left[\int_{\Omega^{(e)}} c\rho[N]^T[N]d\Omega^{(e)} \right]\{\dot{d}_\tau\} = \int_{\Omega^{(e)}} [N]^T d\Omega^{(e)} - \oint_{\Gamma_f^{(e)}} [N]^T f_B\, d\Gamma^{(e)} \tag{10.28}$$

This equation can be written in the following compact form,

$$[k_k]^{(e)}\{d_T\} + [c]^{(e)}\{\dot{d}_T\} = \{r_Q\}^{(e)} + \{r_b\}^{(e)} \tag{10.29}$$

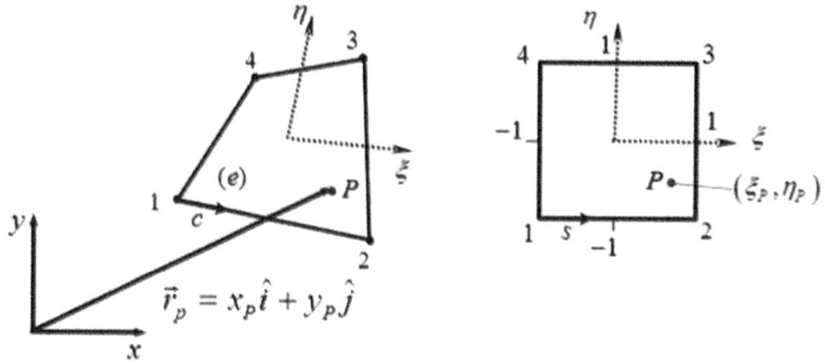

FIGURE 10.3 The Q4 element in physical and mapped coordinate systems.

with

$$
\begin{aligned}
&\text{Conductivity matrix:} && [k_k]^{(e)} = \int\limits_{\Omega^{(e)}} [B]^T[\kappa][B]\, d\Omega^{(e)} && \text{(a)}\\
&\text{Specific heat capacity matrix:} && [c]^{(e)} = \int\limits_{\Omega^{(e)}} c\rho[N]^T[N]\, d\Omega^{(e)} && \text{(b)}\\
&\text{Internal heat generation vector:} && \{r_Q\}^{(e)} = \int\limits_{\Omega^{(e)}} [N]^T Q\, d\Omega^{(e)} && \text{(c)}\\
&\text{Boundary heat flow vector:} && \{r_b\}^{(e)} = -\oint\limits_{\Gamma_f^{(e)}} [N]^T f_B\, d\Gamma^{(e)} && \text{(d)}
\end{aligned}
\tag{10.30}
$$

Note that Eq. (10.30d) should be modified depending on the specified heat flux boundary condition (Eq. (10.9)). This modification will be handled in the next sections.

10.5.1 Finite element form with linear, quadrilateral (Q4) element

Details of the derivation of this element are very similar to the Q4 elements used for two-dimensional elasticity problems (Fig. 10.3). In this formulation, coordinate mapping between the physical and mapped coordinate systems is provided by,

$$
\begin{aligned}
x &= N_1x_1 + N_2x_2 + N_3x_3 + N_4x_4 && \text{(a)}\\
y &= N_2y_1 + N_2y_2 + N_3y_3 + N_4y_4 && \text{(b)}
\end{aligned}
\tag{10.31}
$$

where N_i are the Q4 shape functions given in Eq. (7.61), (x_i, y_i) are the coordinates of the corners of the element. The nodes $i = 1 - 4$ are shown in Fig. 10.3. The temperature variation inside the element is interpolated as follows:

$$
T = \sum_{i=1}^{4} N_i T_i = [N]\{d_T\}
\tag{10.32}
$$

The shape function matrix $[N]$ and the shape functions for this element are as follows:

$$[N] = \begin{bmatrix} N_1 & N_2 & N_3 & N_4 \end{bmatrix} \quad \text{(a)}$$

$$N_1 = \frac{1}{4}(1-\xi)(1-\eta),\ N_2 = \frac{1}{4}(1+\xi)(1-\eta), \quad \text{(b, c)}$$

$$N_3 = \frac{1}{4}(1+\xi)(1+\eta),\ N_4 = \frac{1}{4}(1-\xi)(1+\eta) \quad \text{(d, e)}$$

$$(10.33)$$

The temperature gradient vector is evaluated by using the isoparametric element formulation presented in Section 7.7.1, as follows:

$$\begin{Bmatrix} \dfrac{\partial T}{\partial \xi} \\ \dfrac{\partial T}{\partial \eta} \end{Bmatrix} = \begin{bmatrix} \dfrac{\partial x}{\partial \xi} & \dfrac{\partial y}{\partial \xi} \\ \dfrac{\partial x}{\partial \eta} & \dfrac{\partial y}{\partial \eta} \end{bmatrix} \begin{Bmatrix} \dfrac{\partial T}{\partial x} \\ \dfrac{\partial T}{\partial y} \end{Bmatrix} \tag{10.34}$$

The matrix in this relationship is the transpose of the *Jacobian matrix* defined in Eq. (7.85), which can be calculated by using the mapping defined in Eq. (10.31) as follows:

$$[J]^T = \begin{bmatrix} J_{11} & J_{21} \\ J_{12} & J_{22} \end{bmatrix} = \begin{bmatrix} \dfrac{\partial x}{\partial \xi} & \dfrac{\partial y}{\partial \xi} \\ \dfrac{\partial x}{\partial \eta} & \dfrac{\partial y}{\partial \eta} \end{bmatrix} = \begin{bmatrix} \sum_{i=1}^{4} \dfrac{\partial N_i}{\partial \xi} x_i & \sum_{i=1}^{4} \dfrac{\partial N_i}{\partial \xi} y_i \\ \sum_{i=1}^{4} \dfrac{\partial N_i}{\partial \eta} x_i & \sum_{i=1}^{4} \dfrac{\partial N_i}{\partial \eta} y_i \end{bmatrix} \tag{10.35}$$

The temperature gradient with respect to the global coordinates can be found by inverting Eq. (10.34),

$$\begin{Bmatrix} \dfrac{\partial T}{\partial x} \\ \dfrac{\partial T}{\partial y} \end{Bmatrix} = [J]^{-T} \begin{Bmatrix} \dfrac{\partial T}{\partial \xi} \\ \dfrac{\partial T}{\partial \eta} \end{Bmatrix} = \frac{1}{|J|} \begin{bmatrix} J_{22} & -J_{21} \\ -J_{12} & J_{11} \end{bmatrix} \begin{Bmatrix} \dfrac{\partial T}{\partial \xi} \\ \dfrac{\partial T}{\partial \eta} \end{Bmatrix} \tag{10.36}$$

where $|J|$ is the determinant of the Jacobian matrix. The temperature gradient with respect to the mapped coordinates can be found from Eq. (10.32) as follows:

$$\begin{Bmatrix} \dfrac{\partial T}{\partial \xi} \\ \dfrac{\partial T}{\partial \eta} \end{Bmatrix} = \begin{bmatrix} \dfrac{\partial N_1}{\partial \xi} & \dfrac{\partial N_2}{\partial \xi} & \dfrac{\partial N_3}{\partial \xi} & \dfrac{\partial N_4}{\partial \xi} \\ \dfrac{\partial N_1}{\partial \eta} & \dfrac{\partial N_2}{\partial \eta} & \dfrac{\partial N_3}{\partial \eta} & \dfrac{\partial N_4}{\partial \eta} \end{bmatrix} \begin{Bmatrix} T_1 \\ T_2 \\ T_3 \\ T_4 \end{Bmatrix} \tag{10.37}$$

Finally, we can find the finite element form of the temperature gradient vector by combining Eqs. (10.36) and (10.37) as follows:

$$\begin{Bmatrix} \dfrac{\partial T}{\partial x} \\ \dfrac{\partial T}{\partial y} \end{Bmatrix} = \frac{1}{|J|} \begin{bmatrix} J_{22} & -J_{21} \\ -J_{12} & J_{11} \end{bmatrix} \begin{bmatrix} \dfrac{\partial N_1}{\partial \xi} & \dfrac{\partial N_2}{\partial \xi} & \dfrac{\partial N_3}{\partial \xi} & \dfrac{\partial N_4}{\partial \xi} \\ \dfrac{\partial N_1}{\partial \eta} & \dfrac{\partial N_2}{\partial \eta} & \dfrac{\partial N_3}{\partial \eta} & \dfrac{\partial N_4}{\partial \eta} \end{bmatrix} \begin{Bmatrix} T_1 \\ T_2 \\ T_3 \\ T_4 \end{Bmatrix} \tag{10.38}$$

The gradient operator matrix $[B]$ introduced in Eq. (10.25) can now be expressed as follows:

$$[B] = \frac{1}{|J|}\begin{bmatrix} J_{22} & -J_{21} \\ -J_{12} & J_{11} \end{bmatrix}\begin{bmatrix} \dfrac{\partial N_1}{\partial \xi} & \dfrac{\partial N_2}{\partial \xi} & \dfrac{\partial N_3}{\partial \xi} & \dfrac{\partial N_4}{\partial \xi} \\ \dfrac{\partial N_1}{\partial \eta} & \dfrac{\partial N_2}{\partial \eta} & \dfrac{\partial N_3}{\partial \eta} & \dfrac{\partial N_4}{\partial \eta} \end{bmatrix} \tag{10.39}$$

and the temperature gradient vector becomes,

$$\{T_\partial\} = [B]\{d_T\} \tag{10.40}$$

The element matrices and vectors defined in Eq. (10.30) can now be evaluated for the Q4 element by using the $[N]$ and $[B]$ matrices defined in Eqs. (10.33) and (10.39). Note that the integrals are evaluated in the mapped coordinate system with the substitutions defined in Section 7.6,

$$d\Omega^{(e)} = t|J|d\xi d\eta \qquad \text{and} \qquad d\Gamma^{(e)} = tdc = tJds$$

where J is the Jacobian along the edge as given in Eq. (7.92) and s is the coordinate axis running along the edge of the nondimensional (mapped) element as given in Eq. (7.88). With these variables defined the components of the element equilibrium equations are expressed as follows:

$$\begin{aligned}
&\text{Conductivity matrix:} && [k_k]^{(e)} = \int_{-1}^{1}\int_{-1}^{1} [B]^T[\kappa][B]t|J|d\xi d\eta && \text{(a)} \\
&\text{Specific heat capacity matrix:} && [c]^{(e)} = \int_{-1}^{1}\int_{-1}^{1} c\rho[N]^T[N]t|J|d\xi d\eta && \text{(b)} \\
&\text{Internal heat generation vector:} && \{r_Q\}^{(e)} = \int_{-1}^{1}\int_{-1}^{1} [N]^T Q t|J|d\xi d\eta && \text{(c)} \\
&\text{Boundary heat flow vector:} && \{r_b\}^{(e)} = -\oint_{\Gamma^{(e)}} [N]^T f_B\, t\, J\, ds && \text{(d)}
\end{aligned} \tag{10.41}$$

Eq. (10.41) is evaluated numerically by using Gauss quadrature, introduced in Section 7.8. Before the numerical integration of this equation is introduced we will present the effects of natural boundary conditions on the equilibrium equations of the element.

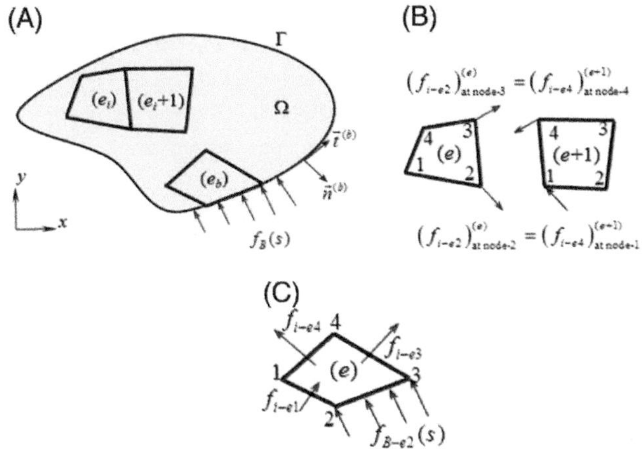

FIGURE 10.4 (A) Heat flux $f_B(s)$ is specified on the boundary of the solution domain. Elements e_i and $e_i + 1$ are internal to Ω, and element e_b shares a node with the boundary. (B) The heat flow through the adjacent edges of two internal elements balances out. (C) The heat flux acting on the boundaries of the element consist of f_{B-e2} on edge-e2 that coincides with the external boundary and the internal heat flux values f_{i-e1}, f_{i-e3}, and f_{i-e4} through edges-e1, –e3, and –e4.

10.6 Natural boundary conditions

The finite element form of the heat Eq. (10.29) can be expressed as follows:

$$[k_k]^{(e)}\{d_T\} + [c]^{(e)}\{\dot{d}_T\} = \{r_Q\}^{(e)} + \{r_b\}^{(e)} \tag{10.42}$$

Heat transfer through the boundary of an element is expressed through the $\{r_b\}$ vector in Eq. (10.41d). Similar to the treatment in Section 7.7, each edge of the element is assessed and integrated separately to give,

$$\{r_b\}^{(e)} = -\oint_{\Gamma^{(e)}} [N]^T f_B\, t\, ds = \{r_{b-ei}\}^{(e)} + \sum_{ej=1, ej\neq ei}^{4} \{r_{i-ej}\}^{(e)} \tag{10.43}$$

where $\{r_{b-ei}\}^{(e)}$ is the *consistent nodal heat flow vector* due to heat flow along an external edge-*ei* and $\{r_{i-ej}\}^{(e)}$ is the *consistent nodal heat flow vector* due to heat flow along an internal edge-*ej*.

If an element is internal to the solution domain, all of its edges are of type $\{r_{i-ej}\}^{(e)}$ (Fig. 10.4B). Heat flow through these edges is balanced with heat flow through the corresponding edges of the adjacent elements. And, as mentioned several times in the previous chapters, this balance forms the basis of element assembly.

If an element shares one or more of its edges with the external boundary Γ of the solution domain, the edges could be exposed to natural boundary conditions, that is, external heat flux. Such an edge gives rise to $\{r_{b-ei}\}^{(e)}$ vector (Fig. 10.4C).

Unlike the solid mechanics problem (Section 7.7) the finite element representation of the $\{r_{b-ei}\}^{(e)}$ depends strongly on the type of flux condition (Eq. (10.9)) specified on the boundary. Edges-1 through -4 of the element are defined in the mapped coordinates as follows:

$$\begin{aligned} &\text{edge–1, e1: } (\xi, -1) && \text{(a)} \\ &\text{edge–2, e2: } (1, \eta) && \text{(b)} \\ &\text{edge–3, e3: } (\xi, 1) && \text{(c)} \\ &\text{edge–4, e4: } (-1, \eta) && \text{(d)} \end{aligned} \tag{10.44}$$

On the boundary the heat flux vector will take different forms depending on the boundary conditions.

10.6.1 Internal edges

In case all the edges of an element are internal to the solution domain, heat flowing out from an edge will be balanced by the heat flowing into the next element that is sharing the same edge. As mentioned before this energy balance forms the basis of element assembly. The internal heat flow will ultimately disappear in the global equilibrium equations upon assembly. Here we present the *nodal heat flow vector* for heat flow along an internal edge-ei as $\{\Gamma_{i-ei}\}^{(e)}$. The finite element form of the heat equation can be simply expressed for an internal element as follows:

$$[k_k]^{(e)}\{d_T\} + [c]^{(e)}\{\dot{d}_T\} = \{r_Q\}^{(e)} + \{r_i\}^{(e)} \tag{10.45}$$

where

$$\{r_i\}^{(e)} = \sum_{ei=1}^{4} \{r_{i-ei}\}^{(e)} \tag{10.46}$$

10.6.2 External edges subjected to prescribed heat flux

In case one of the edges of an element is coincident with a boundary on which a constant heat flux is prescribed, Eq. (10.41d) should be used to compute the consistent nodal heat flow vector. The prescribed heat flux can be constant or variable along a given edge. By assuming that it varies linearly along the edge we can find the following relationships to represent $f_{B-ei}(s)$:

$$\begin{aligned} &\text{edge–e1}, \eta = -1: && f_{B-e1}(\xi, -1) = [N]\{f_{B1} \quad f_{B2} \quad 0 \quad 0\}^T && \text{(a)} \\ &\text{edge–e2}, \xi = 1: && f_{B-e2}(1, \eta) = [N]\{0 \quad f_{B2} \quad f_{B3} \quad 0\}^T && \text{(b)} \\ &\text{edge–e3}, \eta = 1: && f_{B-e3}(\xi, 1) = [N]\{0 \quad 0 \quad f_{B3} \quad f_{B4}\}^T && \text{(c)} \\ &\text{edge–e4}, \xi = -1: && f_{B-e4}(-1, \eta) = [N]\{f_{B1} \quad 0 \quad 0 \quad f_{B4}\}^T && \text{(d)} \end{aligned} \tag{10.47}$$

where f_{Bi} are the nodal values of the heat flux vector (Fig. 10.5). The *consistent nodal heat flow vector* for edge-*ei* when it is exposed to a prescribed heat flux

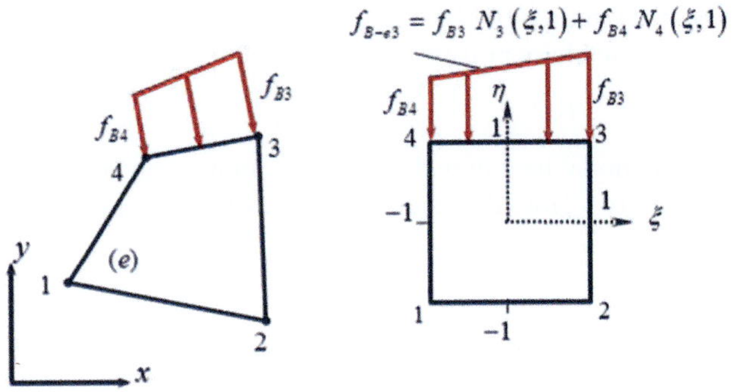

FIGURE 10.5 Linearly varying heat flux f_{B-e3} prescribed along edge-e3 of the element. The linear variation can be expressed by using the values of the prescribed heat flux at nodes-3 and -4 and the interpolation as shown in the figure.

can be then computed by using Eq. (10.41d),

$$\{r_{b-ei}\}^{(e)} = -\int_{-1}^{1} [N]^T f_{B-ei} tJds \tag{10.48}$$

where s and J are defined in Eqs. (7.87) and (7.92), respectively. For example, Fig. 10.5 shows an element subjected to linearly varying prescribed heat flux along its edge-e3. In this case we will find,

$$f_{B-e3}(\xi,1) = \begin{bmatrix} N_1(\xi,1) & N_2(\xi,1) & N_3(\xi,1) & N_4(\xi,1) \end{bmatrix}\{r_{B-e3}\} \tag{10.49}$$

where $\{0 \;\; 0 \;\; f_{B3} \;\; f_{B4}\}^T$ and $f_{B-e3}(\xi,1) = f_{B3}N_3(\xi,1) + f_{B4}N_4(\xi,1)$.

The $\{r_{b-e3}\}^{(e)}$ vector for this example can be found as follows, by using $(\xi,\eta) = (\xi,1)$:

$$\{r_{b-e3}\}^{(e)} = -\int_{-1}^{1} \{0 \quad 0 \quad N_3(f_{B3}N_3 + f_{B4}N_4) \quad N_4(f_{B3}N_3 + f_{B4}N_4)\}^T tJd\xi \tag{10.50}$$

In practice, this equation should be *integrated numerically*.

By using Eqs. (10.42) and (10.43) the finite element form of the heat equation for element-e is found as follows:

$$[k_k]^{(e)}\{d_T\} + [c]^{(e)}\{\dot{d}_T\} = \{r_Q\}^{(e)} + \{r_{b-ei}\}^{(e)} + \{r_i\}^{(e)} \tag{10.51}$$

with

$$\{r_i\}^{(e)} = \sum_{ej=1,(ej\neq ei)}^{4} \{r_{i-ej}\}^{(e)} \tag{10.52}$$

10.6.3 External edges subjected to convection

In case the edge of the element is coincident with a boundary where heat transfer takes place through convection, the heat flux through the boundary follows $f_B =$

$h(T - T_\infty)$ as given in Eq. (10.9b). The convective heat flux on the boundary can be expressed in finite element form as follows:

$$f_B = h([N]\{d_T\} - T_\infty) \tag{10.53}$$

The consistent nodal heat flow due to convection along an edge can be found by using Eq. (10.41d). Note that each edge needs to be considered separately.

$$\begin{aligned}
&\text{edge–e1}, \ \eta - 1: && \{r_{b-ei}\}^{(e)} = -\int_{-1}^{1} h\,[N]^T\,([N]\{d_T\} - T_\infty)\,J\,t\,d\xi && \text{(a)}\\
&\text{edge–e2}, \ \xi = 1: && \{r_{b-ei}\}^{(e)} = -\int_{-1}^{1} h\,[N]^T\,([N]\{d_T\} - T_\infty)\,J\,t\,d\eta && \text{(b)}\\
&\text{edge–e3}, \ \eta = 1: && \{r_{b-ei}\}^{(e)} = -\int_{-1}^{1} h\,[N]^T\,([N]\{d_T\} - T_\infty)\,J\,t\,d\xi && \text{(c)}\\
&\text{edge–e4}, \ \xi = -1: && \{r_{b-ei}\}^{(e)} = -\int_{-1}^{1} h\,[N]^T\,([N]\{d_T\} - T_\infty)\,J\,t\,d\eta && \text{(d)}
\end{aligned} \tag{10.54}$$

where J is given in Eq. (7.92).

In general, the *consistent nodal heat flow vector* due to *convection heat flux* along edge-*ei* (Eq. (10.54)) is expressed in finite element form as follows:

$$\{r_{b-ei}\}^{(e)} = -[k_{h-ei}]^{(e)}\{d_T\} + \{r_{h-ei}\}^{(e)} \tag{10.55}$$

where the boundary convection matrix and boundary convection vectors on the appropriate boundaries are computed as follows:

$$\begin{aligned}
&\text{edge–e1}, \ \eta = -1: && [k_{h-e1}]^{(e)} = \int_{-1}^{1} h\,[N]^T\,[N]\,J\,t\,d\xi && \text{and} && \{r_{h-e1}\}^{(e)} = \int_{-1}^{1} hT_\infty\,[N]^T\,J\,t\,d\xi\\
&\text{edge–e2}, \ \xi = 1: && [k_{h-e2}]^{(e)} = \int_{-1}^{1} h\,[N]^T\,[N]\,J\,t\,d\eta && \text{and} && \{r_{h-e2}\}^{(e)} = \int_{-1}^{1} hT_\infty\,[N]^T\,J\,t\,d\eta\\
&\text{edge–e3}, \ \eta = 1: && [k_{h-e3}]^{(e)} = \int_{-1}^{1} h\,[N]^T\,[N]\,J\,t\,d\xi && \text{and} && \{r_{h-e3}\}^{(e)} = \int_{-1}^{1} hT_\infty\,[N]^T\,J\,t\,d\xi\\
&\text{edge–e4}, \ \xi = -1: && [k_{h-e4}]^{(e)} = \int_{-1}^{1} h\,[N]^T\,[N]\,J\,t\,d\eta && \text{and} && \{r_{h-e4}\}^{(e)} = \int_{-1}^{1} hT_\infty\,[N]^T\,J\,t\,d\eta
\end{aligned} \tag{10.56}$$

By using Eq. (10.6.1) the finite element form of the heat equation for element-*e* is found as follows:

$$\left([k_k]^{(e)} + [k_{h-ei}]^{(e)}\right)\{d_T\} + [c]^{(e)}\{\dot{d}_T\} = \{r_q\}^{(e)} + \{r_{h-ei}\}^{(e)} + \{r_i\}^{(e)} \tag{10.57}$$

where the *internal heat flow* vector is,

$$\{r_i\}^{(e)} = \sum_{ej=1,(ej\neq ei)}^{4} \{r_{i-ej}\}^{(e)} \tag{10.58}$$

This vector will disappear upon element assembly due to energy conservation. Eq. (10.57) shows that the element stiffness matrix is augmented by the $[k_{h-ei}]^{(e)}$

when convection boundary condition is applied on edge-ei. For example for edge-2 the $[k_{h-e2}]^{(e)}$ matrix can be computed as follows:

$$[k_{h-e2}]^{(e)} = \int_{-1}^{1} ht[N(1,\eta)]^T[N(1,\eta)]Jd\eta = \frac{1}{6}htL_{23}\begin{bmatrix} 0 & 0 & 0 & 0 \\ 0 & 2 & 1 & 0 \\ 0 & 1 & 2 & 0 \\ 0 & 0 & 0 & 0 \end{bmatrix} \tag{10.59}$$

It can easily be shown that $J = L_{23}/2$ by using Eqs. (7.88) and (7.92), where L_{23} is the length of edge-e2 in the Cartesian representation of the domain. Note that the zero columns and rows of Eq. (10.59) indicate that the inside nodes (1 and 4) have no contribution to boundary convection as expected. In practice, for a nonrectangular quad-element the $[k_{h-ei}]^{(e)}$ matrix should be computed numerically.

10.6.4 External edges subjected to radiation

In case the edge of the element is coincident with a boundary that is exposed to radiation, the heat flux through the boundary will be $f_B = \psi\sigma(T^4 - T_2^4)$ as given in Eq. (10.9). Note that the first term of this equation depends on the fourth power of the unknown temperature T on the boundary. Therefore, radiative heat flux gives rise to a nonlinear equation. There are several techniques to find the solution of nonlinear system of equations. These are primarily iterative techniques which advance the solution and minimize the error.

In order to express the radiative heat flux equation in finite element form, let's assume that the T^4 can be decomposed into a cubic and a linear term, which are computed at a known iteration level, and an unknown iteration level, respectively: $T^4 = (T^{(l-1)})^3T^{(l)}$, where l represents the iteration level at which the temperature is yet unknown. Then we can express the radiative heat flux along edge-ei, in finite element form, as follows:

$$f_{B-ei} = \sigma\psi T^3[N]\{d_T\} - \sigma\psi T_2^4 \tag{10.60}$$

Note that the iteration level l has been dropped for sake of notational simplification. By using Eq. (10.60) the consistent nodal heat flux vector can be calculated as follows:

$$\{r_{b-ei}\}^{(e)} = -\int_{-1}^{1}[N]^T f_{B-ei}\,t\,J\,ds = -\int_{-1}^{1}(\sigma\psi T^3)[N][N]^T\,t\,ds + \int_{-1}^{1}(\sigma\psi T_2^4)[N]^T t\,J\,ds \tag{10.61}$$

where ds is one of $Jd\xi$ or $Jd\eta$ as described in Eqs. (7.87) and (7.91). This equation can be represented in vector form as follows:

$$\{r_{b-ei}\}^{(e)} = -[k_{r-ei}]^{(e)}\{d_T\} + \{r_{r-ei}\}^{(e)} \tag{10.62}$$

The nodal values of the T^3 term are introduced by using $[I]\{d_T\}^3$ in the computations, where $[I]$ is the 4 × 4 identity matrix. Thus the *radiation matrix* and *radiation vector* for edge-ei are defined, respectively, as follows:

$$[k_{r-ei}(T)]^{(e)} = \int_{-1}^{1} (\sigma\psi[T]\{d_T\}^3)[N][N]^T\, t\, J\, ds \qquad \text{(a)}$$
$$\{r_{r-ei}\}^{(e)} = \int_{-1}^{1} (\sigma\psi T_2^4)[N]^T\, t\, J\, ds \qquad \text{(b)} \qquad (10.63)$$

It is common to represent the radiation matrix as $[k_r(T)]$ to emphasize that the elements of this matrix depend on the yet unknown temperature values, and that the matrix is nonlinear for this reason. By substituting Eq. (10.62) into Eq. (10.43) we find the finite element form of the heat equilibrium if one of the element boundaries is exposed to radiation,

$$\left([k_k]^{(e)} + [k_{r-ei}(T)]^{(e)}\right)\{d_T\} + [c]^{(e)}\{\dot{d}_T\} = \{r_Q\}^{(e)} + \{r_{r-ei}\}^{(e)} + \{r_i\}^{(e)} \qquad (10.64)$$

where $\{r_i\}^{(e)}$ represents the internal heat flow vector defined in Eq. (10.58).

The nonlinearity in Eq. (10.64) is carried into the global stiffness matrix during the assembly process. The resulting nonlinear system of equations can be solved by one of many available iterative solution (e.g., Newton–Raphson) methods if the problem is formulated at steady state, or by an explicit time integration method if the problem is formulated for transient analysis.

10.7 Summary of finite element form of the heat equation and natural boundary conditions

Summary of the finite element form of the heat equation presented in Section 10.6 is given below.

Internal elements: $[k_k]^{(e)}\{d_T\} + [c]^{(e)}\{\dot{d}_T\} = \{r_Q\}^{(e)} + \{r_i\}^{(e)}$ (10.65)

where all the edges are internal and we have,

$$\{r_i\}^{(\theta)} = \sum_{4}^{ei=1} \{r_{i-ei}\}^{(\theta)} \qquad (10.66)$$

Boundary elements:

Element exposed to *prescribed heat flux* along edge-ei:

$$[k_k]^{(e)}\{d_T\} + [c]^{(e)}\{\dot{d}_T\} = \{r_Q\}^{(e)} + \{r_{b-ei}\}^{(e)} + \{r_i\}^{(e)} \qquad (10.67)$$

Element exposed to *convection* along edge-ei:

$$[k_k]^{(e)}[k_{h-ei}]^{(e)}\{d_T\} + [c]^{(e)}\{\dot{d}_T\} = \{r_Q\}^{(e)} + \{r_{h-ei}\}^{(e)} + \{r_i\}^{(e)} \tag{10.68}$$

Elements exposed to *radiation* along edge-ei:

$$\left([k_k]^{(e)} + [k_{r-ei}(T)]^{(e)}\right)\{d_T\} + [c]^{(e)}\{\dot{d}_T\} = \{r_Q\}^{(e)} + \{r_{r-ei}\}^{(e)} + \{r_i\}^{(e)} \tag{10.69}$$

where the internal flow vector is,

$$\{r_i\}^{(e)} = \sum_{ej=1,(ej\neq ei)}^{4} \{r_{i-ej}\}^{(e)}, \tag{10.70}$$

and

Conductivity matrix:	$[k_k]^{(e)}$ is given in Eq. (10.41),
Heat capacity matrix:	$[c]^{(e)}$ is given in Eq. (10.41),
Internal heat generation vector:	$\{r_Q\}^{(e)}$is given in Eq. (10.41),
Prescribed heat flow vector:	$\{r_{b-ei}\}^{(e)}$ is given in Eq. (10.48),
Convection matrix:	$[k_{h-ei}]^{(e)}$ is given in Eq. (10.56),
Convection vector:	$\{r_{h-ei}\}^{(e)}$ is given in Eq. (10.56),
Radiation matrix:	$[k_{r-ei}]^{(e)}$ is given in Eq. (10.63),
Radiation vector:	$\{r_{r-ei}\}^{(e)}$ is given in Eq. (10.63).

10.8 Numerical integration of element equilibrium equations

The element equilibrium equations summarized in Section 10.7 are integrated numerically by using Gauss-quadrature introduced in Section 7.8. By using this method, an integral of the form,

$$I = \int_{-1}^{1} \phi(\xi)d\xi \tag{10.71}$$

can be evaluated approximately as follows:

$$I \cong I_n = \sum_{i=1}^{n} W_i\phi(\xi_i) \tag{10.72}$$

where W_i are scalar values known as *weight factors*, ξ_i are predetermined *Gauss-quadrature points*, and n is the *order of quadrature*. The weights and quadrature points are tabulated values for a given order of quadrature (Table 7.3). Gaussian-quadrature method is designed to evaluate polynomials exactly, as demonstrated in Section 7.8. An integral of the type given in Eq. (10.71) can be integrated exactly by the quadrature formula, that is, Eq. (10.72), with degree n, if the integrand $\phi(\xi)$ is a polynomial of order $2n-1$ [1].

The details of integration of matrices and vectors by Gauss-quadrature are provided in Section 7.9 for the case of solid elements. The implementation of these methods to heat transfer equations is a straightforward extension of the Gauss-quadrature method. Here we will give the Gauss integration formulas for the matrices and vectors summarized in Section 10.7.

Conductivity matrix $[k_k]^{(e)} = \int_{-1}^{1}\int_{-1}^{1} [B]^T[\kappa][B]t|J|d\xi d\eta$

define, $[C]^{(e)} = [B]^T[E][B]|J|t.$ (10.73)

Gauss-quadrature: $k_{k_{ij}}^{(e)} = \sum_{l=1}^{n}\sum_{m=1}^{n} W_l W_m C_{ij}^{(e)}(\xi_l, \eta_m)$

Specific heat capacity matrix $[C]^{(e)} = \int_{-1}^{1}\int_{-1}^{1} c\rho[N]^T[N]t|J|d\xi d\eta$

define, $[C]^{(e)} = c\rho[N]^T[N]|J|t.$ (10.74)

Gauss-quadrature: $c_{ij}^{(e)} = \sum_{l=1}^{n}\sum_{m=1}^{n} W_l W_m C_{ij}^{(e)}(\xi_l, \eta_m)$

Boundary convection matrix $[k_{h-e1}]^{(e)} = \int_{-1}^{1} [N]^T[N]t ds$

where $ds = J\,d\xi$ or $ds = J\,d\eta$

define, $[C]^{(e)} = h[N]^T[N]J\,t.$

Gauss-quadrature: $k_{h_{ij}}^{(e)} = \sum_{l=1}^{n}\sum_{m=1}^{n} W_l W_m C_{ij}^{(e)}(\xi_l, \eta_m)$

(10.75)

Boundary radiation matrix $[k_{r-ei}(T)]^{(e)} = \int_{-1}^{1} (\sigma\psi[T]\{d_T\}^3)[N][N]^T t\,ds$

where $ds = J\,d\xi$ or $ds = J\,d\eta$

define, $[C]^{(e)} = (\sigma\psi[T]\{d_T\}^3)[N][N]^T J\,t$

Gauss-quadrature: $k_{r_{ij}}^{(e)} = \sum_{l=1}^{n}\sum_{m=1}^{n} W_l W_m C_{ij}^{(e)}(\xi_l, \eta_m)$

(10.76)

Internal heat generation vector: $\{r_Q\}^{(e)} = \int_{-1}^{1}\int_{-1}^{1} [N]^T Qt|J|d\xi d\eta$

define, $\{g_Q\}^{(e)} = [N]^T Q|J|t$

Gauss-quadrature: $r_{Qi}^{(e)} = \sum_{l=1}^{n}\sum_{m=1}^{n} W_l W_m g_{Qi}^{(e)}(\xi_l, \eta_m)$

(10.77)

Boundary heat flow vector: $\{r_b\}^{(e)} = -\oint_{\Gamma_f^{(e)}} [N]^T f_B ds^e$ where $\{r_b\}^{(e)}$ is one of $\{f_{b-ei}\}^{(e)}$, $\{r_{h-ei}\}^{(e)}$ or $\{\eta_{r-ei}\}^{(e)}$ given in Eqs. (10.48), (10.56), and (10.63), respectively. As this is a line integral, the appropriate variable ξ or η should be chosen according to the edge-ei on which the integration is evaluated, and similarly, the following substitution should be used $ds = J\,d\xi$ or $ds = J\,d\eta$. Then we can define,

$$\{g_b\}^{(e)} = [N]^T f_B t$$

$$\text{Gauss-quadrature:} \quad r_{bi}^{(e)} = \sum_{l=1}^{n} W_l g_{bi}^{(e)} \tag{10.78}$$

More detail on this integral is provided in in Section 7.9.

As also stated in Section 7.9, the highest order Gauss quadrature should be used for numerical integration to obtain the most accurate results. But, it should also be noted that increased order of quadrature comes at a cost of increased computational expense. For rectangular elements, the Jacobian is a constant and some element types can be integrated exactly. *Full integration* is possible for an eight- or nine-node plane element by using a 3×3 Gauss integration rule, and for a four-node plane element by 2×2 Gauss rule.

10.9 Element assembly

Detailed explanation of the element assembly process is given in Section 4.6 for a general second-order ordinary differential equation and in Section 9.5 for the one-dimensional heat transfer problem. In both of these sections, we demonstrate that in order to represent the original boundary value problem in the global solution domain Ω the temperature and heat flow must be continuous across element boundaries. In fact, we can state that element assembly represents the equilibrium of the internal heat flow. To this end, let's note that internal nodal heat flow vector for an element can be found from the element equilibrium equations presented for different natural boundary conditions in Eqs. (10.65–10.70),

$$\{r_i\}^{(e)} = [k_k]^{(e)}\{d_T\} + [c]^{(e)}\{\dot{d}_T\} - \{r_Q\}^{(e)} \quad \text{(a)}$$

$$\{r_i\}^{(e)} = [k_k]^{(e)}\{d_T\} + [c]^{(e)}\{\dot{d}_T\} - \left(\{r_Q\}^{(e)} + \{r_{b-ei}\}^{(e)}\right) \quad \text{(b)}$$

$$\{r_i\}^{(e)} = \left([k_k]^{(e)} + [k_{h-ei}]^{(e)}\right)\{d_T\} + [c]^{(e)}\{\dot{d}_T\} - \left(\{r_Q\}^{(e)} + \{r_{h-ei}\}^{(e)}\right) \quad \text{(c)}$$

$$\{r_i\}^{(e)} = \left([k_k]^{(e)} + [k_{r-ei}(T)]^{(e)}\right)\{d_T\} + [c]^{(e)}\{\dot{d}_T\} - \left(\{r_Q\}^{(e)} + \{r_{r-ei}\}^{(e)}\right) \quad \text{(d)}$$

$$\tag{10.79}$$

Eq. (10.79) can be expressed in a general element equilibrium form as follows,

$$\{r_i\}^{(e)} = [k]^{(e)}\{d_T\} + [c]^{(e)}\{\dot{d}_T\} - \{r\}^{(e)} \tag{10.80}$$

where the element stiffness matrix $[k]^{(e)}$ and the external loading vector $\{r\}$ can be deduced from Eq. (10.79).

As shown in Section 4.6, the sum of heat flow at the internal nodes must be zero (Fig. 10.4B). This can be represented by the following summation over the elements:

$$\sum_{e=1}^{N} \{r_i\}^{(e)} = 0 \tag{10.81}$$

Note that at the boundaries of the solution domain Γ there will be a net heat flow in to the domain. Therefore, a more strict interpretation of this equation would be,

$$\sum_{e=1}^{N} \{r_i\}^{(e)} = \{R_B\} \tag{10.82}$$

where $\{R_B\}$ is the *global external heat flow vector.* But, if the elements are adjusted appropriately to include the boundary effects, as shown in this chapter in equations Sections 10.6 and 10.7, then Eq. (10.81) can be used. By using Eqs. (10.79) and (10.81) we find,

$$\sum_{e=1}^{N} [k]^{(e)}\{d_T\} + [c]^{(e)}\{\dot{d}_T\} - \{r\}^{(e)} = 0 \tag{10.83}$$

This summation results in the assembly of the global equilibrium equation.

$$[C]\{\dot{D}_T\} + [K]\{D_T\} = \{R\} \tag{10.84}$$

where the *global heat capacity matrix, global stiffness matrix, global heat load vector* are,

$$[C] = \sum_{e=1}^{N} [c]^{(e)},\ [K] = \sum_{e=1}^{N} [k]^{(e)},\ \{R\} = \sum_{e=1}^{N} \{r\}^{(e)} \tag{10.85}$$

and $\{D_T\}$ is the *global degree of freedom vector* which contains the *nodal temperatures* for the entire solution domain.

10.10 Imposing the Essential boundary conditions

Constant temperature boundary condition on the surface has the same effect as the fixed-displacement (prescribed displacement) condition in solid mechanics problems. Since, the temperature is known on that part of the boundary those degrees of freedom are eliminated from the global stiffness matrix. Of course the heat flux associated with these boundary locations can be calculated once the solution has been obtained.

10.10.1 Symbolic representation of essential boundary conditions

In order to visually grasp the implementation of the essential boundary conditions it is helpful to partition the global temperature vector as,

$$\{D_T\} = \begin{Bmatrix} \{D_x\} \\ \{D_B\} \end{Bmatrix} \tag{10.86}$$

where the vector $\{D_x\}$ represents all the unknown nodal temperatures and vector $\{D_B\}$ represents the prescribed temperature values on the boundary. By interchanging the appropriate rows and columns the global stiffness matrix can be repartitioned, and the global equilibrium equation can be expressed as follows:

$$\begin{bmatrix} [K_{XX}][K_{XB}] \\ [K_{BX}][K_{BB}] \end{bmatrix} \begin{Bmatrix} \{D_X\} \\ \{D_B\} \end{Bmatrix} = \begin{Bmatrix} \{R_B\} \\ \{R_X\} \end{Bmatrix} \tag{10.87}$$

where $\{R_B\}$ is the vector of externally applied thermal load and $\{R_X\}$ is the thermal load at the nodes where the temperature is prescribed. The unknown temperature and thermal load vectors can be computed as shown in Eqs. (7.178) and (7.179),

$$\{D_X\} = [K_{XX}]^{-1}(\{R_B\} - [K_{XB}]\{D_B\}) \tag{10.88}$$

$$\{R_X\} = [K_{BX}]\{D_X\} + [K_{BB}]\{R_B\} \tag{10.89}$$

10.10.2 Numerical implementation of essential boundary conditions

In Section 5.6, several techniques are introduced for the numerical implementation of the essential boundary conditions. These include direct assembly of only the active degrees of freedom (Section 5.6.2), row and column manipulations (Section 5.6.3.1), and the penalty method (Section 5.6.3.2). All of these methods can be used to impose the essential boundary conditions on the global stiffness matrix.

Example 10.1 Consider the concrete block ($k = 0.5\,\mathrm{W/m{\cdot}K}$) shown in the figure. The base of this block is kept at 27°C, the two lateral surfaces and the top surface are subjected to constant heat flux with magnitudes of $f_{Bs} = -5\mathrm{W/m^2}$ and $f_{Bt} = 2\,\mathrm{W/m^2}$, respectively. Dimensions of the block are $W_b = 0.5$ m, $W_t = 0.3$ m, $H = 0.5$ m, and $t = 1$ m.

a) Find the steady state temperature distribution in this block by using a 2×2 mesh.

b) Perform a mesh convergence study. Monitor and report the temperature at the top right corner, by using a hierarchically increasing mesh density that keeps the nodes of the less dense meshes in the analysis.

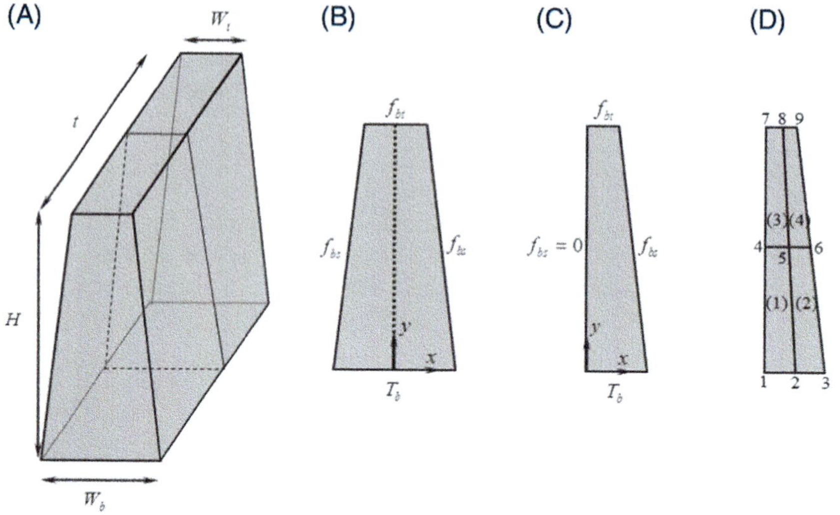

FIGURE 10.6 (A) Dimensions of the rectangular block are shown in the figure. The block is sufficiently deep in the z-direction so that we can assume a two-dimensional heat transfer analysis can be used as shown in (B). The essential and natural boundary conditions applied on the block are also shown in figure (B). As the problem is symmetric in geometry and boundary conditions we will simulate only one half of the problem. The symmetry boundary condition, $f_{bs} = 0$, is shown along the $x = 0$ line in figure (C). In part (A) of the question we use the four elements shown in figure (D).

Solution

a) At steady state, we would expect most of the block's volume to have the same temperature distribution except in a narrow zone at the front and back edges. We will neglect the effects near to front and back edges and use the two-dimensional heat equation, that is representative of a mid-section as shown in Fig. 10.6A.

The geometry and the boundary conditions of the problem are symmetric with respect to the axis $(0, y)$ as shown in Fig. 10.6B. Boundary conditions of the original problem are,

$$\begin{aligned}
&\text{edge–1: } T(x,0) = T_b = 300\,\text{K} && \text{(a)}\\
&\text{edge–2: } -k\nabla T \cdot \vec{n}^{(b)} = f_B = -5\text{W/m}^2 && \text{(b)}\\
&\text{edge–3: } -k\nabla T \cdot \vec{n}^{(b)} = f_B = 2\text{W/m}^2 && \text{(c)}\\
&\text{edge–4: } -k\nabla T \cdot \vec{n}^{(b)} = f_B = -5\,\text{W/m}^2 && \text{(d)}
\end{aligned} \qquad (10.90)$$

Considering the geometric symmetry and the symmetry of the two boundary conditions along edges-2 and -4, we will simulate only one-half of the problem and move the edge-4 to the symmetry axis $(0, y)$. On the symmetry axis the net heat flow must be zero. Therefore, we assign the symmetry boundary condition (Fig. 10.6C) as follows:

$$\text{edge-4: } -k\nabla T \cdot \vec{n}^{(b)} = f_B = 0 \qquad (10.91)$$

TABLE 10.1 Element connectivity table.

(e)	Node-1, 2, 3, 4	$t^{(e)}$ (m)	$k^{(e)}$ (W/m^2 K)	$(f_{B1}^{(e)}, f_{B2}^{(e)}), (f_{B2}^{(e)}, f_{B3}^{(e)})$ $(f_{B3}^{(e)}, f_{B4}^{(e)}), (f_{B4}^{(e)}, f_{B1}^{(e)})$(W/m^2)
1	1, 2, 5, 4	1	0.5	(0, 0), (0, 0), (0, 0), (0, 0)
2	2, 3, 6, 5	1	0.5	(0, 0), (-5,-5), (0, 0), (0, 0)
3	4, 5, 8, 7	1	0.5	(0, 0), (0, 0), (2,2), (0, 0)
4	5, 6, 9, 8	1	0.5	(0, 0), (-5,-5), (2,2), (0, 0)

TABLE 10.2 The positions of the nodes for the 2×2 discretization of the domain. Node locations are shown in Fig. 10.6.

Global node no.	Position (x, y)
1	(0,0)
2	$(W_b/4_b, 0)$
3	$(W_b/2, 0)$
4	$(0, H/2)$
5	$((W_b/4 + W_t/4)/2, H/2)$
6	$((W_b/2 + W_t/2)/2, H/2)$
7	$(0, H)$
8	$(W_t/4, H)$
9	$(W_t/2, H)$

The right half of the two-dimensional domain is divided into four elements as shown in Fig. 10.6D. The nodes and the elements of the domain are numbered as shown in the figure. The element connectivity table and the nodal coordinates are provided in Table 10.1 and 10.2.

Equilibrium equations for each element are found by using the relationships provided in Section 10.6. Noting that we are interested in the steady state solution and no internal heat source is specified in this problem Equation (10.50) becomes:

$$[k_k]^{(e)}\{d_T\} = \{r_{b-ei}\}^{(e)} + \{r_i\}^{(e)} \tag{10.92}$$

where $\{r_{b-ei}\}^{(e)}$ and $\{r_i\}^{(e)}$ are defined in Eqs. (10.48) and (10.52), respectively. The element conductivity matrices are found by integrating Eq. (10.41a) numerically, as described in Section 10.5. The heat flux vectors are found by using Eqs. (10.46) and (10.47). For example, for element-4, by using Eq. (10.46) and the element connectivity table, we find the consistent nodal heat flux vectors along its edges-e2 and -3 in the following.

Element-4:

edge-e2: The external heat flux is constant, $f_{B-82}(1,\eta) = -5 \times 10^3$. The consistent nodal heat flux vector can be found by using Eq. (10.48), where $s = \eta$ and $\xi = 1$,

$$\{r_{b-e2}\}^{(e)} = -\int_{-1}^{1} [N]^T f_{B-e2} t J d\eta \tag{10.93}$$

The Jacobian along the edge-2, can be computed by using Eq. (7.92) as follows:

$$J = \left[\left(\frac{\partial x}{\partial s}\right)^2 + \left(\frac{\partial y}{\partial s}\right)^2\right]^{1/2} \tag{10.94}$$

where

$$x = N_1x_1 + N_2x_2 + N_3x_3 + N_4x_4 \text{ and } y = N_1y_1 + N_2y_2 + N_3y_3 + N_4y_4 \tag{10.95}$$

Along edge-e2, where $\xi = 1$, the shape functions become, as shown in Section 7.6.1.3,

$$N_1 = N_4 = 0, N_2 = \frac{1}{2}(1-s) \text{ and } N_3 = \frac{1}{2}(1+s) \tag{10.96}$$

Thus the Jacobian along edge-2 can then be found as follows:

$$J = \frac{1}{2}\left[(x_3-x_2)^2 + (y_3-y_2)^2\right]^{1/2} = \frac{1}{2}L_{23} = 1.2748 \times 10^{-1} \tag{10.97}$$

Eq. (10.93) then becomes,

$$\{r_{b-e2}\}^{(4)} = -(5 \times 10^3)t \int_{-1}^{1} \begin{bmatrix} 0 \\ N_2(1,\eta) \\ N_3(1,\eta) \\ 0 \end{bmatrix} 1.2748 \times 10^{-1} d\eta \tag{10.98}$$

This equation is evaluated by using Gauss-quadrature with the quadrature points $p_1 = -1/\sqrt{3}$, $p_2 = 1/\sqrt{3}$, and weight factors $W_1 = W_2 = 1$ obtained from Table 7.3, as follows:

$$\{r_{p-\varepsilon 2}\}^{(4)} = 5\times10^3 tJ \begin{Bmatrix} 0 \\ W_1N_2(1,p_1) \\ W_1N_3(1,p_1) \\ 0 \end{Bmatrix} + 5\times10^3 tJ \begin{Bmatrix} 0 \\ W_2N_2(1,p_2) \\ W_2N_3(1,p_2) \\ 0 \end{Bmatrix} = -t \begin{Bmatrix} 0 \\ 0.6374 \\ 0.6374 \\ 0 \end{Bmatrix} \tag{10.99}$$

edge-e3: We note that $f_{B-e3}(\xi, 1) = 2$ and compute the Jacobian determinant along edge-3 as $J = 3.75 \times 10^{-2}$. Numerical integration gives,

$$\{r_{p-e3}\}^{(4)} = -2t\int_{-1}^{1}\begin{bmatrix}0\\0\\N_3(\xi,1)\\N_4(\xi,1)\end{bmatrix}3.75\times10^{-2}d\xi = -t\begin{Bmatrix}0\\0\\3.75\times10^{-2}\\3.75\times10^{-2}\end{Bmatrix} \tag{10.100}$$

Details are omitted but the equilibrium equations in the element coordinate system are found as follows:

$$\text{Element-1:}\begin{bmatrix}0.4357 & -0.3107 & -0.2366 & 0.1116\\ -0.3107 & 0.4107 & 0.1116 & -0.2116\\ -0.2366 & 0.1116 & 0.4855 & -0.3605\\ 0.1116 & -0.2116 & -0.3605 & 0.4605\end{bmatrix}\begin{Bmatrix}T_1\\T_2\\T_3\\T_4\end{Bmatrix} = \begin{Bmatrix}0\\0\\0\\0\end{Bmatrix} \tag{10.101}$$

$$\text{Element-2:}\begin{bmatrix}0.4678 & -0.3178 & -0.2653 & 0.1153\\ -0.3178 & 0.3928 & 0.1153 & -0.1903\\ -0.2653 & 0.1153 & 0.5184 & -0.3684\\ 0.1153 & -0.1903 & -0.3684 & 0.4434\end{bmatrix}\begin{Bmatrix}T_1\\T_2\\T_3\\T_4\end{Bmatrix} = -\begin{Bmatrix}0\\0.6374\\0.6374\\0\end{Bmatrix} \tag{10.102}$$

$$\text{Element-3:}\begin{bmatrix}0.5134 & -0.4134 & -0.2822 & 0.1822\\ -0.4134 & 0.4884 & 0.1822 & -0.2572\\ -0.2822 & 0.1822 & 0.5904 & -0.4904\\ 0.1822 & -0.2572 & -0.4904 & 0.5654\end{bmatrix}\begin{Bmatrix}T_1\\T_2\\T_3\\T_4\end{Bmatrix} = \begin{Bmatrix}0\\0\\0.0750\\0.0750\end{Bmatrix} \tag{10.103}$$

$$\text{Element-4:}\begin{bmatrix}0.5473 & -0.4223 & -0.3120 & 0.1870\\ -0.4223 & 0.4723 & 0.1870 & -0.2370\\ -0.3120 & 0.1870 & 0.6257 & -0.5007\\ 0.1870 & -0.2370 & -0.5007 & 0.5507\end{bmatrix}\begin{Bmatrix}T_1\\T_2\\T_3\\T_4\end{Bmatrix} = \begin{Bmatrix}0\\-0.6374\\-0.5624\\0.0750\end{Bmatrix} \tag{10.104}$$

Note that the relation between the element nodes numbers 1–4 and the global node numbers can be deduced from the element connectivity table (Table 10.1). The active degrees of freedom in this problem are $T_4 - T_9$. The element equilibrium equations are first rearranged switching from the local node numbering to global node numbering. Then the global equilibrium equation is found by summing the element stiffness matrices (10.101)–(10.104),

$$\sum_{e=1}^{4}\left([k]^{(e)}\{d\} - \left\{r_{p-ei}\right\}^{(e)}\right) = 0 \tag{10.105}$$

This results in the following global equilibrium equation:

$$\begin{bmatrix} 0.9739 & -0.7739 & 0 & 0.1822 & -0.2822 & 0 \\ & 1.9645 & -0.7906 & -0.2572 & 0.3692 & -0.3120 \\ & & 0.9906 & 0 & -0.2370 & 0.1870 \\ & & & 0.5654 & -0.4904 & 0 \\ & \text{sym} & & & 1.1411 & -0.5007 \\ & & & & & 0.6257 \end{bmatrix} \begin{Bmatrix} T_4 \\ T_5 \\ T_6 \\ T_7 \\ T_8 \\ T_9 \end{Bmatrix} = \begin{Bmatrix} 0 \\ 0 \\ 1.2748 \\ -0.0750 \\ -0.1500 \\ 0.5624 \end{Bmatrix} \tag{10.106}$$

Solution of this system of equations gives the following global temperature vector for the 2×2 mesh:

$$\{D\} = \left\{303.1651 \quad 303.3935 \quad 304.1125 \quad 304.0762 \quad 304.2487 \quad 304.7617\right\}^T \tag{10.107}$$

b) A mesh convergence study is performed by increasing the number of elements hierarchically. To accomplish this, each element edge is divided by two at each level of mesh refinement. This results in $2^r \times 2^r$ mesh at each level r of the mesh refinement. The resulting mesh scheme is presented in Fig. 10.7 along with the predicted temperature variations in the domain.

In order to quantify the solution convergence we monitor the temperature at each mesh refinement level. The corner temperature T_{corner} changes at each level of mesh refinement are shown in Table 10.3. The relative error between each refinement level is also computed. The rate of change gradually diminishes with increasing refinement, r. About 0.81% change in the decimal point of the solution is achieved when the mesh size changes from 8 × 8 to 16 × 16. Further refinement results in an increasingly smaller change in the result.

Temperature contours for the first four levels of mesh refinement are plotted in Fig. 10.7. We note that the essential boundary condition of $T_b = 300$ K is met at the lower boundary for all mesh levels. The boundary along the left edge shows the temperature gradient to be zero along the symmetry boundary, as expected. The temperature distribution in the solution domain is far from trivial and the contour levels converge toward a final distribution with increasing number of elements.

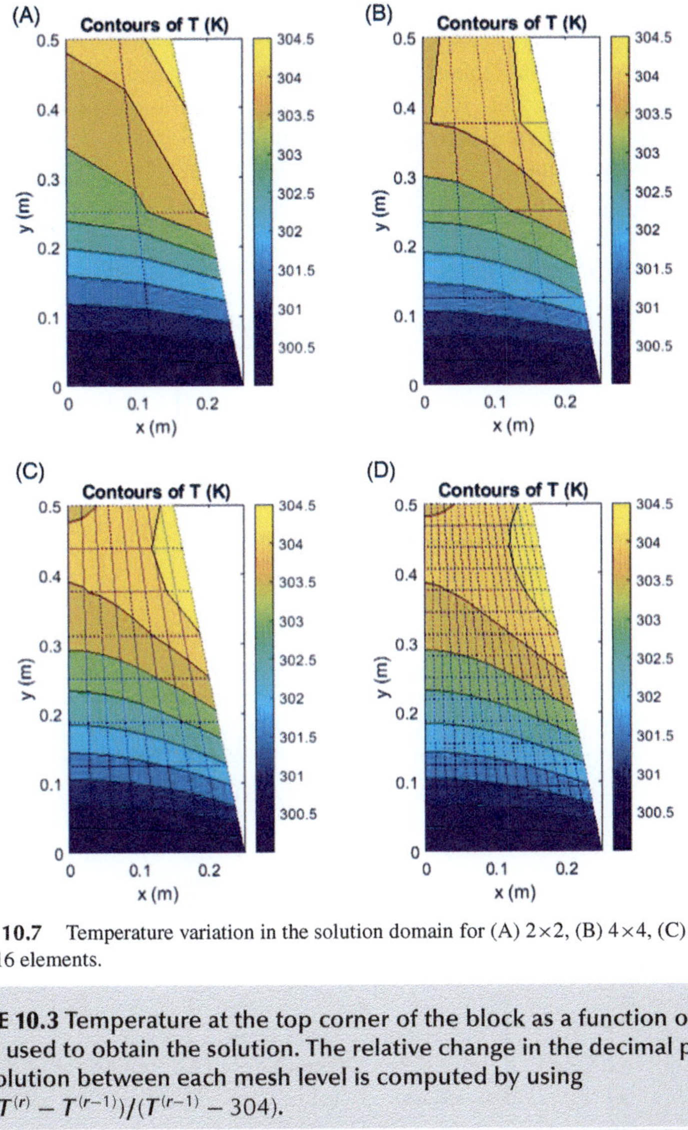

FIGURE 10.7 Temperature variation in the solution domain for (A) 2×2, (B) 4×4, (C) 8×8, and (D) 16×16 elements.

TABLE 10.3 Temperature at the top corner of the block as a function of the mesh used to obtain the solution. The relative change in the decimal part of the solution between each mesh level is computed by using $\cdot 100(T^{(r)} - T^{(r-1)})/(T^{(r-1)} - 304)$.

	T_{corner}	Change (%)
2 × 2	304.7617	
4 × 4	304.7046	7.50E+00
8 × 8	304.6837	2.96E+00
16 × 16	304.6781	8.13E−01
32 × 32	304.6767	2.13E−01
64 × 64	304.6763	5.50E−02

Problems

10.1 A Q4-element is subjected to an external convection along its edge-4 as shown in Fig. 10.8. The other edges of this element are internal. Calculate the element equilibrium equation for this element. Use the following values: h =5 W/m^2 K, $k = 900$ W/m K, $T_\infty = 20$°C, $t =$ 10 cm, $(x_1, y_1) = (10, 10)$, $(x_2, y_2) = (15, 8)$, $(x_3, y_3) = (14, 15)$, and $(x_4, y_4) = (11, 13)$ cm.

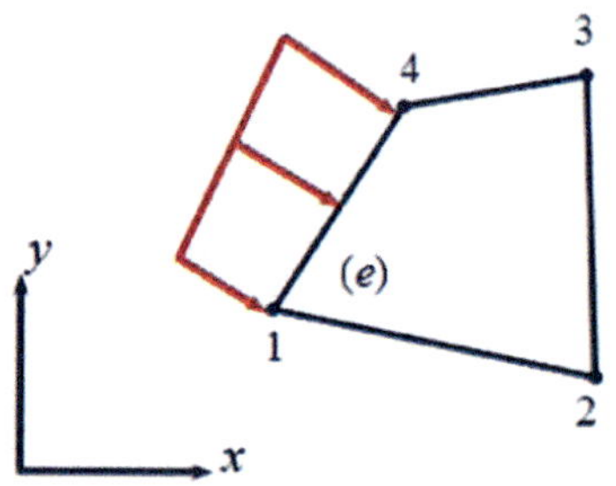

FIGURE 10.8 Problem 10.1.

10.2 A Q4-element is subjected to external radiation along its edge-1 (Fig. 10.9). The other edges of this element are internal. Calculate the element equilibrium equation for this element. Use the following values: Use the following values, $\sigma = 5.669 \times 10^{-8}$ W/K^4 m^2, $\psi = 1$, $T_2 = 1000$ K, $k = 900$ W m K, $t = 10$ cm, $(x_1, y_1) = (10, 10)$, $(x_2, y_2) = (15, 8)$, $(x_3, y_3) = (14, 15)$, and $(x_4, y_4) = (11, 13)$ cm.

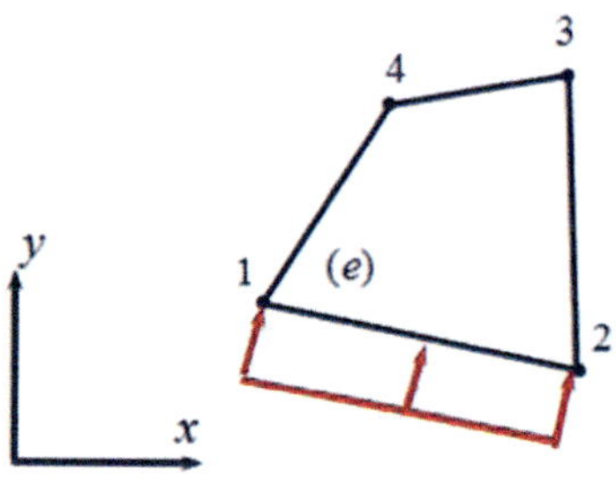

FIGURE 10.9 Problem 10.2.

10.3 The rectangular domain shown in Fig. 10.10 has an internal point heat source Q, located at $\frac{3}{4}(L_x, L_y)$. Its edges along $(0, y)$ and (x, L_y) are kept at constant temperature, T_c, and its other two edges are insulated. Find the nodal temperatures in this problem if the domain is discretized by using four geometrically identical Q4-elements. Use the following values: $k = 900$ W m K, $T_c = 100$°C, $Q = 200$ W/m^3, $L_x = 1.5$ m, $L_y = 1$ m, $t = 0.2$ m.

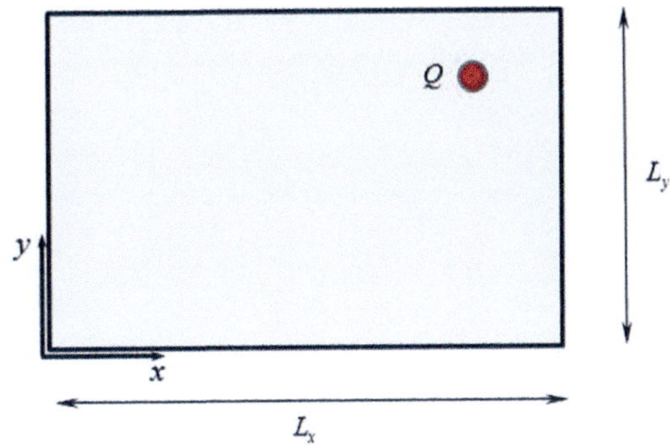

FIGURE 10.10 Problem 10.3.

10.4 Two opposite edges of the rectangular domain shown in Fig. 10.10 along $(0, y)$ and (L_x, y) are kept at constant temperature, T_4 and T_2, respectively. The top and bottom edges are subjected to convective effects. The heat source shown in the figure has a zero value. Find the nodal temperatures in this problem if the domain is discretized by using 4 geometrically identical Q4-elements. Use the following values: $k = 1000$ W.m.K, $T_2 = 50°C$, $T_4 = 100°C$, $Q = 0$, $L_x = 1.5$ m, $L_y = 1$ m, $h = 50$ W/m^{2}K, $T_\infty = 20°C$, $t = 0.2$ m.

10.5 By using a three node, linear interpolation element (Fig. 10.11), defined in Section 8.2.2, develop the finite element form of the heat transfer equations for the case of constant heat flux applied along its edge-e1 located between nodes-1 and -2.

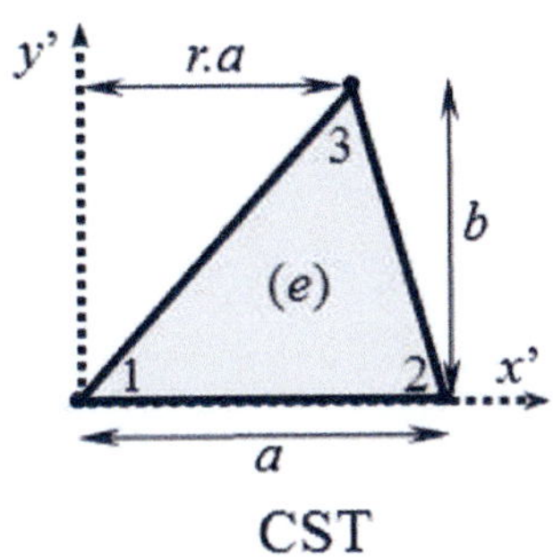

FIGURE 10.11 Problem 10.4.

Reference

[1] R.D. Cook, D.S. Malkus, M.E. Plesha, R.J. Witt, Concepts and Applications of Finite Element Analysis, fourth ed., John Wiley & Sons, New York, NY, 2002.

Chapter 11

Transient thermal analysis

11.1 Introduction

The *semidiscrete* form of the heat transfer problem is given in Eq. (10.84),

$$[C]\{\dot{D}_T\} + [K]\{D_T\} = \{R\} \tag{11.1}$$

where $[C]$ is the global heat capacity matrix, $[K]$ is the global conductivity matrix, $\{R\}$ is the loading vector, and $\{D_T\}$ and $\{\dot{D}_T\}$ represent the nodal temperatures and time rate of change of the nodal temperatures. This equation is subjected to *essential* and/or *natural boundary* conditions along its boundary as given in Eqs. (10.6) and (10.7),

$$\text{Essential boundary condition:} \qquad T(\vec{r}_P, t) = T_b(t) \quad \text{for } \vec{r}_P \in \Gamma_T \tag{11.2}$$

$$\text{Natural boundary condition:} \quad -k\nabla T(\vec{r}_P, t) \cdot \vec{n}^{(b)} = f_B(t) \quad \text{for } \vec{r}_P \in \Gamma_f \tag{11.3}$$

where $\vec{r}_P$ is the position vector for all the points on the boundary Γ (Fig. 10.2). The segment of the boundary on which the temperature is prescribed is the *temperature boundary* Γ_T and the segment over which the heat flux is prescribed is the *flux boundary* Γ_f. In general, Γ_T and Γ_f are equivalent to the Γ_E and Γ_N which represent the boundaries on which the essential and natural boundary conditions are prescribed, respectively. The heat flux on the boundary can take on different forms (Eq. (10.9)) depending on the physics of the problem. Eq. (11.1) is also subjected to an initial condition for the entire domain,

$$T(\vec{r}, 0) = T^{(0)}(\vec{r}) \text{ for } \vec{r} \in \Omega \tag{11.4}$$

where $\vec{r}$ represents all the points that are in the solution domain Ω. The global conductivity matrix includes contributions from the conduction and the boundary convection matrices, as described in Sections 9.6 and 10.6, for one- and two-dimensional domains, respectively.

Note that the semidiscrete form of the heat transfer Eq. (11.1) is discretized in space, but continuous in time. Therefore, discretization in time is still needed. Variation of temperatures at the nodes of the finite element mesh as a function of time is obtained by the *direct time integration* method.

Finite Element Method: Physics and Solution Methods. DOI: https://doi.org/10.1016/B978-0-12-821127-4.00010-4
Copyright © 2022 Elsevier Inc. All rights reserved.

Direct time integration methods are based on the finite difference (FD) approximations to partial derivative operators. In this chapter, time discretization is demonstrated for the one-dimensional heat transfer problem. The concepts developed here can be easily implemented to find the transient solution of the two- and three-dimensional problems. In Section 11.3 an overview of the derivation of the finite difference formulas is provided. This is followed by time discretization of Eq. (11.1) by the *forward, backward,* and *central* finite difference approximations. The generalized trapezoidal method, which is a one parameter algorithm that encompasses these three and other time discretization methods, is also introduced. The solution algorithm including the explicit and implicit time integration approaches are described in Section 11.5. The concepts of stability, accuracy, and convergence are presented in Section 11.6. Two examples are provided to demonstrate the solution methods and the stability conditions developed in the chapter.

11.2 Transient heat transfer equation

In this chapter, time discretization is demonstrated for the one-dimensional heat transfer problem. The derivation for a two-dimensional element is similar and is provided in Chapter 10, in detail.

11.2.1 Boundary/initial value problem

The transient heat transfer equation in one-dimensional domain is given as follows:

$$\frac{\partial}{\partial x}\left(kA\frac{\partial T}{\partial x}\right)+QA-\rho cA\frac{\partial T}{\partial t}=0 \text{ in } 0<x<L \text{ for } t>0 \tag{11.5}$$

with boundary conditions,

$$\text{at } x=0 \text{ either } T(0,t)=T_0(t) \text{ or } n\left(-k\frac{\partial T}{\partial x}\right)=f_{B0}(t) \text{ for } t>0 \tag{11.6a}$$

$$\text{at } x=L \text{ either } T(L,t)=T_L(t) \text{ or } n\left(-k\frac{\partial T}{\partial x}\right)=f_{BL}(t) \text{ for } t>0 \tag{11.6b}$$

and the initial condition at $t=t_0$,

$$T(x,t_0)=T^{(0)} \text{ in } 0\le x\le L \tag{11.7}$$

where temperature $T=T(x,t)$ is a function of the spatial coordinate x and time t, Q is the internal rate of heat generation, k is the conductivity, ρ is the density, c is the heat capacity, and A is the cross-sectional area of the material. n represents the unit outward normal at the boundaries, with $n=-1$ at $x=0$ and $n=1$ at $x=L$. Heat can flow in or out of the boundaries through different physical mechanisms such as convection and radiation, or the boundary can be insulated as described in Eq. (2.176). In most cases the initial time stamp is set as $t_0 = 0$ If prescribed, the heat flux at the ends of domain can vary in time; hence, $f_{B0}(t)$ and $f_{BL}(t)$ are known functions of time. Similarly, if prescribed $T_0(t)$ and $T_L(t)$ can also vary in time.

11.2.2 Element equilibrium equation of one-dimensional, transient heat transfer

The one-dimensional solution domain is spatially discretized in to N_E elements. Each element is numbered consecutively as $1 \leq e \leq N_E$. For $e = j$, it is easy to see that: element-j is connected to nodes j and $j + 1$. element-j spans the range $x_j < x < x_{j+1}$, and its length is,

$$L^{(j)} = x_{j+1} - x_j \tag{11.8}$$

position of node $j + 1$ can be found by using the relationship,

$$x_{j+1} = \sum_{e=1}^{j} L^{(e)} \tag{11.9}$$

Next, we briefly review the procedure to obtain the semidiscrete form of the heat transfer equation (2.171) specified over a one-dimensional element defined in a local coordinate system that spans the range $0 < x' < L^{(e)}$. We will assume that this is an internal element with element number $e = j$, attached to nodes j and $j+1$, on its left and right boundaries, respectively.

The transient heat transfer equation for this element becomes,

$$\frac{\partial}{\partial x'}\left(kA\frac{\partial T}{\partial x'}\right) + QA - \rho cA\frac{\partial T}{\partial t} = 0 \text{ in } 0 < x' < L^{(e)} \tag{11.10}$$

Heat flux conditions are imposed on the element boundaries,

$$n\left(-k\frac{\partial T}{\partial x'}\right)_{x'=0} = f_{i0} = f_{i(j)} \quad \text{and} \quad n\left(-k\frac{\partial T}{\partial x'}\right)_{x=L^{(e)}} = f_{iL^{(e)}} = f_{i(j+1)} \tag{11.11}$$

The heat flux on boundary of the element are given as f_{i0} and $f_{iL^{(e)}}$, or by using the node-numbering convention $f_{i(j)}$ and $f_{i(j+1)}$, respectively.

We use the weak form of Eq. (11.10) to find its spatially discretized form. To this end, the weighted residual integral form of this equation is expressed as follows:

$$\int_0^{L^{(e)}} \delta T\left(\frac{\partial}{\partial x'}\left(kA\frac{\partial T}{\partial x'}\right) + QA - \rho cA\frac{\partial T}{\partial t}\right)dx' = 0 \tag{11.12}$$

where δT is the variation of temperature with the characteristics of a weight function w as described in Chapter 3. Integrating the first term of Eq. (11.12) by parts, and using the natural boundary conditions given in Eq. (11.11) results in the following relationship:

$$-\left(\delta T A_{j+1} f_{i(j+1)}\right) - \left(\delta T A_j f_{i(j)}\right) - \int_0^{L^{(e)}} \left(\frac{\partial \delta T}{\partial x'} kA \frac{\partial T}{\partial x'} - QA\delta T + \rho c A \delta T \frac{\partial T}{\partial t} \right) dx' = 0 \tag{11.13}$$

The temperature and its *spatial derivative* in the element are interpolated as follows:

$$T = [N]\{d_T\} \quad \text{and} \quad \frac{\partial T}{\partial x'} = [B]\{d_T\} \tag{11.14}$$

The shape function matrix $[N]$ and the temperature-gradient matrix $[B]$ are defined in the usual manner as described in Section 9.3. The nodal temperature vector is,

$$\{d_T\} = \left\{T_j \quad T_{j+1}\right\}^T \tag{11.15}$$

where the unknown nodal temperatures vary in time, $T_j = T_j(t)$ and $T_{j+1} = T_{j+1}(t)$. The *time derivative* of the nodal temperatures can be found as follows:

$$\frac{\partial T}{\partial t} = [N]\left\{\dot{d}_T\right\} \tag{11.16}$$

Substituting Eqs. (11.14)–(11.16) in to the weak form and eliminating the arbitrary temperature variation $\{\delta T\}$ in the same manner as described in Chapter 3, we find the finite element form of the energy balance equation,

$$\left[\int_0^{L^{(e)}} kA[B]^T[B]dx'\right]\{d_T\} + \left[\int_0^{L^{(e)}} \rho c A[N]^T[N]dx'\right]\left\{\dot{d}_T\right\} = \int_0^{L^{(e)}} QA[N]^T dx' + \{r_i\}^{(e)} \tag{11.17}$$

where,

$$\text{Conductivity matrix: } [k_k]^{(e)} = \int_0^{L^{(e)}} kA[B]^T[B]dx' \tag{11.18a}$$

$$\text{Heat capacity matrix: } [c]^{(e)} = \int_0^{L^{(e)}} \rho c A[N]^T[N]dx' \tag{11.18b}$$

$$\text{Internal heat vector: } \left\{r_q\right\}^{(e)} = \int_0^{L^{(e)}} Q[N]^T A dx' \tag{11.18c}$$

$$\text{Boundary flux vector: } \{r_i\}^{(e)} = -\left\{\left(A_j f_{i(j)}\right) \quad \left(A_{j+1} f_{i(j+1)}\right)\right\}^T \tag{11.18d}$$

Thus, the semidiscrete form of the transient heat transfer equation for an internal element is expressed in matrix form as follows:

$$[k_k]^{(e)}\{d_T\} + [c]^{(e)}\{\dot{d}_T\} = \{r_q\}^{(e)} + \{r_i\}^{(e)} \tag{11.19}$$

Note that the heat flux boundary conditions, $f_b(0)$ and $f_b(L^{(e)})$ could substantially modify Eq. (11.19) for the elements that share a node with one of the physical boundaries. Details of this are presented in Section 9.3. Here, we simply add the transient term into Eqs. (9.25) to find the finite element form of the one-dimensional, transient heat transfer equation,

$$\begin{aligned}
&\text{Boundary element } (e=1): && ([k_k]^{(1)} + [k_{h-b0}]^{(1)})\{d_T\} + [c]^{(1)}\{\dot{d}_T\} = \{r_q\}^{(1)} + \{r_{h-b0}\}^{(1)}\\
&\text{Internal elements } (2 \le e \le N_E - 1): && [k_k]^{(e)}\{d_T\} + [c]^{(e)}\{\dot{d}_T\} = \{r_q\}^{(e)} + \{r_i\}^{(e)}\\
&\text{Boundary element } (e=N_E): && ([k_k]^{(N_E)} + [k_{h-bL}]^{(N_E)})\{d_T\} + [c]^{(N_E)}\{\dot{d}_T\} = \{r_q\}^{(N_E)} + \{r_{h-bL}\}^{(N_E)}
\end{aligned} \tag{11.20}$$

The boundary convection matrices $[k_{h-b0}]$ and $[k_{h-bL}]$ and the boundary heat flux vectors $\{r_{h-b0}\}$ and $\{r_{h-bL}\}$ are given in Eqs. (9.21) and (9.23), respectively.

11.2.2.1 Heat capacity matrix

Eq. (11.18b) represents the finite element form of the heat capacity matrix. For a linear element, with uniform cross-sectional area, it can be shown that the *consistent heat capacity matrix* has the following form:

$$[c]^{(e)} = \frac{1}{6}(c\rho AL)^{(e)}\begin{bmatrix} 2 & 1 \\ 1 & 2 \end{bmatrix} \tag{11.21}$$

The shape of this matrix shows that heat flux at a given node due to transient effects is affected by both nodes of the element,

$$\begin{Bmatrix} f_j \\ f_{j+1} \end{Bmatrix} = \frac{1}{6}(c\rho AL)^{(e)}\begin{bmatrix} 2 & 1 \\ 1 & 2 \end{bmatrix}\begin{Bmatrix} \dot{T}_j \\ \dot{T}_{j+1} \end{Bmatrix} \tag{11.22}$$

In Section 11.5 we show that using a heat capacity matrix in diagonal form has some algorithmic advantages. A *lumped heat capacity matrix* for the one-dimensional heat transfer problem can be obtained simply by lumping the total heat capacity into each node as follows:

$$[c]^{(e)} = \frac{1}{2}(c\rho AL)^{(e)}\begin{bmatrix} 1 & 0 \\ 0 & 1 \end{bmatrix} \tag{11.23}$$

In this case the heat flux at a given node depends only on the transient rate of change of temperature of the same node.

11.2.3 Global equilibrium equation of one-dimensional, transient heat transfer

Detailed explanation of the element assembly process is given in Sections 4.6 and 9.5. There, we demonstrate that in order to represent the original boundary value problem in the global solution domain Ω defined in $0 < x < L$, the temperature and heat flow must be continuous across element boundaries, that is, at the nodes. Thus, we show that the global equilibrium equation can be found by summing the contributions from each element as follows:

$$\sum_{e=1}^{N_E} = \{r_i\}^{(e)} = \sum_{e=1}^{N_E}\left([k_k]^{(e)}\{d_T\} + [c]^{(e)}\{\dot{d}_T\} - \{r_q\}^{(e)}\right) = 0 \tag{11.24}$$

This summation results in the global equilibrium equation given in the form of Eq. (11.1).

11.2.4 Global boundary conditions

The techniques to impose the natural and essential boundary conditions on the heat transfer problem are discussed in Section 9.6 for the steady state problem. These methods do not change for the transient problem. However, here we need to keep in mind that the boundary conditions can vary in time as indicated by Eqs. (11.6a). Therefore, this time variation should be incorporated in to the computer codes. Example 11.1 addresses such a problem.

11.3 Finite difference approximations to derivatives

The time derivative of the nodal temperatures is evaluated by using the finite difference approximations.

11.3.1 Temporal discretization of a continuous function

Consider a function $f(t)$ that is continuous and differentiable in the interval $t_0 \leq t \leq t_F$. Before we describe the time derivatives in this interval, we will evaluate the function at uniformly[1] spaced subintervals, such that,

$$t^{(n)} = t_0 + n\Delta t \quad \text{for} \quad n = 0 - N_T \tag{11.25}$$

where $^{(n)}$ represents a time step, N_T represents the number of subintervals between t_0 and t_F, and Δt is the size of the *time step* defined as follows:

$$\Delta t = \frac{1}{N_T}(t_F - t_0) \tag{11.26}$$

1. In general, the subintervals (or the time step size) does not have to be uniform. But, variable time step is beyond the scope of this text.

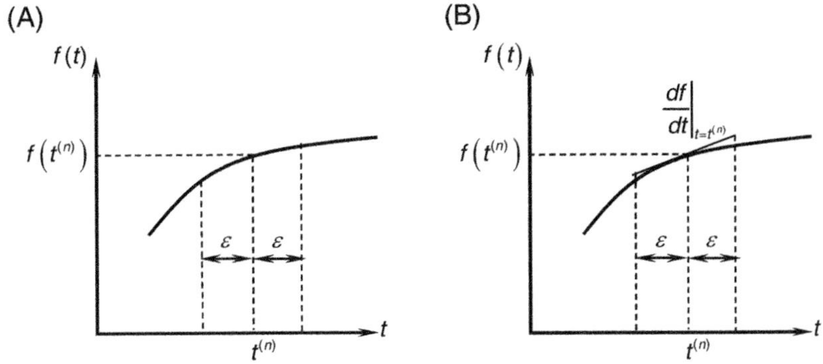

FIGURE 11.1 Geometric depiction of the construction of the Taylor series approximation to a function near $t = t^{(n)}$. (A) The exact function is $f(t)$. This figure shows that $f(t^{(n)} \pm \varepsilon) = f(t^{(n)})$ is a poor approximation in the range $t^{(n)} - \mathcal{E} < t < t^{(n)} + \varepsilon$, even though it is exact when $\varepsilon = 0$. (B) Shows that the linear approximation to $f(t) f(t^{(n)} \pm \varepsilon) = f(t^{(n)}) \pm \varepsilon df/dt$ in the mentioned range could be reasonable.

In a *temporal approximation* scheme, a function's values $f(t^{(n)})$ at instances $t^{(n)}$ are represented by discretely computed, approximate $f^{(n)}$ values. As the accuracy of the time integration method is improved $f^{(n)}$ should approach $f(t^{(n)})$.

The finite difference method is one of the methods to approximate the derivative operator at a given time-step $t^{(n)}$ by using the discretized values of the function $(\ldots, f^{(n-1)}, f^{(n)}, f^{(n+1)}, \ldots)$ near this time instance. The finite difference approximations to derivative operators are derived by using the Taylor series expansion.

11.3.2 Taylor series expansion

Let's consider a function $f(t)$ which is continuously differentiable at $t = t^{(n)}$ (Fig. 11.1). This function can be evaluated at $t^{(n+\varepsilon)} = t^{(n)} + \varepsilon$ by using *Taylor series expansion* as follows:

$$f\left(t^{(n+\varepsilon)}\right) = f\left(t^{(n)}\right) + \sum_{i=1}^{\infty} \frac{\varepsilon^i}{i!} \left.\frac{d^i f}{dt^i}\right|_{t=t^{(n)}} \tag{11.27}$$

ε is a small variation of time around $t^{(n)}$.

11.3.3 Approximations to the first derivative of a function

In order to obtain the finite difference approximations of the derivatives of a function at $f(t)$ at $t = t^{(n)}$ the Taylor series expansion of the function is considered with $\varepsilon = \Delta t$. Thus Eq. (11.27) can be expressed as,

$$f\left(t^{(n+1)}\right) = f\left(t^{(n)}\right) + \Delta t \frac{df}{dt}\bigg|_{t=t^{(n)}} + \Delta t^2 \frac{d^2 f}{dt^2}\bigg|_{t=t^{(n)}} + \sum_{i=3}^{\infty} \frac{\Delta t^i}{i!} \frac{d^i f}{dt^i}\bigg|_{t=t^{(n)}} \tag{11.28}$$

Note that in Eq. (11.28) as the size of the time step Δt is reduced, the effects of the second and the third term, which are proportional to Δt^2 and Δt^3, respectively, will diminish rapidly. By considering the positive and negative Δt values, Eq. (11.28) becomes,

$$f\left(t^{(n+1)}\right) = f\left(t^{(n)}\right) + \Delta t \frac{df}{dt}\bigg|_{t=t^{(n)}} + \frac{\Delta t^2}{2} \frac{d^2 f}{dt^2}\bigg|_{t=t^{(n)}} + \sum_{i=3}^{\infty} \frac{\Delta t^i}{i!} \frac{d^i f}{dt^i}\bigg|_{t=t^{(n)}} \tag{11.29a}$$

$$f\left(t^{(n-1)}\right) = f\left(t^{(n)}\right) - \Delta t \frac{df}{dt}\bigg|_{t=t^{(n)}} + \frac{\Delta t^2}{2} \frac{d^2 f}{dt^2}\bigg|_{t=t^{(n)}} + \sum_{i=3}^{\infty} \frac{(-\Delta t)^i}{i!} \frac{d^i f}{dt^i}\bigg|_{t=t^{(n)}} \tag{11.29b}$$

The *forward difference* approximation to the derivative operator can be obtained by rearranging Eq. (11.29a) as follows:

$$\frac{df}{dt}\bigg|_{t=t^{(n)}} = \frac{f\left(t^{(n+1)}\right) - f\left(t^{(n)}\right)}{\Delta t} - \tau(O(\Delta t)) \tag{11.30}$$

where the *residual or truncation error* τ is,

$$\tau = \frac{\Delta t}{2} \frac{d^2 f}{dt^2}\bigg|_{t=t^{(n)}} + \sum_{i=3}^{\infty} \frac{\Delta t^{i-1}}{i!} \frac{d^i f}{dt^i}\bigg|_{t=t^{(n)}} \tag{11.31}$$

Note that as $\Delta t \rightarrow 0$ the first term of Eq. (11.30) approaches df/dt, while $\tau \rightarrow 0$ in linear proportion to Δt. Therefore, the highest order of truncation error in (11.30) is said to be $O(\Delta t)$.

The *backward difference approximation* for the first derivative of $f(t)$ at $t^{(n)}$ can be found by using Eq. (11.29b) as follows:

$$\frac{df}{dt}\bigg|_{t=t^{(n)}} = \frac{f\left(t^{(n)}\right) - f\left(t^{(n-1)}\right)}{\Delta t} + \tau(O(\Delta t)) \tag{11.32}$$

A higher order approximation to the first derivative can be obtained by considering the Taylor series expansion at $t^{(n)} \pm \Delta t$ simultaneously. Subtracting Eq. (11.29a) from (11.29b) the following expression is obtained,

$$f\left(t^{(n-1)}\right) - f\left(t^{(n-1)}\right) = 2\Delta t \frac{df}{dt}\bigg|_{t=f^{(n)}} + \frac{2\Delta t^3}{3!} \frac{d^3 f}{dt^3}\bigg|_{t=t^{(n)}} \tag{11.33}$$

The *central difference approximation* for the first derivative of $f(t)$ at $t^{(n)}$ is found as follows:

$$\left.\frac{df}{dt}\right|_{t=t^{(n)}} = \frac{f\left(t^{(n+1)}\right) - f\left(t^{(n-1)}\right)}{2\Delta t} - \tau\left(O\left(\Delta t^2\right)\right) \tag{11.34}$$

In this case the highest truncation error is proportional to Δt^2,

$$\tau = \left.\frac{\Delta t^2}{3!}\frac{d^3 f}{dt^3}\right|_{t=t^{(n)}} \tag{11.35}$$

Summary of finite difference approximations to the first-order derivative operator:

Forward difference formula: $$\left.\frac{df}{dt}\right|_{t=t^{(n)}} = \frac{f\left(t^{(n+1)}\right) - f\left(t^{(n)}\right)}{\Delta t} - \tau(O(\Delta t)) \tag{11.36a}$$

Backward difference formula: $$\left.\frac{df}{dt}\right|_{t-n^{(n)}} = \frac{f\left(t^{(n)}\right) - f\left(t^{(n-1)}\right)}{\Delta t} + \tau(O(\Delta t)) \tag{11.36b}$$

Central difference formula: $$\left.\frac{df}{dt}\right|_{t=t^{(n)}} = \frac{f\left(t^{(n+1)}\right) - f\left(t^{(n-1)}\right)}{2\Delta t} - \tau\left(O\left(\Delta t^2\right)\right) \tag{11.36c}$$

11.4 Direct time integration of the heat transfer equation

The time derivative in Eq. (11.1) can be approximated by either one of backward, forward, or central time finite difference formulas. Each approach yields valid methods for temporal discretization of Eq. (11.1) with different stability and accuracy characteristics. In this section, we develop forward difference, backward, difference, central difference methods, and also the generalized trapezoidal method that can yield any one of these methods.

Eq. (11.1) represents the semidiscrete heat equation. By using the finite difference approximations this equation will be transformed to a temporally discrete form in the time domain, where the *current, future*, and *previous* time steps are indicated by the superscripts (n), $(n+1)$, and $(n-1)$, respectively. Thus, for example, the temperature at position $x = x_j$ and at time $t = t^{(n)}$, becomes $T(x_j, t^{(n)}) = T_j^{(n)}$ (Fig. 11.2).

11.4.1 Forward difference or Euler method

In order to develop the *forward difference* (or *Euler*) *method* we will express Eq. (11.1) at the time current step $t^{(n)}$ where the temperature vector $\{D_T^{(n)}\}$ is known,

$$[C]\left\{\dot{D}_T^{(n)}\right\} + [K]\left\{D_T^{(n)}\right\} = \left\{R^{(n)}\right\} \tag{11.37}$$

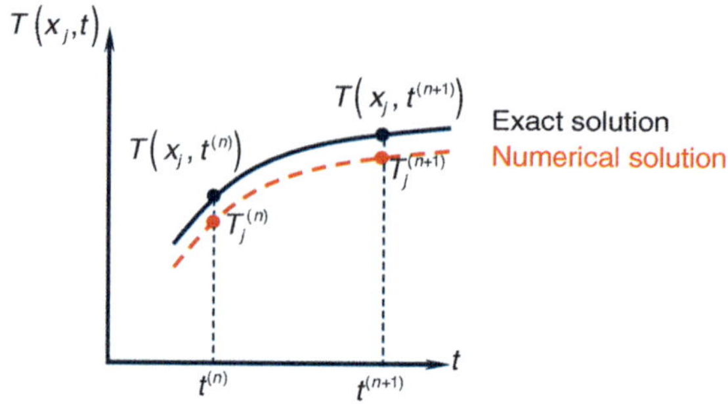

FIGURE 11.2 Schematic depiction of the exact and numerical solution to the transient response of temperature at node-j.

Note that the loading vector $\{R^{(n)}\}$ represents prescribed values of the external or internal effects and thus it can be easily evaluated at all-time steps.

By using the forward difference formula (11.36a) in the semidiscrete heat transfer Eq. (11.37) and separating the terms that belong to the known $t^{(n)}$ and unknown $t^{(n+1)}$ time steps, the following recursive relationship can be established:

$$\frac{1}{\Delta t}[C]\left\{D_T^{(n+1)}\right\} = \left\{R_T^{(n)}\right\} - \left([K] - \frac{1}{\Delta t}[C]\right)\left\{D_T^{(n)}\right\} \tag{11.38}$$

The temperature vector for the time step $t^{(n+1)}$ can then be obtained by inverting Eq. (11.38). This method of temporal discretization is known as the *forward difference* or *Euler method*. This method is conditionally stable and only $0(\Delta t)$ accurate. The concepts of accuracy and stability in the context of time integration methods are discussed in Section 11.6.

11.4.2 Backward difference method

In order to develop a time integration scheme with the backward difference method, we express the governing equation at the unknown time step $t^{(n+1)}$,

$$[C]\left\{\dot{D}_T^{(n+1)}\right\} + [K]\left\{D_T^{(n+1)}\right\} = \{R^{(n+1)}\} \tag{11.39}$$

By using the backward difference formula (11.36b) in the semidiscrete heat transfer Eq. (11.37) and separating the terms that belong to the known $t^{(n)}$ and unknown $t^{(n+1)}$ time steps, the following recursive relationship can be established:

$$\left(\frac{1}{\Delta t}[C] + [K]\right)\left\{D_T^{(n+1)}\right\} = \left\{R_T^{(n)}\right\} + \frac{1}{\Delta t}[C]\left\{D_T^{(n)}\right\} \tag{11.40}$$

The method described above is known as the *backward difference method*. This method is unconditionally stable and only $0(\Delta t)$ accurate. As mentioned before, the concepts of accuracy and stability are introduced in Section 11.6.

11.4.3 Central difference or Crank–Nicholson method

In order to use the central finite difference formula to approximate the time derivative in Eq. (11.37), let's express the governing equation at time step $t^{(n+1/2)}$ which is halfway between $t^{(n)}$ and $t^{(n+1)}$, defined as $t^{(n+1/2)} = t^{(n)} + \frac{1}{2}\Delta t$,

$$[C]\left\{\dot{D}_T^{(n+1/2)}\right\} + [K]\left\{D_T^{(n+1/2)}\right\} = \left\{R^{(n+1/2)}\right\} \tag{11.41}$$

We evaluate $\{\dot{D}_T^{(n+1/2)}\}$ using the central difference formula,

$$\left\{\dot{D}_T^{(n+1/2)}\right\} = \frac{1}{2\frac{\Delta t}{2}}\left(\left\{D_T^{(n+1)}\right\} - \left\{D_T^{(n)}\right\}\right) \tag{11.42}$$

and for $\{D_T^{(n+1/2)}\}$ and $\{R_T^{(n+1/2)}\}$ use the average values,

$$\left\{D_T^{(n+1/2)}\right\} = \frac{1}{2}\left(\left\{D_T^{(n)}\right\} + \left\{D_T^{(n+1)}\right\}\right) \text{ and } \left\{R^{(n+1/2)}\right\} = \frac{1}{2}\left(\left\{R^{(n)}\right\} + \left\{R^{(n+1)}\right\}\right) \tag{11.43}$$

Substituting Eqs. (11.42) and (11.43) into (11.41) and keeping the unknown temperatures at time step $t^{(n+1)}$ on the left hand side of the equation gives,

$$\left(\frac{1}{\Delta t}[C] + \frac{1}{2}[K_T]\right)\left\{T^{(n+1)}\right\} = \frac{1}{2}\left(\left\{R_T^{(n+1)}\right\} + \left\{R_T^{(n)}\right\}\right) + \frac{1}{\Delta t}[C]\left\{T^{(n)}\right\} - \frac{1}{2}[K_T]\left\{T^{(n)}\right\} \tag{11.44}$$

The recursive relationship presented in Eq. (11.44) is known as the *Crank-Nicholson formula*. This formula leads to an unconditionally stable and $O(\Delta t)^2$ accurate time integration method. The concepts of accuracy and stability are discussed in Section 10.6.

11.4.4 Generalized trapezoidal method

Note that the rate of change of temperatures at time step $t^{(n+1)}$ can be approximated by using the backward time difference formula as follows:

$$\left\{\dot{D}_T^{(n+1)}\right\} = \frac{1}{\Delta t}\left(\left\{D_T^{(n+1)}\right\} - \left\{D_T^{(n)}\right\}\right) \tag{11.45}$$

This variable can also be expressed as the weighted average of the $\{\dot{D}_T^{(n+1)}\}$ and $\{\dot{D}_T^{(n)}\}$ as follows:

$$\left\{\dot{D}_T^{(n+1)}\right\} = \alpha\left\{\dot{D}_T^{(n+1)}\right\} + (1-\alpha)\left\{\dot{D}_T^{(n)}\right\} \tag{11.46}$$

where α is a scalar variable with values in the range $0 \le \alpha \le 1$ (see Fig. 11.3). Combining Eqs. (11.45) and (11.46) the following relationship is obtained:

$$\left\{D_T^{(n+1)}\right\} = \left\{D_T^{(n)}\right\} + \left[\alpha\left\{\dot{D}_T^{(n+1)}\right\} + (1-\alpha)\left\{\dot{D}_T^{(n)}\right\}\right]\Delta t \tag{11.47}$$

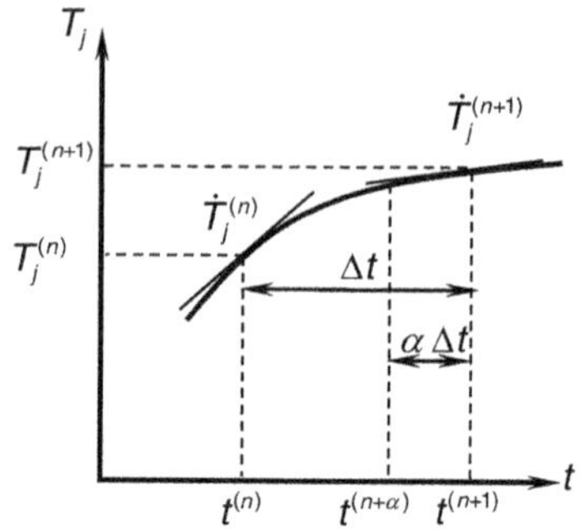

FIGURE 11.3 Definition of the intermediate time step $t^{(n+\alpha)}$, where α is a scalar parameter in the range $0 \leq \alpha \leq 1$. Setting $\alpha = \frac{1}{2}$ leads to the Crank–Nicholson method.

This relationship is used to eliminate $\{\dot{D}_T\}$ from Eq. (11.37). To this end Eq. (11.1) is evaluated at time step $t^{(n)}$ and multiplied by $(1-\alpha)$, and evaluated at time step $t^{(n+1)}$ and multiplied by α. This results in,

$$\begin{aligned} &(1-\alpha)\Big[[C]\Big\{\dot{D}_T^{(n)}\Big\} + [K]\Big\{D_T^{(n)}\Big\} - \big\{R^{(n)}\big\}\Big] = 0 \\ &\alpha\Big[[C]\Big\{\dot{D}_T^{(n+1)}\Big\} + [K]\Big\{D_T^{(n+1)}\Big\} - \big\{R^{(n+1)}\big\}\Big] = 0 \end{aligned} \tag{11.48}$$

Adding these two equations gives,

$$\begin{aligned} &[C]\Big((1-\alpha)\Big\{\dot{D}_T^{(n)}\Big\} + \alpha\Big\{\dot{D}_T^{(n+1)}\Big\}\Big) + [K]\Big((1-\alpha)\Big\{D_T^{(n)}\Big\} + \alpha\Big\{D_T^{(n+1)}\Big\}\Big) \\ &\quad = (1-\alpha)\big\{R^{(n)}\big\} + \alpha\big\{R^{(n+1)}\big\} \end{aligned} \tag{11.49}$$

Note that Eq. (11.47) can be used to replace the time derivatives from this equation,

$$\begin{aligned} &\frac{1}{\Delta t}[C]\Big(\Big\{D_T^{(n+1)}\Big\} - \Big\{D_T^{(n)}\Big\}\Big) + [K]\Big((1-\alpha)\Big\{D_T^{(n)}\Big\} + \alpha\Big\{D_T^{(n+1)}\Big\}\Big) \\ &\quad = (1-\alpha)\big\{R^{(n)}\big\} + \alpha\big\{R^{(n+1)}\big\} \end{aligned} \tag{11.50}$$

This equation can be rearranged by collecting $\{D_T\}$ at different time steps as follows:

$$\left(\frac{1}{\Delta t}[C] + \alpha[K]\right)\Big\{D_T^{(n+1)}\Big\} = \left(\frac{1}{\Delta t}[C] - (1-\alpha)[K]\right)\Big\{D_T^{(n)}\Big\} + (1-\alpha)\big\{R^{(n)}\big\} + \alpha\big\{R^{(n+1)}\big\} \tag{11.51}$$

The temperature vector at the unknown time step $\{D_T^{(n+1)}\}$ can be obtained by inverting Eq. (11.51). Details of the solution algorithm are introduced in Section 10.5.

Note that the parameter α in Eq. (11.51) controls the contribution of the time derivatives at the known and unknown time steps (Eq. (11.46)). If $\alpha = 0$, it can easily be shown that the generalized method given (11.57) becomes the forward difference method (11.54). Similarly, $\alpha = 1/2$ gives the Crank–Nicholson method and $\alpha = 1$ gives the backward difference method.

11.5 Solution algorithm

In the previous section, recursive relationships for computation of the temperature vector in the unknown time step $t^{(n+1)}$ are developed by using the backward, forward, central difference schemes and the generalized trapezoidal method. In particular, these are given by Eqs. (11.38), (11.40), (11.44), and (11.51), respectively. These four equations can also be expressed in the following general form:

$$[K_{eL}]\left\{D_T^{(n+1)}\right\} = \{R_e\} \tag{11.52}$$

The temperature vector for the time step $t^{(n+1)}$ can then be obtained by inverting Eq. (11.52),

$$\left\{D_T^{(n+1)}\right\} = [K_{eL}]^{-1}\{R_e\} \tag{11.53}$$

The effective stiffness matrices and the load vector for each method are given below.

Forward difference or Euler method:

$$\begin{gathered}[K_{eL}] = \frac{1}{\Delta t}[C], \qquad [K_{eR}] = -\left([K] - \frac{1}{\Delta t}[C]\right) \\ \{R_e\} = \left\{R^{(n)}\right\} + [K_{eR}]\left\{D_T^{(n)}\right\}\end{gathered} \tag{11.54}$$

Backward difference method:

$$\begin{gathered}[K_{eL}] = \frac{1}{\Delta t}[C] + [K], \qquad [K_{eR}] = \frac{1}{\Delta t}[C] \\ \{R_\varepsilon\} = \left\{R^{(n)}\right\} + [K_{\varepsilon R}]\left\{D_T^{(n)}\right\}\end{gathered} \tag{11.55}$$

Central difference or Crank–Nicholson method:

$$\begin{gathered}[K_{eR}] = -\left(\frac{1}{2}[K] - \frac{1}{\Delta t}[C]\right) \qquad [K_{eL}] = \frac{1}{\Delta t}[C] + \frac{1}{2}[K] \\ \{R_e\} = \left\{R^{(n+1/2)}\right\} + [K_{eR}]\left\{D_T^{(n)}\right\} \qquad \left\{R^{(n+1/2)}\right\} = \frac{1}{2}\left(\left\{R^{(n)}\right\} + \left\{R^{(n+1)}\right\}\right)\end{gathered} \tag{11.56}$$

Generalized trapezoidal method:

$$\begin{gathered}[K_{eL}] = \frac{1}{\Delta t}[C] + \alpha[K], \qquad [K_{eR}] = -\left((1-\alpha)[K] - \frac{1}{\Delta t}[C]\right) \\ \{R_e\} = \left\{R^{(n+\alpha)}\right\} + [K_{eR}]\left\{D_T^{(n)}\right\} \qquad \left\{R^{(n+\alpha)}\right\} = (1-\alpha)\left\{R^{(n)}\right\} + \alpha\left\{R^{(n+1)}\right\}\end{gathered} \tag{11.57}$$

Any one of the time integration equations given above can be integrated in to the following *time integration algorithm* to find the transient response of the temperature.

```
Given:                  {D_T^(0)} and {R(t)}, Δt, N_t = (t_F − t_0)/Δt
Initialize for n = 0: {D_T^(0)}, {R^(0)} and {R^(1)}
for n = 1: Nt
   Evaluate:            [K_eL], [K_eR], {R_e}
   Find:                {D_T^(n+1)} = [K_eL]^(-1){R_e}
   Update:              {D_T^(n)} = {D_T^(n+1)}
end
```

In most (linear) problems, the effective stiffness matrix does not change as a function of time. In that case $[K_{eL}]$ can be computed and inverted only once before the recursive solution algorithm is started. If the problem is nonlinear, $[K_{eL}]$ will need to be recomputed and inverted at each time step.

11.5.1 Explicit and implicit time integration methods

11.5.1.1 Explicit time integration method

Close inspection of Eq. (11.57) shows that if $\alpha = 0$ the effective stiffness matrix becomes,

$$[K_{\varepsilon L}] = \frac{1}{\Delta t}[C] \tag{11.58}$$

Furthermore, if $[C]$ is a diagonal matrix, for example (11.23), the matrix inversion operation to obtain $\{D_T^{(n+1)}\}$ is replaced by a division operation. A diagonal form of the heat capacity matrix can be obtained if we use the *lumped matrix* approximation to find $[C]$ instead of using the consistent heat capacity matrix. In general, a method that advances the solution by a division operation is known as an *explicit time integration method* [1].

As an example consider element matrices, with identical properties, that is, $(c\rho AL)^{(e)}$ and $\alpha = 0$. In this case Eq. (11.57) becomes,

$$\frac{1}{2\Delta t}(c\rho AL)^{(e)}[I]\left\{D_T^{(n+1)}\right\} = \left\{R^{(n)}\right\} - \left([K] - \frac{1}{\Delta t}[C]\right)\left\{D_T^{(n)}\right\} \tag{11.59}$$

where $[I]$ is the identity matrix. The recursive relationship to obtain $\{D_T^{(n+1)}\}$ becomes,

$$\left\{D_T^{(n+1)}\right\} = \frac{2\Delta t}{(c\rho AL)^{(e)}}\left(\left\{R^{(n)}\right\} - \left([K] - \frac{1}{\Delta t}[C]\right)\left\{D_T^{(n)}\right\}\right) \tag{11.60}$$

Note that all terms on the right hand side of this equation are evaluated in the current/known time step. Thus Eq. (11.60) provides a particularly effective method. But, as in many good things in life, this advantage comes with some conditions.

TABLE 11.1 The stability and accuracy characteristics of the generalized trapezoidal method of time integration. α is a parameter that controls the stability and accuracy of the time integration, and it is in the range, $0 \le \alpha \le 1$. λ_{max} is the maximum eigenvalue of the system.

α		Stability	Δt_{cr}	Accuracy
0	Forward difference, Euler	Conditional	$\frac{2}{(1-2\alpha)\lambda_{max}}$	$O(\Delta t)$
$< \frac{1}{2}$		Conditional	$\frac{2}{(1-2\alpha)\lambda_{max}}$	
$< \frac{1}{2}$		Unconditional		
$\frac{1}{2}$	Crank–Nicholson, trapezoidal	Unconditional		$O(\Delta t^2)$
$\frac{2}{3}$	Galerkin	Unconditional		
1	Backward difference	Unconditional		$O(\Delta t)$

We will show in the next section that the explicit time integration method that results from the generalized trapezoidal method is only $O(\Delta t)$ accurate and it is conditionally stable. This means that the time step must be chosen according to a stability criterion (Table 11.1 and Section 11.6) in order to obtain a stable and accurate solution method.

11.5.1.2 Implicit time integration method

In case, the parameter α is not equal to zero, the effective stiffness matrix remains as given in Eq. (11.57),

$$[K_{eL}] = \frac{1}{\Delta t}[C] + \alpha[K]$$

We see that both the conductivity and heat capacity matrices contribute to the effective stiffness matrix. The solution is advanced in time by using the approach, which uses the inverse of $[K_{eL}]$ as defined in Eq. (11.57). In general, a method that advances the solution by a matrix inversion operation is known as an *implicit time integration method* [1].

For systems with large degrees of freedom, matrix inversion could increase the computational effort. But, an advantage of the implicit time integration methods is that most are unconditionally stable (Table 11.1). This means that a relatively large time step size Δt can be chosen to advance the solution. Using a large time step size with an implicit time integration method would guarantee a solution, but not necessarily an accurate one. The concepts of stability, accuracy, and convergence are discussed in Section 11.6.

11.5.1.3 Explicit time integration for nonlinear problems

In nonlinear problems, the state of the material depends on the variables of the problem. Several different types of nonlinearities are typically encountered. For example, in the case of *material nonlinearity* the material properties depend on strain and/or temperature; in *geometric nonlinearity*, the deformations are so large that small strain approximations are not valid; in *load nonlinearity*, the external loading is a nonlinear function of the degrees of freedom. In Section 10.6.5 we showed the effect of the radiation heat flux on a boundary and how it introduces a nonlinearity into the global conductivity matrix.

Let's describe a generic first-order nonlinear, time-dependent matrix equation,

$$[C(D_T)]\{\dot{D}_T\} + [K(D_T)]\{D_T\} = \{R\} \tag{11.61}$$

This equation appears to be particularly challenging as the $[C]$ and the $[K]$ matrices can only be computed if the nodal degrees of freedom D_T are known. If we compute these matrices at the known time step, with $\{D_T^{(n)}\}$ the general solution method can be expressed as follows:

$$\left[K_{eL}\left(D_T^{(n)}\right)\right]\left\{D_T^{(n+1)}\right\} = \{R_e\} \tag{11.62}$$

Thus we see that for nonlinear problems, the effective stiffness matrix should be computed at each time step. As discussed above, if we use an implicit time integration method, the solution can only be advanced if the $[K_{eL}(D_T^{(n)})]$ is inverted at each time step. For nonlinear problems, the implicit method could be computationally expensive.

The matrix inversion can be avoided, however, by using an explicit time integration method. The ability to avoid matrix inversion provided by this approach is advantageous for systems with large degrees of freedom as it results in reduction in computational effort.

11.6 Convergence, stability, and accuracy of time integration methods

Ideally, we would like a numerical time integration method to predict the exact result. A time integration algorithm is said to be *convergent* if for a fixed time step $t^{(n)}$ the numerical solution approaches the exact solution as the time step size is reduced toward zero, that is,

$$\left\{D^{(n)}\right\} \to \left\{D\left(t^{(n)}\right)\right\} \text{ as } \Delta t \to 0 \tag{11.63}$$

Convergence of a time integration algorithm depends on its *stability* and *consistency*. A *stable* time integration algorithm behaves in the same manner as the analytical solution. *Consistency* and *rate of convergence* ensure that the truncation error after temporal discretization remains below the following threshold,

$$|\tau| < c\Delta t^k \tag{11.64}$$

where c is a constant independent of Δt, k is a positive number named the *rate of convergence*.

The *Lax equivalence theorem* ensures that a consistent and stable time integration algorithm is also convergent [1–3].

There are several techniques to study the stability characteristics of a time integration algorithms. Here we follow two of these approaches. The *modal approach* described by Hughes for the generalized trapezoidal method [3] gives results applicable to all first-order semidiscrete equations and identifies a critical time step for the stability in terms of the system's eigenvalues. We supplement this specifically for the heat equation by using the *Fourier–von Neumann stability analysis* of the generalized trapezoidal method in order to interpret the stability conditions from physical and numerical perspectives.

11.6.1 Modal expansion of the semidiscrete first-order equation

The homogeneous form of the first-order semidiscrete matrix Eq. (11.1) and its initial condition are given as follows:

$$[C]\{\dot{D}\} + [K]\{D\} = 0 \tag{11.65}$$

$$\{D(0)\} = \left\{D^{(0)}\right\} \tag{11.66}$$

where $\{D^{(0)}\}$ represents the initial values of the nodal degrees of freedom at $t = 0$. The temperature vector can be expressed by using the method of eigenvector expansion as follows:

$$\{D\} = \sum_{m=1}^{N} T_m(t)\{S_m\} \tag{11.67}$$

$$\left\{\dot{D}\right\} = \sum_{m=1}^{N} \dot{T}_m(t)\{S_m\} \tag{11.68}$$

where N is the number of active degrees of freedom in the system, $T_m(t)$ are the time-dependent coefficients, and $\{S_m\}$ are the eigenvectors of the system. The eigenvectors are orthogonal with respect to the $[C]$ matrix [4],

$$\{S_l\}^T[C]\{S_m\} = \delta_{lm} \tag{11.69}$$

and have the property,

$$\{S_l\}^T[K]\{S_m\} = \lambda_l\delta_{lm} \tag{11.70}$$

where δ_{lm} is the Kronecker delta[2] and λ_1 is an eigenvalue. Substituting Eqs. (11.67) and (11.68) into Eq. (11.65) we find,

$$\sum_{m=1}^{N} \left([C]\dot{T}_m(t) + [K]T_m(t)\right)\{S_m\} = 0 \tag{11.71}$$

2. $\delta_{lm} = 1$ if $l = m$ and $\delta_{lm} = 0$ if $l \neq m$.

Recalling that the eigenvectors $\{S_m\}$ are nonzero, we note that the term in the bracket must be equal to zero for each m value. Multiplying each such term with $\{S_l\}^T$ and invoking the orthogonality conditions given in Eqs. (11.69) and (11.70), we find,

$$\dot{T}_l(t) + \lambda_l T_l = 0 \tag{11.72}$$

The time-dependent coefficients can then be found by solving Eq. (11.72) as follows:

$$T_l(t) = c_l e^{-\lambda_l t} \tag{11.73}$$

where c_l are scalar constants determined from the initial conditions.

In order to find an equation for the eigenvalues of the system, we combine Eq. (11.73) with (11.67) and (11.68). Substituting the result into Eq. (11.65) we find,

$$\sum_{m=1}^{N} ([K] - \lambda_m [C])\{S_m\} = 0 \tag{11.74}$$

This expression must be correct for each m value and defines the following eigenvalue problem:

$$([K] - \lambda_m [C])\{S_m\} = 0 \tag{11.75}$$

A unique eigenvector λ_m must be identified for each eigenvector $\{S_m\}$, and the following equation must be valid for Eq. (11.75) to have a nontrivial solution:

$$\det([K] - \lambda_m [C]) = 0 \tag{11.76}$$

This equation is known as the *characteristic equation* of the system and results in an N-term polynomial. Solution of this equation gives the eigenvalues of the system. In the heat transfer problem the eigenvalues are positive and ordered as follows $0 \le \lambda_1 \le \lambda_2 \le \cdots \le \lambda_N$.

11.6.2 Stability of the semidiscretized first-order equation

The stability of the time response of the semidiscretized heat equation can be determined from Eq. (11.72). To this end, let's consider a time interval $t^{(n)} \le t \le t^{(n+1)}$. If the value of $T_l(t^{(n)})$ is taken as the initial condition $T_l(t^{(n)}) = c_l(t^{(n)})$, the solution of Eq. (11.73) becomes,

$$T_l(t) = c_l\left(t^{(n)}\right)e^{\lambda_l\left(t^{(n)} - t\right)} \tag{11.77}$$

By using Eq. (11.77) at $t = t^{(n+1)}$, the coefficient $c_l(t^{(n+1)})$ can be found as follows,

$$c_l\left(t^{(n+1)}\right) = c_l\left(t^{(n)}\right)e^{\lambda_l\left(t^{(n)} - t^{(n+1)}\right)} \tag{11.78}$$

We can define an *amplification factor* A that represents the ratio of the temporal coefficients as follows,

$$A = \frac{c_l\left(t^{(n+1)}\right)}{c_l\left(t^{(n)}\right)} = e^{-\lambda_1 \Delta t} \tag{11.79}$$

We conclude only one of the two following conditions will hold, based on Eq. (11.79) and by noting that eigenvalues of this system and Δt are all positive:

$$c_l\left(t^{(n+1)}\right) < c_l\left(t^{(n)}\right) \text{ for } \lambda_l > 0 \tag{11.80a}$$

$$c_l\left(t^{(n+1)}\right) = c_l\left(t^{(n)}\right) \text{ for } \lambda_l = 0 \tag{11.80b}$$

This result indicates that the solution either becomes smaller in time or does not change, for first-order systems with positive eigenvalues. We seek the same type of behavior for the generalized trapezoidal algorithm.

11.6.3 Modal expansion of the generalized trapezoidal algorithm

Let's express the temperature vectors at time steps n and $n+1$ by using eigenfunction expansion,

$$\left\{D^{(n)}\right\} = \sum_{m=1}^{N} c_m^{(n)}\{S_m\} \text{ and } \left\{D^{(n+1)}\right\} = \sum_{m=1}^{N} c_m^{(n+1)}\{S_m\} \tag{11.81}$$

where $\{S_m\}$ are the eigenvectors and $c_m^{(n)}$ and $c_m^{(n+1)}$ are the coefficients that represent the temporal component of the response similar to Eq. (11.67). Substituting Eqs. (11.81) in to the generalized trapezoidal form of the fully discretized heat Eq. (11.57) we find,

$$\sum_{m=1}^{N} [K_{eL}]c_m^{(n+1)}\{S_m\} = \sum_{m=1}^{N} [K_{eR}]c_m^{(n)}\{S_m\} + \left\{R^{(n+\alpha)}\right\} \tag{11.82}$$

Multiplying both sides of this equation by $\{S_l\}^T$, using Eqs. (11.69) and (11.70) and after some algebra, we find the modal form of the fully discretized generalized trapezoidal method,

$$(1 + \alpha\Delta t\lambda_l)c_l^{(n+1)} = \Delta t F_l^{(n+\alpha)} + (1 - \Delta t(1-\alpha)\lambda_l)c_l^{(n)} \tag{11.83}$$

where $F_l^{(n+\alpha)} = \{S_l\}^T\{R^{(n+\alpha)}\}$ and representing,

$$F_l^{(n+\alpha)} = (1-\alpha)F_l^{(n)} + \alpha F_l^{(n+1)} \tag{11.84}$$

11.6.4 Stability of the generalized trapezoidal algorithm

The stability characteristics of the generalized trapezoidal algorithm can be determined from the homogeneous form of Eq. (11.83). The amplification factor

A in this case becomes,

$$A_l = \frac{c_l^{(n+1)}}{c_l^{(n)}} = \frac{1 - \Delta t(1-\alpha)\lambda_l}{1 + \alpha \Delta t \lambda_l} \tag{11.85}$$

We note that A_l should behave in the same manner as the amplification factor of the semidiscrete equation given in Eq. (11.79). Therefore, we can see that in order for the fully discretized system to behave as the semidiscretized system, the following bounds must be imposed on the amplification factor,

$$-1 < A_l < 1 \tag{11.86}$$

The left side of this inequality leads to the condition,

$$\Delta t < \frac{2}{(1-2\alpha)\lambda_l} \tag{11.87}$$

By recalling that $0 \le \alpha \le 1$ and $\Delta t > 0$ we observe that if $0 \le \alpha < 1/2$, the right hand side of Eq. (11.87) will be positive. Thus in this range Δt will have to be smaller than the limiting value of $2/(1-2\alpha)\lambda$. If $\alpha = 1/2$ we find Δt must be a less than 1/0, which is true for all practical problems; therefore, no condition on Δt develops. Finally, if $1/2 < \alpha \le 1$ we find that $\Delta t \lambda_1 > 2/(2\alpha - 1)$, which is true as all eigenvalues are positive. Thus, based on this analysis we conclude that the generalized trapezoidal algorithm is,

$$\begin{aligned} &\text{Conditionally stable for} \quad 0 \le \alpha < \frac{1}{2} \text{ with } \Delta t < \Delta t_{cr} = \frac{2}{(1-2\alpha)\lambda_{\max}} \\ &\text{Unconditionally stable for } \frac{1}{2} \le \alpha \le 1 \end{aligned} \tag{11.88}$$

Note that for the conditionally stable case the largest eigenvalue $\lambda_{\max}$ has been used to ensure that the strictest condition is met. Eq. (11.88) gives the stability analysis of the semidiscrete first-order matrix Eq. (11.65) provided that all eigenvalues are positive.

11.6.5 Fourier–von Neumann stability analysis of the generalized trapezoidal method

We next present an analysis of the stability characteristics of the fully discretized heat equation by using the Fourier–von Neumann method. This analysis will give an insight to the stability limit in terms of the physical and numerical parameters of the system. To this end we will investigate the finite element form of the one-dimensional heat transfer problem. We will assume that the one-dimensional domain has uniform properties and it is discretized by using equal-length, linear elements. We will analyze the problem for both consistent and lumped heat capacity matrices.

11.6.5.1 Consistent heat capacity matrix

The recursive formula for the generalized trapezoidal method is given in Eq. (11.57). The conductivity and heat capacity matrices in this case can be determined from Eq. (9.30) and Eq. (11.21) respectively. We then find the finite element form of the heat equation for element-j as follows:

$$\frac{1}{6}\left(c^{(j)}\rho^{(j)}A^{(j)}L^{(j)}\right)\begin{bmatrix} 2 & 1 \\ 1 & 2 \end{bmatrix}\begin{Bmatrix} \dot{T}_j \\ \dot{T}_{j+1} \end{Bmatrix}+\left(\frac{k^{(j)}A^{(j)}}{L^{(j)}}\right)\begin{bmatrix} 1 & -1 \\ -1 & 1 \end{bmatrix}\begin{Bmatrix} T_j \\ T_{j+1} \end{Bmatrix}=\begin{Bmatrix} R_j \\ R_{j+1} \end{Bmatrix} \tag{11.89}$$

Assembly of all the element equations leads to the global equilibrium equation given in Eq. (11.65).

The load vector does not affect the stability of the system, hence we omit it from following analysis. The homogeneous form of the generalized trapezoidal method can be obtained from Eqs. (11.52) and (11.57),

$$\left(\frac{1}{\Delta t}[C]+\alpha[K]\right)\left\{D_T^{(n+1)}\right\}=-\left((1-\alpha)[K]-\frac{1}{\Delta t}[C]\right)\left\{D_T^{(n)}\right\} \tag{11.90}$$

By using Eqs. (11.89) and (11.90), the jth row of the global heat transfer equation is obtained by assembling elements-j and $j-1$,

$$\begin{aligned}&\frac{1}{6}\left(T_{j-1}^{(n+1)}+4T_j^{(n+1)}+T_{j+1}^{(n+1)}\right)+\alpha s\left(-T_{j-1}^{(n+1)}+2T_j^{(n+1)}-T_{j+1}^{(n+1)}\right)\\&=\frac{1}{6}\left(T_{j-1}^{(n)}+4T_j^{(n)}+T_{j+1}^{(n)}\right)-(1-\alpha)s\left(-T_{j-1}^{(n)}+2T_j^{(n)}-T_{j+1}^{(n)}\right)\end{aligned} \tag{11.91}$$

where

$$s=\frac{k}{c\rho}\frac{\Delta t}{L^{(e)2}} \tag{11.92}$$

We will assume that the spatial and temporal components of the temperature response are separable, and the temperature at time step-n and node-j is given as follows [5]:

$$T(x,t)\approx\tilde{T}_j^{(n)}=e^{i\beta x}Q^{t/\Delta t} \tag{11.93}$$

We will recall that in the discretized problem we have $x=j\Delta x$ and $t=n\Delta t$. Using these variables in Eq. (11.93) we find,

$$\tilde{T}_j^{(n)}=e^{i\beta\Delta x j}Q^n \tag{11.94}$$

In what follows, we drop the approximation sign $\sim$ for the sake of notational simplicity. Substituting Eq. (11.94) into (11.91) and dividing the resulting equation by $e^{i\beta\Delta xj}$ we find[3],

$$\left[\frac{1}{3}(2+\cos\beta\Delta x)+2\alpha s(1-\cos\beta\Delta x)\right]Q^{n+1}=\left[\frac{1}{3}(2+\cos\beta\Delta x)-2s(1-\alpha)(1-\cos\beta\Delta x)\right]Q^{n} \tag{11.95}$$

Thus an amplification ratio A can be expressed as follows:

$$A=\frac{Q^{n+1}}{Q^{n}}=\frac{2+\cos\beta\Delta x-6s(1-\alpha)(1-\cos\beta\Delta x)}{2+\cos\beta\Delta x+6\alpha s(1-\cos\beta\Delta x)} \tag{11.96}$$

We expect the amplitude of the response $Q^{t/\Delta}$ to be bounded due to the nature of the problem as shown in Section 11.6.3. Therefore, the amplification ratio should lie between the following limits:

$$-1\le A\le 1 \tag{11.97}$$

The case $A=1$ represents the steady state condition and $0\le A<1$ represent monotonically decaying response amplitude. If A is in the range $-1<A<0$ the amplitude decays in an oscillatory manner. While we do not expect this result from the exact solution, this type of numerical solution indicates a stable but perhaps inaccurate time response. Finally, $A=-1$ indicates stable oscillations.

We will next analyze the two ends of the range defined in Eq. (11.97). It can easily be shown that the case $A\le 1$ leads to trivial identities that do not impose any restrictions between s and α. But, in the case $-1\le A$, Eq. (11.96) reduces to,

$$s<\frac{(2+\cos\beta\Delta x)}{3(1-2\alpha)(1-\cos\beta\Delta x)} \tag{11.98}$$

Thus we see that the relationship between s and α depends on $\cos\beta\Delta x$. This term is limited to be in the range $-1\le\cos\beta\Delta x\le 1$. Using the upper and lower limits of this variable should be sufficient to analyze the behavior of Eq. (11.98).

In the case, $\cos\beta\Delta x=1$, Eq. (11.98) becomes $s\le 1/0$, which is always true. But, in the case, $\cos\beta\Delta x=-1$, Eq. (11.98) becomes,

$$s<\frac{1}{6(1-2\alpha)} \tag{11.99}$$

Here we see two conditions on α develop. Recalling the definition of s given in Eq. (11.92), we find that Eq. (11.99) can be reduced to,

$$\text{for } 0<\alpha<\frac{1}{2}:\ \Delta t\le\Delta t_{cr}=\frac{1}{6}\frac{1}{(1-2\alpha)}\left(\frac{c\rho L^{(e)2}}{k}\right) \tag{11.100a}$$

$$\text{for } \frac{1}{2}\le\alpha<1:\ \ \Delta t\ge-\frac{1}{6}\frac{1}{(2\alpha-1)}\left(\frac{c\rho L^{(e)2}}{k}\right) \tag{11.100b}$$

The condition in part (a) of this equation indicates that for values of α in the range $0\le\alpha<1/2\Delta t$ must be smaller than the *critical time step* Δt_{cr}. The

3. Also using the relationship $\cos\beta\Delta x=(e^{i\beta\Delta xj}+e^{-i\beta\Delta xj})/2$.

condition in part (b), on the other hand is always true; and, it is interpreted as the algorithm will be stable for all values of Δt, if α is in the range $1/2 \le \alpha \le 1$.

11.6.5.2 Lumped heat capacity matrix

The finite element form of the heat equation by using the lumped heat capacity matrix given in Eq. (11.23) is as follows,

$$\frac{1}{2}\left(c^{(j)}\rho^{(j)}A^{(j)}L^{(j)}\right)\begin{bmatrix}1 & 0\\ 0 & 1\end{bmatrix}\begin{Bmatrix}\dot{T}_j\\ \dot{T}_{j+1}\end{Bmatrix}+\left(\frac{k^{(j)}A^{(j)}}{L^{(j)}}\right)\begin{bmatrix}1 & -1\\ -1 & 1\end{bmatrix}\begin{Bmatrix}T_j\\ T_{j+1}\end{Bmatrix}=\begin{Bmatrix}R_j\\ R_{j+1}\end{Bmatrix} \tag{11.101}$$

By using Eqs. (11.101) and (11.90), the jth row of the global heat equation is obtained by assembling elements-j and $j-1$,

$$T_j^{(n+1)}+\alpha s\left(-T_{j-1}^{(n+1)}+2T_j^{(n+1)}-T_{j+1}^{(n+1)}\right)=T_j^{(n)}-(1-\alpha)s\left(-T_{j-1}^{(n)}+2T_j^{(n)}-T_{j+1}^{(n)}\right) \tag{11.102}$$

Upon substituting Eq. (11.94) in Eq. (11.102) we find the amplification ratio for the generalized trapezoidal method with lumped heat capacity matrix as follows:

$$A=\frac{Q^{n+1}}{Q^n}=\frac{1-2(1-\alpha)s(1-\cos\beta\Delta x)}{1+2\alpha s(1-\cos\beta\Delta x)} \tag{11.103}$$

Analysis of this amplification ratio as presented above leads to the following stability condition:

$$\text{for } 0<\alpha<\frac{1}{2}:\ \Delta t<\Delta t_{cr}=\frac{1}{2}\frac{1}{(1-2\alpha)}\left(\frac{c\rho L^{(e)2}}{k}\right) \tag{11.104a}$$

$$\text{for } \frac{1}{2}\le\alpha<1:\ \text{no condition on } \Delta t \text{ develops} \tag{11.104b}$$

11.6.5.3 Stability of the heat equation discretized with generalized trapezoidal algorithm

The results presented in the two preceding subsections show that the generalized trapezoidal algorithm is, unconditionally stable for $1/2 \le \alpha \le 1$, conditionally stable for $0 \le \alpha \le 1/2$ with $\Delta t < \Delta t_{cr}$.

The critical time step is,

$$\Delta t_{cr}=\frac{1}{2}\frac{1}{(1-2\alpha)}\left(\frac{c\rho L^{(e)2}}{k}\right)\ \text{for the lumped heat capacity matrix} \tag{11.105a}$$

$$\Delta t_{cr}=\frac{1}{6}\frac{1}{(1-2\alpha)}\left(\frac{c\rho L^{(e)2}}{k}\right)\ \text{for the consistent heat capacity matrix.} \tag{11.105b}$$

Therefore, we see that the critical time step is larger when the lumped heat capacity matrix is used. For a large number of time steps this could help reduce computational effort.

Analytical approaches to the solution of the heat transfer equation show that the rate of temperature propagation is infinite. By using one of Eqs. (11.105), we can see that the numerical speed of temperature propagation is proportional to the thermal diffusivity $(k/c\rho)$ of the problem,

$$\frac{L^{(e)}}{\Delta t} \propto \frac{k}{c\rho}\frac{1}{L^{(e)}} \tag{11.106}$$

Eq. (11.106) shows that while the numerical propagation speed is finite, it increases with decreasing element size. Thus we see that mesh-size refinement should improve the numerical temporal response.

Note that the preceding analysis is limited to the assumptions we made at the onset: use of linear interpolation functions and analysis of row-j of the stiffness matrix. It is reasonable to assume that the boundary conditions would modify these results. In fact, the more generalized method involving eigenvalue expansion presented in Section 11.6.4, resulting in the stability condition given in Eq. (11.88) includes the effects of the boundary conditions. Comparison of Eqs. (11.88) and Eq. (11.105) helps interpret the physical content of the eigenvalues of the heat transfer problem.

11.6.6 Consistency and rate of convergence

Let's recall the modal form of the generalized trapezoidal algorithm given in Eq. (11.83),

$$(1+\alpha\Delta t\lambda_l)c_l^{(n+1)} = \Delta t F_l^{(n+\alpha)} + (1-\Delta t(1-\alpha)\lambda_l)c_l^{(n)} \tag{11.107}$$

Recalling the amplification factor A derived for this method in Eq. (11.85) we can express Eq. (11.107) as follows,

$$c_l^{(n+1)} - Ac_l^{(n)} - \frac{\Delta t}{1+\alpha\Delta t\lambda_l}F_l^{(n+\alpha)} = 0 \tag{11.108}$$

If we replace the numerical values of $c_l^{(n)}$ and $c_l^{(n+1)}$ with the exact values $c_l(t^{(n)})$ and $c_l(t^{(n+1)})$, respectively, we need to take into account the *truncation error*, $\tau(t^{(n)})$. In this case Eq. (11.108) becomes,

$$c_l\left(t^{(n+1)}\right) - Ac_l\left(t^{(n)}\right) - \frac{\Delta t}{1+\alpha\Delta t\lambda_l}F_l^{(n+\alpha)} = \Delta t\tau\left(t^{(n)}\right) \tag{11.109}$$

It can be shown that the truncation error is of the following form [3],

$$\tau\left(t^{(n)}\right) = (1-2\alpha)O(\Delta t) + O\left(\Delta t^2\right) \tag{11.110}$$

where $O(\Delta t^n)$ indicates the order of the time-dependent terms in this equation.

This shows that the truncation error is,

$$\tau = O(\Delta t^2) \text{ for } \alpha = \frac{1}{2} \tag{11.111a}$$

$$\tau = O(\Delta t) \text{ for all of } \alpha \text{ in the range } 0 \le \alpha < \frac{1}{2} \text{ and } \frac{1}{2} < \alpha \le 1 \tag{11.111b}$$

Table 11.1 gives a summary of the stability and accuracy of the generalized trapezoidal method and all the methods that it represents.

Example 11.1 Write a finite element program to solve the following boundary/initial value problem that describes transient heat transfer in one-dimensional domain:

$$\frac{\partial}{\partial x}\left(kA\frac{\partial T}{\partial x}\right) - \rho c A\frac{\partial T}{\partial t} = 0 \text{ in } 0 < x < L \text{ and } t > 0 \tag{11.112}$$

$$\begin{aligned} \text{Boundary conditions: at } x = 0, &\quad T(0,t) = T_0 \text{ and } t > 0 \\ \text{at } x = L, &\quad T(L,t) = T_L \text{ and } t > 0 \end{aligned} \tag{11.113}$$

$$\text{Initial condition:} \qquad \text{at } t = 0, T(x,0) = T^{(0)} \text{ in } 0 \le x \le L \tag{11.114}$$

where

$$T^{(0)} = \begin{cases} T_0(0), & 0 \le x \le L/2 \\ T_0 + 4(T_M - T_0)(x/L - 1/2), & L/2 \le x \le 3L/4 \\ T_M + 4(T_L - T_M)(x/L - 3/4), & 3L/4 \le x \le L \end{cases} \tag{11.115}$$

with $T_0(t) = 5(1 + 2.5 \times 10^{-2}\sin(\pi t/500))$ °C, $T_M = 25$°C and $T_L = 15$°C. Use uniform length, linear elements and the following parameters to solve this problem: $L = 1$ m, $k = 220$ W/mK, $\rho = 2,700$ kg/m^3, $c = 920$ J/kgK, and total duration of $t_F = 5 \times 10^3$ s.

a) By using the generalized trapezoidal method, carry out an analysis of the critical time step Δt_{cr} for diagonal and consistent mass matrices.
Use $N_E = 25, 50, 100, 200$, and $\alpha = 0, 0.25, 0.5, 0.75$, and 1.

b) Demonstrate the effect of critical time step for the conditionally stable algorithms by using the consistent heat capacity matrix, for $\alpha = 0.25$ and $N_E = 100$.

Solution:

We use the C^0 linear interpolation elements in this solution. The conductivity and heat capacity matrices for such elements are given in Eqs. (9.30), (11.21) and (11.23), respectively. For demonstration purpose, let's assume the total number of elements is $N_E = 4$. The element length for a uniform mesh size is $L^{(e)} = L/N_E$ and the problem has $N_E + 1$ degrees of freedom. The semidiscrete form of the global heat equilibrium equation upon element assembly is represented

as follows:

$$[C]\{\dot{D}_T\} + [K]\{D_T\} = 0 \tag{11.116}$$

Note that the load vector due to internal heat source is zero in this problem. Eq. (11.37) becomes by using the consistent heat capacity matrix, and four elements,

$$\frac{c\rho L^{(e)}}{6}\begin{bmatrix} 2 & 1 & 0 & 0 & 0 \\ 1 & 4 & 1 & 0 & 0 \\ 0 & 1 & 4 & 1 & 0 \\ 0 & 0 & 1 & 4 & 1 \\ 0 & 0 & 0 & 1 & 2 \end{bmatrix}\begin{Bmatrix} \dot{T}_1 \\ \dot{T}_2 \\ \dot{T}_3 \\ \dot{T}_4 \\ \dot{T}_5 \end{Bmatrix} + \frac{k}{L^{(e)}}\begin{bmatrix} 1 & -1 & 0 & 0 & 0 \\ -1 & 2 & 1 & 0 & 0 \\ 0 & -1 & 2 & -1 & 0 \\ 0 & 0 & -1 & 2 & -1 \\ 0 & 0 & 0 & -1 & 1 \end{bmatrix}\begin{Bmatrix} T_1 \\ T_2 \\ T_3 \\ T_4 \\ T_5 \end{Bmatrix} = 0 \tag{11.117}$$

Let's use the generalized trapezoidal method (11.57) to discretize Eq. (11.37). Noting the internal heat load vector is zero we find,

$$\left(\frac{1}{\Delta t}[C] + \alpha[K]\right)\left\{D_T^{(n+1)}\right\} = \left(\frac{1}{\Delta t}[C] - (1-\alpha)[K]\right)\left\{D_T^{(n)}\right\} \tag{11.118}$$

In this problem, both boundaries are subjected to essential boundary conditions, such that $T_1 = T_0(t)$ and $T_5 = T_L$. These two essential boundary conditions are implemented as described in Section 4.7.2. The fully discretized equation becomes,

$$\begin{aligned}
&\left(\frac{c\rho L^{(\varepsilon)}}{6\Delta t}\begin{bmatrix} 4 & 1 & 0 \\ 1 & 4 & 1 \\ 0 & 1 & 4 \end{bmatrix} + \frac{\alpha k}{L^{(e)}}\begin{bmatrix} 2 & -1 & 0 \\ -1 & 2 & -1 \\ 0 & -1 & 2 \end{bmatrix}\right)\begin{Bmatrix} T_2 \\ T_3 \\ T_4 \end{Bmatrix}^{(n+1)} \\
&= \left(\frac{c\rho L^{(e)}}{6\Delta t}\begin{bmatrix} 4 & 1 & 0 \\ 1 & 4 & 1 \\ 0 & 1 & 4 \end{bmatrix} - \frac{(1-\alpha)k}{L^{(e)}}\begin{bmatrix} 2 & -1 & 0 \\ -1 & 2 & -1 \\ 0 & -1 & 2 \end{bmatrix}\right)\begin{Bmatrix} T_2 \\ T_3 \\ T_4 \end{Bmatrix} \\
&- \begin{Bmatrix} \left(\frac{c\rho L^{(\varepsilon)}}{6\Delta t} + \frac{\alpha k}{L^{(e)}}\right)T_0^{(n+1)} \\ 0 \\ \left(\frac{c\rho L^{(e)}}{6\Delta t} + \frac{\alpha k}{L^{(e)}}\right)T_L \end{Bmatrix} + \begin{Bmatrix} \left(\frac{c\rho L^{(e)}}{6\Delta t} - \frac{(1-\alpha)k}{L^{(\varepsilon)}}\right)T_0^{(n)} \\ \left(\frac{c\rho L^{(e)}}{6\Delta t} - \frac{(1-\alpha)k}{L^{(e)}}\right)T_L \end{Bmatrix}
\end{aligned} \tag{11.119}$$

Extension of this method to lumped mass matrix is straight forward. Note that as the boundary condition at $x = 0$ varies with time $T_0^{(n)}$ and $T_0^{(n+1)}$ can be easily computed for each time step. The initial condition, $T^{(0)}$ for nodes-2, -3, and -4 can be computed by using Eq. (11.115). Eq. (11.119) represents the recursive equation to solve the problem with $N = 4$. This approach is generalized in to a computer code by using the general algorithm outlined in Section 11.5, and the rest of the problem is thus solved.

a) Analysis of the critical time step

The eigenvalue problem associated with the semidiscretized form of the problem is given in Eq. (11.76) as follows:

$$([K] - \lambda[C])\{S\} = 0 \tag{11.120}$$

where the global conductivity matrix $[K]$, the consistent $[C_c]$, and diagonal $[C_d]$ heat capacity matrices are,

$$[K] = \frac{k}{L^{(e)}}\begin{bmatrix} 1 & -1 & 0 & & \\ & \cdots & & & \\ & -1 & 2 & -1 & \\ & & & \cdots & \\ & & 0 & -1 & 1 \end{bmatrix}, \quad [C_c] = \frac{c\rho L^{(e)}}{6}\begin{bmatrix} 2 & 1 & 0 & & \\ & \cdots & & & \\ & 1 & 4 & 1 & \\ & & & \cdots & \\ & & 0 & 1 & 2 \end{bmatrix}, \quad [C_d] = \frac{c\rho L^{(e)}}{2}[I] \tag{11.121}$$

with $[I]$ as the identity matrix. The size of all the matrices is $(N_E + 1) \times (N_E + 1)$. In this solution, the eigenvalues and the eigenvectors of this system are found by using Matlab's eigenfunction,

```
[S, lambda] = eig(K, C)
```

where S is a matrix whose columns represent the eigenvectors, *lambda* is a diagonal matrix that contains the eigenvalues of the problem. Table 11.2 shows the maximum eigenvalue of the system for different mesh sizes, α values and choice of heat capacity matrix. The critical time step determined by using the stability condition (11.88), and $\Delta t_{cr(\lambda)} = 2/(1 - 2\alpha)\lambda_{max}$ is calculated. This critical time step is compared to the critical time step developed for the heat Eq. (11.105a).

Table 11.2 shows that the maximum eigenvalue λ_{max} of the system increases with decreasing element size. Consequently, the critical time step size $\Delta t_{cr(\lambda)}$ also decreases in inverse proportion to λ_{max}. Increasing the α value in the conditionally stable range $(0 \leq \alpha < 1/2)$ helps to increase the critical time step size. It is also worth noting that the critical time step size determined for the heat equation matches the one obtained from eigenvalue analysis $\Delta t_{cr(\lambda)}$.

b) Conditionally stable algorithms

As we are using $\alpha = 0.25$, we expect the solution algorithm to be conditionally stable. Analysis provided Table 11.2 shows that for $\alpha = 0.25$ and $N_E = 100$ the critical time step for a stable solution is $\Delta t_{cr(\lambda)} = 0.3764$ seconds for the case of using the consistent heat capacity matrix. Fig. 11.4 shows the solution of this transient problem obtained by using $\Delta t_{cr}= 0.36$ seconds. In this case we see the solution behaves in a stable manner with a bounded response. The temperature distribution along the solution domain experiences an initial transition period following the rather abrupt looking initial temperature distribution marked by

TABLE 11.2 Analysis of critical time step for the generalized trapezoidal algorithm by using diagonal or consistent mass matrices, for different values of N_E and α. Note that $\Delta t_{cr(\lambda)} = 2/(1-2\alpha)\lambda_{max}$ represents the critical time step determined by the eigenvalue analysis of the system and Δt_{cr} is the critical time step determined from Eq. (11.105).

N_E	λ_{max}	$\Delta t_{cr(\lambda)}(s)$			$\Delta t_{cr}(s)$	
		$\alpha = 0$	$\alpha = 1/4$	$\alpha = 1/2$	$\alpha = 0$	$\alpha = 1/4$
Diagonal $[C_d]$ matrix						
25	0.2214	9.0327	18.0655	–	9.0327	18.0655
50	0.8857	2.2582	4.5164	–	2.2582	4.5164
100	3.5427	0.5645	1.1291	–	0.5645	1.1291
200	14.1707	0.1411	0.2823	–	0.1411	0.2823
Consistent $[C_c]$ matrix						
25	0.6643	3.0109	6.0218	–	3.0109	6.0218
50	2.6570	0.7527	1.5055	–	0.7527	1.5055
100	10.6280	0.1882	0.3764	–	0.1882	0.3764
200	42.5121	0.0470	0.0941	–	0.0470	0.0941

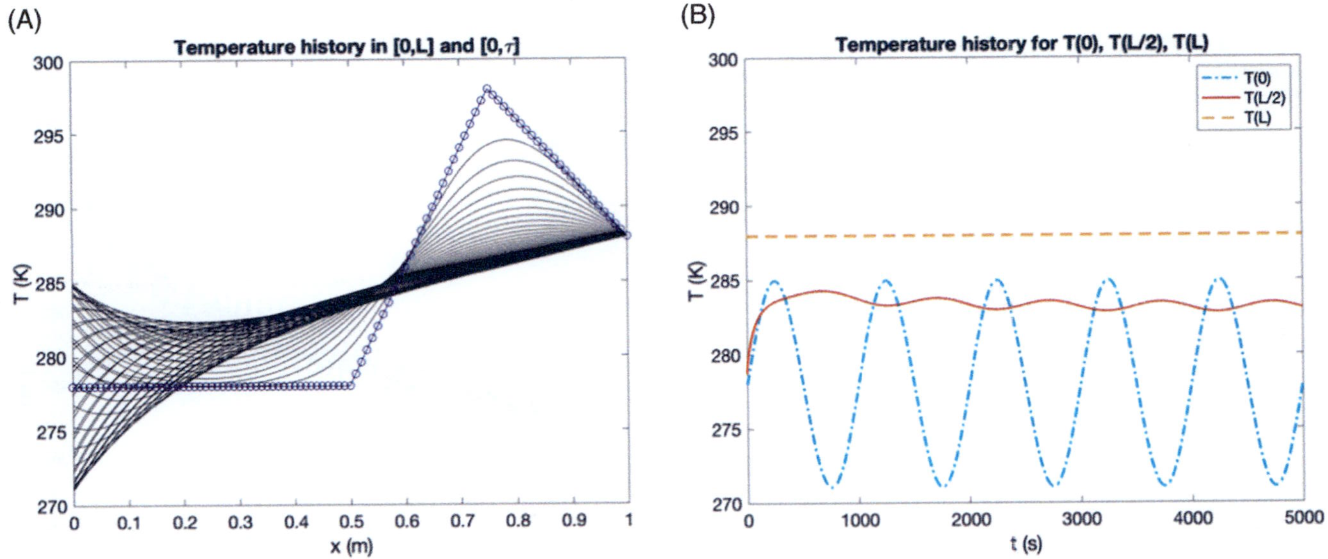

FIGURE 11.4 (A) Time history of the solution for $\tau = 5000$ seconds, printed with 100-second intervals. (B) Time history of the temperature variation at the left boundary $T(0)$, middle $T(L/2)$, and the right boundary $T(L)$ of the one-dimensional solution domain. The parameters used in this problem are $L = 1$ m, $k = 220$ W/m K, $\rho = 2,700$ kg/m^3, $c = 920$ J/kg K, $\tau = 5 \times 10^3$ seconds, $N = 100$, $\alpha = 0.25$, $\Delta t = 0.36$ seconds.

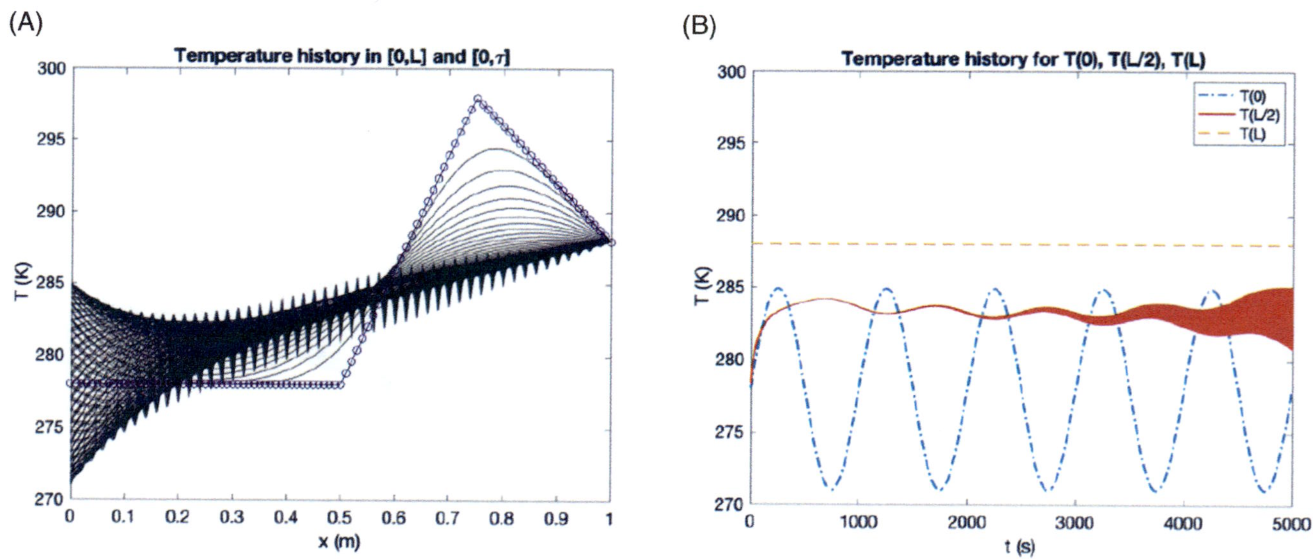

FIGURE 11.5 Demonstration of the effect of time step size on the stability of the transient solution. $\Delta t = 0.3768$ seconds was used to obtain this solution whereas the critical time step is $\Delta t_{cr(\lambda)} = 0.3764$ seconds for $\alpha = 0.25$. (A) Time history of the solution. (B) Time history of the temperature variation for $T(0)$, $T(L/2)$, and $T(L)$. Except for Δt, the parameters used in this problem are identical to those used in Fig. 11.4.

blue circles. In less than 500 seconds, the response becomes periodic due to the transient boundary condition prescribed on the left boundary.

Whether or not the solution provided in Fig. 11.4 is converged is not certain. In order to probe this question we would need to conduct a study of mesh and time step size concurrently. An example of such a study is provided in Example 11.2. But, in Fig. 11.5 where we used Δt_{cr}= 0.3768 seconds, we can certainly see choosing the time step size slightly larger than the critical time step size of $\Delta t_{cr(\lambda)} = 0.3764$ seconds, leads to unbounded oscillations after 1000 seconds of simulation. Clearly we would not pursue this approach.

Example 11.2 Solve the initial value problem described in Example 11.1 with the following boundary and initial conditions

$$\begin{aligned} \text{Boundary conditions: at } x = 0 \quad & T(0,t) = T_0 \quad \text{and } t > 0 \\ \text{at } x = L \quad & T(L,t) = T_L \quad \text{and } t > 0 \end{aligned} \tag{11.122}$$

$$\text{Initial condition:} \quad \text{at } t = 0, \quad T(x,0) = T^{(0)} \tag{11.123}$$

where

$$T^{(0)} = \begin{cases} T_0 + 2(T_M - T_0)(x/L), & 0 \le x \le L/2 \\ T_M + 2(T_L - T_M)(x/L - 1/2), & L/2 \le x \le L \end{cases} \tag{11.124}$$

with $T_0 = 0$, $T_M = 27$ and $T_L = 0$. Use the same material properties and element given in Example 11.1. Perform a mesh ($N = 4, 8, 16, 32, 64$) and time step ($\Delta t = 0.04, 0.4, 4, 40$ seconds) size convergence study for $\alpha = 0.5$. Compare the results to the analytical solution of the problem.

Solution:
Analytical solution of Eq. (11.112) can be obtained by using separation of variables as follows [6]:

$$T(x,\, t) = \sum_{n=1}^{\infty} b_n \sin\left(\frac{n\pi x}{L}\right) e^{-\left(\frac{n\pi}{L}\right)^2 \alpha_T^2 t} \tag{11.125}$$

where $\alpha_T^2 = k/c\rho$ is the *thermal diffusivity* of the material. The initial condition can be expanded in Fourier series,

$$T^{(0)}(x) = \sum_{n=1}^{\infty} b_n \sin\left(\frac{n\pi x}{L}\right) \tag{11.126}$$

and the constant coefficients b_n are found as follows:

$$b_n = \frac{2}{L} \int_0^L T^{(0)} \sin\left(\frac{n\pi x}{L}\right) dx \tag{11.127}$$

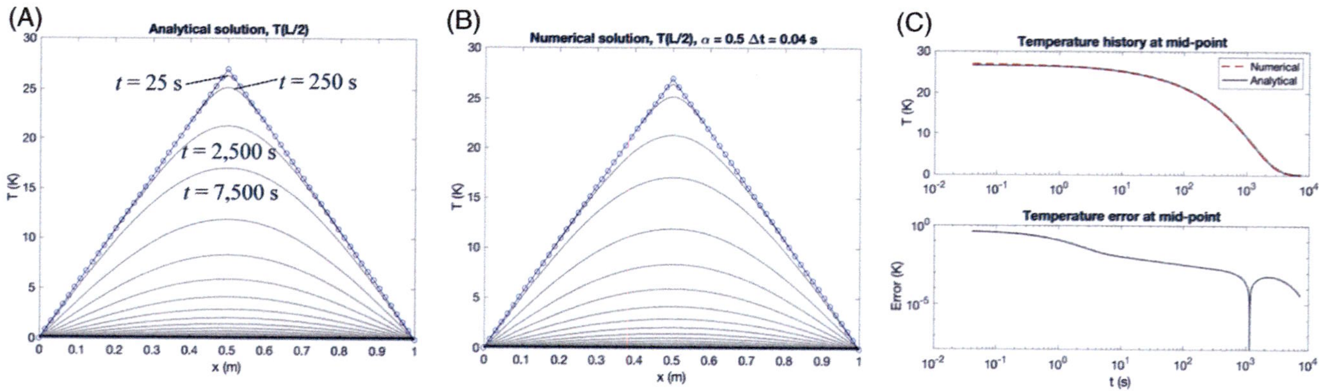

FIGURE 11.6 (A, B) Transient temperature history obtained with analytical and numerical solution approaches. Time steps shown are 25, 250, 2500, 7500, and increments of 10^4 thereafter. (C) History of the midpoint temperature $T(L/2)$ and the error between the two solution approaches. Parameters used in this solution are $\alpha = 0.5$, $\Delta t = 0.04$ seconds, $N = 64$.

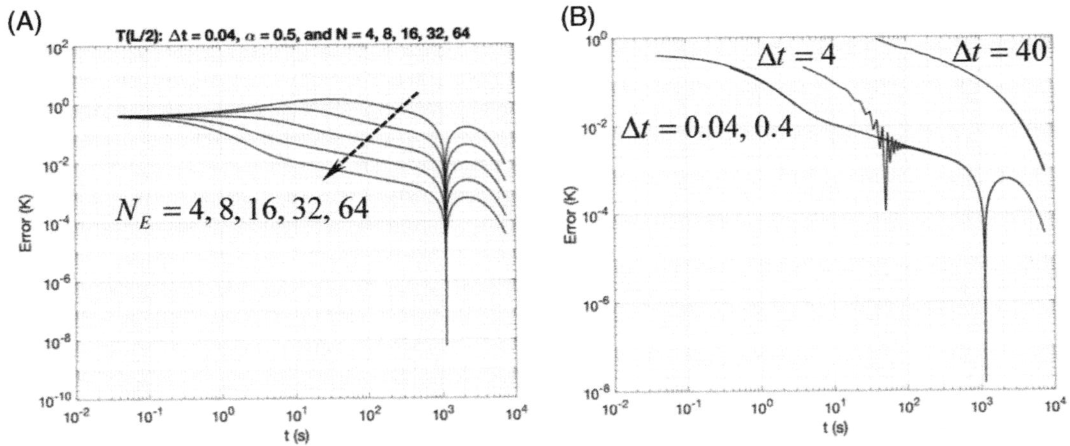

FIGURE 11.7 Effects of (A) mesh size and (B) time step size on the numerical solution error. Parameters used in this solution are $\alpha = 0.5$, (A) $N_E = 4, 8, 16, 32$, and 64, (B) $\Delta t = 0.04, 0.4, 4$, and 40 seconds.

By using Eqs. (11.124) and (11.127) we find the analytical solution as follows:

$$T(x,t) = 16T_M \sum_{n=1}^{\infty} \frac{1}{n^2\pi^2} \sin^2\left(\frac{n\pi}{4}\right) \sin\left(\frac{n\pi}{2}\right) \sin\left(\frac{n\pi x}{L}\right) e^{-\left(\frac{n\pi}{L}\right)^2 \alpha_T^2 t} \tag{11.128}$$

Numerical solution is carried out by using the program developed for Example 11.1 with the boundary and initial conditions given in Eqs. (11.122)–(11.124). Figs. 11.6A and 11.6B show the progressive cooling of the medium for α=0.5, N_E= 64, and Δt = 0.04 seconds. An upper limit of 25 terms provided sufficient accuracy for the series in Eq. (11.128). The first four curves are plotted at t = 25, 250, 2500, 7500 seconds, and the rest are plotted with 10^4-second intervals for a total of 37,500 seconds.

Comparison of the analytical (Fig. 11.6A) and numerical (Fig. 11.6B) results shows that they are virtually identical. The mid-point temperature and the error between the two solution approaches are compared in Fig. 11.6C. This figure shows that the *initiation error* in the solution is diminished as time evolves. There is a rapid drop in the error near the inflection point of the temperature. Temperature approaches its steady state value asymptotically, for both solution methods.

Effects of mesh and time step size on the transient solution are presented in Fig. 11.7 for N_E = 4, 8, 16, 32, and 64 and Δt = 0.04, 0.4, 4, and 40 seconds. As expected the error becomes smaller, in general, for smaller element size. However, the way the error is reduced is also different, and for smaller elements the solution starts to become more accurate earlier in the solution. Similarly, we see that Δt = 0.04 and 0.4 seconds have virtually the same transient response, but solution accuracy deteriorates for the large time steps of Δt = 4 and 40 seconds. In none of these four cases the error affects the steady state solution, but the earlier response of the system suffers with the larger time steps sizes.

References

[1] R.D. Richtmyer, K.W. Morton, Difference Methods for Initial-Value Problems, second ed., Interscience Publishers, New York, 1967.

[2] W.L. Wood, Practical Time-Stepping Schemes, Clarendon Press, Oxford, 1990.

[3] T.J.R. Hughes, The Finite Element Method, Linear Static and Dynamic Finite Element Analysis, Prentice-Hall, Englewood Cliffs, NJ, 1987.

[4] R.D. Cook, D.S. Malkus, M.E. Plesha, R.J. Witt, Concepts and Applications of Finite Element Analysis, fourth ed., Wiley, New York, NY, 2002.

[5] R. Haberman, Elementary Applied Partial Differential Equations, second ed., Prentice-Hall, Englewood Cliffs, NJ, 1987.

[6] W.E. Boyce, R.C. DiPrima, Elementary Differential Equations and Boundary Value Problems, John Wiley & Sons, New York, 1977.

Chapter 12

Transient analysis of solids and structures

12.1 Introduction

Deformation of a structure becomes time-dependent if it is subjected to time-dependent boundary conditions and/or body forces. In case the loading is cyclical, the need to carryout dynamic analysis depends on the comparison of the loading frequency and the natural frequencies of the system. The problem can be classified as *quasistatic* if the dynamic response is hardly larger than the static response. On the other hand, a fully *transient* solution may be required if the external loads acting on the structure are transient with no particular periodicity.

Mechanical vibrations is a rich and interconnected area in engineering science that spans from analysis of single degree of freedom, mass–spring–damper systems to wave propagation in solids. The goal of this chapter is to present the foundational principles of vibration analysis in finite element method in a step-by-step manner, staring from the single degree of freedom systems in Section 12.2 to full transient analysis of general solids in Section 12.10. The concepts introduced in Section 12.2, *natural frequency, damping ratio, critical damping, free* and *forced vibrations* form the foundations of advanced topics introduced in the rest of the chapter.

The boundary/initial value problems that represent the dynamics of a two-dimensional elastic solid, a deformable bar and an Euler–Bernoulli beam are presented in Section 12.3. An analytical solution to the free vibration of an Euler–Bernoulli beam is presented in Section 12.4. This solution not only presents basis for comparisons with the numerical solutions in the subsequent sections, but it also serves as an example to highlight the similarities between the analytical and numerical solution approaches. The semidiscrete versions of the boundary/initial value problems introduced in Section 12.3 are obtained by using the finite element approach in Section 12.5. The mass and damping matrices that are associated with the inertial and damping forces are discussed in Sections 12.6 and 12.7, respectively. In Section 12.8, the element stiffness matrices are assembled to represent the physics of the global problem in the same way that was introduced in the previous chapters.

Finite Element Method: Physics and Solution Methods. DOI: https://doi.org/10.1016/B978-0-12-821127-4.00007-4
Copyright © 2022 Elsevier Inc. All rights reserved.

The semidiscrete form of the equation of motion contains a wealth of information useful to assess the dynamic response of dynamical systems. Semidiscrete systems can be encountered in modeling multidegree of freedom lumped–mass systems, or by using finite difference or finite element discretization of continuous systems. Semidiscrete systems have as many critical frequencies as their degrees of freedom, and they vibrate with specific vibration modes when excited with one of these critical frequencies. In Section 12.9, we present analytical analysis of discrete systems. Eigenvalue analysis provides information on *critical frequencies* (*eigenvalues*) and *mode-shapes* (*eigenvectors*) of the system. By using the orthogonality property of the eigenvalues, the degrees of freedom can be decoupled. This provides an opportunity to introduce damping, specific to each mode and can be used to reduce the effect of discretization error associated with high frequencies. The analysis of single degree of freedom systems becomes handy when we can deal with each mode separately, thanks to this decoupling. Dynamic response of the system can be obtained by modal superposition.

Direct time integration methods introduce discretization to the temporal variation of the degrees of freedom (Section 12.10). This approach is particularly useful for the solution of nonlinear problems or problems with general transient loading. Temporal discretization results in explicit and implicit solution algorithms which have their own advantages and drawbacks. In fact stability, accuracy, and convergence of the direct time integration methods have to be established (Section 12.11) before these methods can be used reliably.

12.2 Vibration of single degree of freedom systems

The system shown in Fig. 12.1 is a single degree of freedom system that consists of a mass m that is anchored to a wall by a linear spring and a damper (dashpot). The mass slides on frictionless rollers and it is subjected to a time-dependent force $F(t)$. The stiffness of the spring is k and the damping coefficient is c. The position $y = y(t)$ of the mass is a function of time, and it is measured with respect to a *reference state* where the spring and the damper apply no forces on the system. The equation of motion and initial conditions of this system can be stated

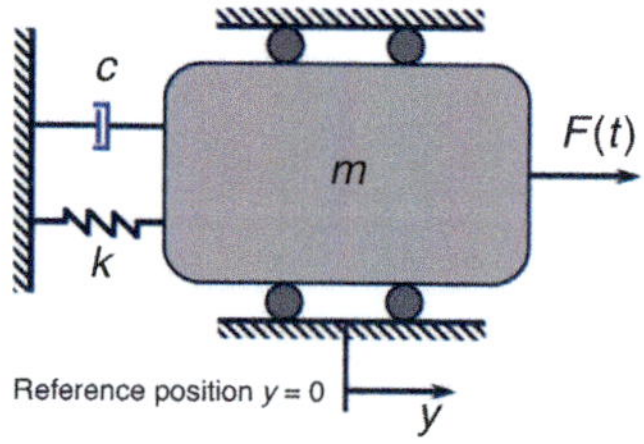

FIGURE 12.1 A single degree of freedom system consist of a spring with stiffness k, a damper with damping constant c and a mass m. The system is subjected to a time-dependent force $F(t)$.

as follows:

$$m\ddot{y} + c\dot{y} + ky = F \tag{12.1}$$

$$y(0) = y^{(0)} \tag{12.2a}$$

and

$$\dot{y}(0) = v^{(0)} \tag{12.2b}$$

where $\dot{y} = dy/dt$ and $\ddot{y} = d^2y/dt^2$.

We can reduce the number of system parameters from three to two by dividing Eq. (12.1) through m resulting in,

$$\ddot{y} + 2\xi\omega_n\dot{y} + \omega_n^2 y = \frac{F}{m} \tag{12.3}$$

where $\omega_n = \sqrt{\frac{k}{m}}$ is the *natural frequency* and $\xi = \frac{c}{2\sqrt{km}}$ is the *damping factor*. It is important for the following analysis to note that the damping factor is a positive number, $\xi \geq 0$.

The system of Eqs. (12.1) and (12.2) represents an *initial value problem*. In particular, Eq. (12.1) is a second-order ordinary differential equation whose solution consists of the superposition of complementary y_c and particular y_p parts.

12.2.1 Free vibrations: complementary solution

The complementary solution satisfies the homogenous form of Eq. (12.1), and it can be found by seeking a solution of the form $y_c = Ce^{rt}$, where C is a constant. Substituting this trial solution into the homogenous form of Eq. (12.1) we find the *characteristic equation*,

$$r^2 + 2\,\xi\,\omega_n r + \omega_n^2 = 0 \tag{12.4}$$

which has the roots $r_{1,2} = (-\xi \pm \sqrt{\xi^2 - 1})\omega_n$. The form of the solution and, therefore, the response of the system depends on the damping factor, ξ. In what follows we provide a brief analysis of four types of responses that depend on the amount of damping present in the system.

Overdamped systems, $\xi > 1$: $r_{1,2} = -\xi\omega_n \pm \sqrt{\xi^2 - 1}\omega_n$

These roots are both negative ($r_1, r_2 < 0$), real and distinct. Therefore, the solution is of the form,

$$y_c(t) = A_1 e^{r_1 t} + A_2 e^{r_2 t} \tag{12.5}$$

where A_1 and A_2 are constants that can be determined by engaging the initial conditions. Because the roots are both negative numbers, the system response is an exponential decay and no periodic motion can be observed. Using the initial

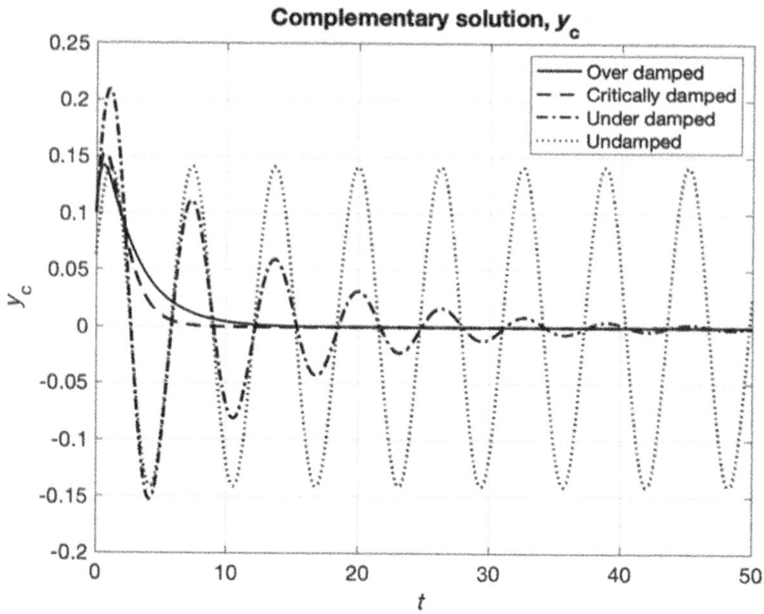

FIGURE 12.2 The complementary solution of the one degree of freedom system for the overdamped ($\xi = 1.5$), critically damped ($\xi = 1$), underdamped ($\xi = 0.1$), and undamped cases. Other parameters of the problem are $\omega_n = 1$, $y^{(0)} = 0.1$, $v^{(0)} = 0.2$.

conditions in Eq. (12.2) the system response can be shown to be [4],

$$y_c(t) = e^{-\xi\omega_n t}\left(\frac{\xi\omega_n y^{(0)} + v^{(0)}}{\sqrt{\xi^2 - 1}\omega_n}\sinh\sqrt{\xi^2 - 1}\omega_n t + y^{(0)}\cosh\sqrt{\xi^2 - 1}\omega_n t\right) \tag{12.6}$$

A typical overdamped system response can be seen in Fig. 12.2.

Critically damped systems, $\xi = 1$: $r_{1,2} = -\omega_n$

Due to the repeated roots, the complementary solution is of the form,

$$y_c(t) = (A_1 + A_2 t)e^{-\omega_v t} \tag{12.7}$$

It can be shown that this equation represents a system response that will diminish to zero, with no periodic characteristics. The *critical damping coefficient* can be found as $c_c = 2\sqrt{km}$ when $\xi = 1$. The system response can be obtained by using the initial conditions [4]:

$$y_c(t) = \left[y^{(0)} + \left(y^{(0)}\omega_n + v^{(0)}\right)t\right]e^{-\omega_n t} \tag{12.8}$$

A typical critically damped system response can be seen in Fig. 12.2.

Underdamped systems $\xi < 1$: $r_{1,2} = -\xi\omega_n \pm i\sqrt{1 - \xi^2}\omega_n$

The roots are complex conjugates with $i = \sqrt{-1}$. The solution can be shown to be of the form,

$$y_c(t) = e^{-\xi\omega_k t}(A_1 \sin(\omega_d t) + A_2 \cos(\omega_d t)) \tag{12.9}$$

where $\omega_d = \omega_n\sqrt{1-\xi^2}$ is the damped natural frequency of the system. Eq. (12.9) can also be expressed in the following form:

$$y_c(t) = Y_c e^{-\xi\omega_t t} \cos(\omega_d t - \varphi) \tag{12.10}$$

where $Y_c = \sqrt{A_1^2 + A_2^2}$ is the initial *amplitude of vibration* and $\varphi = \tan^{-1}(A_1/A_2)$ is the *phase angle*. Thus, we see that the complementary solution has a periodic nature with a *response (vibration) frequency* ω_d and with a decaying amplitude $Y_c e^{-\xi\omega_\pi t}$. Note that frequency of free vibrations is always lower than the critical frequency $\omega_d = \omega_n\sqrt{1-\xi^2} < \omega_n$.

The complementary solution depends on the initial conditions, form which the constants Y_e and φ can be found as follows [4]:

$$Y_c = \sqrt{y^{(0)2} + \left(\frac{\xi\omega_n y^{(0)} + v^{(0)}}{\omega_d}\right)^2} \text{ and } \varphi = \tan^{-1}\left(\frac{\xi\omega_n y^{(0)} + v^{(0)}}{\omega_d y^{(0)}}\right) \tag{12.11}$$

The complementary solution will decay in the long term as shown in Fig. 12.2.

Undamped systems $\xi = 0$: $r_{1,2} = \pm\omega_n i$

In this case the characteristic equation becomes,

$$r^2 + \omega_n = 0 \tag{12.12}$$

The roots of the corresponding characteristic equation are both complex with no real parts. Thus it can be shown that the solution is of the form [4],

$$y_c(t) = A_1 \sin(\omega_n t) + A_2 \cos(\omega_n t) \tag{12.13}$$

The constants A_1 and A_2 can be found from the initial conditions and the solution becomes,

$$y_c(t) = y^{(0)} \cos(\omega_n t) + \frac{v^{(0)}}{\omega_n} \sin(\omega_n t) \tag{12.14}$$

By using appropriate trigonometric relationships this equation can also be expressed as follows:

$$y_c(t) = Y_c \cos(\omega_n t - \varphi) \tag{12.15}$$

where the amplitude Y_e and the phase angle $\phi\varphi$ are determined from the initial conditions [4],

$$Y_c = \sqrt{y^{(0)2} + \left(\frac{v^{(0)}}{\omega_n}\right)^2} \text{ and } \varphi = \tan^{-1}\left(\frac{v^{(0)}}{y^{(0)}\omega_n}\right) \tag{12.16}$$

A typical undamped system response is shown in Fig. 12.2.

We note that the undamped systems vibrate with the natural frequency, ω_n and once set into motion, the system will vibrate indefinitely at this frequency (Fig. 12.2). In practice, all real systems are subjected to some level of dissipation and truly undamped systems are hard to realize.

12.2.2 Response to harmonic excitations: particular solution

Let's assume that the single degree of freedom system shown in Fig. 12.1 is subjected to an external, harmonic excitation of the form $F(t) = kY_p \cos \omega t$, where kY_p is the amplitude and ω is the frequency of excitation. The equation of motion in this case becomes,

$$m\ddot{y} + c\dot{y} + ky = kY_p \cos \omega t \tag{12.17}$$

The solution of this equation consists of the *complementary solution* y_c and the *particular solution* y_p. The complimentary solution develops due to the initial conditions imposed on the system, as shown in the previous section. No complementary solution develops if the initial conditions are zero.

Here we focus on the particular solution for the *harmonic forcing* function of the type given in Eq. (12.17) [4],

$$y_p(t) = \frac{Y_p}{\sqrt{\left(1 - r^2\right)^2 + (2\xi r)^2}} \cos(\omega t - \phi) \tag{12.18}$$

where $r = \omega/\omega_n$ is the *frequency ratio*, and $\phi = \tan^{-1}(\frac{2\xi r}{1-r^2})$ is the *phase angle* of the response.

Eq. (12.18) shows that if an undamped system ($\xi = 0$) is excited with the natural frequency, ($\omega = \omega_n$ or $r = 1$), the system response will be infinite. On the other hand, if the system is damped ($\xi > 0$) even if the system is critically excited the system response will be bounded. This can be observed by plotting the non-dimensional amplitude y/Y_p as a function of the frequency r and damping ratio ξ as shown in Fig. 12.3A. For undamped systems, the response amplitude can grow in an unbounded manner if the forcing frequency ω is equal to the *natural frequency* ω_n of the system. For damped systems, on the other hand, the response amplitude is large but does not grow in an unbounded manner.

Eq. (12.18) also shows that the system response will always have a phase shift with respect to the forcing function. The phase shift is determined by the phase angle ϕ, which is plotted as a function of the frequency r and damping ratio ξ in Fig. 12.3B. For low excitation frequencies ($r < 1$) the phase shift tends toward zero with decreasing damping ($\xi \to 0$). If the system is excited with the natural frequency ($r = 1$) the phase shifts is $\pi/2$ for all positive damping values. For large frequencies the phase shift lies in the range, $\pi/2 < \phi < \pi$. For undamped systems the phase shift is zero for $0 \leq r < 1$ and jumps to π for $r > 1$ [4].

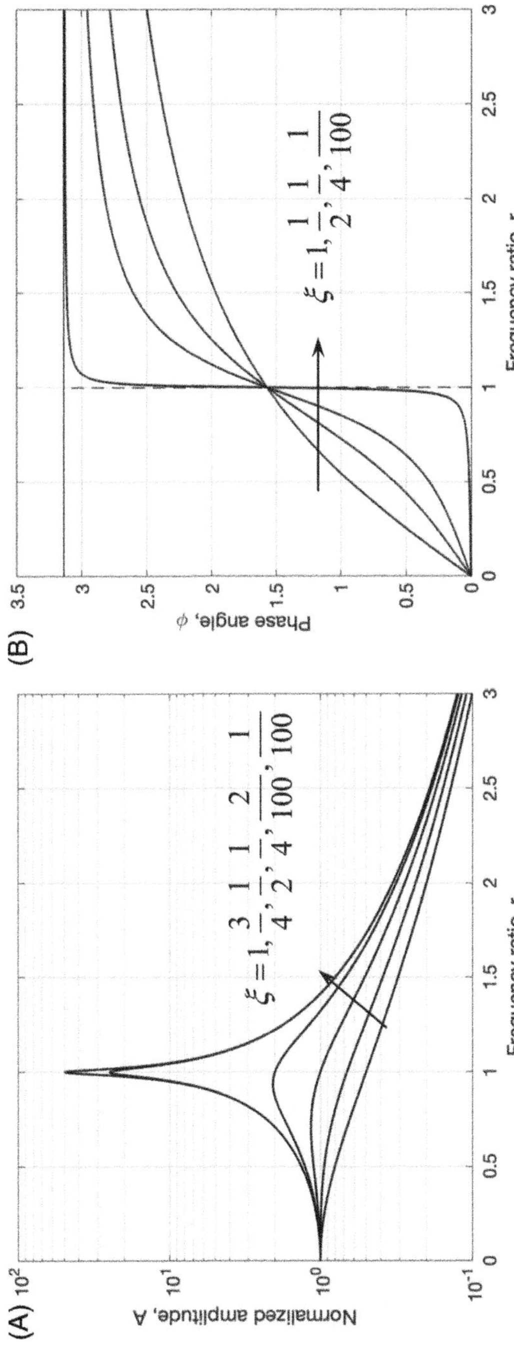

FIGURE 12.3 (A) Normalized amplitude y/Y_p and (B) phase angle ϕ of the response as a function of frequency ratio r and damping coefficient ξ.

12.2.3 Combined response: complimentary and particular solutions

When a linear system is subjected to both nonzero initial conditions and a harmonic excitation, the solution can be obtained by the superposition of the complementary and particular solutions ($y = y_c + y_p$). Most mechanical systems have moderate damping. Taking this into account, the *response of an underdamped system to harmonic excitation* of the type used in Eq. (12.17) becomes,

$$y = y_c + y_p = Y_c e^{-\xi\omega_n t}\cos(\omega_d t - \varphi) + \frac{Y_p}{\sqrt{(1-r^2)^2 + (2\xi r)^2}}\cos(\omega t - \phi) \tag{12.19}$$

where the constants Y_c and φ depend on the initial conditions and they are given in Eq. (12.11). Let's note that the response has two distinct frequencies $\omega_d = \omega_n\sqrt{1-\xi^2}$ and ω, and two distinct phase angles φ and ϕ for the free vibration and the harmonic loading response, respectively. Note that the contribution of the complementary solution of a damped system decays in time as it is only energized by the initial conditions. The particular solution, on the other hand, is energized continually by the harmonic loading, and thus it moves with the same frequency as the load. Therefore, we can see that initial system response has contributions from both components, but the steady state response is dominated by the particular solution as shown in Fig. 12.4.

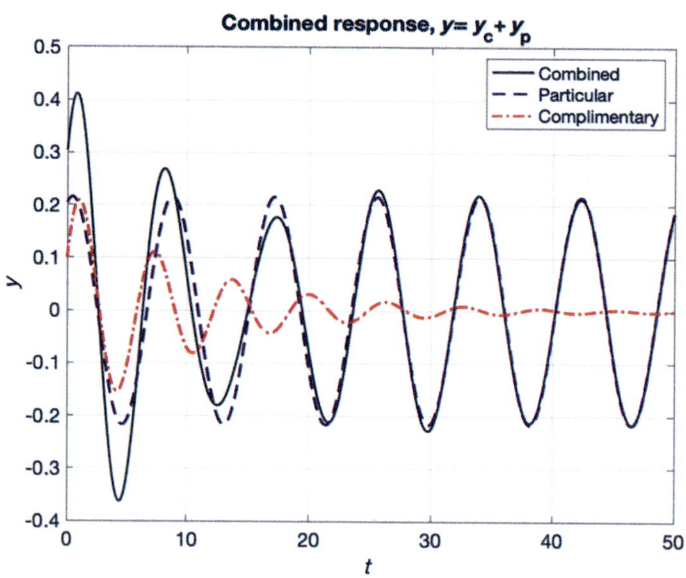

FIGURE 12.4 The general system response to harmonic loading $F(t)$ given in Eq. (12.17). The parameters of the problem are $\omega_n = 1$, $\xi = 0.1$, $Y_p = 0.1$, $\omega = 0.75$, $y^{(0)} = 0.1$, $v^{(0)} = 0.2$.

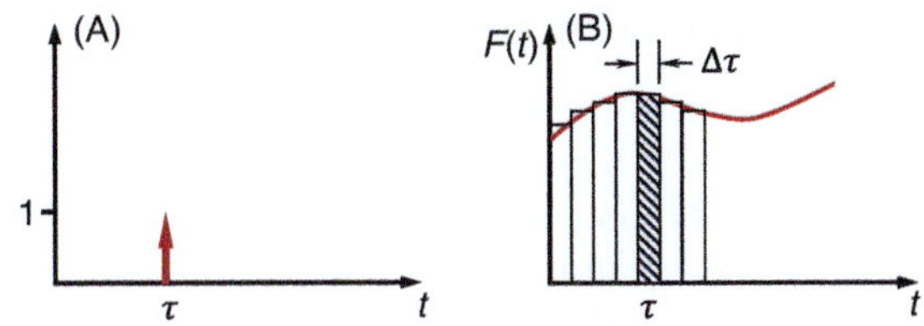

FIGURE 12.5 Approximation of a transient force by impulses of magnitude $F(t)\Delta\tau$.

12.2.4 Transient vibration

In the previous sections we presented vibrations of systems that are energized by initial conditions and by harmonically varying external loads. *Arbitrary excitation* $F(t)$ is another type of loading that can act on a system. *Transient vibrations* of a single degree of freedom system subjected to arbitrary loading can be calculated by modeling the external load as superposition of the series of pulses. To this end it is important to realize that the response of a damped, single degree of freedom system to a *unit impulse* that occurs at $t = \tau$ is given by [7],

$$h(t-\tau) = \frac{1}{\omega_d m} e^{-\xi\omega_n(t-\tau)} \sin \omega_d(t-\tau) \text{ for } t > \tau \tag{12.20}$$

As mentioned above, the transient excitation can be approximated by a collection of impulses of strength $F(\tau)\Delta\tau$, as shown in Fig. 12.5. The system's response to an impulse of magnitude $F(\tau)\Delta\tau$ that occurs at $t = \tau$ should be $h(t-\tau)F(\tau)\Delta\tau$. Summing up all the impulses that make up the transient loading and letting $\Delta\tau$ approach zero, we can find the particular solution to the transient load as follows:

$$y_p(t) = \int_0^t h(t-\tau)F(\tau)d\tau = \frac{1}{\omega_d m}\int_0^t e^{-\xi\omega_n(t-\tau)} \sin \omega_d(t-\tau)F(\tau)d\tau \tag{12.21}$$

If impulse occurs on a system with nonzero initial conditions, then the solution can be found by superposition of the appropriate complementary solution y_c and the particular solution y_p presented in Eq. (12.21). For example, for an underdamped system with nonzero initial conditions the solution becomes,

$$\begin{aligned} y &= y_c + y_p \\ &= Y_c e^{-\xi\omega_n t}\cos(\omega_d t - \varphi) + \frac{1}{\omega_d m}\int_0^t e^{-\xi\omega_n(t-\tau)} \sin \omega d(t-\tau)F(\tau)d\tau \end{aligned} \tag{12.22}$$

where the constants y_c and φ which depend on the initial conditions $y^{(0)}$ and $v^{(0)}$ are given in Eq. (12.11).

12.3 Initial/boundary value problems for deformable solids

The displacement, strain, and stress in deformable solids become functions of time, if the solid is subjected to *time-dependent* essential and/or natural boundary conditions, and/or if the body force acting on the system is time dependent. In this section, we summarize the equations of motion of two-dimensional deformable solids, deformable bars, and Euler–Bernoulli beams. Details of the derivations of these equations are presented in Chapter 2.

12.3.1 Two-dimensional deformable solid

The boundary/initial value problem associated with loading response of a two-dimensional elastic domain, described in Section 2.2, can be summarized as follows:

Equations of motion:

$$\frac{\partial \sigma_{xx}}{\partial x} + \frac{\partial \tau_{xy}}{\partial y} + F_{B_x} = \rho \frac{\partial^2 u_x}{\partial t^2} + c\frac{\partial u_x}{\partial t} \text{ in } \Omega \text{ for } t > 0 \tag{12.23a}$$

$$\frac{\partial \tau_{yx}}{\partial x} + \frac{\partial \sigma_{yy}}{\partial y} + F_{B_y} = \rho \frac{\partial^2 u_y}{\partial t^2} + c\frac{\partial u_y}{\partial t} \text{ in } \Omega \text{ for } t > 0 \tag{12.23b}$$

Boundary conditions:

$$\text{for } (x, y) \text{ on } \Gamma_E \text{ and for } t > 0: \quad \vec{u} = \vec{u}^{(b)}(x, y, t) \tag{12.24a}$$

$$\text{for } (x, y) \text{ on } \Gamma_N \text{ and for } t > 0: \quad \vec{q} = \vec{q}^{(b)}(x, y, t) \tag{12.24b}$$

Initial conditions:

$$\text{for } (x, y) \text{ in } \Omega \cup \Gamma \text{ and at } t = 0: \vec{u} = \vec{u}^{(0)}(x, y, 0) \tag{12.25a}$$

$$\frac{\partial \vec{u}}{\partial t} = \dot{\vec{u}}^{(0)}(x, y, 0) \tag{12.25b}$$

The vectors $\vec{u}^{(b)} = u_x^{(b)}\hat{i} + u_y^{(b)}\hat{j}$ and $\vec{q}^{(b)} = q_x^{(b)}\hat{i} + q_y^{(b)}\hat{j}$ are the displacements and external tractions prescribed on the different (non-intersecting) parts of the boundary Γ. The initial condition of the system is described by the initial displacement $\vec{u}^{(0)}$ and velocity states $\dot{\vec{u}}^{(0)}$ of the system. The last terms of Eqs. (12.23a) and (12.23b) represent the *dissipative forces* acting on the domain. There is strong empirical evidence showing that for most physical systems dissipation is proportional to the velocity of vibration. The dissipative forces are added to the equations of motion derived in Chapter 2 by using d'Alembert's principle. The coefficient c is generally referred to as the *damping coefficient*, though its physical origins depend on the dissipation mechanism acting on the system (Section 12.7). The strain–displacement relations for a two-dimensional solid are given in Eq. (2.43) and the constitutive behavior of the material is described in Section 2.2.5.

12.3.2 One-dimensional bar

The boundary/initial value problem associated with loading response of a one-dimensional bar is described in Section 2.2.7 and it is given as follows,
Equation of motion:

$$\frac{\partial \sigma_{xx} A}{\partial x} + q_x + F_B A = \rho A \frac{\partial^2 u_x}{\partial t^2} + cA \frac{\partial u_x}{\partial t} \text{ in } 0 < x < L \text{ for } t > 0 \tag{12.26}$$

Boundary conditions:

$$\text{at } x = 0 \text{ for } t > 0\text{: either } u_x = u_0(t) \text{ or } (\sigma_{xx} A)_{x=0} = F_0(t) \tag{12.27a}$$

$$\text{at } x = 0 \text{ for } t > 0\text{: either } u_x = u_L(t) \text{ or } (\sigma_{xx} A)_{x=L} = F_L(t) \tag{12.27b}$$

Initial conditions:

$$\text{in } 0 \le x \le L \text{ at } t = 0\text{: } u_x(x, 0) = u^{(0)}(x) \tag{12.28a}$$

$$\frac{\partial u_x}{\partial t}(x, 0) = \dot{u}^{(0)}(x) \tag{12.28b}$$

Note that the boundary conditions $u_0(t)$, $u_L(t)$, $F_0(t)$ and $F_L(t)$ can vary in time, the initial conditions $u^{(0)}$ and $\dot{u}^{(0)}$ are defined over the entire solution domain. *Small normal strain* is defined in Eq. (2.36) as$\varepsilon_{xx} = \partial u_x / \partial x$, and the *constitutive behavior* of a linearly elastic material is given as $\sigma_{xx} = E\varepsilon_{xx}$, as described in Section 2.2.5.

12.3.3 Euler–Bernoulli beam

Flexural vibrations of an Euler–Bernoulli beam can be posed as a boundary/initial value problem. The equation of motion and the admissible boundary conditions are derived in Eq. (2.154) and Example 3.7, respectively. Note that the effects of dissipative forces are added to Eq. (2.154) in an ad hoc manner as described in Section 12.3.1. It is common to use the symbol v to represent the transverse deflection component u_y of the beam, and we ask the reader to make the distinction between the velocity and transverse beam deflections based on the context.
Equation of motion:

$$\rho A \frac{\partial^2 v}{\partial t^2} + cA \frac{\partial v}{\partial t} + \frac{\partial^2}{\partial x^2}\left(EI \frac{\partial^2 v}{\partial x^2} \right) = q_y \qquad \text{in } 0 < x < L, \qquad \text{for } t > 0 \tag{12.29}$$

Boundary conditions:

$$\text{at } x = 0,\ \text{for } t > 0\text{: either } v = v_0(t) \text{ or } V(0) = F_{y0}(t),\ \text{and} \tag{12.30a}$$

$$\text{either } \frac{\partial v}{\partial x} = \theta_0(t) \text{ or } M(t) = M_0(t) \tag{12.30b}$$

$$\text{at } x = L,\ \text{for } t > 0\text{: either } v = v_L(t) \text{ or } V(L) = F_{yL}(t),\ \text{and} \tag{12.30c}$$

$$\text{either } \frac{\partial v}{\partial x} = \theta_L(t) \text{ or } M(L) = M_L(t) \tag{12.30d}$$

Initial conditions:

$$\text{in } 0 \le x \le L \text{ at } t = 0\text{: } v(x,0) = v^{(0)}(x) \tag{12.31a}$$

$$\frac{\partial v}{\partial t}(x,0) = \dot{v}^{(0)}(x) \tag{12.31b}$$

The admissible boundary conditions given in Eq. (12.30a) indicate that at each boundary, $x=0$ or $x=L$, two boundary conditions must be specified. These can be either the prescribed displacement, v_0, v_L, or the prescribed transverse force, F_{y0}, F_{yL} and the prescribed slope, θ_0, θ_L or the prescribed moment, M_0, M_L. Note that the boundary conditions can vary with time. The initial conditions $v^{(0)}$ and $\dot{v}^{(0)}$ are defined over the entire solution domain. The internal moment and shear force resultants are given in Eqs. (2.140) and (2.143) as $M(x,t) = -EI\partial^2 v/\partial x^2$ and $V(x,t) = \partial M/\partial x$, respectively.

12.4 Vibration response of an Euler–Bernoulli beam

In this section we present an analytical solution for the vibration of a simply supported, undamped Euler–Bernoulli beam that is released from an initially deformed state $v^{(0)}$ as shown in Fig. 12.6. The homogenous form of the equation of motion, the boundary and the initial conditions for this problem are expressed as follows:

$$\frac{\partial^2 v}{\partial t^2} + a^2 \frac{\partial^4 v}{\partial x^4} = 0 \tag{12.32a}$$

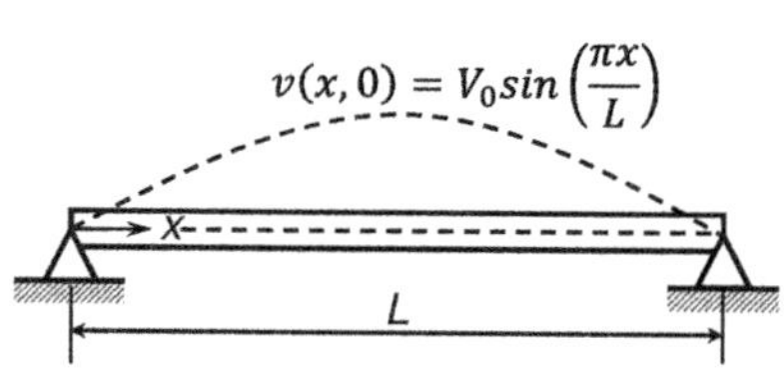

FIGURE 12.6 A simply supported beam subjected to an initial displacement $v(x,0) = v^{(0)}(x) = V^{(0)}\sin(\pi x/L)$ is released with zero initial velocity.

$$\text{at } x = 0 \text{ and for } t > 0\text{: } v(0,t) = 0 \text{ and } \left.\frac{\partial^2 v}{\partial x^2}\right|_0 = 0 \tag{12.32b}$$

$$\text{at } x = L \text{ and for } t > 0\text{: } v(L,t) = 0 \text{ and } \left.\frac{\partial^2 v}{\partial x^2}\right|_L = 0 \tag{12.32c}$$

$$\text{for } 0 \le x \le L \text{ and at } t = 0\text{: } v(x,0) = v^{(0)}(x) \tag{12.32d}$$

$$\frac{\partial v}{\partial t}(x,0) = \dot{v}^{(0)}(x) \tag{12.32e}$$

where $a^2 = EI/\rho A$ and $v^{(0)}$ and $\dot{v}^{(0)}$ are the functions that describe the initial shape and initial velocity of the beam. We will first solve the eigenvalue problem associated with the equation of motion and then find the free vibration response to the initial conditions. We will assume that the beam has uniform propertiesE, I, ρ and A along its length L.

12.4.1 Eigenvalue problem

In order to find a solution to Eq. (12.32a), the spatial and temporal parts of the response will be separated as $V(x)$[1] and $T(t)$, respectively, and the solution will be represented as the product of these two functions,

$$v(x,t) = V(x)T(t) \tag{12.33}$$

Substituting this relationship into Eq. (12.32a) we find [4]:

$$-\frac{1}{T}\frac{\partial^2 T}{\partial t^2} = \frac{a^2}{V}\frac{\partial^4 V}{\partial x^4} = \lambda \tag{12.34}$$

Note that the left and right sides of Eq. (12.34) are functions of time and space, respectively. Therefore, both of these equations should be equal to the same scalar parameter λ for such an equality to hold. The parameter λ is known as the *eigenvalue* of the system. It is matematically convenient to define the eigenvalue in terms of the expected vibration frequency of the system as $\lambda = \omega_n^2$

By using this definition, we can see that Eq. (12.34) represents the following ordinary differential equations in space and time,

$$\frac{\partial^2 T}{\partial t^2} + \omega_n^2 T = 0 \tag{12.35a}$$

$$\frac{\partial^4 V}{\partial x^4} - \beta^4 V = 0 \tag{12.35b}$$

where $\beta^4 = (\omega_n/a)^2$. The solutions of Eqs. (12.35a) and (12.35b) can be found

1. Note that, here, the symbol V only represents the spatial part of the response, and it should not be confused with the internal shear force resultant V of the beam theory.

as follows [10]:

$$T(t) = A_1 \sin \omega_n t + A_2 \cos \omega_n t \tag{12.36a}$$

$$V(x) = C_1 \sin \beta x + C_2 \cos \beta x + C_3 \sinh \beta x + C_4 \cosh \beta x \tag{12.36b}$$

where the integration constants $C_1 - C_4$ can be found by using the boundary conditions. The process of finding $C_1 - C_4$ leads to the eigenvalue problem described in the rest of this section. The integration constants A_1 and A_2 can be found by using the initial conditions. The process of finding A_1 and A_2 is described in the next section. Note that ω_n represents the *frequency of temporal variations* and β represent the frequency of spatial variations of the solution. Borrowing the idea of *natural frequency* from the undamped single degree of freedom systems introduced in Section 12.2, we will note that ω_n represents the *natural frequency* of the system.

Eq. (12.36b) and the boundary conditions given in Eqs. (12.32b,c) describe an eigenvalue problem from which the *characteristic spatial frequency* $\beta = (\omega_n/a)^{1/2}$ and the *natural vibration frequency* ω_n can be determined. In order to express this eigenvalue problem let's use Eq. (12.36b) and the boundary conditions Eq. (12.32a). This gives,

$$V(0) = C_2 + C_4 = 0 \text{ and } \left.\frac{\partial^2 V}{\partial x^2}\right|_0 = -C_2 + C_4 = 0 \tag{12.37}$$

and indicate that $C_2 = C_4 = 0$. Similarly, the boundary conditions Eqs. (12.32c) become,

$$V(L) = C_1 \sin \beta L + C_3 \sinh \beta L = 0 \text{ and } \left.\frac{\partial^2 V}{\partial x^2}\right|_L = -C_1 \sin \beta L + C_3 \sinh \beta L = 0 \tag{12.38}$$

Close examination of Eq. (12.38) shows that $C_3 = 0$. Thus we find that the constant β must satisfy the following equation,

$$\sin \beta L = 0 \tag{12.39}$$

Eq. (12.39) is known as the *characteristic equation* of the system. The characteristic equation will take on different forms for different boundary conditions. See reference [7] for a summary of the characteristic equations of the Euler–Bernoulli beam for different types of boundary conditions.

Eq. (12.39) has multiple solutions,

$$\beta_k = \frac{k\pi}{L} \text{ for } k = 1 - \infty \tag{12.40}$$

Consequently, the spatial component represented by Eq. (12.36b) should take into account these multiple solutions. Thus the spatial component of the solution is represented by modifying Eq. (12.36b) as follows:

$$V_k(x) = C_k \sin (\beta_k x) \tag{12.41}$$

The spatial components of the solution given in Eq. (12.41) are known as the *mode shapes* or *eigenfunctions* of the system. As we defined the relationship $\beta^4 = (\omega_n/a)^2$ between the spatial and natural frequencies, we find the *natural frequencies* of vibration associated with the mode shapes from Eq. (12.40) as follows,

$$\omega_{n_k} = a\beta_k^2 = \left(\frac{k\pi}{L}\right)^2 \left(\frac{EI}{\rho A}\right)^{1/2} = (k\pi)^2 \left(\frac{EI}{mL^3}\right)^{1/2} \tag{12.42}$$

where $m = \rho AL$ is the total mass of the beam. Note that each natural frequency ω_{n_k} is associated with a mode shape $V_k(x)$ and there are infinitely many natural frequency and mode shape pairs. The first four mode shapes $V_1 - V_4$ of the simply supported beam are presented in Fig. 12.7A.

The temporal component of the solution for mode-k can be obtained from Eq. (12.36a),

$$T_k(t) = A_{1k} \sin a\beta_k^2 t + A_{2k} \cos a\beta_k^2 t \tag{12.43}$$

The vibration response of mode-k of the system can be found by combining Eqs. (12.41) and (12.43) in Eq. (12.33),

$$v_k(x,t) = C_k \sin(\beta_k x)\left(A_{1k} \sin a\beta_k^2 t + A_{2k} \cos a\beta_k^2 t\right) \tag{12.44}$$

As there are infinitely many modes they should be superimposed to find the total response to the free vibration problem,

$$v(x,t) = \sum_{k=1}^{\infty} \left(D_{1k} \sin a\beta_k^2 t + D_{2k} \cos a\beta_k^2 t\right) \sin(\beta_k x) \tag{12.45}$$

where constant coefficients are combined as $D_k = C_k A_k$. These coefficients can be found by using the initial conditions as described next.

12.4.2 Free vibration problem

The constant coefficients of Eq. (12.45) can be found by using the initial conditions. By using Eq. (12.45) we can express the initial conditions as follows:

$$v(x,0) = \sum_{k=1}^{\infty} D_{2k} \cos\left(a\beta_k^2 t\right) \sin(\beta_k x) = v^{(0)}(x) \tag{12.46a}$$

$$\frac{\partial v}{\partial t}(x,0) = \sum_{n=1}^{\infty} D_{1k} a\beta_k^2 \cos\left(a\beta_k^2 t\right) \sin(\beta_k x) = \dot{v}^{(0)}(x) \tag{12.46b}$$

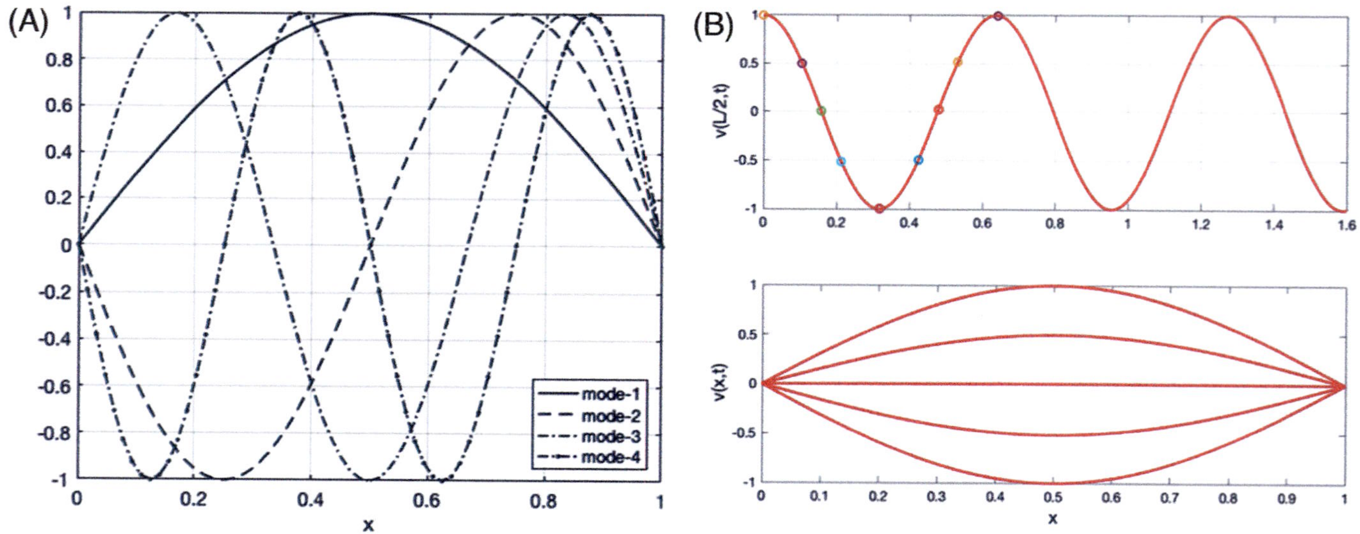

FIGURE 12.7 (A) The first four mode shapes (eigenvectors) of the simple supported Euler–Bernoulli beam. (B) Transient response of the mid-point (top) and the whole beam in response to the initial condition given in Eq. (12.50).

Let's next multiply these equations by $\sin(\beta_m x)$ and integrate through the domain. This gives,

$$\int_0^L \left(\sum_{k=1}^{\infty} D_{2k} \sin(\beta_m x) \sin(\beta_k x) \right) dx = \int_0^L \sin(\beta_m x) v^{(0)}(x) dx \tag{12.47a}$$

$$\int_0^L \left(\sum_{k=1}^{\infty} D_{1k} a \beta_k^2 \sin(\beta_m x) \sin(\beta_k x) \right) dx = \int_0^L \sin(\beta_m x) \dot{v}^{(0)}(x) dx \tag{12.47b}$$

Before proceeding to evaluate these integrals let's recall the *orthogonality condition* for the sine function [2],

$$\int_0^L \sin(\beta_m x) \sin(\beta_k x) dx = \begin{cases} \frac{L}{2} \text{ for } k = m \\ 0 \text{ for } k \neq m \end{cases} \tag{12.48}$$

Thus by using the orthogonality condition in Eq. (12.47) we can find the coefficients D_{2k} and D_{1k} as follows:

$$D_{2k} = \frac{2}{L} \int_0^L v^{(0)} \sin(\beta_k x) dx \tag{12.49a}$$

$$D_{1k} = \frac{2}{L} \int_0^L \frac{\dot{v}^{(0)}}{a \beta_k^2} \sin(\beta_k x) dx \tag{12.49b}$$

In this example the simple-support conditions lead to relatively simple mode shapes given in Eq. (12.39). It can be shown that the mode shapes that arise from more general boundary conditions are also orthonormal [4].

Let's next consider the initial conditions shown in Fig. 12.6,

$$\text{at } t = 0,\ v^{(0)} = V^{(0)} \sin \frac{\pi x}{L} \text{ and } \dot{v}^{(0)} = 0 \tag{12.50}$$

where $V^{(0)}$ is the amplitude of the initial displacement of the beam. By recalling $\beta_k = k\pi x/L$ the coefficients D_{1k} and D_{2k} of the solution Eq. (12.45) can be found from Eq. (12.49a) as follows:

$$D_{1k} = 0 \text{ and } D_{2k} = \frac{2}{L} \int_0^L V^{(0)} \sin \frac{\pi x}{L} \sin\left(\frac{k\pi x}{L}\right) dx \tag{12.51}$$

Integration of this equation shows that only the case $k = 1$ is nonzero due to the orthogonality condition given in Eq. (12.48) above. Thus, we find the coefficients of Eq. (12.45) as follows:

$$D_{1k} = 0, \text{ for } k \geq 1 \tag{12.52a}$$

$$D_{2(1)} = V^{(0)}, \text{ and } D_{2k} = 0 \text{ for } k \geq 2 \tag{12.52b}$$

The free vibration of an undamped Euler–Bernoulli beam subjected to an initial displacement of the half-sine shape can then be stated by using Eqs. (12.45) and (12.52a) as follows:

$$v(x,t) = V^{(0)} \sin(\beta_1 x) \cos\left(a\beta_1^2 t\right) \tag{12.53}$$

where $\beta_1 = \pi/L$ and $a = (EI/\rho A)^{1/2}$. The analytical solution given by Eq. (12.53) is plotted in Fig. 12.7B.

12.5 Semidiscrete equations of motion

Steady state finite element formulation of the equations of equilibrium for the deformable bars, Euler–Bernoulli beams, and two-dimensional solids are presented in Sections 5.2.4, 6.3.4, and, 7.4, respectively. For steady state problems, deflections depend only on the spatial coordinates. In the dynamic problems considered in this chapter, deflection is a function of both space and time, for example, $u_x = u_x(x, y, t)$. In this section we use the finite element method to discretize the spatial component of the equations of motion. This results in the *semidiscrete equation of motion*. Fully discretized equation of motion is addressed in Section 12.10.

12.5.1 Two-dimensional deformable element

To develop the semidiscrete equation of motion for an element, the initial/boundary value problem defined in Section 12.3.1 is described in the element domain $\Omega^{(e)}$ in the same manner as described in Section 7.4. The boundaries of the element are assumed to be subjected to natural boundary conditions as presented in Section 7.4. The weighted residual integral associated with the problem described by Eqs. (12.23a,b) is expressed as follows:

$$\int_{\Omega^{(e)}} \left[\delta u_x \left(\frac{\partial \sigma_{xx}}{\partial x} + \frac{\partial \tau_{xy}}{\partial y} + F_{B_x} - \rho \frac{\partial^2 u_x}{\partial t^2} - c \frac{\partial u_x}{\partial t} \right) + \delta u_y \left(\frac{\partial \tau_{yx}}{\partial x} + \frac{\partial \sigma_{yy}}{\partial y} + F_{B_y} - \rho \frac{\partial^2 u_y}{\partial t^2} - c \frac{\partial u_y}{\partial t} \right) \right] d\Omega^{(e)} \tag{12.54}$$

where δu_x and δu_y represent the variation of the displacement components in the sense defined in Chapter 3. Eq. (12.54) can be expressed in the following vector form, after using the Green–Gauss theorem and noting that the natural boundary conditions are given in the form described in Eq. (7.28). The details of this derivation are given in Section 7.4. What is new here is the addition of the effects of the inertial and dissipative forces acting on the system, represented in the second and third terms of the left side of the following equation:

$$\int_{\Omega^{(e)}} \{\delta\varepsilon\}^T\{\sigma\}d\Omega^{(e)} + \int_{\Omega^{(e)}} \rho\{\delta u\}^T\{\ddot{u}\}d\Omega^{(e)} + \int_{\Omega^{(e)}} c\{\delta u\}^T\{\dot{u}\}d\Omega^{(e)}$$
$$= \oint_{\Gamma^{(e)}} \{\delta u\}^T\{q^{(b)}\}d\Gamma^{(e)} + \int_{\Omega^{(e)}} \{\delta u\}^T\{F_B\}d\Omega^{(e)} + \{\delta u\}^T\{F\}^{(e)} \tag{12.55}$$

where

$$\{\delta u\} = \{\delta u_x \quad \delta u_y\}^T,$$
$$\dot{u} = \{\dot{u}_x \quad \dot{u}_y\}^T,$$
$$\{\ddot{u}\} = \{\ddot{u}_x \quad \ddot{u}_y\}^T$$
$$\{\delta\varepsilon\} = \{\varepsilon_{xx} \quad \varepsilon_{yy} \quad \gamma_{xy}\}^T$$
$$\{\sigma\} = \{\sigma_{xx} \quad \sigma_{yy} \quad \tau_{xy}\}^T$$

Eq. (12.55) is spatially discretized by using the finite element representation of the displacement vector,

$$\{u\} = [N]\{d\} \tag{12.56}$$

where $\{u\} = \{u_x \quad u_y\}^T$ is the displacement vector and $[N]$ and$\{d\}$ are the shape function matrix and the nodal displacement vector as given in Eqs. (7.34) and (7.35), respectively. The strain vector in finite element form is expressed as follows:

$$\{\varepsilon\} = \{\varepsilon_{xx} \ \varepsilon_{yy} \ \gamma_{xy}\}^T = [\partial]\{u\} \tag{12.57}$$

where $[\partial]$ is the derivative operator matrix defined in Eq. (7.38). The variation of the displacement and strain vectors is found as follows:

$$\{\delta u\} = [N]\{\delta d\} \text{ and } \{\delta\varepsilon\} = [\partial][N]\{\delta d\} = [B]\{\delta d\} \tag{12.58}$$

where the strain-displacement operator matrix $[B] = [\partial][N]$ is defined in Eq. (7.40).

The time derivatives of the displacement matrix are found from Eq. (12.56),

$$\{\dot{u}\} = [N]\{\dot{d}\} + \text{ and } \{\ddot{u}\} = [N]\{\ddot{d}\} \tag{12.59}$$

Substituting Eqs. (12.58) and (12.59) in Eq. (12.55), and eliminating the $\{\delta u\}$ vector in the same manner as described in Section 7.4 we find,

$$\int_{\Omega^{(e)}} [B]\{\sigma\}d\Omega^{(e)} + \left(\int_{\Omega^{(e)}} \rho[N]^T[N]d\Omega^{(e)}\right)\{\ddot{d}\} + \left(\int_{\Omega^{(e)}} c[N]^T[N]d\Omega^{(e)}\right)\{\dot{d}\}$$
$$= \oint_{\Gamma^{(e)}} [N]^T\{q^{(b)}\}d\Gamma^{(e)} + \int_{\Omega^{(e)}} [N]^T\{F_B\}d\Omega^{(e)} + \{F\}^{(e)} \tag{12.60}$$

Eq. (12.60) represents the equation of motion for an element and it can be represented in vector form as follows:

$$[m]^{(e)}\{\ddot{d}\}^{(e)} + [c]^{(e)}\{\dot{d}\} + \{r_{\text{int}}\}^{(e)} = \{r_B\}^{(e)} + \{r_q\}^{(e)} + \{F\}^{(e)} \tag{12.61}$$

where

Consistent mass matrix: $$[m]^{(e)} = \int_{\Omega^{(e)}} \rho[N]^T[N]d\Omega^{(e)} \tag{12.62a}$$

Consistent damping matrix: $$[c]^{(e)} = \int_{\Omega^{(e)}} c[N]^T[N]d\Omega^{(e)} \tag{12.62b}$$

Internal load vector: $$\{r_{\text{int}}\}^{(e)} = \int_{\Omega^{(e)}} [B]^T\{\sigma\}d\Omega^{(e)} \tag{12.62c}$$

Body force vector: $$\{r_B\}^{(e)} = \int_{\Omega^{(e)}} [N]^T\{F_B\}d\Omega^{(e)} \tag{12.62d}$$

External traction vector: $$\{r_q\}^{(e)} = \oint_{\Gamma^{(e)}} [N]^T\{q^{(b)}\}d\Gamma^{(e)} \tag{12.62e}$$

External nodal force vector: $\{F\}^{(e)}$ (12.62f)

Note that in Eq. (12.61) the internal forces are specified for a material with a general constitutive behavior. For a linearly elastic material, Hooke's law given in its general form in Eq. (7.22) should be used to replace the $\{\sigma\}$ vector in Eq. (12.62c),

$$\{r_{\text{int}}\}^{(e)} = \left(\int_{\Omega(e)} [B]^T[E][B]d\Omega^{(e)}\right)\{d\} - \int_{\Omega(e)} [B]^T[E]\{\varepsilon^{(0)}\}d\Omega^{(e)}$$

Defining,

Stiffness matrix: $$[k]^{(e)} = \int_{\Omega^{(e)}} [k]^T[E][B]d\Omega^{(e)} \quad \text{(a)}$$

Initial strain load vector: $$\{r_{\varepsilon 0}\}^{(e)} = \int_{\Omega^{(e)}} [B]^T[E]\{\varepsilon^{(0)}\}d\Omega^{(e)} \quad \text{(b)}$$

The equation of motion for an element becomes,

$$[m]^{(e)}\{\ddot{d}\} + [c]^{(e)}\{\dot{d}\} + [k]^{(e)}\{d\} = \{r_B\}^{(e)} + \{r_q\}^{(e)} + \{r_{\varepsilon 0}\}^{(e)} + \{F\}^{(e)} \tag{12.63}$$

12.5.2 One-dimensional elastic bar element

The initial/boundary value problem defined in Section 12.3.2 is described in the element domain $\Omega^{(e)}, 0 < x' < L^{(e)}$ in the same manner as described in

Chapter 5. The local coordinate system and the axial deformation of the bar are indicated by x' and $u'(x', t)$ and the ends of the bar are subjected to the natural boundary conditions as described in Eq. (5.3). The weak form of the equation of motion given in Eq. (12.26) is then found as follows:

$$\int_0^{L(e)} \left(\delta\varepsilon_{x'}(E\varepsilon_{x'}) + \rho\delta u'\ddot{u}' + c\delta u'\dot{u}'\right)Adx' - \int_0^{L(e)} \delta u'(F_{Bx'}A + q_{x'})dx' - \delta u_1' F_{x'1} - \delta u_2' F_{x'2} = 0 \tag{12.64}$$

Eq. (12.64) can be spatially discretized by using the finite element representation of the axial displacement of the bar,

$$u'(x', t) = [N]\{d_m'\} \tag{12.65}$$

where $\{d_m'\}$ is the nodal displacement vector and $[N]$ is the shape function matrix described in Section 5.2.4. The strain vector in finite element form is expressed as follows:

$$\varepsilon_{x'} = \frac{\partial u'}{\partial x'} = [B]\{d_m'\} \tag{12.66}$$

where strain-displacement operator matrix $[B] = \partial([N])/\partial x'$ is defined in Eq. (5.15).

The variation of the displacement and strain is expressed as follows:

$$\delta u' = [N]\{\delta d_m'\} \text{ and } \delta\varepsilon_{x'} = [B]\{\delta d_m'\} \tag{12.67}$$

The time derivatives of the deformation are found from Eq. (12.65),

$$\dot{u}' = [N]\{\dot{d}_m'\} \text{ and } \ddot{u}' = [N]\{\ddot{d}_m'\} \tag{12.68}$$

Substituting Eqs. (12.67) and (12.68) in Eq. (12.64) and eliminating the $\{\delta d\}$ vector in the same manner as described in Section 5.2.4 we find,

$$\begin{aligned}\left(\int_0^{L(e)} EA[B]^T[B]dx'\right)\{d_m'\} + \left(\int_0^{L(e)} \rho A[N]^T[N]dx'\right)\{\ddot{d}_m'\} + \left(\int_0^{L(e)} cA[N]^T[N]dx'\right)\{\dot{d}_m'\}\\ = \int_0^{L(e)} q_{x'}[N]^T dx' + \int_0^{L(e)} q_{Bx'}[N]^T A dx'\end{aligned} \tag{12.69}$$

This is the equation of motion for an elastic bar element and in vector form it can be expressed as follows:

$$[m_m']^{(e)}\{\ddot{d}_m\}' + [c_m']^{(e)}\{\dot{d}_m'\} + [k_m']^{(e)}\{d_m\}'^{(e)} = \{r_B'\}^{(e)} + \{r_q'\}^{(e)} + \{r_F\}^{(e)} \tag{12.70}$$

where

$$\text{Consistent mass matrix:} \quad [m'_m]^{(e)} = \int_0^{L(e)} \rho A[N]^T[N]dx' \tag{12.71a}$$

$$\text{Consistent damping matrix:} \quad [c'_m]^{(e)} = \int_0 cA[N]^T[N]dx' \tag{12.71b}$$

$$\text{Stiffness matrix:} \quad [k'_m]^{(e)} = \int_0^{L(e)} EA[B]^T[B]dx' \tag{12.71c}$$

$$\text{Body force vector:} \quad \{r'_q\}^{(e)} = \int_0^{L(e)} F_{Bx'}[N]^T A dx' \tag{12.71d}$$

$$\text{External traction force vector:} \quad \{r'_q\}^{(e)} = \int_0^{L(e)} q_{x'}[N]^T dx' \tag{12.71e}$$

$$\text{External nodal force vector:} \quad \{r_F\}'^{(e)} = \{F_{x'1} \quad F_{x'2}\}^T \tag{12.71f}$$

12.5.3 Euler–Bernoulli beam element

The initial/boundary value problem defined in Section 12.3.3 is described in the element domain $\Omega^{(e)}$, $0 < x' < L^{(e)}$ in the same manner as described in Chapter 6. The local coordinate system and the transverse deflection of the beam are indicated by x' and $v'(x', t)$ and the ends of the beam are subjected to the natural boundary conditions as described in Eq. (6.14). The weak form of the equation of motion (12.29) is found as follows:

$$\int_0^{L^{(e)}} \left(EI\frac{\partial^2(\delta v')}{\partial x'}\frac{\partial^2 v'}{\partial x'} + \rho A\delta v'\,\ddot{v}' + cA\delta v'\,\dot{v}' \right)dx' \int_0^{L^{(e)}} \delta u' f_{y'} dx'$$
$$- F_{y'0}\delta v'(0) - F_{y'L^{(e)}}\delta v'(L^{(e)}) - M_0\frac{\partial(\delta v')}{\partial x}\bigg|_0 - M_{L^{(e)}}\frac{\partial(\delta v)}{\partial x}\bigg|_{L^{(e)}} = 0 \tag{12.72}$$

Eq. (12.72) is spatially discretized by using the finite element representation of the transverse deflection of the beam,

$$v'(x', t) = [N^{(b)}]\{d'_b\} \tag{12.73}$$

where $\{d'_b\} = \{v'_1\ \theta_1\ v'_2\ \theta_2\}^T$ is the nodal displacement vector where the subscripts 1 and 2 refer to the end points of the beam, $[N^{(b)}]$ is the shape function matrix described in Eq. (6.29). The curvature of the deformed beam in finite element form is expressed as follows:

$$\frac{\partial^2 v'}{\partial x'^2} = [B^{(b)}]\{d'_b\} \tag{12.74}$$

where curvature–displacement operator matrix $[B^{(b)}] = \partial^2([N^{(b)}])/\partial x'^2$ is defined in Eq. (6.30).

The variation of the transverse deflection and curvature is expressed as follows:

$$\delta v' = \left[N^{(b)}\right]\left\{\delta d_b'\right\} \text{ and } \frac{\partial^2 \delta v'}{\partial x'^2} = \left[B^{(b)}\right]\left\{\delta d_b'\right\} \tag{12.75}$$

The time derivatives of the deflection are found from Eq. (12.73),

$$\dot{v}' = \left[N^{(b)}\right]\left\{\dot{d}_b'\right\} \text{ and } \ddot{v}' = \left[N^{(b)}\right]\left\{\ddot{d}_b'\right\} \tag{12.76}$$

Substituting Eqs. (12.75) and (12.76) in Eq. (12.72) and eliminating the $\{\delta d_b'\}$ vector in the same manner as described in Section 6.3.4 we find,

$$\left[\int_0^{L^{(e)}} EI\left[B^{(b)}\right]^T\left[B^{(b)}\right]dx'\right]\{d\} + \left[\int_0^{L^{(e)}} \rho A\left[N^{(b)}\right]^T\left[N^{(b)}\right]dx'\right]\{\ddot{d}\} + \left[\int_0^{L^{(e)}} cA\left[N^{(b)}\right]^T\left[N^{(b)}\right]dx'\right]\{\dot{d}\}$$
$$= \int_0^{L^{(e)}} q_{y'}\left[N^{(b)}\right]^T dx' + \int_0^{L^{(e)}} F_{By'}\left[N^{(b)}\right]^T A dx' + \left\{r_F'\right\}^{(e)}$$

This is the equation of motion for an Euler–Bernoulli beam element. It can be expressed in vector form as follows:

$$\left[m_b'\right]^{(e)}\left\{\ddot{d}_b'\right\} + \left[c_b'\right]^{(e)}\left\{\dot{d}_b'\right\} + \left[k_b'\right]^{(e)}\left\{d_b'\right\}^{(e)} = \left\{r_B'\right\}^{(e)} + \left\{r_q'\right\}^{(e)} + \left\{r_F'\right\}^{(e)} \tag{12.77}$$

where

Consistent mass matrix: $$\left[m_b'\right]^{(e)} = \int_0^{L^{(e)}} \rho A\left[N^{(b)}\right]^T\left[N^{(b)}\right]dx' \tag{12.78a}$$

Consistent damping matrix: $$\left[c_b'\right]^{(e)} = \int_0^{L^{(e)}} cA\left[N^{(b)}\right]^T\left[N^{(b)}\right]dx' \tag{12.78b}$$

Stiffness matrix: $$\left[k_b'\right]^{(e)} = \int_0^{L^{(e)}} EI\left[B^{(b)}\right]^T\left[B^{(b)}\right]dx'. \tag{12.78c}$$

Body force vector: $$\left\{r_B'\right\}^{(e)} = \int_0^{L^{(e)}} F_{By'}\left[N^{(b)}\right]^T A dx' \tag{12.78d}$$

External traction force vector: $$\left\{r_q'\right\}^{(e)} = \int_0^{L^{(e)}} q_{y'}\left[N^{(b)}\right]^T dx' \tag{12.78e}$$

External nodal force vector: $$\left\{r_F'\right\}^{(e)} = \left\{F_{y1} \quad M_1 \quad F_{y2} \quad M_2\right\}^T \tag{12.78f}$$

12.6 Mass matrix

12.6.1 Consistent mass matrices

The mass matrix is a discrete representation of a continuous mass distribution. Close inspection of Eqs. (12.62a), (12.71a) , and (12.78a) shows that the general relationship for a consistent mass matrix is of the form,

$$[m_C]^{(e)} = \int_{\Omega^{(e)}} \rho\,[T]^T\,[N] d\,\Omega^{(e)} \tag{12.79}$$

The size and content of the mass matrix depends on the element type. We introduce the subscript $_C$ to indicate the mass matrix is computed consistently with the interpolations used for the element.

12.6.1.1 *Consistent mass matrix for two-dimensional elements*

Two-dimensional elements are either triangular or quadrilateral with a minimum of three or four nodes, respectively, for linear interpolation of the variables. Higher order interpolation formulas add additional nodes and degrees of freedom to the problem, as described in Section 7.5. Integration of Eq. (12.79) is carried out in the mapped domain by using isoparametric mapping of the coordinates, as described in Sections 7.7 and 7.8. Thus, for two-dimensional elements Eq. (12.79) is evaluated as,

$$[m_C]^{(e)} = \int_{-1}^{1}\int_{-1}^{1} [N]^T\,[N]\rho t|J| d\xi d\eta \tag{12.80}$$

The shape functions for *rectangular* elements are given in Section 8.2.1. Using these relationships, the following mass matrix can be found for a rectangular Q4-element ($2a\times2b$):

$$\left[m_C^{(Q4)}\right]^{(e)} \int_{-a}^{a}\int_{-b}^{b} [N]^T\,[N]\rho t dx dy = \frac{1}{9}\rho abt \begin{bmatrix} 4 & 0 & 2 & 0 & 1 & 0 & 2 & 0 \\ 0 & 4 & 0 & 2 & 0 & 1 & 0 & 2 \\ 2 & 0 & 4 & 0 & 2 & 0 & 1 & 0 \\ 0 & 2 & 0 & 4 & 0 & 2 & 0 & 1 \\ 1 & 0 & 2 & 0 & 4 & 0 & 2 & 0 \\ 0 & 1 & 0 & 2 & 0 & 4 & 0 & 2 \\ 2 & 0 & 1 & 0 & 2 & 0 & 4 & 0 \\ 0 & 2 & 0 & 1 & 0 & 2 & 0 & 4 \end{bmatrix} \tag{12.81}$$

where $2a$ and $2b$ are the edge lengths and t is the thickness of the rectangular element (Fig. 8.2). For general quadrilateral elements the element mass matrix should be computed by using Eq. (12.80).

12.6.1.2 Consistent mass matrix for bar elements

The shape function matrices for two-node and three-node bar elements are given in Eqs. (4.41) and (4.43) as,

$$[N^{(m)}] = \left[1 - \frac{x'}{L^{(e)}} \quad \frac{x'}{L^{(e)}}\right] \tag{12.82a}$$

$$[N^{(m)}] = \left[\left(1 - 2\frac{x'}{L^{(e)}}\right)\left(1 - \frac{x'}{L^{(e)}}\right) \quad 4\left(\frac{x'}{L^{(e)}}\right)\left(1 - \frac{x'}{L^{(e)}}\right) \quad -\left(\frac{x'}{L^{(e)}}\right)\left(1 - 2\frac{x'}{L^{(e)}}\right)\right] \tag{12.82b}$$

respectively. Using the general form of the consistent mass matrix given in Eq. (12.71a) we find:

$$\text{Two-node element:} \quad [m_{Cm}'] = \frac{1}{6}\rho A L^{(e)}\begin{bmatrix} 2 & 1 \\ 1 & 2 \end{bmatrix} \tag{12.83a}$$

$$\text{Three-node element:} \quad [m_{Cm}']^{(e)} = \frac{1}{15}\rho A L^{(e)}\begin{bmatrix} 2 & 1 & -1/2 \\ 1 & 8 & 1 \\ -1/2 & 1 & 2 \end{bmatrix} \tag{12.83b}$$

12.6.1.3 Consistent mass matrix for Euler–Bernoulli beam element

The shape function matrix for an Euler–Bernoulli beam element is given in Eq. (6.29),

$$[N^{(b)}] = [N_1^{(b)} N_2^{(b)} N_3^{(b)} N_4^{(b)}]$$

where the shape functions are given in Eq. (6.5) as,

$$N_1^{(b)}(x') = 1 - 3\left(\frac{x'}{L^{(e)}}\right)^2 + 2\left(\frac{x'}{L^{(e)}}\right)^3 \tag{12.84a}$$

$$N_2^{(b)}(x') = x'\left[1 - 2\left(\frac{x'}{L^{(e)}}\right) + \left(\frac{x'}{L^{(e)}}\right)^2\right] \tag{12.84b}$$

$$N_3^{(b)}(x') = 3\left(\frac{x'}{L^{(e)}}\right)^2 - 2\left(\frac{x'}{L^{(e)}}\right)^3 \tag{12.84c}$$

$$N_4^{(b)}(x') = x'\left[-\left(\frac{x'}{L^{(e)}}\right) + \left(\frac{x'}{L^{(e)}}\right)^2\right] \tag{12.84d}$$

Using the general form of the consistent mass matrix given in Eq. (12.78a) we find,

$$[m_{Cb}']^{(e)} = \frac{1}{420}\rho A L^{(e)}\begin{bmatrix} 156 & 22L^{(e)} & 54 & -13L^{(e)} \\ 22L & 4L^{(e)2} & 13L^{(e)} & -3L^{(e)2} \\ 54 & 13L^{(e)} & 156 & -22L^{(e)} \\ -13L^{(e)} & -3L^{(e)2} & -22L^{(e)} & 4L^{(e)2} \end{bmatrix} \tag{12.85}$$

It should be noted that when consistent mass matrix is used the inertial forces at a given node are affected by the accelerations of all of the nodes. This can be seen, for example, by computing the inertial forces acting at node-2 of a beam element as follows:

$$F_2 = \frac{1}{420}\rho A L^{(e)}\left(54\ddot{v}_1 + 13L^{(e)}\ddot{\theta}_1 + 156\ddot{v}_2 - 22L^{(e)}\ddot{\theta}_2\right) \tag{12.86}$$

The inertial force at node-2 is affected by the acceleration of both nodes of the element. Thus inertia is distributed across the element when we use the consistent mass matrices in a manner consistent with the interpolation used in the element. This is not the case for the lumped mass matrix.

12.6.2 Lumped mass matrix

An alternative approach to determine the inertial forces acting on the nodes is to *lump* the mass/inertia to each node and neglect the interconnected nature of inertial forces across the elements. This approach results in a diagonal mass matrix and as such it has advantages in reducing the storage requirements and improving the computational speed for the explicit methods as described in Section 12.10.

The simplest method to obtain a lumped mass matrix is to assume that the mass is distributed to the nodes evenly. For a linear, one-dimensional element of length $L^{(e)}$ and constant cross-sectional area A this simply gives,

$$\left[m_{Lm}{}'\right]^{(e)} = \frac{1}{2}\rho A L^{(e)}\begin{bmatrix} 1 & 0 \\ 0 & 1 \end{bmatrix} \tag{12.87}$$

For a Q4-element of size ($2a \times 2b$) and thickness t the lumped mass matrix becomes,

$$\left[m_L^{(Q4)}\right]^{(e)} = \rho abt[I] \tag{12.88}$$

where $[I]_{8\times 8}$ is the 8×8 diagonal identity matrix.

For more general mass matrices, such as a general quadrilateral element (Chapter 7), or elements that involve higher order interpolation polynomials, how to lump the mass to the nodes is not very clear. One of the approaches commonly used for such elements is the *row sum method* [1,9]. In this method the diagonal values of the lumped mass matrix, m_{Lii} are found by adding the values of the corresponding row of the consistent mass matrix $[m_C]$. For an element with N degrees of freedom, the diagonal values are found by using,

$$m_{Lii} = \sum_{N}^{j=1} m_{Cij} \ \text{for}\ i = 1 - N \tag{12.89}$$

During this process we must ensure that the total mass of the element is preserved.

By using this approach, Eq. (12.87) can easily be obtained from (12.83a) and Eq. (12.88) from (12.81). We can also show that lumped mass matrix for the three node bar element becomes,

$$\left[m_m{}'\right]^{(e)} = \frac{1}{6}\rho AL^{(e)}\begin{bmatrix} 1 & 0 & 0 \\ 0 & 4 & 0 \\ 0 & 0 & 1 \end{bmatrix} \tag{12.90}$$

by using the row sum method with Eq. (12.83b).

In the case of the Euler–Bernoulli beam elements the lumped mass matrix can be given as follows,

$$\left[m_{Lb}{}'\right]^{(\varepsilon)} = \rho AL^{(\varepsilon)}\begin{bmatrix} 1/2 & 0 & 0 & 0 \\ 0 & \phi L^{(\varepsilon)2} & 0 & 0 \\ 0 & 0 & 1/2 & 0 \\ 0 & 0 & 0 & \phi L^{(\varepsilon)2} \end{bmatrix} \tag{12.91}$$

where ϕ is a scalar parameter that determines whether or not rotary inertia will be included in the problem. In case rotary inertia can be neglected we use $\phi = 0$. An ad hoc way to determine ϕ is described in [1] where $\phi = 1/24$. is derived based on simple rigid body dynamics principles. The reader is referred to reference [1] for description of more sophisticated methods such as the *HRZ lumping* and *optimal lumping*.

12.7 Damping matrix

Some internal or external physical effects that act on solids or structures could cause the kinetic energy to be dissipated in an irreversible manner. While the causes of damping may be due to different physical mechanisms, acting alone or in combination, the final effect is typically a dissipative (damping) force that is proportional to the velocity of the vibrating structure. In general, damping reduces the response amplitude of a structure. In *free vibration*, damping causes a terminal decay of the vibration amplitude, whereas in *forced vibration* it causes a lower amplitude response as shown for the single degree of freedom system presented in Section 12.2. The effects that cause a damped response can be a due to vastly different physical mechanisms, including viscous, hysteresis, Coulomb and algorithmic damping.

Viscous damping is observed in submerged structures due to their interaction with a surrounding fluid. For example, vibration of thin, light, and flexible structures would be affected by the surrounding air. The viscous damping force is affected by the viscosity of the fluid and proportional to the vibration velocity of the fluid, $c\dot{u}$. Energy dissipated by viscous damping per each vibration cycle is proportional to $A^2\omega$ where A is the response amplitude and ω is the vibration frequency. The general form of the viscous damping matrix is given in

Eq. (12.62b),

$$[c]^{(e)} = \int\limits_{\Omega^{(e)}} c[N]^T[N]d\Omega^{(e)} \tag{12.92}$$

where c is the damping coefficient. Note that the form of this matrix is very similar to the mass matrix. Therefore, the expressions developed in Section 12.6 can be used to compute the damping matrix by replacing ρ with c.

Hysteresis (or *solid*) *damping* is observed as a result of the internal dissipation mechanisms in a solid. Most materials have many internal imperfections such as small cracks. During dynamic response of the structure, the cracks can be opening or sliding depending on the local stresses they experience and their tips could be plastically deforming. If the crack opening/sliding remains under a critical threshold, the overall response of the structure will be in elastic range while some amount of the external work and internal kinetic energy is dissipated in plastic work at the crack tips. In hysteresis damping, the energy dissipated per cycle is independent of vibration frequency.

Coulomb damping is associated with dry friction. As two surfaces in contact slide past each other, a frictional shear stress τ that is proportional to the normal stress σ develops and points in the direction opposite to the sliding motion. The proportionality constant between τ and σ is the coefficient of friction μ_f, and *Coulomb's friction law* states $|\tau| = \mu_f|\sigma|$. A system that experiences sliding friction has to dissipate some of the available energy to overcome the resistance to motion/deformation imposed by the frictional shear stress. Coulomb damping and hysteresis damping are generally modeled similarly.

With direct time integration methods *algorithmic* (*artificial, numerical*) *damping* is introduced into the system. More on this type of damping is introduced later in this chapter. Another theoretical tool that is used to introduce frequency dependence is *proportional* (*Rayleigh*) *damping*. The damping matrix is computed as a linear combination of the mass and stiffness matrices,

$$[c]^{(e)} = a[m]^{(e)} + b[k]^{(e)} \tag{12.93}$$

where a and b are two model parameters that can be determined based on the frequency range of interest. It can be shown that the first term of Eq. (12.93) damps the lowest vibration modes while the second term damps the highest modes [1].

Modal damping is a convenient theoretical tool that can be used to introduce damping into a select range of vibration frequencies if required. To this end, the semidiscrete equation of motion needs to be represented as a set of decoupled differential equations. The semidiscrete equation of motion can be decoupled by using eigenvalue analysis, and the original system can be represented in terms of unknown modal coordinates and known eigenvalues of the system, as described in Section 12.9. Each mode can then be treated separately and customized level of damping ratio can be applied to each mode. *Modal analysis* can thus be used to selectively damp out response associated with specific vibration frequencies.

12.8 Global equation of motion

The equation of motion of the system is assembled by noting that the internal forces are continuous across element boundaries as described in Section 7.10. The global equation of equilibrium for nonlinear systems is found as follows:

$$[M]\{\ddot{D}\} + [C]\{\dot{D}\} + \{R_{\text{int}}\} = \{R\} \tag{12.94}$$

For linear elastic systems the global equation of equilibrium becomes,

$$[M]\{D\} + [C]\{\dot{D}\} + [K]\{D\} = \{R\} \tag{12.95}$$

where $[M]$ is the global mass matrix, $[C]$ is the global damping matrix, $[K]$ is the global stiffness matrix, $\{D\}$ is the global displacement vector, $\{\dot{D}\}$ is the global velocity vector, $\{\ddot{D}\}$ is the global acceleration vector, $\{R\}$ is the external load vector, and $\{R_{\text{int}}\}$ is the internal load vector encountered in nonlinear problems.

Eqs. (12.94) and (12.95) represent the *semidiscrete equation of motion.* These equations are subjected to the boundary and initial conditions applied on the original problem. The external tractions, that is, natural boundary conditions, are represented in the external force vector $\{R\}$ and they can vary in time. Imposed displacements, that is, the essential boundary conditions, can also be time dependent. Time-dependent essential boundary conditions generally affect the stiffness matrix and the loading vector as described in Section 7.10.2.

12.9 Analytical analysis of vibration of semidiscrete systems

In classical vibration analysis, the equations of motion of multidegree of freedom, lumped mass systems result in *discrete equations of motion* represented in vector form identical to Eq. (12.95). In addition, finite difference and finite element discretization of the spatial components of the boundary/initial value problems also result in semidiscrete system equations. Regardless of the origin of derivation, semidiscrete systems are represented by mass, damping, and stiffness matrices which have important common properties. Among these are the *positive definiteness* and *positive semidefiniteness* of these matrices.

In order to define these concepts, first consider the *kinetic energy* of the system,

$$T = \frac{1}{2}\{\dot{D}\}^T[M]\{\dot{D}\} \tag{12.96}$$

where $[M]$ is the mass matrix and $\{\dot{D}\}$ is an arbitrary velocity vector of the system. It should be noted that physical systems result in mass matrices that have fully populated diagonals, thus it is easy to realize that the kinetic energy is always a positive variable for nonzero velocities. Thus it is said that kinetic energy is a *positive-definite* variable and the mass matrix is a *positive definite matrix.*

Next consider the *strain energy* in the system,

$$U = \frac{1}{2}\{D\}^T[K]\{D\} \tag{12.97}$$

where $[K]$ is the stiffness matrix. If the structure is not anchored, any displacement vector $\{D\}$ applied on the system will cause a rigid body motion and strain energy cannot be stored in such a system. Therefore, there is a physical condition under which the system can have a zero strain energy for a nonzero displacement vector. However, if the structure is properly anchored, the strain energy will be always positive. Thus, considering both of these possibilities it is said that strain energy is a *positive semidefinite variable*, and the stiffness matrix is a *positive semidefinite matrix*. Note that the mass diagonalization process to obtain lumped mass matrices mentioned in Section 12.6.2 can occasionally result in a zero diagonal entry. In such a case the mass matrix becomes positive semidefinite. But, these occurrences are rare and should be avoided.

Vibration of semidiscrete systems can be analyzed by analytical approaches much like the continuous systems (Section 12.4). These approaches also involve an eigenvalue problem and modal expansion. In the next few sections we provide a brief introduction to such solutions.

12.9.1 Eigenvalue problem for the semidiscrete equation of motion

Free vibrations of an undamped system can be investigated by using the following homogenous equation of motion:

$$[M]\{\ddot{D}\} + [K]\{D\} = 0 \tag{12.98}$$

We assume that similar to the one-dimensional problem the undamped system will experience a simple harmonic motion. However, unlike the single degree of freedom system which only has a single natural frequency ($\omega_n = \sqrt{k/m}$) a semidiscrete system with N degrees of freedom has as many natural frequencies, ω_n. A critical mode of the system will vibrate synchronously at the corresponding natural frequency. To find such a harmonic solution we assume the response to be of the form,

$$\{D(t)\} = \{\underline{D}\}\sin\omega_n t \tag{12.99}$$

where $\{\underline{D}\}$ represents the vector of instantaneous nodal amplitudes. For the three element model of the simply supported beam shown in Fig. 12.8, the nodal amplitude vector is $\{\underline{D}\} = \{\theta_1 \quad v_2 \quad \theta_2 \quad v_3 \quad \theta_3 \quad \theta_4\}^T$. Using Eq. (12.99) in (12.98) we find,

$$\left([K] - \omega_n^2[M]\right)\{\underline{D}\}\sin\omega_n t = 0 \tag{12.100}$$

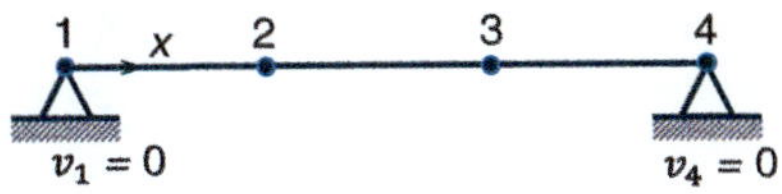

FIGURE 12.8 Finite element model of a simply supported beam.

Note that, in general, the $\sin \omega_n t$ term cannot be zero, and therefore the following term must be zero:

$$([K] - \lambda[M])\{\underline{D}\} = 0 \tag{12.101}$$

where we replaced $\lambda = \omega_n^2$ for convenience. Eq. (12.101) is the *eigenvalue problem* for the semidiscrete, homogeneous equation of motion. In order find a unique $\{\underline{D}\}$ from Eq. (12.101) the determinant of the term in the bracket must be zero. This gives the *characteristic equation* for the system,

$$\det([K] - \lambda[M]) = 0 \tag{12.102}$$

The characteristic equation is a polynomial equation, solution of which gives the eigenvalues λ_i for $i = 1 - N$ of the system. If the stiffness matrix $[K]$ and the mass matrix $[M]$ are positive definite, then the system will have N positive eigenvalues ordered as $\lambda_1 \leq \lambda_2 \leq \ldots \leq \lambda_N$. The *eigenvalues* $\lambda$of the system are related to the *natural frequencies* of vibration ω_n and $\{\underline{D}\}$ the *eigenvectors* of the system. An eigenvector or *mode-shape* represents the unique shape the structure vibrates with when it is excited with a specific natural frequency.

In a positive definite stiffness matrix all rigid body modes and mechanisms are suppressed. A stiffness matrix can be rendered positive definite by imposing essential boundary conditions and suppressing the mechanism modes. In a positive definite mass matrix (consistent or lumped) all diagonal entries are positive. The system will have zero eigenvalues if the stiffness matrix $[K]$ is positive semidefinite. If the mass matrix $[M]$ is positive semidefinite then the eigenvalues will be infinite for each zero diagonal. The zero or infinite eigenvalues can be suppressed and the remaining positive ones can be obtained as described in [7].

Returning back to positive definite systems, we conclude that the free vibration response of the undamped system is obtained by a superposition of all the modes, as the system possesses N eigenvalues, ω_{ni} for $i = 1 - N$. Note that this assertion is similar to the analytical solution presented in Section 12.4. The harmonic solution given in Eq. (12.99) becomes,

$$\{D(t)\} = \sum_{i=1}^{N} \{\underline{D}_i\} \sin \omega_{ni} t \tag{12.103}$$

Example 12.1 Find the eigenvalues and eigenvectors of a simply supported beam and a cantilevered beam by using finite element discretization.

a) Use a single element and demonstrate the effects of using consistent and lumped mass matrices on the predicted eigenvalues.

b) Carryout a mesh convergence study by using 4, 8, 16, 32, and 64 elements, and compare the numerically predicted eigenvalues to the analytically determined ones.

Solution:

a) Simply supported beam with consistent mass matrix

The homogenous form of the equation of motion of an *undamped* Euler–Bernoulli beam can be obtained from Eq. (12.77),

$$\left[m_b{}'\right]^{(e)}\left\{\ddot{d}_b'\right\} + \left[k_b'\right]^{(e)}\left\{d_b{}'\right\}^{(e)} = 0 \tag{12.104}$$

Consistent mass $[m_{Cb}{}']^{(e)}$ and stiffness $[k_b']^{(e)}$ matrices for the Euler–Bernoulli beam element are given in Eqs. (12.85) and (6.40), respectively. The essential boundary conditions for the simply supported beam are,

$$\text{at } x = 0 : v(0) = 0 \text{ and at } x = L : v(L) = 0 \tag{12.105}$$

For the case where a single element is used, these boundary conditions become $v_1 = v_2 = 0$. The global equation of motion can be found as follows:

$$\frac{EI}{L^3}\begin{bmatrix} 4L^2 & 2L^2 \\ 2L^2 & 4L^2 \end{bmatrix}\begin{Bmatrix} \theta_1 \\ \theta_2 \end{Bmatrix} + \frac{\rho AL}{420}\begin{bmatrix} 4L^2 & -3L^2 \\ -3L^2 & 4L^2 \end{bmatrix}\begin{Bmatrix} \ddot{\theta}_1 \\ \ddot{\theta}_2 \end{Bmatrix} = 0 \tag{12.106}$$

The eigenvalue problem associated with this equation of motion can be established from Eq. (12.101),

$$\left(\begin{bmatrix} 4L^2 2L^2 \\ 2L^2 4L^2 \end{bmatrix} - \lambda\frac{1}{420}\frac{mL^3}{EI}\begin{bmatrix} 4L^2 & -3L^2 \\ -3L^2 & 4L^2 \end{bmatrix}\right)\begin{Bmatrix} \theta_1 \\ \theta_2 \end{Bmatrix} = 0 \tag{12.107}$$

where $m = \rho AL$. The characteristic equation is established by the determinant of the term in the bracket as defined in Eq. (12.102),

$$\det([K] - \lambda[M]) = \frac{L^4}{25200}\left(\frac{mL^3}{EI}\lambda - 120\right)\left(\frac{mL^3}{EI}\lambda - 2520\right) = 0$$

Considering the natural frequencies are related to the eigenvalues as $\lambda = \omega^2$ we find,

$$\bar{\omega}_1 = \sqrt{120} = 10.954, \quad \bar{\omega}_2 = \sqrt{2520} = 50.120 \tag{12.108}$$

where the nondimensional frequency $\bar{\omega} = \omega/(EI/mL^3)^{1/2}$ is used for notational simplicity. The eigenvalues calculated by the analytical approach in Eq. (12.42) give,

$$\bar{\omega}_1 = \pi^2 = 9.867, \quad \bar{\omega}_2 = 4\pi^2 \tag{12.109}$$

Thus, we see that the numerically determined natural frequencies are about 11% and 27% off from the exact values. This will improve with mesh refinement as shown below. Note that for notational simplicity we assume $\rho = E = I = L = 1$ in this problem. The mode shapes can be calculated by using Eq. (12.139) with $\lambda_1 = \omega_1^2$

$$\left(\begin{bmatrix} 4 & 2 \\ 2 & 4 \end{bmatrix} - \frac{120}{420}\begin{bmatrix} 4 & -3 \\ -3 & 4 \end{bmatrix}\right)\begin{Bmatrix} \underline{\theta}_1 \\ \underline{\theta}_2 \end{Bmatrix} = 0 \tag{12.110}$$

It can be shown that this relationship results in $\underline{\theta}_1 = -\underline{\theta}_2$. Thus, we see that one of the degrees of freedom is arbitrary. We choose $\underline{\theta}_1 = 1$. Repeating the same procedure for ω_2 the eigenvector associated with natural frequency ω_1 and ω_2 are found as follows:

$$\{\underline{D}_{\omega_1}\} = \begin{Bmatrix} 1 \\ -1 \end{Bmatrix} \text{ and } \{\underline{D}_{\omega_1}\} = \begin{Bmatrix} 1 \\ 1 \end{Bmatrix} \tag{12.111}$$

Simply supported beam with lumped mass matrix
The eigenvalue problem when the lumped mass matrix given in Eq. (12.91) is used becomes

$$\left(\begin{bmatrix} 4L^2 & 2L^2 \\ 2L^2 & 4L^2 \end{bmatrix} - \lambda\frac{mL^3}{EI}\begin{bmatrix} \phi L^2 & 0 \\ 0 & \phi L^2 \end{bmatrix}\right)\begin{Bmatrix} \theta_1 \\ \theta_2 \end{Bmatrix} = 0 \tag{12.112}$$

Using $\phi = 1/24$, solution of this eigenvalue problem is found as follows:

$$\bar{\omega}_1 = 4\sqrt{3}, \bar{\omega}_2 = 12 \tag{12.113}$$

These predictions are significantly lower than the exact values given in Eq. (12.109) or the values found by using the consistent mass matrix in Eq. (12.108).

Cantilevered beam with consistent mass matrix
The boundary conditions for a cantilevered beam are,

$$\text{at } x = 0: \quad v(0) = 0 \tag{12.114a}$$

$$\theta(0) = 0 \tag{12.114b}$$

$$\text{at } x = L: EI\frac{\partial^2 v}{\partial x^2} = 0 \tag{12.114c}$$

$$\frac{\partial}{\partial x}\left(EI\frac{\partial^2 v}{\partial x^2}\right) = 0 \tag{12.114d}$$

We only impose the essential boundary conditions on the global stiffness matrix. For the case where a single element is used, these boundary conditions become $v_1 = \theta_1 = 0$. It can be shown by using Eqs. (6.41) and (12.85) that the eigenvalue problem when the consistent mass matrix is used, becomes,

$$\left(\begin{bmatrix} 12 & -6L \\ -6L & 4L^2 \end{bmatrix} - \lambda\frac{1}{420}\frac{mL^3}{EI}\begin{bmatrix} 156 & -22L \\ -22L & 4L^2 \end{bmatrix}\right)\begin{Bmatrix} v_2 \\ \theta_2 \end{Bmatrix} = 0 \tag{12.115}$$

The corresponding characteristic equation is found as follows,

$$\det([K]-\lambda[M]) = \frac{L^2}{1260}\left(15120 + \frac{mL^3}{EI}\lambda\left(\frac{mL^3}{EI}\lambda - 1124\right)\right) = 0$$

The natural frequencies of this system are thus found as follows,

$$\bar{\omega}_1 = 3.5327, \bar{\omega}_2 = 34.8069 \tag{12.116}$$

Comparing these values to the exact values $\bar{\omega}_1 = 3.5160$ and $\bar{\omega}_2 = 22.0345$ in Table 12.1 shows that the first natural frequency is closer to the exact value than the second one. As expected these predications improve with a refined mesh as shown below. By following the same procedure described above the eigenvectors, or the mode shapes can be found as follows:

$$\{\underline{D}_{\omega_1}\} = \begin{Bmatrix} 1 \\ 1.38 \end{Bmatrix} \text{ and } \{\underline{D}_{\omega_2}\} = \begin{Bmatrix} 1 \\ 7.62 \end{Bmatrix} \tag{12.117}$$

Cantilevered beam with lumped mass matrix

By using Eq. (12.91), the eigenvalue problem with the lumped mass matrix becomes,

$$\left(\begin{bmatrix} 12 & -6L \\ -6L & 4L^2 \end{bmatrix} - \lambda\frac{mL^3}{EI}\begin{bmatrix} 1/2 & 0 \\ 0 & \phi L^2 \end{bmatrix}\right)\begin{Bmatrix} v_2 \\ \theta_2 \end{Bmatrix} = 0 \tag{12.118}$$

The characteristic equation associated with this equation is,

$$\det([K]-\lambda[M]) = \frac{L^2}{48}\left(576 + \frac{mL^3}{EI}\lambda\left(\frac{mL^3}{EI}\lambda - 120\right)\right) = 0 \tag{12.119}$$

Solution of this equation gives the following natural frequencies:

$$\bar{\omega}_1 = 2.2381, \bar{\omega}_2 = 10.7234 \tag{12.120}$$

Using a lumped mass matrix and a single element, the predicted natural frequencies are even farther off from the exact values (Table 12.1) than those predicted by using the consistent mass matrix.

b) Effect of solution mesh on natural frequencies of the semidiscrete system

A computer program to compute the natural frequencies and the mode shapes is written by using the Euler–Bernoulli beam elements given in Eqs. (6.41), (12.85), and (12.91).

The first four nondimensional natural frequencies of the simply supported and cantilevered beam problems are presented in Table 12.1. For reference, the exact natural frequencies of the simply supported the cantilevered beam problems are obtained from Eq. (12.42) and from reference [7], respectively. The first four mode shapes for both problems are plotted in Figs. 12.9 and 12.10.

Inspection of the results presented in Table 12.1 shows that for both types of support conditions, the consistent mass matrix converges to the exact values with fewer elements. Thus, for this problem, we conclude that the consistent mass

TABLE 12.1 First four natural frequencies of the simply supported (top) and cantilevered beams (bottom) computed by using different number of elements. The analytical solution for the cantilevered beam can be found in reference [7].

Simply supported Euler–Bernoulli beam									
Consistent mass matrix					Lumped mass matrix				
N_E	$\bar{\omega}_1$	$\bar{\omega}_2$	$\bar{\omega}_3$	$\bar{\omega}_4$	N_E	$\bar{\omega}_1$	$\bar{\omega}_2$	$\bar{\omega}_3$	$\bar{\omega}_4$
4	9.8722	39.6342	90.4495	175.2712	4	9.6228	35.8096	72.0655	110.8513
8	9.8698	39.4887	88.9407	158.5369	8	9.8066	38.4912	83.9863	143.2384
16	9.8696	39.4791	88.8338	157.9547	16	9.8538	39.2265	87.5621	153.9647
32	9.8696	39.4785	88.8269	157.9163	32	9.8656	39.4151	88.5067	156.9061
64	9.8696	39.4784	88.8265	157.9138	64	9.8686	39.4626	88.7463	157.6604
Exact	**9.8696**	**39.4784**	**88.8264**	**157.9137**	**Exact**	**9.8696**	**39.4784**	**88.8264**	**157.9137**
Cantilevered Euler–Bernoulli beam									
Consistent mass matrix					**Lumped mass matrix**				
4	3.5161	22.0602	62.1749	122.6576	4	3.3791	18.8338	47.4806	83.7574
8	3.5160	22.0363	61.7347	121.1728	8	3.4806	21.0810	56.7312	106.1225
16	3.5160	22.0346	61.6997	120.9202	16	3.5071	21.7849	60.3333	116.5856
32	3.5160	22.0345	61.6974	120.9031	32	3.5138	21.9714	61.3482	119.7804
64	3.5160	22.0345	61.6972	120.9020	64	3.5155	22.0187	61.6095	120.6189
Exact	**3.5160**	**22.0345**	**61.6972**	**120.9019**	**Exact**	**3.5160**	**22.0345**	**61.6972**	**120.9019**

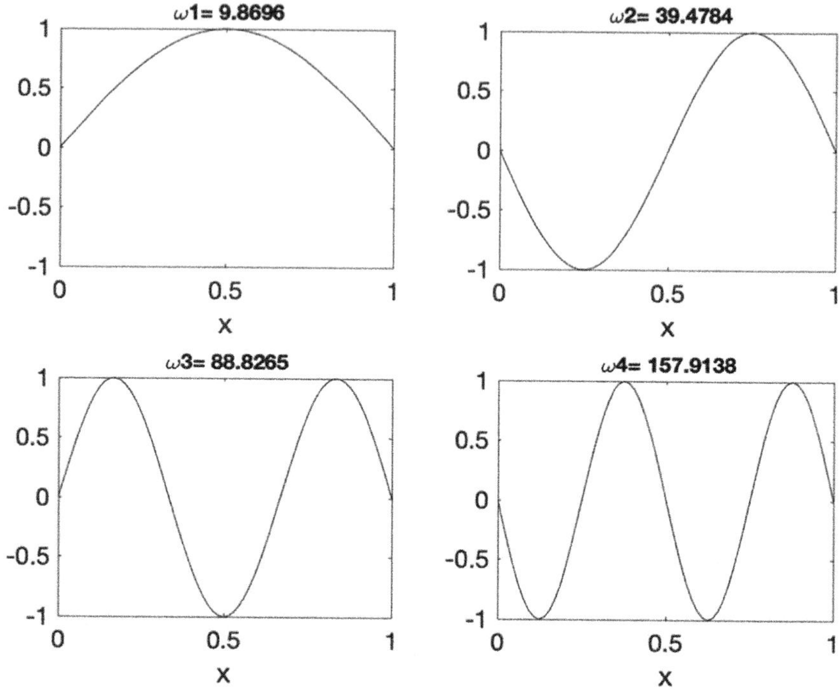

FIGURE 12.9 The first four mode shapes (eigenvectors) of the simply supported beam, calculated by using 64 elements and the consistent mass matrix.

matrix is a better representation of the original system than the lumped mass matrix. We also note that the first natural frequency $\bar{\omega}_1$ converges very quickly toward the exact solution with mesh refinement when the consistent mass matrix is used.

The effect of mesh refinement on natural frequencies is presented in Fig. 12.11 by using the simply supported beam and the consistent mass matrix. This figure shows the natural frequency $\bar{\omega}_1$ and the percent error as a function of mode numbers. The percent error is computed based on the exact solution and the numerically predicted natural frequency. Further mesh refinement improves the higher values of the computed frequencies. Results are shown for mesh sizes $N_E = 4$, 8, 16, 32, and 64. Note that the predictions become unreliable for higher mode numbers as increasingly more severe aliasing error would be expected. This figure shows that as the mesh is refined the error in the predicted natural frequency drops by approximately an order of magnitude. Generally speaking, high frequencies have a lower share of the dynamic response of the system. Thus, the error can be tolerated for systems with low-frequency loading and response. Nevertheless, a similar error analysis could be conducted to determine the error in the predicted natural frequencies.

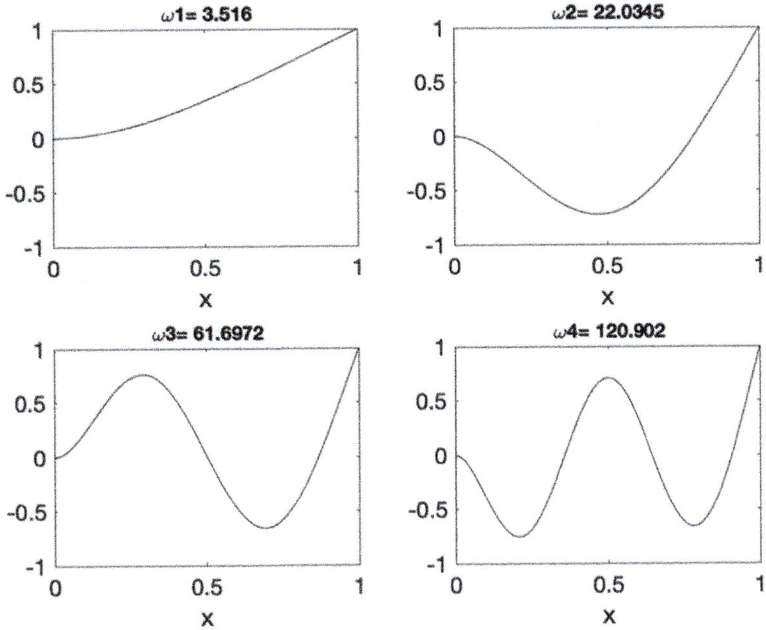

FIGURE 12.10 The first four mode shapes (eigenvectors) of the cantilevered beam calculated by using 64 elements and the consistent mass matrix.

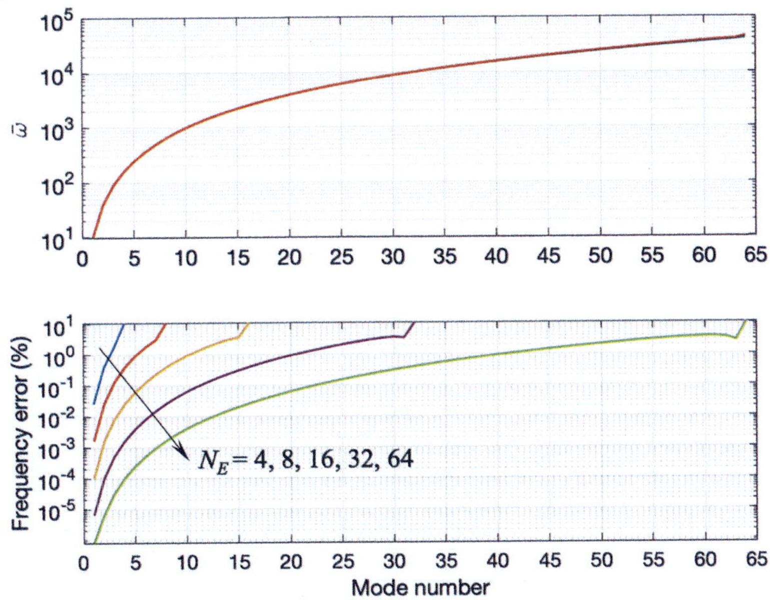

FIGURE 12.11 The exact and numerically predicted natural frequencies for $N_E = 4, 8, 16, 32$, and 64. The error grows for higher frequencies, but mesh refinement can improve the error to a certain extent.

12.9.2 Orthogonality of the eigenvectors

The *eigenfunctions* that represent the mode shapes of a continuous system are orthogonal to one another as mentioned in Section 12.3.3. A similar property can be shown for the eigenvectors of the semidiscrete equation of motion. However, in contrast to the eigenfunctions of the continuous systems which are orthogonal with respect to one another, the eigenvectors of the semidiscretized dynamical systems are orthogonal with respect to the mass and stiffness matrices. A detailed derivation of this type of orthogonality property is straightforward and it can be found in reference [4] among others. Here we simply state the resulting relationships.

Consider two different eigenvectors $\{\underline{D}_r\}$ and $\{\underline{D}_s\}$, and positive definite stiffness and mass matrices, $[K]$ and $[M]$, respectively. Orthogonality of these eigenvectors with respect to the these matrices means that,

$$\{\underline{D}_r\}^T[K]\{\underline{D}_s\} = 0 \text{ if } r \neq s \tag{12.121a}$$

$$\{\underline{D}_r\}^T[M]\{\underline{D}_s\} = 0 \text{ if } r \neq s \tag{12.121b}$$

Two different eigenvectors ($r \neq s$) will always yield zero through the multiplications given in Eq. (12.121a). But, using the same eigenvector ($r = s$) must give nonzero values, due to the definition of the positive definiteness, as described at the beginning of this section. This condition is expressed as follows:

$$\{\underline{D}_r\}^T[K]\{\underline{D}_r\} = k_{rr} \tag{12.122a}$$

$$\{\underline{D}_r\}^T[M]\{\underline{D}_r\} = m_r \tag{12.122b}$$

where k_{rr} and m_{rr} are positive scalar numbers.

This property can be used to identify a normalization and decoupling for the dynamical system. To demonstrate this, let's use the rth eigenvalue ω_{nr} and the corresponding eigenvector $\{\underline{D}_r\}$ in the solution given in Eq. (12.99) to find, $\{D_r(t)\} = \{\underline{D}_r\}\sin\omega_{nr}t$. Substituting this in Eq. (12.98) and premultiplying the resulting equation through $\{\underline{D}_r\}^T$ we find,

$$\{\underline{D}_r\}^T\left([K] - \omega_{nr}^2[M]\right)\{\underline{D}_r\}\sin\omega_{nr}t = 0 \tag{12.123}$$

By using the orthogonality property defined in Eq. (12.122a) we can show that Eq. (12.123) becomes,

$$k_{rr} - \omega_{nr}^2 m_{rr} = 0 \text{ for } r = 1 - N \tag{12.124}$$

This property leads to a convenient normalization of the system equations,

$$\{\underline{D}_r\}^T[K]\{\underline{D}_s\} = \delta_{rs}\omega_r^2 \tag{12.125a}$$

$$\{\underline{D}_r\}^T[M]\{\underline{D}_s\} = \delta_{rs} \tag{12.125b}$$

where δ_{rs} is the Dirac delta function.

12.9.3 Response to initial excitations by modal analysis

Free vibration of an undamped system is represented in semidiscrete form by the equation of motion and the initial conditions,

$$[M]\{\ddot{D}\} + [K]\{D\} = 0 \tag{12.126a}$$

$$\{D(0)\} = \{d^{(0)}\} \tag{12.126b}$$

and

$$\{\dot{D}(0)\} = \{v^{(0)}\} \tag{12.126c}$$

The solution $\{D(t)\}$ can be expressed by using the superposition of the normal modes $\{\underline{D}_r\}$,

$$\{D(t)\} = \sum_{r=1}^{N} c_r\{\underline{D}_r\} \tag{12.127}$$

where c_r are the unknown time-dependent coefficients known as the *modal coordinates*. To explain this term, note that the normal modes $\{\underline{D}_r\}$ represent the axes of an N-dimensional space because they are orthogonal with respect to the $[M]$ and $[K]$ matrices as described in Eq. (12.122). The coefficients c_r, thus, can be thought of as the *coordinates* on the corresponding axis $\{\underline{D}_r\}$.

In order to obtain the modal coordinates, let's premultiply Eq. (12.127) by $\{\underline{D}_s\}^T[M]$ and use the orthogonality condition given in Eq. (12.125b) on the resulting expression,

$$\{\underline{D}_s\}^T [M]\{D(t)\} = \sum_{r=1}^{N} c_r \{\underline{D}_s\}^T [M]\{\underline{D}_r\} \tag{12.128}$$

This gives,

$$c_r = \{\underline{D}_r\}^T [M]\{D(t)\} \quad \text{for} \quad r = 1 - N \tag{12.129}$$

Similarly, premultiplying Eq. (12.127) by $\{\underline{D}_s\}^T[K]$ and using the condition given in Eq. (12.125a) we can find,

$$\omega_{rnr}^2 c_r = \{\underline{D}_r\}^T [K]\{D(t)\} \quad \text{for} \quad r = 1 - N \tag{12.130}$$

Eqs. (12.127), (12.129), and (12.130) constitute the *expansion theorem*. Substituting Eq. (12.127) into equation of motion (12.126a) we can find,

$$[M]\sum_{r=1}^{N} \ddot{c}_r\{\underline{D}_r\} + [K]\sum_{r=1}^{N} c_r\{\underline{D}_r\} = 0 \tag{12.131}$$

Premultiplying this equation by $\{\underline{D}_s\}^T$ and using the orthogonality conditions in Eq. (12.125a), we can find the following ordinary differential equation for the modal coordinates,

$$\ddot{c}_r + \omega_{nr}^2 c_r = 0 \text{ for } r = 1 - N \tag{12.132}$$

Thus, we can see that the equation of motion (12.126a) is decoupled into N independent ordinary differential equations. Note that Eq. (12.132) is in the same form as the equation of motion of a single degree of freedom *simple harmonic oscillator* in Eq. (12.12). Therefore, it has the same solution,

$$c_r(t) = c_r(0)\cos\omega_{nr}t + \frac{\dot{c}_r(0)}{\omega_{nr}}\sin\omega_{nr}t \quad \text{for} \quad r = 1 - N \tag{12.133}$$

where the initial modal coordinates $c_r(0)$ and velocities $\dot{c}_r(0)$ can be obtained from the initial conditions in Eq. (12.126b) (12.126c) and by using Eq. (12.129) as follows:

$$c_r(0) = \{\underline{D}_r\}^T[M]\{d^{(0)}\} \text{ for } r = 1 - N \tag{12.134a}$$

$$\dot{c}_r(0) = \{\underline{D}_r\}^T[M]\{v^{(0)}\} \text{ for } r = 1 - N \tag{12.134b}$$

Once each modal coordinate is found by using Eq. (12.133), the response of entire system can be obtained by using Eqs. (12.127) and (12.133) as follows,

$$\{D(t)\} = \sum_{r=1}^{N}\left(c_r(0)\cos\omega_{nr}t + \frac{\dot{c}_r(0)}{\omega_{nr}}\sin\omega_{nr}t\right)\{\underline{D}_r\} \tag{12.135}$$

Modal analysis presented above is a powerful method to find analytical solutions for the free vibration of a semidiscrete equation of motion that represents an undamped dynamical system. This method can be expanded to damped systems, harmonic and transient loading cases as well. Interested readers can find the details of these implementations in references [4,7] among others.

Example 12.2 Find the free vibration response of the simply supported Euler–Bernoulli beam presented in Section 12.4, subjected to the following boundary conditions, by using the modal superposition method:

$$v(x,0) = A\sin\frac{\pi x}{L}, \quad \dot{v}(x,0) = 0 \tag{12.136}$$

a) Use the semidiscrete equation of motion with consistent mass matrix. For demonstration purposes use three elements evaluate the system response in the range, $0 \le t \le 2$.

b) Perform a mesh convergence study using 4, 8, 16, 32, and 64 elements and compare the results to the analytical solution given in Eq. (12.53).

Solution to part (a):

Element level equation of motion for an undamped system with no external loading can be found by using Eqs. (6.41) and (12.85) as follows:

$$\frac{m^{(e)}}{420}\begin{bmatrix} 156 & 22L^{(e)} & 54 & -13L^{(e)} \\ 22L & 4L^{(e)2} & 13L^{(e)} & -3L^{(e)2} \\ 54 & 13L^{(e)} & 156 & -22L^{(e)} \\ -13L^{(e)} & -3L^{(e)2} & -22L^{(e)} & 4L^{(e)2} \end{bmatrix}\begin{Bmatrix} \ddot{v}_1 \\ \ddot{\theta}_1 \\ \ddot{v}_2 \\ \ddot{\theta}_2 \end{Bmatrix} + k_b\begin{bmatrix} 12 & 6L^{(e)} & -12 & 6L^{(e)} \\ 6L^{(e)} & 4L^{(e)2} & -6L^{(e)} & 2L^{(e)2} \\ -12 & -6L^{(e)} & 12 & -6L^{(e)} \\ 6L^{(e)} & 2L^{(e)2} & -6L^{(e)} & 4L^{(e)2} \end{bmatrix}\begin{Bmatrix} v_1 \\ \theta_1 \\ v_2 \\ \theta_2 \end{Bmatrix} = 0 \tag{12.137}$$

where $m^{(e)} = \rho A^{(e)} L^{(e)}$ is the mass of the beam and $k_b = EI/L^{(e)3}$ is the bending rigidity. In this problem we use $\rho = A = E = I = 1$ for notational simplicity. For the case of three elements we find $L^{(e)} = 1/3$ Thus Eq. (12.137) becomes,

$$10^{-4}\begin{bmatrix} 1238.1 & 58.201 & 428.57 & -34.392 \\ 58.201 & 3.5273 & 34.392 & -2.6455 \\ 428.57 & 34.392 & 1238.1 & -58.201 \\ -34.392 & -2.6455 & -58.201 & 3.5273 \end{bmatrix}\begin{Bmatrix} \ddot{v}_1 \\ \ddot{\theta}_1 \\ \ddot{v}_2 \\ \ddot{\theta}_2 \end{Bmatrix} + \begin{bmatrix} 324 & 54 & -324 & 54 \\ 54 & 12 & -54 & 6 \\ -324 & -54 & 324 & -54 \\ 54 & 6 & -54 & 12 \end{bmatrix}\begin{Bmatrix} v_1 \\ \theta_1 \\ v_2 \\ \theta_2 \end{Bmatrix} = 0 \tag{12.138}$$

After assembling the three elements and imposing the boundary conditions $v_1 = v_3 = 0$ the equation of motion becomes,

$$[M]\{\ddot{D}\} + [K]\{D\} = 0 \tag{12.139}$$

where,

$$\{\ddot{D}\} = \begin{Bmatrix} \ddot{\theta}_1 & \ddot{v}_2 & \ddot{\theta}_2 & \ddot{v}_3 & \ddot{\theta}_3 & \ddot{\theta}_4 \end{Bmatrix}^T \tag{12.140a}$$

$$[M] = 10^{-4}\begin{bmatrix} 3.5273 & 34.392 & -2.6455 & 0 & 0 & 0 \\ & 2476 & 0 & 42.875 & -34.392 & 0 \\ & & 7.0547 & 34.392 & -2.6455 & 0 \\ & & & 2476 & 0 & -34.392 \\ & & & & 7.0547 & -2.6455 \\ & & & & & 3.5273 \end{bmatrix} \tag{12.140b}$$

$$[K] = \begin{bmatrix} 12 & -54 & 6 & 0 & 0 & 0 \\ & 648 & 0 & -324 & 54 & 0 \\ & & 24 & -54 & 6 & 0 \\ & & & 648 & 0 & 54 \\ & & & & 24 & 6 \\ & & & & & 12 \end{bmatrix} \tag{12.140c}$$

Note that only the symmetric parts of the mass and stiffness matrices are shown. The eigenvalue problem associated with Eq. (12.139) is solved in Example 12.1. The characteristic equation of this problem is given, in its general form, in Eq. (12.102),

$$\det([K] - \lambda[M]) = 0 \tag{12.141}$$

where $\lambda = \omega^2$. We solve this equation by using Matlab and find the natural frequencies as follows:

$$\begin{aligned} &\omega_1 = 9.8776, \quad &\omega_2 = 39.9451, \quad &\omega_3 = 98.5901, \\ &\omega_4 = 183.3209, \quad &\omega_5 = 328.015, \quad &\omega_6 = 451.7964 \end{aligned} \tag{12.142}$$

Note that a comparison with Table 12.1, where the exact eigenvalues for this problem are listed, shows that with the exception of ω_1 the values given above have some error. But, this is expected as the size of the mesh ($N_E = 3$) is kept low for this demonstration.

The eigenvectors corresponding to the eigenvalues given in Eq. (12.142) are found by using Eq. (12.101) as follows:

$$\begin{aligned} \{\underline{D}_1\} &= \{4.4501 \quad 1.2267 \quad 2.2250 \quad 1.2267 \quad -2.2250 \quad -4.4501\}^T \\ \{\underline{D}_2\} &= \{9.0761 \quad 1.2251 \quad -4.5381 \quad -1.2251 \quad -4.5381 \quad 9.0761\}^T \\ \{\underline{D}_3\} &= \{-16.4317 \quad 0 \quad 16.4317 \quad 0 \quad -16.4317 \quad -16.4317\}^T \\ \{\underline{D}_4\} &= \{27.4660 \quad -1.1823 \quad -13.7330 \quad 1.1823 \quad -13.7330 \quad 27.4660\}^T \\ \{\underline{D}_5\} &= \{45.5151 \quad -0.9360 \quad 22.7576 \quad -0.9360 \quad -22.7576 \quad -45.5151\}^T \\ \{\underline{D}_6\} &= \{43.4741 \quad 0 \quad 43.4741 \quad 0 \quad 43.4741 \quad 43.4741\}^T \end{aligned} \tag{12.143}$$

It is important to realize that these eigenvectors contain the variables in the order of the vector defined in Eq. (12.140a),

$$\{\bar{D}_r\} = \{\theta_{r1} \; v_{r2} \; \theta_{r2} \; v_{r3} \; \theta_{r3} \; \theta_{4r}\}^T \tag{12.144}$$

As mentioned before the magnitudes of these eigenvectors are arbitrary. For plotting convenience we normalize the displacements and slopes separately,

$$\begin{aligned} \{\underline{D}_1\} &= \{1.0 \quad 1.0 \quad 0.5 \quad 1.0 \quad -0.50 \quad -1.0\}^T \\ \{\underline{D}_2\} &= \{1.0 \quad 1.0 \quad -0.5 \quad -1.0 \quad -0.5 \quad -1.0\}^T \\ \{\underline{D}_3\} &= \{-1.0 \quad 0 \quad 1.0 \quad 0 \quad -1.0 \quad 1.0\}^T \\ \{\underline{D}_4\} &= \{1.0 \quad -1.0 \quad -0.5 \quad 1.0 \quad -0.5 \quad 1.0\}^T \\ \{\underline{D}_5\} &= \{1.0 \quad -1.0 \quad 0.5 \quad -1.0 \quad -0.5 \quad -1.0\}^T \\ \{\underline{D}_6\} &= \{1.0 \quad 0 \quad 1.0 \quad 0 \quad 1.0 \quad 1.0\}^T \end{aligned} \tag{12.145}$$

In order to visualize the deflections and slopes, the appropriate subvectors should be extracted from the vectors given in Eq. (12.143) accordingly. Fig. 12.12A shows the first two mode shapes extracted from Eq. (12.143),

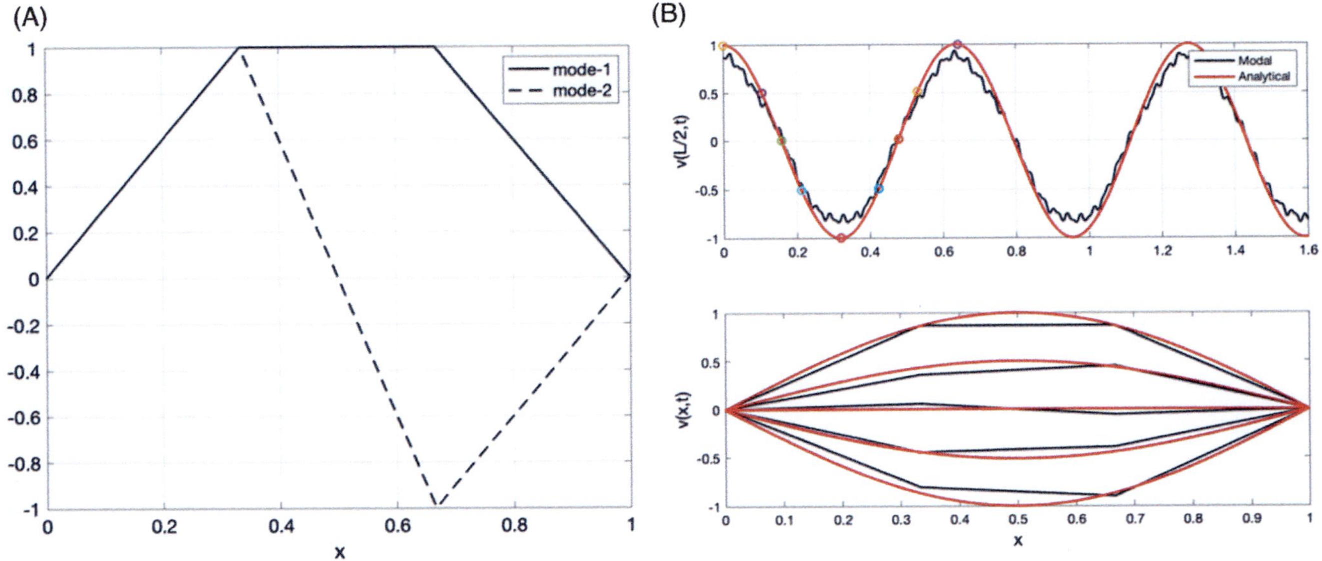

FIGURE 12.12 (A) The first two mode shapes (eigenvectors) of the simple supported Euler–Bernoulli beam when the beam is discretized by three elements. (B) Transient response of the mid-point (top) and the whole beam in response to the initial condition given in Eq. (12.136). The mode shapes are poor representations of $\sin(\pi x/L)$ and $\sin(2\pi x/L)$. With mesh refinement this situation improves significantly.

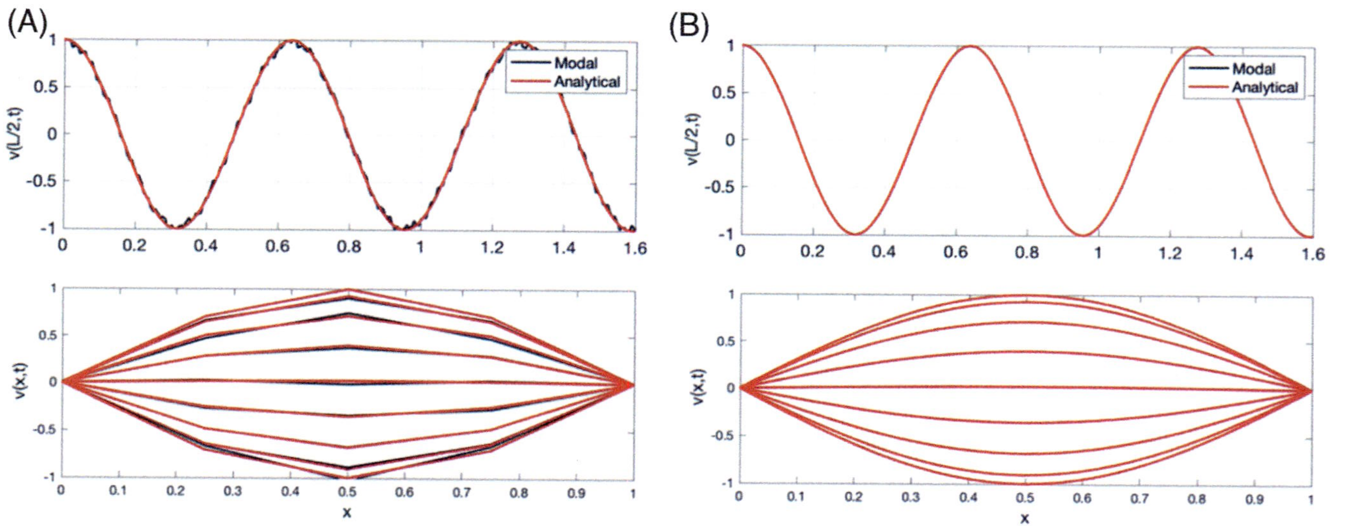

FIGURE 12.13 Free vibration response of an Euler–Bernoulli beam in response to a half-sine-shaped initial condition. Analytical and modal responses are superimposed. The beam is modeled by using (A) 4 and (B) 32 elements. The high-frequency error is much more pronounced for the solution when $N_E = 4$.

$$\begin{aligned}\{\underline{D}_1\} &= \{\underline{v}_1 \quad \underline{v}_2 \quad \underline{v}_3 \quad \underline{v}_4\}^T = \{0 \quad 1 \quad 1 \quad 0\}^T \\ \{\underline{D}_2\} &= \{0 \quad 1 \quad -1 \quad 0\}^T\end{aligned} \tag{12.146}$$

These vectors are representative of the first two eigenvectors predicted analytically in Section 12.4, $\sin(\pi x/L)$ and $\sin(2\pi x/L)$. The mesh refinement study presented in part (b) of this problem shows that the computed eigenvectors approach the exact values.

With the eigenvectors determined, the dynamic response of the system to initial conditions can be found from Eq. (12.135),

$$\{D(t)\} = \sum_{r=1}^{N}\left(c_r(0)\cos\omega_r t + \frac{\dot{c}_r(0)}{\omega_r}\sin\omega_r t\right)\{\underline{D}_r\} \tag{12.147}$$

where the initial values of the modal coordinates $c_r(0)$ and $\dot{c}_r(0)$ are found from Eq. (12.134a),

$$c_r(0) = \{\underline{D}_r\}^T[M]\{d^{(0)}\} \quad \text{and} \quad \dot{c}_r(0) = \{\underline{D}_r\}^T[M]\{v^{(0)}\} \tag{12.148}$$

By using the eigenvectors given in Eq. (12.143) and the initial conditions given in (12.136) the model coordinates are found as follows:

$$\begin{aligned}&c_1(0) = 0.6896, \quad c_2(0) = 0.0274, \quad c_3(0) = 0.0319, \\ &c_4(0) = 0.0291, \quad c_5(0) = -0.0214, \quad c_6(0) = 0.012\end{aligned} \tag{12.149a}$$

and

$$\dot{c}_1(0) = \dot{c}_2(0) = \dot{c}_3(0) = \dot{c}_4(0) = \dot{c}_5(0) = \dot{c}_6(0) = 0. \tag{12.149b}$$

Note that the initial values are dominated by the first mode. This is not surprising as initial condition $v(x,0)$ has the same shape as the first eigenfunction of the problem given in Section 12.4. The vibration of the beam can be computed by using Eq. (12.147),

$$D(t) = \sum_{r=1}^{6} c_r(0)\cos(\omega_{nr}t)\{\underline{D}_r\} \tag{12.150}$$

Free vibration response of the beam to the initial condition in Eq. (12.136) is plotted in Fig. 12.12B. The response is dominated by the first mode, but the effect of the higher modes can also be seen. As the solution is obtained with a coarse mesh, high frequencies introduce large errors, and the result plotted in Fig. 12.12 is not an accurate representation of the solution. A mesh convergence study is presented next.

Solution to part (b):

A mesh convergence study is performed by using $(N_E) = 4$, 8, 16, 32, and 64 elements. The first three natural frequencies of this problem and the convergence characteristics can be found in Table 12.1 of Example 12.1. There we showed that the error between the eigenvalues of the semidiscrete equation of motion and the analytical solution improves with mesh refinement. Here we compare the free vibration response to the initial condition given in Eq. (12.136), computed by modal analysis of the semidiscrete equation of motion (12.139) and the analytical solution provided by Eq. (12.53).

Fig. 12.13 shows the free vibration response for two different meshes: $N_E = 4$ and $N_E = 32$ computed by using modal analysis and the analytical solution. The overall solution is very similar to the analytical solution in the case $N_E = 4$, but there is clearly a high-frequency error that is prominent in the response. This is expected as the natural frequencies of the higher modes are predicted poorly, in general. In this particular problem, we can eliminate the high-frequency error by computing the response with only mode-1 of the solution. But, this is not a good strategy in general, as different initial conditions might require input from all modes of vibration. Therefore, increasing the mesh density presents a better approach to reducing the effects of the high-frequency error as demonstrated in Fig. 12.13B.

12.10 Direct time integration of the equation of motion of a solid

Direct time integration refers to calculation of response history by using step-by-step integration of the equation of motion in time. Deflection of the structure $\{D\}$ is evaluated at discrete time instances,

$$\{D\}^0, \{D\}^{\Delta t}, \{D\}^{2\Delta t}, \ldots \{D\}^{n\Delta t}, \ldots \{D\}^{N\Delta t} \tag{12.151}$$

The time domain is discretized as described in Section 11.3.1,

$$t^{(n)} = t_0 + n\Delta t \qquad \text{for} \qquad n = 0 - N_T \tag{12.152}$$

where $^{(n)}$ represents a time step, N_T represents the number of subintervals between t_0 and t_F, and Δt is the size of the *time step* defined as follows:

$$\Delta t = \frac{1}{N_T}(t_F - t_0) \tag{12.153}$$

Note that in most problems $t^{(0)} = 0$.

The semidiscrete equation of motion for a dynamical system given in Eqs. (12.94) and (12.95) for nonlinear and linear systems, respectively, are expressed for spatially discrete, temporally continuous displacement, velocity and acceleration vectors $\{D^{(n\Delta t)}\}$, $\{\dot{D}^{(n\Delta t)}\}$, and $\{\ddot{D}^{(n\Delta t)}\}$, respectively. In what follows we will replace these vectors with the temporally discrete vectors $\{x^{(n)}\}$, $\{v^{(n)}\}$, and $\{a^{(n)}\}$, respectively. Let's also consider that these vectors are known at the *current time step* $^{(n)}$ and but unknown at the next (future) time step $^{(n+1)}$. Using this notation the equation of motion for the nonlinear problems, (12.94), can be expressed as follows:

$$[M]\{a^{(n)}\} + [C]\{v^{(n)}\} + \left\{R_{\text{int}}^{(n)}\right\} = \{R^{(n)}\} \tag{12.154}$$

Similarly, we find the equation of motion for linear problems, (12.95) as,

$$[M]\{a^{(n)}\} + [C]\{v^{(n)}\} + [K]\{x^{(n)}\} = \{R^{(n)}\} \tag{12.155}$$

The initial conditions can be expressed as follows:

$$\{x^{(0)}\} = \{D^{(0)}\} \tag{12.156b}$$

$$\{v^{(0)}\} = \{\dot{D}^{(0)}\} \tag{12.156c}$$

Next, we introduce different approaches to approximate the velocity and acceleration vectors. These approaches lead to different *direct time integration methods* with different stability and convergence characteristics.

12.10.1 Central finite difference approximations: explicit time integration

The velocity and acceleration vectors can be approximated by using the central finite difference formulas as follows:

$$\{v^{(n)}\} = \frac{1}{2\Delta t}\left(\{x^{(n+1)}\} - \{x^{(n-1)}\}\right) \tag{12.157a}$$

$$\{a^{(n)}\} = \frac{1}{\Delta t^2}\left(\{x^{(n+1)}\} - 2\{x^{(n)}\} + \{x^{(n-1)}\}\right) \tag{12.157b}$$

Note that both of these formulas are second-order accurate in Δt, that is, $O(\Delta t^2)$. Substituting these approximations into the equation of motion (12.154), we find,

$$\begin{aligned}&\left[\frac{1}{\Delta t^2}[M] + \frac{1}{2\Delta t}[C]\right]\{x^{(n+1)}\}\\ &= \{R^{(n)}\} - \left\{R^{(n)}_{\text{int}}\right\} + \frac{2}{\Delta t^2}[M]\{x^{(n)}\} - \left[\frac{1}{\Delta t^2}[M] - \frac{1}{2\Delta t}[C]\right]\{x^{(n-1)}\}\end{aligned} \tag{12.158}$$

This equation provides a recursive relationship to solve for the unknown displacement vector at time step $^{(n+1)}$. Note that all of the variables on the right hand side of this equation are at the previous time steps where the displacement vectors have already been computed. In case the mass and damping matrices are both diagonal matrices, the solution of Eq. (12.158) can be obtained by a relatively straight forward division operation. A diagonal mass matrix can be obtained by using lumped element matrices as described in Section 12.6.2. However, a diagonal damping matrix may not always be compatible with the physics of damping. In case the damping matrix cannot be assumed to be diagonal the system can still be diagonalized by taking a penalty in accuracy. To this end, instead of using the $O(\Delta t^2)$ accurate central difference formula, we use the $O(\Delta t)$ accurate backward difference formula,

$$\left\{v^{(n)}\right\} = \frac{1}{\Delta t}\left(\left\{x^{(n)}\right\} - \left\{x^{(n-1)}\right\}\right) \tag{12.159}$$

Using Eq. (12.159) instead of (12.157a) in Eq. (12.154) we find the following approximation to the equation of motion:

$$\begin{aligned}\frac{1}{\Delta t^2}[M]\left\{x^{(n+1)}\right\} &= \left\{R^{(n)}\right\} - \left\{R_{\text{int}}^{(n)}\right\} \\ &+ \left[\frac{2}{\Delta t^2}[M] - \frac{1}{\Delta t}[C]\right]\left\{x^{(n)}\right\} - \left[\frac{1}{\Delta t^2}[M] - \frac{1}{\Delta t}[C]\right]\left\{x^{(n-1)}\right\}\end{aligned} \tag{12.160}$$

If the system has linear stiffness characteristics, the internal force vector can be replaced with $\{R_{\text{int}}^{(n)}\} = [K]\{x^{(n)}\}$. The mass matrix in Eq. (12.160) can be diagonalized by using lumped elements very easily, and leads to an efficient solution algorithm. The effectiveness of this algorithm is particularly significant for systems with large numbers of degrees of freedom. However, because we mixed first- and second-order accurate approximations to the temporal derivatives, the accuracy of this algorithm is $O(\Delta t)$. Note that the explicit time integration algorithm is a conditionally stable as shown in Section 12.10.3.

In order to start the recursive solution at $t = 0$ we need the vectors $\{x^{(0)}\}$ and $\{x^{(-1)}\}$. These vectors can be found by using the initial conditions given in Eq. (12.156b). In particular, Taylor series expansion of $\{D^{(n)\Delta t}\}$ with $-\Delta t$ gives,

$$\left\{D^{(n-1)\Delta t}\right\} = \left\{D^{(n)\Delta t}\right\} - \Delta t\left\{\dot{D}^{(n)\Delta t}\right\} + \frac{\Delta t^2}{2}\left\{\ddot{D}^{(n)\Delta t}\right\} - \frac{\Delta t^3}{3!}\left\{\dddot{D}^{(n)\Delta t}\right\} + \ldots \tag{12.161}$$

Discretized version of this expansion at $n = 0$ gives,

$$\left\{x^{(-1)}\right\} \approx \left\{x^{(0)}\right\} - \Delta t\left\{v^{(0)}\right\} + \frac{\Delta t^2}{2}\left\{a^{(0)}\right\} \tag{12.162}$$

where we neglected the terms higher than Δt^2 in the expansion.

12.10.2 Linear and average acceleration methods: implicit time integration

Let's consider the time interval $t^{(n)} \leq t \leq t^{(n+1)}$ in the progression of a transient solution. Let's also assume that the vectors $\{x^{(n)}\}$ and $\{v^{(n)}\}$are known. We can easily shift the time interval to $0 \leq t \leq \Delta t$ to analyze the events in this time interval, where $\Delta t = t^{(n+1)} - t^{(n)}$.

12.10.2.1 Average acceleration method

Let's assume that the acceleration within the time interval $0 \le t \le \Delta t$ is approximated by the average of the accelerations,

$$\{\ddot{D}(t)\} = \frac{1}{2}\left(\{\ddot{D}^{(n+1)}\} + \{\ddot{D}^{(n)}\}\right) \tag{12.163}$$

Integrating this relationship and using the initial conditions at $t = 0$, we find,

$$\{\dot{D}(t)\} = \{\dot{D}^{(n)}\} + \frac{t}{2}\left(\{\ddot{D}^{(n+1)}\} + \{\ddot{D}^{(n)}\}\right) \tag{12.164a}$$

$$\{D(t)\} = \{D^{(n)}\} + t\{\dot{D}^{(n)}\} + \frac{t^2}{4}\left(\{\ddot{D}^{(n+1)}\} + \{\ddot{D}^{(n)}\}\right) \tag{12.164b}$$

Evaluating Eq. (12.164a) at the end of the interval, $t = \Delta t$ and replacing the continuous variables with the numerical ones, we find,

$$\{v^{(n+1)}\} = \{v^{(n)}\} + \frac{\Delta t}{2}\left(\{a^{(n+1)}\} + \{a^{(n)}\}\right) \tag{12.165a}$$

$$\{x^{(n+1)}\} = \{x^{(n)}\} + \Delta t\{v^{(n)}\} + \frac{\Delta t^2}{4}\left(\{a^{(n+1)}\} + \{a^{(n)}\}\right) \tag{12.165b}$$

Solving for $\{a^{(n+1)}\}$ from Eq. (12.165b) and substituting the result into Eq. (12.165a) we find,

$$\{a^{(n+1)}\} = \frac{4}{\Delta t^2}\left(\{x^{(n+1)}\} - \{x^{(n)}\} - \Delta t\{v^{(n)}\}\right) - \{a^{(n)}\} \tag{12.166a}$$

$$\{v^{(n+1)}\} = -\{v^{(n)}\} + \frac{2}{\Delta t}\left(\{x^{(n+1)}\} - \{x^{(n)}\}\right) \tag{12.166b}$$

Substituting these equations into the equation of motion evaluated at time step $(n+1)$

$$[M]\{a^{(n+1)}\} + [C]\{v^{(n+1)}\} + \left\{R_{\text{int}}^{(n+1)}\right\} = \{R^{(n+1)}\} \tag{12.167}$$

we find,

$$\begin{aligned}&\frac{4}{\Delta t^2}[M]\{x^{(n+1)}\} + \frac{2}{\Delta t}[C]\{x^{(n+1)}\} + \left\{R_{\text{int}}^{(n+1)}\right\}\\ &= \{R^{(n+1)}\}+[M]\left(\frac{4}{\Delta t^2}\{x^{(n)}\}+\frac{4}{\Delta t}\{v^{(n)}\}+\{a^{(n)}\}\right)+[C]\left(\frac{2}{\Delta t}\{x^{(n)}\}+\{v^{(n)}\}\right)\end{aligned}$$

In case the system has a linear stiffness, we can use $\{R_{\text{int}}^{(n+1)}\} = [K]\{x^{(n+1)}\}$. Then we can find the recursive relationship for the *average acceleration method*

$$[K_e]\{x^{(n+1)}\} = \{R_e^{(n+1)}\} \tag{12.168}$$

where effective stiffness matrix and effective loading vector are as follows:

$$[K_e] = \frac{4}{\Delta t^2}[M] + \frac{2}{\Delta t}[C] + [K] \tag{12.169a}$$

$$\{R_e^{(n+1)}\} = \{R^{(n+1)}\} + [M]\left(\frac{4}{\Delta t^2}\{x^{(n)}\} + \frac{4}{\Delta t}\{v^{(n)}\} + \{a^{(n)}\}\right) + [C]\left(\frac{2}{\Delta t}\{x^{(n)}\} + \{v^{(n)}\}\right) \tag{12.169b}$$

12.10.2.2 *Linear acceleration method*

Let's assume that the acceleration within the time interval $0 \leq t \leq \Delta t$ is varying linearly. Therefore, the acceleration can be approximated as follows:

$$\{\ddot{D}(t)\} = \{\ddot{D}^{(n)}\} + \frac{t}{\Delta t}\left(\{\ddot{D}^{(n+1)}\} - \{\ddot{D}^{(n)}\}\right) \tag{12.170}$$

Integrating this relationship and using the initial conditions at $t = 0$, we find,

$$\{\dot{D}(t)\} = \{\dot{D}^{(n)}\} + t\{\ddot{D}^{(n)}\} + \frac{t^2}{2\Delta t}\left(\{\ddot{D}^{(n+1)}\} - \{\ddot{D}^{(n)}\}\right) \tag{12.171a}$$

$$\{D(t)\} = \{D^{(n)}\} + t\{\dot{D}^{(n)}\} + \frac{t^2}{2}\{\ddot{D}^{(n)}\} + \frac{t^3}{6\Delta t}\left(\{\ddot{D}^{(n+1)}\} - \{\ddot{D}^{(n)}\}\right) \tag{12.171b}$$

Following the same steps as above, we find the effective stiffness matrix and load vector for the *linear acceleration method* as follows:

$$[K_\varepsilon] = \frac{6}{\Delta t^2}[M] + \frac{3}{\Delta t}[C] + [K] \tag{12.172a}$$

$$\begin{aligned}\{R_e^{(n+1)}\} = \{R^{(n+1)}\} &+ [M]\left(\frac{6}{\Delta t^2}\{x^{(n)}\} + \frac{6}{\Delta t}\{v^{(n)}\} + 2\{a^{(n)}\}\right) \\ &+ [C]\left(\frac{3}{\Delta t}\{x^{(n)}\} + 2\{v^{(n)}\} + \frac{\Delta t}{2}\{a^{(n)}\}\right)\end{aligned} \tag{12.172b}$$

12.10.3 Newmark's method for direct time integration

The average and linear acceleration methods make assumptions on how the acceleration varies within a given time interval $0 \leq t \leq \Delta t$. *Newmark's method* generalizes this variation by introducing the parameters γ and β into the calculation of the velocity and displacement vectors within a time step,

$$\{v^{(n+1)}\} = \{v^{(n)}\} + \Delta t\left(\gamma\{a^{(n+1)}\} + (1-\gamma)\{a^{(n)}\}\right) \tag{12.173a}$$

$$\{x^{(n+1)}\} = \{x^{(n)}\} + \Delta t\{v^{(n)}\} + \frac{\Delta t^2}{2}\left(2\beta\{a^{(n+1)}\} + (1-2\beta)\{a^{(n)}\}\right) \tag{12.173b}$$

Here we see that the contributions of accelerations at time steps $^{(n)}$ and $^{(n+1)}$ can be tuned by using the parameters γ and β. Note that Eq. (12.173a) can also be expressed in a *predictor-corrector* format,

$$\{v^{(n+1)}\} = \{\tilde{v}^{(n+1)}\} + \Delta t\ \gamma\{a^{(n+1)}\} \tag{12.174a}$$

$$\{x^{(n+1)}\} = \{\tilde{x}^{(n+1)}\} + \beta\Delta t^2\{a^{(n+1)}\} \tag{12.174b}$$

where the *predictors* are variables computed at the known time step $^{(n)}$,

$$\{\tilde{v}^{(n+1)}\} = \{v^{(n)}\} + \Delta t(1-\gamma)\{a^{(n)}\} \tag{12.175a}$$

$$\{\tilde{x}^{(n+1)}\} = \{x^{(n)}\} + \Delta t\{v^{(n)}\} + (1-2\beta)\frac{1}{2}\Delta t^2\{a^{(n)}\} \tag{12.175b}$$

Substituting Eq. (12.174a) into the equation of motion at time step $^{(n+1)}$ we can find the recursive relationship for the unknown acceleration vector,

$$[K_e]\{a^{(n+1)}\} = \{R_e^{(n+1)}\} \tag{12.176}$$

where

$$[K_e] = [M] + \gamma\Delta t[C] + \beta\Delta t^2[K] \tag{12.177a}$$

$$\{R_e^{(n+1)}\} = \{R^{(n+1)}\} - [K]\{\tilde{x}^{(n+1)}\} - [C]\{\tilde{v}^{(n+1)}\} \tag{12.177b}$$

The *acceleration-form* of the Newmark's algorithm described above has the advantage of a compact notation.

A *displacement-form* of the algorithm can also be obtained. In order to find this form, we first solve for the acceleration vector $\{a^{(n+1)}\}$ from Eq. (12.173b) and substitute the result into Eq. (12.173a) to get,

$$\{a^{(n+1)}\} = \frac{1}{\beta\Delta t^2}\left(\{x^{(n+1)}\} - \{x^{(n)}\} - \Delta t\{v^{(n)}\}\right) - \left(\frac{1}{2\beta} - 1\right)\{a^{(n)}\} \tag{12.178a}$$

$$\{v^{(n+1)}\} = \frac{\gamma}{\beta\Delta t}\left(\{x^{(n+1)}\} - \{x^{(n)}\}\right) - \left(\frac{\gamma}{\beta} - 1\right)\{v^{(n)}\} - \Delta t\left(\frac{\gamma}{2\beta} - 1\right)\{a^{(n)}\} \tag{12.178b}$$

Substituting Eq. (12.178) into the equation of motion at time step $^{(n+1)}$ we can find the recursive relationship for the unknown acceleration vector,

$$[K_e]\{x^{(n+1)}\} = \{R_e^{(n+1)}\} \tag{12.179}$$

where

$$[K_\varepsilon] = [M] + \gamma\Delta t[C] + \beta\Delta t^2[K] \tag{12.180a}$$

$$\begin{aligned}\{R_\varepsilon^{(n+1)}\} = {} & \beta\Delta t^2\{R^{(n+1)}\} + [M]\left(\{x^{(n)}\} + \Delta t\{v^{(n)}\} + \frac{1}{2}(1-2\beta)\Delta t^2\{a^{(n)}\}\right) \\ & + \Delta t[C]\left(\gamma\{x^{(n)}\} + (\gamma-\beta)\Delta t\{v^{(n)}\} + \frac{1}{2}(\gamma-2\beta)\Delta t^2\{a^{(n)}\}\right)\end{aligned} \tag{12.180b}$$

12.10.4 α-Method for direct time integration

The *α-method* also known as the *Hilber–Hughes–Taylor method* is a general direct integration method that represents several well-known time integration methods such as the *Newmark, linear-acceleration, average-acceleration,* and *Fox–Goodwin* methods. This method was developed specifically to add controlled numerical damping, without sacrificing the accuracy of time integration.

The *fully discretized equation of motion* is expressed at a time step between $^{(n)}$ and $^{(n+1)}$ as follows:

$$[M]\{a^{(n+1)}\} + [C]\{v^{(n+\alpha)}\} + [K]\{x^{(n+\alpha)}\} = \{R^{(n+\alpha)}\} \tag{12.181}$$

where α is a real number which is known as the *numerical damping parameter* and

$$\{x^{(n+\alpha)}\} = (1+\alpha)\{x^{(n+1)}\} - \alpha\{x^{(n)}\} \tag{12.182a}$$

$$\{v^{(n+\alpha)}\} = (1+\alpha)\{v^{(n+1)}\} - \alpha\{v^{(n)}\} \tag{12.182b}$$

$$\{R^{(n+\alpha)}\} = (1+\alpha)\{R^{(n+1)}\} - \alpha\{R^{(n)}\} \tag{12.182c}$$

Displacements and velocities for time step $^{(n+1)}$ are approximated by using the predictor-corrector form given in Eq. (12.174). The predictors $\{\tilde{x}^{(n+1)}\}$ and $\{\tilde{v}^{(n+1)}\}$ are defined by using the known values at time step $^{(n)}$ given in Eq. (12.175).

12.10.4.1 Acceleration-form

The acceleration-form of the α-method is obtained by substituting Eqs. (12.174), (12.175), and (12.182) into Eq. (12.181),

$$\begin{aligned}&\left([M] + (1+\alpha)\gamma\Delta t[C] + (1+\alpha)\beta\Delta t^2[K]\right)\{a^{(n+1)}\} = \\ &\{R^{(n+\alpha)}\} - (1+\alpha)\left([K]\{\tilde{x}^{(n+1)}\} + [C]\{\tilde{v}^{(n+1)}\}\right) + \alpha\left([K]\{x^{(n)}\} + [C]\{v^{(n)}\}\right)\end{aligned} \tag{12.183}$$

This recursive relationship can be represented by defining the following simplified notation:

$$[K_e] = [M] + (1+\alpha)\gamma\Delta t[C] + (1+\alpha)\beta\Delta t^2[K] \tag{12.184a}$$

$$\{F_e^{(n+1)}\} = \{R^{(n+\alpha)}\} - (1+\alpha)\left([K]\{\tilde{x}^{(n+1)}\} + [C]\{\tilde{v}^{(n+1)}\}\right) + \alpha\left([K]\{x^{(n)}\} + [C]\{v^{(n)}\}\right) \tag{12.184b}$$

12.10.4.2 Displacement-form

The displacement-form of the α-method can be found by following the same steps for the Newmark's method. By solving Eq. (12.174b) for $\{a^{(n+1)}\}$ and substituting the result into Eq. (12.174a), we find,

$$\left\{v^{(n+1)}\right\} = \left\{\tilde{v}^{(n+1)}\right\} + \frac{\gamma}{\beta\Delta t}\left(\left\{x^{(n+1)}\right\} - \left\{\tilde{x}^{(n+1)}\right\}\right) \tag{12.185a}$$

$$\left\{a^{(n+1)}\right\} = \frac{1}{\beta\Delta t^2}\left(\left\{x^{(n+1)}\right\} - \left\{\tilde{x}^{(n+1)}\right\}\right) \tag{12.185b}$$

Substituting Eq. (12.182a), (12.182b) into Eq. (12.181) we find,

$$[M]\left\{a^{(n+1)}\right\}+[C](1+\alpha)\left\{v^{(n+1)}\right\}+(1+\alpha)[K]\left\{x^{(n+1)}\right\} = \left\{R^{(n+\alpha)}\right\}+\alpha[C]\left\{v^{(n)}\right\}+\alpha[K]\left\{x^{(n)}\right\} \tag{12.186}$$

Combining the last two equations we find,

$$\left[\frac{1}{\beta\Delta t^2}[M] + (1+\alpha)\frac{\gamma}{\beta\Delta t}[C] + (1+\alpha)[K]\right]\left\{x^{(n+1)}\right\} = \left\{R^{(n+\alpha)}\right\} + \alpha[C]\left\{v^{(n)}\right\}$$
$$+\alpha[K]\left\{x^{(n)}\right\} + \frac{1}{\beta\Delta t^2}[M]\left\{\tilde{x}^{(n+1)}\right\} - (1+\alpha)[C]\left(\left\{\tilde{v}^{(n+1)}\right\} - \frac{\gamma}{\beta\Delta t}\left\{\tilde{x}^{(n+1)}\right\}\right)$$

Displacement-form of the time-stepping algorithm is expressed, in more compact representation, as follows:

$$[K_e]\left\{x^{(n+1)}\right\} = \left\{F_e^{(n+1)}\right\} \tag{12.187}$$

with

$$[K_e] = \frac{1}{\beta\Delta t^2}[M] + (1+\alpha)\frac{\gamma}{\beta\Delta t}[C] + (1+\alpha)[K] \tag{12.188a}$$

$$\left\{F_\varepsilon^{(n+1)}\right\} = \left\{R^{(n+\alpha)}\right\} + \alpha[K]\left\{x^{(n)}\right\} + \frac{1}{\beta\Delta t^2}[M]\left\{\tilde{x}^{(n+1)}\right\}$$
$$+[C]\left[\alpha\left\{v^{(n)}\right\} - (1+\alpha)\left(\left\{\tilde{v}^{(n+1)}\right\} - \frac{\gamma}{\beta\Delta t}\left\{\tilde{x}^{(n+1)}\right\}\right)\right] \tag{12.188b}$$

12.10.4.3 Explicit version of the α-method

High volume of matrix multiplications involved with the implicit algorithm could slow the solution especially when large numbers of degrees of freedom are involved. Explicit time integration reduces this burden. In order to obtain an explicit algorithm for the α-method Eq. (12.174) is modified as follows [3]:

$$\left\{x^{(n+1)}\right\} = \left\{\tilde{x}^{(n+1)}\right\} \tag{12.189a}$$

$$\left\{v^{(n+1)}\right\} = \left\{\tilde{v}^{(n+1)}\right\} \tag{12.189b}$$

Substituting Eqs. (12.182) and (12.189) into Eq. (12.181), the following recursive relationship is obtained to find $\{a^{(n+1)}\}$:

$$[M]\left\{a^{(n+1)}\right\} = \left\{F_e^{(n+1)}\right\} \tag{12.190}$$

where the equivalent load vector becomes [5],

$$\left\{F_e^{(n+1)}\right\} = \left[R^{(n+\alpha)} - (1+\alpha)\left([C]\left\{\tilde{v}^{(n+1)}\right\} + [K]\left\{\tilde{d}^{(n+1)}\right\}\right) + \alpha\left([C]\left\{v^{(n)}\right\} + [K]\left\{d^{(n)}\right\}\right)\right] \tag{12.191}$$

If a lumped mass matrix is chosen, the acceleration vector $\{a^{(n+1)}\}$ can be obtained by a simple division operation. This has considerable advantages in the speed of the algorithm. Explicit schemes can be made such that matrix inversions are minimized or totally eliminated. But, they are only conditionally stable requiring small Δt for stability.

12.10.5 Initial conditions

The start-up processes for all of the time-stepping methods introduced above are the same,

$$\{x^{(0)}\} = \{D^{(0)}\} \tag{12.192a}$$

$$\{v^{(0)}\} = \{\dot{D}^{(0)}\} \tag{12.192b}$$

$$\{a^{(0)}\} = [M]^{-1}(\{R^{(0)}\} - [C]\{v^{(0)}\} - [K]\{x^{(0)}\}) \tag{12.192c}$$

The effective stiffness matrix $[K_\varepsilon]$ can be computed and inverted once before the time-stepping algorithm starts, if the problem has linear stiffness characteristics and if the mass and damping matrices do not change in time. This results in reduction of computation time. Once the solution for $\{a^{(n+1)}\}$ is found from the chosen form of the time-stepping algorithm the variables $\{x^{(n+1)}\}$ and $\{v^{(n+1)}\}$ can be found by using Eq. (12.174).

12.10.6 Solution algorithm

The following pseudo code represents an algorithm for the acceleration-form of the α-method.

```
%..........................................................
% Set initial conditions
n = 0
        {x(0)},{v(0)} are given
        {a(0)} = [M]^-1({R(0)} - [C] {v(0)} - [K]{x(0)})
%
% Time-stepping algorithm
for n = 1:N_T
        Compute the predictors:                   {x̃(n+1)}, {ṽ(n+1)}
        Compute the effective stiffness
                matrix and load vector:           [Ke],{Fe(n+1)}
        Solve for acceleration:                   {a(n+1)}
        Compute current displacement
                and velocities:                   {x(n+1)},{v(n+1)}
        Update variables:                         {}(n) = {}(n+1)
end
%......................................................
```

12.11 Convergence, stability, and accuracy of time integration methods

The concepts of convergence, stability, and accuracy are introduced in Section 11.6 for first-order systems. In this section, we extend these concepts to second-order systems. The literature of the derivation of these conditions is rich and explained in detail by Hughes [3] and Cook et al. [1]. Here we give a brief overview of the stability conditions for the α-method.

As stated in Section 11.6, the *Lax equivalence theorem* states if a time integration algorithm is *consistent* and *stable* it is also *convergent* [3,6,8]. Thus it is sufficient to determine the conditions that control the consistency and stability of the algorithms. A *stable* time integration algorithm behaves in the same manner as the analytical solution. *Consistency* and *rate of convergence* ensure that the *truncation error* after temporal discretization remains below the following threshold,

$$|\tau| \le C_\tau \Delta t^k \tag{12.193}$$

where C_τ is a constant independent of Δt and k is a positive number named the *rate of convergence*.

12.11.1 Stability of the explicit method

The explicit time integration method introduced in Section 12.10.1 is conditionally stable, similar to its counterpart used in the solution of the heat transfer problem. In order to investigate the stability characteristics of the explicit time integration, let's consider the numerical solution of the transient form of the one-dimensional elasticity problem given in Eq. (12.26). The equation of motion for a bar with uniform cross section, uniform material properties, negligible damping and no body and external forces becomes,

$$\frac{\partial^2 u_x}{\partial t^2} = c^2 \frac{\partial^2 u_x}{\partial x^2} \tag{12.194}$$

where $c = \sqrt{E/\rho}$ is the wave speed. Eq. (12.194) is also known as the *wave equation*. Note that we use symbol u instead of u_x here to simplify the notation.

Let's assume that the one-dimensional solution domain is discretized into elements of equal length $L^{(e)}$. In the previous chapters we have shown that in this simple problem element-j is connected to nodes-j and $j+1$. It can be easily shown that the semidiscrete form of the equation of motion for element-j becomes,

$$\frac{1}{2}\begin{bmatrix} 1 & 0 \\ 0 & 1 \end{bmatrix}\begin{Bmatrix} \ddot{\tilde{u}}_j \\ \ddot{\tilde{u}}_{j+1} \end{Bmatrix} + \left(\frac{c}{L^{(e)}}\right)^2 \begin{bmatrix} 1 & -1 \\ -1 & 1 \end{bmatrix}\begin{Bmatrix} \tilde{u}_j \\ \tilde{u}_{j+1} \end{Bmatrix} = 0 \tag{12.195}$$

where $\tilde{u}$ is represents the numerically approximated value of the displacement. The jth-row of the global equation of motion can be shown to be in the following form after assembling elements-j and $j-1$:

$$\ddot{\tilde{u}}_j - \left(c/L^{(e)}\right)^2\left(\tilde{u}_{j-1} - 2\tilde{u}_j + \tilde{u}_{j+1}\right) = 0 \tag{12.196}$$

For this problem damping is neglected and the stress–strain relationship is linear. Therefore, the central difference method, (Eq. (12.160)), developed to discretize the equation of motion becomes,

$$\frac{1}{\Delta t^2}[M]\left\{x^{(n+1)}\right\} = -[K]\left\{x^{(n)}\right\} + \frac{2}{\Delta t^2}[M]\left\{x^{(n)}\right\} - \frac{1}{\Delta t^2}[M]\left\{x^{(n-1)}\right\} \tag{12.197}$$

This equation can be rearranged as,

$$\frac{1}{\Delta t^2}[M]\left(\left\{x^{(n+1)}\right\} - 2\left\{x^{(n)}\right\} + \left\{x^{(n-1)}\right\}\right) = -[K]\left\{x^{(n)}\right\} \tag{12.198}$$

The jth-row of this equation can be obtained from this equation by using (12.196) as follows:

$$\left(\tilde{u}_j^{(n+1)} - 2\tilde{u}_j^{(n)} + \tilde{u}_j^{(n-1)}\right) - s^2\left(\tilde{u}_{j-1}^{(n)} - 2\tilde{u}_j^{(n)} + \tilde{u}_{j+1}^{(n)}\right) = 0 \tag{12.199}$$

where $s^2 = (c\,\Delta t/L^{(e)})^2$. The stability of the explicit time integration method for the wave equation, described in Eq. (12.199) can be analyzed by using the Fourier–von Neumann method introduced in Section 11.6.5 for the heat equation. We will assume that the spatial and temporal components of the displacement are separable, and the temperature at time step-n and node-j can be given as described in Eq. (11.6.32) follows:

$$\tilde{u}_j^{(n)} = e^{i\beta\,\Delta x\,j}Q^n \tag{12.200}$$

where $i = \sqrt{-1}$. Note that Q is the amplitude of the spatial component of the transient response. In order for the solution to remain stable we see that Q should be in the range of $-1 \leq Q \leq 1$. Outside this range the response will grow unboundedly with increasing n. Substituting Eq. (12.200) into (12.199) we find,

$$\left(Q^{n+1} - 2Q^n + Q^{n-1}\right) = s^2\left(e^{-i\beta\,\Delta x(-1)} + e^{i\beta\,\Delta x} - 2\right)Q^n \tag{12.201}$$

Dividing this equation through $Q^{(n)}$ and using the relationship $\cos\beta\Delta x = (e^{i\beta\Delta xj} + e^{-i\beta\Delta xj})/2$, we find,

$$Q^2 - (\sigma + 2)Q + 1 = 0 \tag{12.202}$$

where

$$\sigma = 2s^2(\cos\beta\Delta x - 1) \tag{12.203}$$

The spatial amplitude of the response can then be found as follows:

$$Q_{1,2} = \frac{1}{2}\left[\sigma + 2 \pm \sqrt{(\sigma + 2)^2 - 4}\right] \tag{12.204}$$

It can be shown that if $-2 < \sigma + 2 < 2$, or $-4 < \sigma < 0$, the roots Q_1 and Q_2 are complex conjugates and $|Q| = 1$. This provides a temporally oscillating solution for a fixed position and ensures stability. If σ is outside the range found above it can be shown that one of the roots will be greater than 1, leading to unstable response [2]. Using the definition of σ and the stability range, we can show that the following stability condition emerges:

$$\frac{c\Delta t}{L^{(e)}} \leq 1 \tag{12.205}$$

This condition is known as the *Courant–Friedrichs–Lewy* (CFL) *stability condition*. The term $L^{(e)}/\Delta t$ can be interpreted as the *numerical wave speed*. Eq. (12.205) can be expressed as a limit on the timestep,

$$\Delta t \leq \Delta t_{cr} = \frac{L^{(e)}}{c} \tag{12.206}$$

which indicates that within one time step the numerically computed solution should stay inside the element.

A generalized treatment of the problem shows that the stability of the explicit time integration method for an undamped system depends on its maximum eigenvalue $\omega_{\max}$ and the stability condition is found as follows [3],

$$\Delta t \leq \Delta t_{cr} = \frac{2}{\omega_{\max}} \tag{12.207}$$

Stability of a damped system, on the other hand, is found from,

$$\Delta t \leq \min\left[\frac{2}{\omega_i}\left(\sqrt{1-\xi_i^2}-\xi_i\right)\right] \tag{12.208}$$

where ω_i and ξ_i are frequency and damping ratio of the ith-mode.

The maximum eigenvalues of different solid mechanics models and element types are summarized in Table 12.2 [3]. Note that the following variables are used in this table:

Bar element:

$c = \sqrt{E/\rho}$,	wave velocity
$L^{(e)}$,	element length

Timoshenko beam element:

A,	cross-sectional area
A_s,	shear area
I,	second moment of area
$c_s = \sqrt{\mu A_s/\rho A}$	shear-wave velocity
μ,	shear modulus.

Rectangular Q4 element:

$2a,\ 2b$,	element dimensions
$c_d = \sqrt{(\lambda+\mu)/\rho}$	dilatational wave velocity
$\lambda = \nu E/(1+\nu)(1-2\nu)$	Lamé constant.

TABLE 12.2 Maximum eigenfrequency and critical timestep for explicit time integration method for different types of elements (based on [3] and references therein). The variables in this table are defined in the text.

Element type	Mass matrix	ω_{max}	Δt_{cr}
2-node, 1D bar element	Lumped	$\frac{2c}{L^{(e)}}$	$\frac{L^{(e)}}{c}$
2-node, 1D bar element	Consistent	$\frac{2\sqrt{3}c}{L^{(e)}}$	$\frac{L^{(e)}}{\sqrt{3}c}$
3-node, 1D bar element	Lumped	$\frac{2\sqrt{6}c}{L^{(e)}}$	$\frac{L^{(e)}}{\sqrt{6}c}$
2-node Timoshenko beam element	Lumped	$\max\left\{\left(\frac{2c}{L^{(e)}}\right)\left(\frac{I}{\alpha A}\right)^{1/2}, \left(\frac{2c_s}{L^{(e)}}\right)\left[1+\frac{1}{\alpha}\left(\frac{L^{(e)}}{2}\right)^2\right]^{1/2}\right\}$	$\frac{2}{\omega_{max}}$
Rectangular-Q4 element			$\left[\frac{c_d}{4}(a^{-2}+b^{-2})\right]^{-1}$

12.11.2 Stability and consistency of the Newmark and α-methods

The α-method is a three parameter method with α, β, and γ as the numerical parameters that provide the user the ability to control the stability, accuracy, and algorithmic damping in the solution algorithm. The α-method represents a family of methods that have been developed over time. By setting $\alpha = 0$ the well-known *Newmark method* is recovered. The Newmark method itself encompasses other well-known methods such as the *average acceleration, linear acceleration*, and the *Fox–Goodwin* methods. Addition of the parameter α to the Newmark's method introduces *algorithmic damping* while keeping second-order accuracy for the algorithm. The stability and accuracy for the time-stepping algorithms introduced in Section 12.10 have been analyzed extensively. This analysis is similar to that presented in Chapter 11 for heat transfer. Here the results of stability and consistency for these methods are briefly summarized, but the interested readers are referred to reference [3] for more on this topic.

Newmark method (Section 12.10.3) is a two parameter method with β and γ which historically preceded the α-method. This method is,

$$\text{unconditionally stable for} \quad 2\beta \geq \gamma \geq \frac{1}{2} \tag{12.209}$$

$$\text{conditionally stable for} \quad \gamma \geq \frac{1}{2}, \beta < \frac{1}{2}\gamma \tag{12.210}$$

$$\text{with the stability condition } \Delta t \leq \frac{\Omega_{\text{crit}}}{\omega_{\max}} \tag{12.211}$$

where $\omega_{\max}$ is the *highest natural frequency* of the system and

$$\Omega_{\text{crit}} = \left(\frac{\gamma}{2} - \beta\right)^{-1}\left[\xi\left(\gamma - \frac{1}{2}\right) + \sqrt{\frac{\gamma}{2} - \beta + \xi^2\left(\gamma - \frac{1}{2}\right)^2}\right] \tag{12.212}$$

where ξ is the damping ratio. This equation shows that when $\gamma = 1/2$ damping has no effect on stability, and also when $\gamma > 1/2$, Ω_{crit} . increases, thus increasing the critical time step. Table 12.3 provides a summary of the stability and accuracy of the Newmark method for some of the well-known algorithms.

Note that the high frequencies in the system response are artifacts of spatial discretization as shown in Example 12.2. Therefore, for direct time integration methods damping of the high-frequency components of the system response, while keeping the low-frequency response intact is desirable. In general, damping can be introduced into the system by increasing the damping ratio ξ. But, this is approach reduces the participation of all of the frequencies. Algorithmic damping can selectively reduce the participation of the high-frequency modes while keeping the low-frequency modes intact. Algorithmic damping can be introduced into Newmark's method by selecting $\lambda > 1/2$ and $\beta > \sqrt{\gamma + 1/2}/4$. This has the drawback to reduce the accuracy of the solution to $O(\Delta t)$.

TABLE 12.3 Stability and accuracy of the Newmark family of methods and the α-method. This table is adopted from [1] and [3]. Note that the stability condition and the order of accuracy are for the undamped problem, $\xi = 0$.

	Type	γ	β	Stability condition	Order of accuracy
			Newmark methods		
Average acceleration	Implicit	1/2	1/4	Unconditional	$O(\Delta t^2)$
Linear acceleration	Implicit	1/2	1/6	$\Omega_{crit} = 2\sqrt{3}$	$O(\Delta t^2)$
Fox–Goodwin	Implicit	1/2	1/12	$\Omega_{crit} = \sqrt{6}$	$O(\Delta t^4)$
Central difference	Explicit	1/2	0	$\Omega_{crit} = 2$	$O(\Delta t^2)$
Algorithmically damped	Implicit	>1/2	$> \sqrt{\gamma + 1/2}/4$	Unconditional	$O(\Delta t)$
			α-method $-1/3 \le \alpha \le 0$		
		$\frac{1}{2}(1 - 2\alpha)$	$\frac{1}{4}(1 - \alpha)^2$	Unconditional	$O(\Delta t^2)$

The α-method introduces algorithmic damping to the time integration while keeping the second-order accuracy $O(\Delta t^2)$, and unconditional stability intact. To achieve these characteristics, the value of the parameter is chosen from the range $-\frac{1}{3} \leq \alpha \leq 0$, while the parameters β and γ are chosen as follows:

$$-\frac{1}{3} < \alpha < 0 \quad \text{and} \quad \gamma = \frac{1}{2}(1 - 2\alpha) \quad \text{and} \quad \beta = \frac{1}{4}(1 - \alpha)^2$$

Note that decreasing α increases numerical dissipation.

Example 12.3 Solve Example 12.2 by using a transient finite element analysis. Use the α-method with consistent mass matrix and 16 Euler–Bernoulli beam elements.

a. Demonstrate the effects of the time step on the solution for a case where $\alpha = 0$, $\beta = 1/2$, $\gamma = 1/4$,
b. Repeat part (a) for a case where $\alpha = 0$, $\beta = 1/2$, $\gamma = 1/12$.

Recall that this is an undamped beam with no external forces. The system is set into motion by the initial conditions that are given in Eq. (12.136),

$$v(x, 0) = A \sin \frac{\pi x}{L}, \; \dot{v}(x, 0) = 0 \tag{12.213}$$

and it is expected to experience free vibration. A finite element program is prepared based on the solution algorithm given in Section 12.10.6. The eigenvalue analysis presented in Example 12.1 for the simply supported beam is used to determine the critical time step Δt_{cr} in part (b).

The free vibration problem described above is analyzed by using the α-method for 16 elements. There are no external forces acting on the system, therefore the force vector is zero for all time steps,

$$\{R\}^{(n)} = 0 \tag{12.214}$$

The active degrees of freedom of the problem at time step-n are,

$$\{x\}^{(n)} = \left\{\theta_1^{(n)} \; v_2^{(n)} \; \theta_2^{(n)} \; \cdots v_{15}^{(n)} \; \theta_{15}^{(n)} \; \theta_{16}^{(n)}\right\}^T \tag{12.215}$$

The initial conditions for the active degrees of freedom are found by using Eq. (12.213) and recalling that the slope at node-i is defined as $\theta_i = dv_i/dx$,

$$\{x\}^{(0)} = \left\{\theta_1^{(0)} v_2^{(0)} \theta_2^{(0)} \cdots v_{15}^{(0)} \theta_{15}^{(0)} \theta_{16}^{(0)}\right\}^T \tag{12.216a}$$

$$\{v\}^{(0)} = 0 \tag{12.216b}$$

$$\{a\}^{(0)} = [M]^{-1}\left(\{R\}^{(0)} - [C]\{v\}^{(0)} - [K]\{x\}^{(0)}\right) \tag{12.216c}$$

Solution of part (a):
In case the numerical integration parameters are chosen as $\alpha = 0$, $\beta = 1/2$, $\gamma = \frac{1}{4}$, the α-method presents an unconditionally stable and implicit solution algorithm (Table 12.3). The transient solutions presented in Fig. 12.14 are

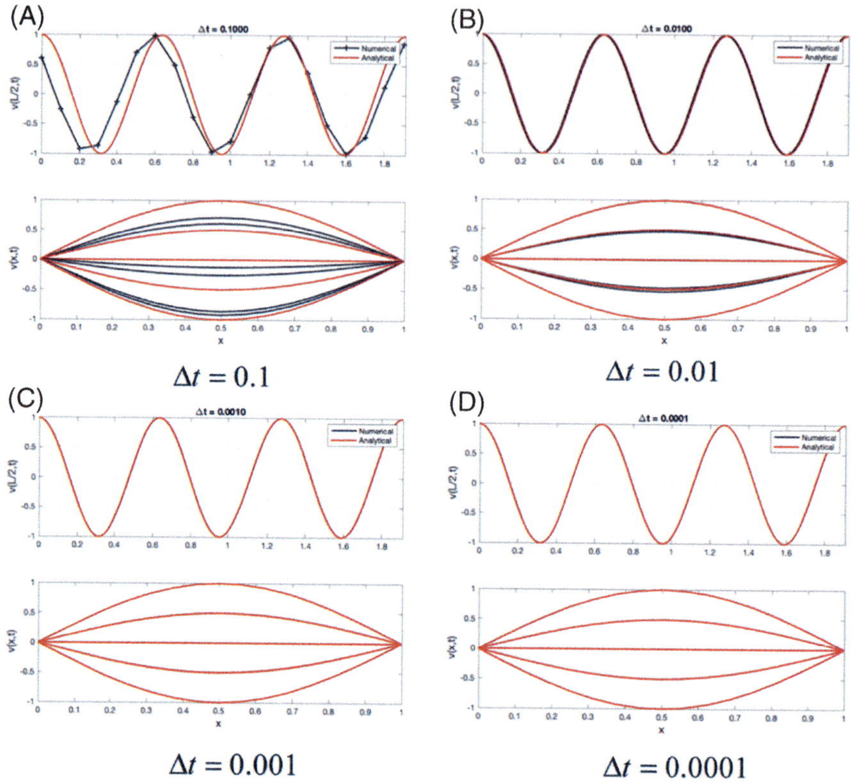

FIGURE 12.14 The effect of the size of the time step Δt used in the solution. This case is run with $\alpha = 0$, $\beta = 1/2$, $\gamma = 1/4$, and $N_E = 16$, Δt as indicated on the captions. This choice of time integration parameter gives an unconditionally stable implicit algorithm. The snapshots of the vibration are plotted at $t = 0,\ T/6,\ T/4,\ 2T/6,\ T/2,\ 4T/6,\ 3T/4,\ 5T/6$, and T where the period of vibration is calculated as $T = 2\pi/\omega_{n1}$.

obtained by using $\Delta t = 10^{-1}$, 10^{-2}, 10^{-3}, and 10^{-4} second-time steps. This figure compares the numerically calculated solution to the analytically predicted solution given in Eq. (12.53); vibration response of the mid-point of the beam (top) and the vibration snapshots (bottom) of the whole beam are plotted as a function of time. We see that the solution remains stable for all four time step sizes, but the coarsest time step, $\Delta t = 10^{-1}$, has significant period and amplitude errors. By using smaller time steps, the error is diminished significantly. For the case of $\Delta t = 10^{-4}$, the numerical and analytical solutions become virtually identical.

Solution of part (b):

In case the numerical integration parameters are chosen as $\alpha = 0$, $\beta = 1/2$, $\gamma = 1/12$, the α-method reduces to the Fox–Goodwin method, which is a conditionally stable and implicit method (Table 12.3). The critical time step size

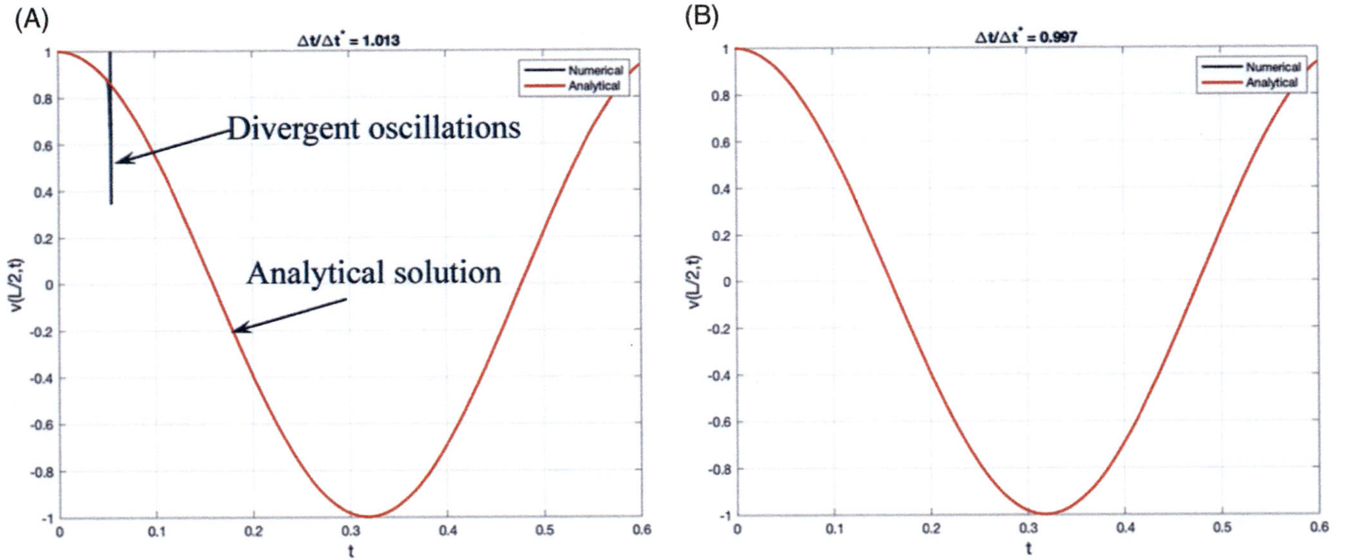

FIGURE 12.15 The effect of time step size for the conditionally stable, Fox–Goodwin method. The critical time step is given as $\Delta t_{cr} = \Omega_{cr}/\omega_{\max}$. For a mesh of $N_E = 16$ elements, $\Delta t_{cr} = 1.9601 \times 10^{-4}$ seconds. The plot in part (A) shows $\Delta t/\Delta t_{cr} = 1.013$ where the numerical solution goes into divergent oscillations before $t = 0.1$. The plot in part (B) shows $\Delta t/\Delta t_{cr} = 0.997$ where the numerical and analytical solutions are very close to one another.

can be deduced from Eq. (12.211) to be,

$$\Delta t_{cr} = \frac{\Omega_{crit}}{\omega_{\max}} \tag{12.217}$$

The maximum frequency of the system depends on the mesh size used. The eigenvalue analysis presented in Example 12.1 shows that for $N_E = 16$ the maximum frequency of the system is 12.851 kHz. The critical time step can then be found by using $\Omega_{crit} = \sqrt{6}$ to be $\Delta t_{cr} = 1.9061 \times 10^{-4}$ seconds.

In order to demonstrate the effect of time step size on this conditionally stable solution algorithm, the problem is solved twice with a $\Delta t > \Delta t_{cr}$ and with $\Delta t < \Delta t_{cr}$. The mid-point deflection history is compared to the analytical solution in Fig. 12.15. In case the time step is $\Delta t = 1.93 \times 10^{-4}$ seconds (Fig. 12.15A), the solution starts to experience unstable oscillations around $t = 0.05$ seconds and it is stopped. Note that the $\Delta t / \Delta t_{cr}$ ratio is 1.013 in this case! On the other hand when we use $\Delta t = 1.9 \times 10^{-4}$ ($\Delta t > \Delta t_{cr} = 0.997$) the solution remains stable and accurate as shown in Fig. E3.2B.

It is also important to note that the free vibration problem is thus solved by two separate approaches: the modal expansion in Example E12.2 and by numerical time integration here. As expected both approaches give results that are very close to the analytical solution, provided that the numerical parameters are chosen appropriately.

Problems

12.1 Find the equation of motion of the general beam element with uncoupled bending and axial deformation capability, presented in Section 6.4. Use the C^1-continuous Hermit polynomials.

12.2 The total potential energy for a tensioned beam is presented in Problem 6.7. Find the equation of motion of a tensioned beam element. Use the C^1-continuous Hermit polynomials.

12.3 Find the equation of motion a two-node tapered bar element.

The following problems require use of a finite element program.

12.4 Consider an Euler–Bernoulli beam supported as shown in Fig. 12.16. Carryout an eigenvalue analysis for this system by using the semidiscrete form obtained by finite element method. Let $L_1 = 1$ m, $L_2 = 0.5$ m, $E = 70$ GPa, $b = 6$ mm, $h = 25$ mm, $\rho = 2500$ kg/m^3.

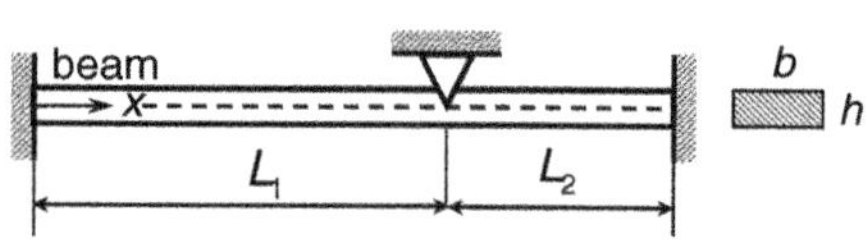

FIGURE 12.16 Problem 12.4.

a. Report the convergence of the first four eigenvalues by conducting a mesh convergence study.

b. Plot the first four eigenvectors of the system.

12.5 Consider a cantilevered, tapered Euler–Bernoulli beam as shown in Fig. 12.17. Carryout an eigenvalue analysis for this system by using the semidiscrete form obtained by finite element method. Let $L = 1.5$ m, $E = 70$ GPa, $b = 12$ mm, $h_1 = 25$ mm, $h_2 = 6$ mm, $\rho = 2500$ kg/m^3.

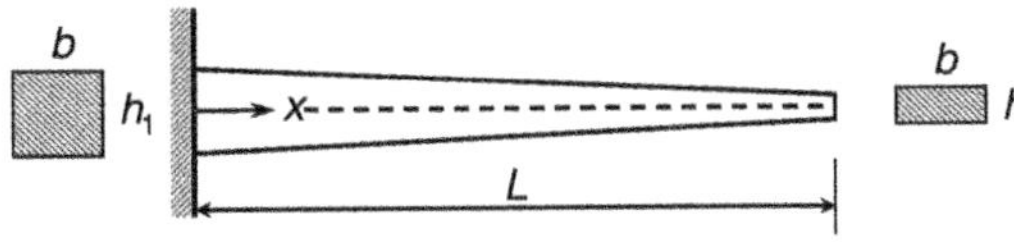

FIGURE 12.17 Problem 12.5.

a. Report the convergence of the first four eigenvalues by conducting a mesh convergence study.
b. Plot the first four eigenvectors of the system.

12.6 Consider a uniform elastic bar fixed between two supports as shown in Fig. 12.18. Carryout an eigenvalue analysis for this system by using the semidiscrete form obtained by finite element method. Let $L = 0.5$ m, $E = 100$ GPa, $b = 12$ mm, $h = 12$ mm, $\rho = 2100$ kg/m^3.

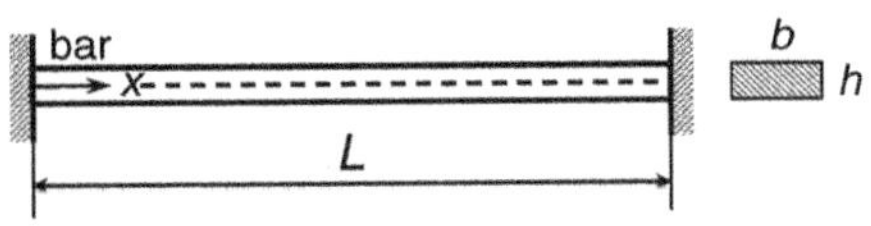

FIGURE 12.18 Problem 12.6.

a. Report the convergence of the first four eigenvalues by conducting a mesh convergence study.
b. Plot the first four eigenvectors of the system.

12.7 Consider a uniform elastic bar fixed only on one side as shown in Fig. 12.19. Carryout an eigenvalue analysis for this system by using the semidiscrete form obtained by finite element method. Let $L = 0.5$ m, $E = 100$ GPa, $b = 12$ mm, $h = 12$ mm, $\rho = 2100$ kg/m^3.

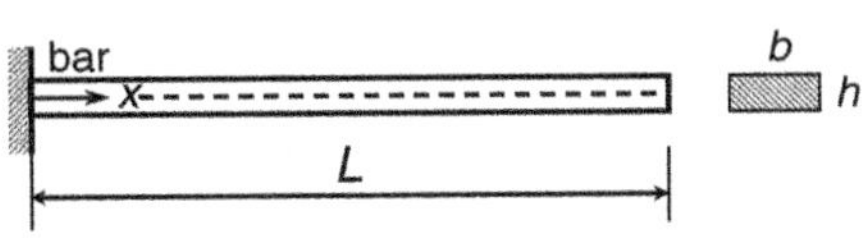

FIGURE 12.19 Problem 12.7.

a. Report the convergence of the first four eigenvalues by conducting a mesh convergence study.
b. Plot the first four eigenvectors of the system.

12.8 Consider the rectangular, elastic domain with uniform material properties, supported on one side as shown in Fig. 12.20. Carryout an eigenvalue analysis for this system by using the semidiscrete form obtained by finite element method. Let $L_x = 0.5$ m, $L_y = 1$ m, $t = 0.1$ m, $E = 100$ GPa, $\rho = 2000$ kg/m^3.

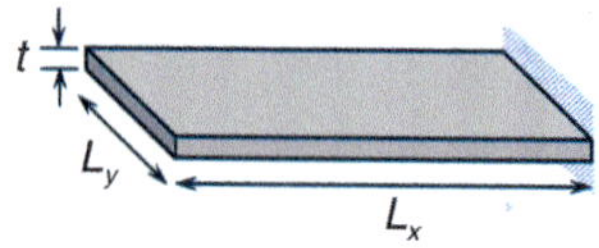

FIGURE 12.20 Problem 12.8.

- **a.** Report the convergence of the first four eigenvalues by conducting a mesh convergence study.
- **b.** Plot the first four eigenvectors of the system.

12.9 Consider the rectangular, elastic domain with uniform material properties and a centrally located circular hole, supported on one side as shown in Fig. 12.21. Carryout an eigenvalue analysis for this system by using the semidiscrete form obtained by finite element method. Let $L_x = 0.5$ m, $L_y = 1$ m, $t = 0.1$ m, $D = 0.15$ m, $E = 100$ GPa, $\rho = 2000$ kg/m^3.

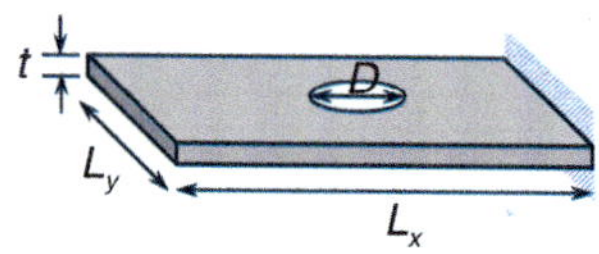

FIGURE 12.21 Problem 12.9.

- **a.** Report the convergence of the first four eigenvalues by conducting a mesh convergence study.
- **b.** Plot the first four eigenvectors of the system.

12.10 Consider a uniform elastic bar initially compressed on both ends by an amount 2δ by Fig. 12.22. The supports are released at time $t = 0$, and the system is allowed to vibrate. Find an analytical solution for the mechanical vibrations of this bar. Present a symbolic solution.

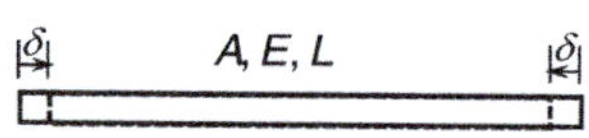

FIGURE 12.22 Problem 12.10.

12.11 Obtain the semidiscretized form of problem 12.10 by using finite element method and then solve this problem by using the modal superposition method. Carryout a mesh convergence study and assume $E = A = L = \rho = 1$ and zero damping.

12.12 Consider a cantilevered beam initially deflected by a concentrated force applied at its free end as shown in Fig. 12.23. At time $t = 0$ the force is released and the beam is allowed to vibrate freely. Find an analytical solution for the beam vibrations in this problem. Present a symbolic solution.

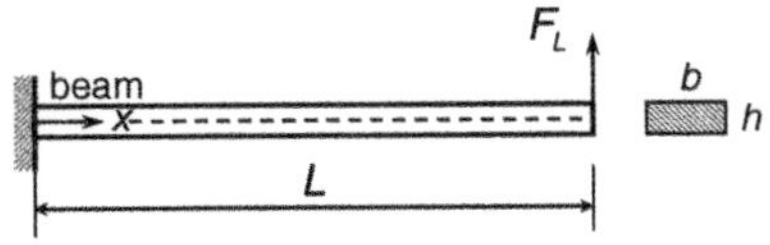

FIGURE 12.23 Problem 12.12.

12.13 Obtain the semidiscretized form of problem 12.12 by using finite element method and then solve this problem by using the modal superposition method. Carryout a mesh convergence study and assume $E = I = L = \rho = 1$ and zero damping.

12.14 Solve problem 12.10 by using direct time integration. Assume $E = A = L = \rho = 1$, use Newmark's method with $\gamma = 1/2$ and $\beta = 1/4$.

a. Carryout a mesh and time step convergence study and report your results.

12.15 Solve problem 12.12 by using direct time integration. Assume $E = I = L = \rho = 1$, use Newmark's method with $\gamma = 1/2$ and $\beta = 1/4$.

a. Carryout a mesh and time step convergence study and report your results.

12.16 Consider an Euler–Bernoulli beam fixed at both ends to be initially at rest, as shown in Fig. 12.24. At time $t = 0$, a point mass m impacts this beam at $x = L/4$ with velocity $v_i = 0.1$ m/s. After the impact the point mass remains stuck on the beam. Let $L = 0.5$ m, $E = 100$ GPa, $b = 12$ mm, $h = 12$ mm, $\rho = 2100$ kg/m^3. Use a damping of the type $[c^{(e)}] = 5 \times 10^{-2}[k^{(e)}]$.

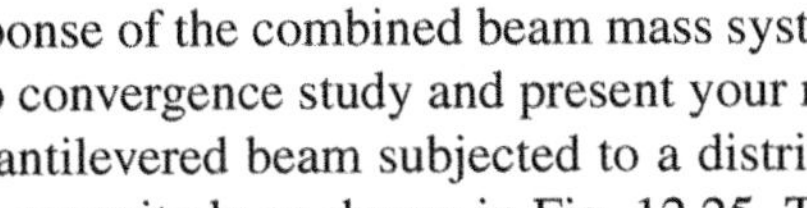

FIGURE 12.24 Problem 12.16.

Use Newmark's method with $\gamma = 1/2$ and $\beta = 1/4$ and determine the transient response of the combined beam mass system. Carryout a mesh and time step convergence study and present your results.

12.17 Consider a cantilevered beam subjected to a distributed force with linearly varying magnitude as shown in Fig. 12.25. The maximum amplitude of this load varies periodically. The loading can be summarized as follows:

$$q_y(x) = q_0(t)\left(1 - \frac{x}{L}\right) \text{ and } q_0(t) = Q_0 \sin \omega t$$

Let $L_1 = 1.5$ m, $E = 70$ GPa, $b = 6$ mm, $h = 25$ mm, $\rho = 2500$ kg/m^3, $\omega = 20$ Hz.

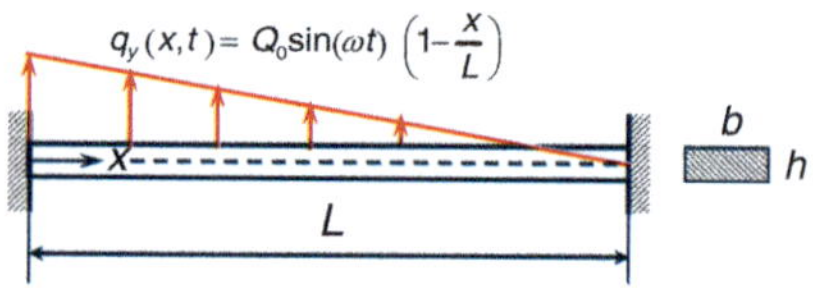

FIGURE 12.25 Problem 12.17.

Use Newmark's method with $\gamma = 1/2$ and $\beta = 1/4$ to find the dynamic response of this beam. Carryout a mesh and time step convergence study and present your results.

12.18 By using the Fourier–von Neumann stability approach, Section 12.11.1, find the stability criterion of the explicit time integration method, for an Euler–Bernoulli beam element with variables given as E, I, ρ, L.

References

[1] R.D. Cook, D.S. Malkus, M.E. Plesha, R.J. Witt, Concepts and Applications of Finite Element Analysis, fourth ed., Wiley, New York, NY, 2002.

[2] R. Haberman, Applied Partial Differential Equations, fifth ed., Pearson Modern Classics, New Jersey, 2019.

[3] T.J.R. Hughes, The Finite Element Method, Prentice-Hall, New Jersey, 1987.

[4] L. Meirovitch, Fundamentals of Vibrations, Waveland Press, Long Grove, IL, 2010.

[5] S. Müftü, The Transient Foil Bearing Problem in Magnetic Recording, Ph.D. thesis, University of Rochester, Rochester, NY, 1994.

[6] R.D. Richtmyer, K.W. Morton, Difference Methods for Initial-Value Problems, second ed., Interscience, New York, 1967.

[7] F.S. Tse, I.E. Morse, R.T. Hinkle, Mechanical Vibrations, second ed., Allyn and Bacon, Inc, Boston, MA, 1978.

[8] W.L. Wood, Practical Time-stepping Schemes, Clarendon Press, Oxford, 1990.

[9] O.C. Zienkiewicz, R.L. Taylor, The Finite Element Method, vol. 1, fourth ed., McGraw-Hill, New York, NY, 1989.

[10] W.E. Boyce, R.C. DiPrima, Elementary Differential Equations and Boundary Value Problems, John Wiley & Sons, New York, NY, 1977.

Appendix A

MATLAB

MATLAB is a programming language made by MathWorks, Inc., Natick, MA. This program provides a practical programming environment that can easily be coupled with many sophisticated functions related to different fields such as signal processing, image processing, data science and statistics, control systems, test, and measurements to name a few. The goal of this appendix is to provide a concise overview of the programming syntax and functions used by MATLAB. MathWorks provides resources to learn MATLAB at all levels [1]. The following are some of the commands and functions that will be useful in creating a simple FEA program. The organization and content of the following sections are in part based on the reference [2]. The *help* function of MATLAB is very useful in learning the syntax.

A.1 Arithmetic

All MATLAB computations are carried out in double precision floating point arithmetic. Each number is represented by a 64-bit number representation.

The number range in MATLAB is approximately between 10^{-308} and 10^{308}. Use the commands `realmax` and `realmin` to find the maximum and minimum numbers that can be represented by MATLAB. If computations result in a number bigger than `realmax` MATLAB produces `inf`. This is an indication of overflow and computations being no longer reliable.

An operation which involves mathematically undefined operations (e.g., `0/0, inf/inf, 0*inf`) produces `NaN` (not a number). NaNs can remain in the computations, but needless to say the results have no determinable meaning.

The accuracy of round-off is 2^{-53} which is approximately 1.11×10^{-16}. Therefore, MATLAB carries up to 16 digit accuracy in its addition, subtraction, multiplication, division, and square root operations. The function `eps` gives twice the unit round-off.

A.1.1 Arithmetic operators

MATLAB's arithmetic operators follow the following precedence rule:

1. ^ exponentiation
2. + – unary plus, unary minus
3. * / multiplication, division
4. + – addition, subtraction

Parentheses should be used to prevent ambiguity.

A.2 Mathematical functions

MATLAB has many elementary and special functions. Typing `help elfun` in the MATLAB command window we can find a brief description of each function as follows:

```
Elementary math functions.
Trigonometric.
    sin         - Sine.
    sind        - Sine of argument in degrees.
    sinh        - Hyperbolic sine.
    asin        - Inverse sine.
    asind       - Inverse sine, result in degrees.
    asinh       - Inverse hyperbolic sine.
    cos         - Cosine.
    cosd        - Cosine of argument in degrees.
    cosh        - Hyperbolic cosine.
    acos        - Inverse cosine.
    acosd       - Inverse cosine, result in degrees.
    acosh       - Inverse hyperbolic cosine.
    tan         - Tangent.
    tand        - Tangent of argument in degrees.
    tanh        - Hyperbolic tangent.
    atan        - Inverse tangent.
    atand       - Inverse tangent, result in degrees.
    atan2       - Four quadrant inverse tangent.
    atan2d      - Four quadrant inverse tangent, result in degrees.
    atanh       - Inverse hyperbolic tangent.
    sec         - Secant.
    secd        - Secant of argument in degrees.
    sech        - Hyperbolic secant.
    asec        - Inverse secant.
    asecd       - Inverse secant, result in degrees.
    asech       - Inverse hyperbolic secant.
    csc         - Cosecant.
    cscd        - Cosecant of argument in degrees.
    csch        - Hyperbolic cosecant.
    acsc        - Inverse cosecant.
    acscd       - Inverse cosecant, result in degrees.
    acsch       - Inverse hyperbolic cosecant.
    cot         - Cotangent.
    cotd        - Cotangent of argument in degrees.
```

```
  coth        - Hyperbolic cotangent.
  acot        - Inverse cotangent.
  acotd       - Inverse cotangent, result in degrees.
  acoth       - Inverse hyperbolic cotangent.
  hypot       - Square root of sum of squares.
  deg2rad     - Convert angles from degrees to radians.
  rad2deg     - Convert angles from radians to degrees.

Exponential.
  exp         - Exponential.
  expm1       - Compute exp(x)-1 accurately.
  log         - Natural logarithm.
  log1p       - Compute log(1+x) accurately.
  log10       - Common (base 10) logarithm.
  log2        - Base 2 logarithm and dissect floating point number.
  pow2        - Base 2 power and scale floating point number.
  realpow     - Power that will error out on complex result.
  reallog     - Natural logarithm of real number.
  realsqrt    - Square root of number greater than or equal to zero.
  sqrt        - Square root.
  nthroot     - Real n-th root of real numbers.
  nextpow2    - Next higher power of 2.

Complex.
  abs         - Absolute value.
  angle       - Phase angle.
  complex     - Construct complex data from real and imaginary parts.
  conj        - Complex conjugate.
  imag        - Complex imaginary part.
  real        - Complex real part.
  unwrap      - Unwrap phase angle.
  isreal      - True for real array.
  cplxpair    - Sort numbers into complex conjugate pairs.

Rounding and remainder.
  fix         - Round toward zero.
  floor       - Round toward minus infinity.
  ceil        - Round toward plus infinity.
  round       - Round toward nearest integer.
  mod         - Modulus (signed remainder after division).
  rem         - Remainder after division.
  sign        - Signum.
```

A.3 Matrices

In general, the entries of two-dimensional matrices and one-dimensional vectors are addressed by their row and column positions.

Matrix: In symbolic notation, the matrix entry at the *i*th-row and the *j*th-column of a matrix [*M*] is stated as M_{ij}. In MATLAB the matrix entry at the location (*i,j*) is indicated as `M(i,j)`.

Row vector: As its name implies a row vector consists of one row and multiple columns. Therefore, to address a position (e.g., *i*) in the row vector we use the notation (1,*i*) such as `v(1,i)`.

Column vector: is the type of vector where the variable values are stored in a single column. To address a position (e.g., i) in the column vector we use the notation $(I,1)$ such as `v(i,1)`.

Matrix vector initialization: It is generally a good idea (in MATLAB) to declare the size of the matrices before using them. The common way to declare a matrix is to use the MATLAB function `zeros(N,M)`, which creates an $N \times M$ matrix of zeros,

```
>> A=zeros(2,3)
A =
      0      0      0
      0      0      0
```

A *column* vector $R_{3\times 1}$ full of zeros could be created by the following command,

`R=zeros(3,1);`

Use of matrices and vectors: In MATLAB values can be assigned to the specific locations in matrices and vectors in several ways.

Enclosing the numbers between `[` and `]` signs can be used to define a matrix. For example, if we want to create the 2×2 square matrix,

$$[B] = \begin{bmatrix} 2 & 3 \\ 22 & 3 \times 10^{-4} \end{bmatrix}$$

We can use one of the following three approaches:

Method 1:

```
B = [ 2  3; 22.0 3e-4 ];
```

Method 2:

```
B =  [2 3
22.03e-4];
```

Method 3:

```
B(1,1) = 2;
B(1,2) = 3;
B(2,1) = 22.0;
B(2,2) = 3e-4;
```

Some useful functions for matrices: MATLAB provides numerous matrix operations and can generate numerous special matrices used in matrix analysis. The reader can consult references [1,2] among others for this wealth of capabilities. The author finds the following functions useful to write MATLAB-based finite element programs: `diag`, `end`, `length`, `min`, `max`, `ones`, and `zeros`.

A.3.1 Subscripting and colon notation

Indexing: Single entries of a matrix can be accessed as `B(i,j)`, as long as: both i and j are integers; $i \geqslant 1$ and $j \geqslant 1$; and, i and j are within the size limits of the matrix for matrices that have been defined before.

For example, for the `B` matrix defined above, by issuing `B(1,2)` command we will get:

```
>> B(1,2)
ans =
      3
```

But, if we issue `B(3,2)`, MATLAB will return:

```
>> B(3,2)
Index in position 1 exceeds array bounds (must not exceed 2)
```

indicating that position (3,2) is not defined for this matrix.

Colon notation: In general, the colon provides a way to generate sequential numbers. For example:

```
>> 2:2.25:13.5
ans =
      2.0000     4.2500     6.5000     8.7500    11.0000    13.2500
```

The first and the last numbers in this command indicate the beginning and ending limits of the series of the numbers, whereas the middle number indicates the increments between each number in the sequence.

By using only integer numbers, MATLAB provides a powerful way to address submatrices in a given matrix by using the : notation. A submatrix that is composed of the rows *i, j* and columns *k, l* of a matrix A can be addressed in MATLAB by `A(i:j,k:l)`.

Example 1: Extraction elements of an array. Consider the 5×5 array $[A]$,

$$[A] = \begin{bmatrix} 1 & 3 & 5 & 7 & 11 \\ 13 & 17 & 19 & 23 & 31 \\ 37 & 2 & 41 & 43 & 47 \\ 53 & 59 & 61 & 71 & 73 \\ 79 & 83 & 89 & 97 & 109 \end{bmatrix}$$

The MATLAB `A(2:3,3:5)` command would result in,

```
>> A(2:3,3:5)
A =
    19 23 31
    41 43 47
```

Example 2: We can use the : notation to insert a smaller array into a larger one. Consider a 3×3 array $[ke]$, and a much larger array $[K]$. One could insert $[ke]$ into $[K]$ as follows:

```
K(59:61,59:61)=ke;
```

If we want to add `ke` into the position of `K(59:61,59:61)` without replacing the contents the following operation would work:

```
K(59:61,59:61)=K(59:61,59:61)+ke;
```

This operation is helpful to assembling the global finite element matrices.

Removing rows and columns from a matrix: On occasion rows and columns must be removed from matrices. This operation should also result in reduction of the matrix size. MATLAB provides a powerful way to accomplish this task. The [ ] notation indicates a 0x0 matrix. Assigning this notation would eliminate the indicated row,

```
A(i,:)= [] ;
```

or column,

```
A(:,j)= [] ;
```

This could be used effectively to apply essential boundary conditions directly on the stiffness matrix.

A.3.2 Matrix and array operations

Matrix product: If $[A]$ is a matrix of dimensions $N \times M$ and $[B]$ is a matrix of dimensions $M \times L$, the matrix product results in a matrix $[C]$ of dimensions $N \times L$. The components of this matrix can be computed by,

$$C_{ij} = \sum_{p=1}^{M} A_{ip} B_{pj} \quad \text{for} \quad i = 1 - N \quad \text{and} \quad j = 1 - L$$

The MATLAB form of this operation is,

```
C=A*B;
```

Vector dot product: consider two column vectors $\vec{a}$ and $\vec{b}$ of length N,

$$\vec{a} = \{a\} = \begin{Bmatrix} a_1 & a_2 & \cdots & a_{N-1} & a_N \end{Bmatrix}^T \text{ and}$$
$$\vec{b} = \{b\} = \begin{Bmatrix} b_1 & b_2 & \cdots & b_{N-1} & b_N \end{Bmatrix}^T$$

The vector dot product of these two vectors is defined as,

$$s = \vec{a}.\vec{b} = \sum_{i=1}^{N} a_i b_i$$

where s is a scalar. The MATLAB form of this operation can be obtained by using any one of the following operations:

```
s = a'*b;
s = transpose(a)*b;
s=dot(a,b);
```

where$'$ indicates the transpose operation.

Cross product: Consider two vectors $\vec{a}$ and $\vec{b}$ of size 3×1

$$\vec{a} = a_1\hat{i} + a_2\hat{j} + a_3\hat{k} \quad \text{and} \quad \vec{b} = b_1\hat{i} + b_2\hat{j} + b_3\hat{k}$$

where $\hat{i}$, $\hat{j}$, and $\hat{k}$ are unit vectors along three mutually orthogonal axes. The cross product of these two vectors yields another vector $\vec{c}$ which is given as,

$$\vec{c} = \vec{a} \times \vec{b} = (a_2b_3 - a_3b_2)\hat{i} + (a_3b_1 - a_1b_3)\hat{j} + (a_1b_2 - a_2b_1)\hat{k}$$

The MATLAB form of this operation is,

```
c=cross(a,b);
```

Elementary matrix and array operations: The following are the elementary operations associated with matrices: addition (+), subtractions (−), multiplication (*), left division (\), right division (/), and power (^). In MATLAB these operations can be carried out in two modes:

- *Matrix sense* according to the rules of matrix algebra,
- *Array sense* by using element wise operations.

Addition and subtraction: are identical in both matrix sense and array sense operations.

Multiplication:

Matrix sense multiplication (`C=A*B`) follows the rules of matrix algebra as described above.

Array sense multiplication (`C=A.*B`) multiplies the corresponding elements of matrices $[A]$ and $[B]$ to find the matrix $[C]$. If $[A]$ and $[B]$ are matrices of the same size, this means $C_{ij}=A_{ij}B_{ij}$.

Division:

Matrix sense division: The left and right division operators in matrix sense define the solutions of linear systems:

for $[A]\{X\}=\{r\}$ the solution is found as $\{X\}=[A]/\{r\}$
for $\{X\}[A]=\{r\}$ the solution is found as $\{X\}=[A]\backslash\{r\}$

Array sense division: (`C=A./B`) divides the corresponding elements of matrices $[A]$ and $[B]$ to find the matrix $[C]$. If $[A]$ and $[B]$ are matrices of the same size, this means $C_{ij}=A_{ij}/B_{ij}$. The assignment `A./B` is equivalent to the assignment `A.\B`.

Power: Consider a square matrix $[A]$.

Matrix sense power: Raising this matrix to power n (`^n`) would trigger n matrix products. For example the MATLAB command `A^2` is equivalent to $[A].[A]$ in matrix algebra `A*A` or in MATLAB.

Array sense power: Raising this matrix to power n in array sense (`.^n`) would be an element-by element operation where the elements of the first matrix are raised to power of the elements of the second matrix ($C_{ij} = A_{ij}\wedge B_{ij}$). Note that for matrices of the same size $[A].\hat{}[B]$ is treated similarly.

Example: Solution of linear set of equations

The fundamental tool for solving a linear system of equations is the backslash operator, \. It handles a linear system of the type `Ax=b`, where the matrix `A` and the vector `b` are given. The solution vector `x` is obtained with the following command, `x=A\b`.

Consider the solution of the system of equations,

$$\begin{bmatrix} 1 & -1 \\ 2 & 3 \end{bmatrix}\begin{Bmatrix} x_1 \\ x_2 \end{Bmatrix}=\begin{Bmatrix} 2 \\ 0 \end{Bmatrix}$$

In MATLAB this system of equations can be solved as follows:

```
>>A = [ 1 -1; 2 3];
>>b = [ 2; 0];
>> x = A\b
x =
    1.2000
   -0.8000
```

A.4 Relational operators and flow control

Relational operators are used in flow control statements such as `if` and `while`, among others.

The relational operators in MATLAB are,

< `less than`
> `greater than`
<= `less than or equal`
<= `greater than equal`
== `equal`
~= `not equal`

Flow control mechanisms in MATLAB are,

`if`-statement

```
if expression
      statements
      elseif
         statements
      else
         statements
      end
```

Note that `elseif` and `else` in the above construct are optional.

`for`-statement

```
      for variable = expression
         statements
      end
```

`while`-statement

```
      while expression
         statements
      end
```

`switch`-statement

```
      switch p
         case value
            statement
         case value
            statement
         otherwise
            statement
      end
```

A.5 Scripts and functions

A.5.1 Script m-files

MATLAB allows the use of a collection of commands from a text file. Typically, the extension of this file is `.m`, such as `my_fea_program.m`. If such a file is saved in a directory whose path is known to MATLAB, then the commands contained in the file can be executed from a MATLAB command window by typing the name of the file:

```
» my_fea_program
```

The default location is the `work` directory typically located at `c:\MATLAB##\work` on a PC.

A.5.2 Function m-files

A *function* is a subprogram that can be called from the MATLAB command line or from a `.m` file. A good example is the sine function whose syntax is `sin(theta)`. The following is a common use of this function:

```
L1=sin(theta)
```

MATLAB allows use of *user defined functions*. The following user defined function `dot_product` can be saved in an m-file named `dot_product.m`. If it is placed in a directory whose path is known to MATLAB, it can be called from any MATLAB script to find the dot product of a matrix and a vector.

```
function r = dot_product(K,d)
% dot_product(K,d) will perform dot product [K].{d}={r}
%               and will return the {r} vector
%
for i=1:length(d)
    for j=1:length(d)
        r(i) = c(i,j)*d(j)   ;
    end
end
```

User defined functions are useful in writing structured programs, such as a FEA program.

The flow of a FEA program could be a collection of the following user defined MATLAB functions:

```
[..., ..., ..., ...]=get_element_properties(..., ..., ..., ...)
[..., ..., ..., ...]=get_boundary_conditions(..., ..., ..., ...)
[..., ..., ..., ...]=perform_assembly(..., ..., ..., ...)
[..., ..., ..., ...]=apply_boundary_conditions(..., ..., ..., ...)
[..., ..., ..., ...]=obtain_displacements(..., ..., ..., ...)
[..., ..., ..., ...]=calculate_stresses(..., ..., ..., ...)
[..., ..., ..., ...]=output_results(..., ..., ..., ...)
```

A.6 Reading and saving files

A.6.1 Reading an input file

MATLAB can read numerous file formats including `mat`, `txt`, `csv`, `xls` among others and provides several functions to this end including `load`, `matfile`, `readmatrix`, `textscan`, `readtable`, `readcell`, `readvars`, `readstruct` among others.

Use of an *input file* is a convenient way of handling data input to a program.

To demonstrate the data reading by the `load` command, consider a text file named `input_1.txt`. Save the following text in file named `input_1.txt` in ascii-text format:

```
1  2  7  48.5e-4
3  4  5  34.6
```

This data can be directly read into MATLAB by using the `load` command as follows:

`A=load ('input_1.txt');`

MATLAB will automatically assign `A` as a 2×4 matrix, and it will assign the numbers in the file into the proper matrix locations according to their order. For example, in order to retrieve `34.6` from `A` and assign it to a variable named `T` one would use,

`T=A(2,4) ;`

Note that each row in the input file `input_1.txt` should have the same number of elements, in order to use the `load` command.

A.6.2 Saving an output file by using a formatted statement

MATLAB provides numerous commands (e.g., `print`, `fprintf`) to save data into different file formats. The `fprintf` command has the following syntax,

`fprintf(formatSpec,A1,...,An)` for displaying formatted data on the screen.

`fprintf(fileID,formatSpec,A1,...,An)` for writing formatted text or numbers into a file.

`formatSpec` consists of keywords used to control the output of the variables, `A1,...,An`.

`formatSpec` can also include text. An example, of printing the elements and associated nodes is as follows:

```
%...........................
E_vec  = zeros(4); EL_vec = zeros(4); ND_vec = zeros(4+1);
EL_vec = 1:1:4;
for i = 1:length(EL_vec)
    ND_vec(i)   = i;
    ND_vec(i+1) = i+1;
    E_vec(i)    = 10.0e9
    fprintf('%2i, (%2i %2i), %10.2e \n', EL_vec(i), ND_vec(i), ND_vec(i+1), E_vec(i))
end
%...........................
```

Executing this block results in the following output to be displayed on the screen:

```
1, ( 1  2), 1.00e+10
1, ( 2  3), 1.00e+10
1, ( 3  4), 1.00e+10
1, ( 4  5), 1.00e+10
```

Each data type has a specific format indicator,

`i`	integer
`e, E, g or G`	exponent
`f, g, G`	real

The format indicators are accompanied by the `%`-sign, the *field width* and *precision*, such as `%12.4e` or `%12.8f` which would print π as,

```
>> fprintf('12.4e \n',pi)
  3.1416e+00
>> fprintf('12.10f \n',pi)
3.1415926536
```

Note that the `\n` operator included in the `formatSpec` implements a carriage return and thus allows for continuation of the rest of the text on the following line.

`fileID` is a file identifier into which the data will be saved. The file identifier is obtained from the `fopen` command. A typical use of this option would look like,

```
%...........................
no_of_elements = 4;
fid01 = fopen('element_information.dat','w');
fprintf(fid01, 'We have %4i elements. \n',no_of_elements);
fclose(fid01);
%...........................
```

This block would create a file named `element_information.dat` in the working folder with the text "`We have 4 elements.`"

More information about the `fprintf, fopen, and fclose` commands can be obtained from the MATLAB help.

A.7 Plotting

MATLAB provides multiple functions for different types of data. These include,

Line Plots

Line plots:	`plot, plot3, stairs, errorbar, area, stackedplot`
Log plots:	`loglog, semilogx, semilogy`
Function plots:	`fplot, fimplicit, fplot3`

Data distribution plots

Histograms and box charts:	`histogram, histogram2, morebins, fewerbins, histcounts, histcounts2, boxchart`
Scatter plots and parallel coordinates:	`scatter, scatter3, binscatter, scatterhistogram, spy, plotmatrix, parallelplot`
Pie charts, heatmaps and word clouds:	`pie, pie3, heatmap, sortx, sorty, wordcloud`

Discrete data plots

Bar graphs:	`bar, barh, bar3, bar3h, pareto`
Stem plots:	`stem, stem3`
Scatter plots:	`scatter, scatter3`
Stairstep plots:	`stairs`

Geographic plots

Plot data:	`geoplot, geoscatter, geobubble, geodensityplot`

Polar plots

Plot data:	`polarplot, polarscatter, polarhistogram, compass, ezpolar`

Contour plots

Plot data:	`contour, contourf, contourc, contour3, contourslice, fcontour`

Vector fields

Plot data:	`feather, quiver, compass, quiver3, streamslice, streamline`

Surfaces, volumes and polygons

Surface and mesh plots:	`surf, surfc, surface, surfl,surfnorm, mesh, meshc, meshz, Hidden, fsurf, fmesh, fimplicit3, waterfall, ribbon, contour3, peaks, cylinder ellipsoid, sphere, pcolor, surf2patch`
Volume visualization:	
Scalar volume data:	`contourslice, flow, isocaps, isocolors, isonormals, sosurface, reducepatch, reducevolume, shrinkfaces, slice, smooth3, subvolume, volumebounds`
Vector volume data:	`coneplot, curl, divergence, interpstreamspeed, stream2, stream3, streamline, streamparticles, streamribbon, streamslice, streamtube`

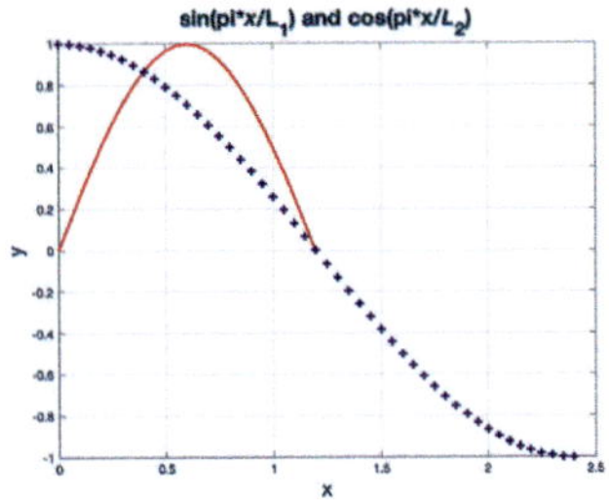

FIGURE A.1 Sample figure generated by the plot function and saved by the print function.

Polygons:

Plot data:	`fill, fill3, patch, surf2patch`

Animation

Animate data:	`movie, getframe, frame2im, im2frame, animatedline, addpoints, getpoints, clearpoints, comet, comet3 drawnow, refreshdata`

A.7.1 Plot function

The function named `plot` is used to create a simple line plot(s) of pair(s) of variables `x` and `y`. These variables need to be supplied to `plot` function as vectors of equal length. The following can be used as simple example to build upon. More on the details of the plot function are available in MATLAB help (Fig. A.1).

```
%................................
x1=0:0.05:1.2;
x2=0:0.05:2.4;
y1 = sin(pi.*x1/x1(end));
y2 = cos(pi.*x2/x2(end));

fig03=figure(3)
   plot(x1,y1,'-r', x2,y2,'b+','linewidth',2)
   grid on;
   xlabel('x','FontSize',16); ylabel('y','FontSize',16);
   title('sin(pi*x/L_1) and cos(pi*x/L_2)','FontSize',18)

 print(fig03,'plot_example','-dtiff') % saves the figure in tiff format
%................................
```

Note that `print` command is used to save the file under the name `plot_example.tif` in the `tiff` format.

References

[1] Mathworks, "Self-Paced Online Courses," https://matlabacademy.mathworks.com, Accessed 1 January 2022.

[2] D.J. Higham, N.J. Higham, MATLAB Guide, SIAM, Philadelphia, PA, 2000.

Appendix B

Guidelines for writing a finite element code in MATLAB

B.1 Structure of a finite element code

In general, all finite element codes have preprocessing, solution, and postprocessing sections. The outline below is for the finite element code `frame2D.m` given in Appendix B4.

Preprocessing

1. *Input and initialization*:
   a. Number of elements (*NUMEL*) and nodes (*number_of_nodes*)
   b. Input the following for each element
      - Element connectivity table (*CT*)
      - Consistent nodal loads (*re*)
      - Material and geometric properties of each element (E, A, L, α_{th},α)
      - Temperature changes (ΔT)
   c. Input the nodal coordinates
   d. Input the boundary conditions
   e. Reserve storage space for global arrays $[K]$ and $\{R\}$ by initializing them to null arrays

Solution

2. *Assemble the global equations of equilibrium*:
   For each element:
   a. Compute element property matrix $[k]^{(e)}$ (*ke*) and
   b. Determine the element load vector $\{r\}^{(e)}$ (*re*)
   c. Add $[k]^{(e)}$ to $[K]$
   d. Add $\{r\}^{(e)}$ to $\{R\}$

3. *Impose displacement boundary conditions*
4. *Solve the equations*: $[K]\{D\} = \{R\}$

Postprocessing

5. *Compute strain and stress variation*
   a. For each element extract $\{d\}^{(e)}$ from $\{D\}$. Compute mechanical strains produced by $\{d\}^{(e)}$.
   b. Calculate the stresses for each element from the strains. Include initial strains, if any.
6. *Report results*
   a. Prepare plots
   b. Prepare output files

B.2 Finite element program for solution of second-order ODEs

In this section a finite element solver is presented to solve a boundary value problem for a second-order ordinary differential equation. This problem is introduced in Section 4.1.1, and repeated here for convenience.

$$\text{BVP: } \frac{d}{dx}\left(a\frac{du}{dx}\right) + pu + q = 0 \text{ in } 0 < x < L \tag{B.1}$$

Boundary conditions:

$$\text{at } x = 0\text{: either } u \text{ is specified or } -\frac{du}{dx} = \alpha_0 u + \beta_0 \tag{B.2a}$$

$$\text{at } x = L\text{: either } u \text{ is specified or } \frac{du}{dx} = \alpha_L u + \beta_L \tag{B.2b}$$

where $u(x)$ is the dependent variable, $a(x)$, $p(x)$, and $q(x)$ are specified functions, $\alpha\_$ and $\beta\_$ are known coefficients on the boundaries. A finite element program based on the process presented in Chapter 4 is given in Section B.2.2.

This program follows the code structure presented in Section B.1. Note that a relatively straightforward input procedure is used, in which the external variables (L, $\alpha_0, \alpha_L, \beta_0, \beta_L$) are directly input through the m-file called `get_inputs.m`, rather than using an *input file*.

B.2.1 Example

Solve the boundary value problem described in Eqs. (B.1) and (B.2a) for the following variables:

$$L = 1 \tag{B.3a}$$

$$N = 64 \tag{B.3b}$$

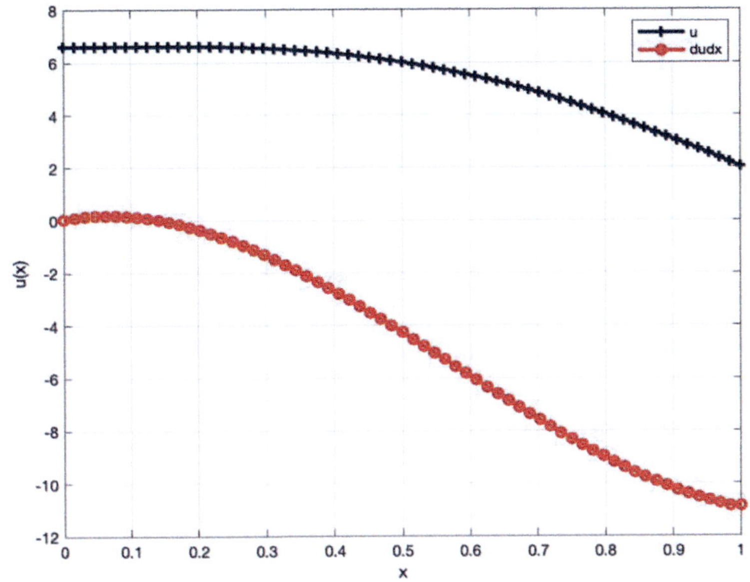

FIGURE B.1 Solution of the boundary value problem defined by Eqs. (B.1)–(B.4a) by using 64 elements (N).

$$a(x) = a_0 + \frac{a_L - a_0}{L}x \tag{B.3c}$$

$$p(x) = p_0 + \frac{p_L - p_0}{L}x \tag{B.3d}$$

$$q(x) = q_0 + \frac{q_L - q_0}{L}x \tag{B.3e}$$

with $a_0 = 1$, $a_L = 2$, $p_0 = -1$, $p_L = 10$, $q_0 = 1$, $q_L = 2$ and,

$$\text{at}\, x = 0\ :\ -\frac{du}{dx} = \alpha_0 u + \beta_0 \tag{B.4a}$$

$$\text{at}\, x = L\ :\ \frac{du}{dx} = \alpha_L u + \beta_L \tag{B.4b}$$

with $\alpha_0 = 1$, $\beta_0 = 0$, $\alpha_L = 0.1$, and $\beta_L = -11.10$.

The variables given in Eqs. (B.3a) and (B.4a) are directly entered into the finite element code through the `get_inputs.m` file. Note that this is a flexibility allowed by MATLAB. For other programming platforms which work with executable files, using hardcoded inputs is not an effective strategy.

The solution of the program is plotted in Fig. B.1.

B.2.2 MATLAB-based ODE solver

```
%.......................................................................
% 2nd Order ODE solver:
%
% ODE: (a(x)*u,x),x + p(x)*u + q(x) = 0 in 0 < x < L
% BCs: at x = 0, either u = u0 or -u,x = alpha0*u(0) + beta0
%      at x = L, either u = uL or  u,x = alphaL*u(L) + betaL
%
% Dependencies: get_inputs.m, prepare_element_CT.m,
assemble_global_variables.m,
%              apply_BCs.m, post_process_solution.m, plot_solution.m
%
% Sinan Muftu, Northeastern University, Boston, MA, 2020
%.......................................................................
clear;clc;                  % clear memory and clear the screen
get_inputs;                 % get problem inputs
K = zeros(dof);             % initialize [K] and {R}
R = zeros(dof,1);
prepare_element_CT          % prepare element connectivity table
assemble_global_variables % assemble [K] and {r}
apply_BCs                   % apply boundary conditions
Dn = Kn\Rn;                 % obtain solution
post_process_solution;      % postprocess the solution
plot_solution;              % plot the solution
%.......................................................................
```

```
%.......................................................................
% 1D-ODE-solver: get_inputs.m
%.......................................................................
L  = 1;  % domain length
% Input variables
A0 =  1e0;  AL = 2e0  ; mA = (AL-A0)/L;
P0 = -1e0;  PL = 10.0; mP = (PL-P0)/L;
q0 =  1e0;  qL = 2    ; mq = (qL-q0)/L;
%BCs
BC_type_0 = 'NBC';
BC_type_L = 'NBC';
% NBC variable values
alpha_0 = 0;    beta_0  = 0;
alpha_L = 0.1;  beta_L  = -11;
% EBC variable values
u0      = -1;   uL      = 2.5;
% number of elements and degrees of freedom
N = 64; dof=N+1;
% prepare mesh (set the x vector)
x = 0:L/N:L;
%.......................................................................
```

```
%.......................................................................
% 1D-ODE-solver: prepare_element_CT.m
% Set the element connectivity table (CT)
%.......................................................................
for e=1:N;
    Le(e)=x(e+1) - x(e);
    A1(e)=A0+mA*x(e); A2(e)=A0+mA*x(e+1);
    P1(e)=P0+mP*x(e); P2(e)=P0+mP*x(e+1);

    q1(e)=q0+mq*x(e); q2(e)=q0+mq*x(e+1);
    CT(e,1)=e;
    CT(e,2)=e+1;
end
%.......................................................................
```

```
%.........................................................................
% 1D-ODE-solver: assemble_global_variables.m
%
% Perform the assembly
%.........................................................................
for e =1:N
    % ka matrix for element-e
    Am          = (A1(e)+A2(e))/(2*Le(e));
    ka          = [Am -Am; -Am Am];
    % kp matrix for element-e
    kp          = (Le(e)/12)*[ (3*P1(e)+P2(e))  (P1(e)+P2(e)) ;...
                               (P1(e)+P2(e))  (P1(e)+3*P2(e)) ];
    % consistent load vector for element-e
    re          = Le(e)/6*[(2*q1(e)+q2(e)); (q1(e)+2*q2(e))];
    %
    kt          = ka - kp;
    % pull out global node numbers from element connectivity table
    i = CT(e,1);
    j = CT(e,2);
    % Assemble the K matrix
    K(i,i)   = K(i,i) + kt(1,1);
    K(i,j)   = K(i,j) + kt(1,2);
    K(j,i)   = K(j,i) + kt(2,1);
    K(j,j)   = K(j,j) + kt(2,2);
    % Assemble the R vector
    R(i)     = R(i)    + re(1);
    R(j)     = R(j)    + re(2);
end
%.........................................................................

%.........................................................................
% 1D-ODE-solver: apply_BCs.m
%
% This file modifies the K matrix and the R vector based on BCs
%
% x = 0: BC_type_0 is either EBC or NBC
% x = L: BC_type_L is either EBC or NBC
% see get_inputs.m for the specific values
%.........................................................................
if BC_type_0 == 'EBC'& BC_type_L == 'EBC'
    Kn    =      K(2:dof-1,2:dof-1) ;
    Rn    =      R(2:dof-1) ;
    Rn(1)=       Rn(1)- u0*K(2,1);
    Rn(dof-2)= Rn(dof-2)- uL*K(dof-1,dof);
end

if BC_type_0 == 'EBC'& BC_type_L == 'NBC'
    Kn         = K(2:dof,2:dof) ;
    Rn         = R(2:dof) ;
    Rn(1)      = Rn(1)- u0*K(2,1);

    Kn(dof-1,dof-1) = Kn(dof-1,dof-1) - alpha_L;
    Rn(dof-1)       = Rn(dof-1) + A2(N)*beta_L;
end

if BC_type_0 == 'NBC'& BC_type_L == 'NBC'
    K(1,1) = K(1,1) - alpha_0 ;
    R(1)   = R(1) + A1(1)*beta_0;

    K(dof,dof) = K(dof,dof) - alpha_L;
    R(dof)     = R(dof) + A2(N)*beta_L

    Kn=K;
    Rn=R;
end
```

```
if BC_type_0 == 'NBC'& BC_type_L == 'EBC'
    K(1,1)  = K(1,1) - alpha_0 ;
    R(1)    = R(1) + A1(1)*beta_0;

    Kn = K(1:dof-1,1:dof-1) ;
    Rn = R(1:dof-1) ;
    Rn(dof-1)=Rn(dof-1)- uL*K(dof-1,dof);
end
%.....................................................................

%.....................................................................
% 1D-ODE-solver: post_process_solution.m
%
% Put the solution back into the full D vector for plotting wrt x
%.....................................................................
if BC_type_0 == 'EBC'& BC_type_L == 'EBC'
    D(1)        = u0;
    D(2:dof-1)= Dn;
    D(dof)      = uL;
end

if BC_type_0 == 'EBC'& BC_type_L == 'NBC'
    D(1)        = u0;
    D(2:dof)    = Dn;
end

if BC_type_0 == 'NBC'& BC_type_L == 'NBC'
    D           = Dn;
end

if BC_type_0 == 'NBC'& BC_type_L == 'EBC'
    D(1:dof-1)  = Dn;
    D(dof)      = uL;
end
%
% Perform postprocessing of the solution (i.e. compute dudx)
%
for e =1:N
    % pull out global node numbers from element connectivity table
    i = CT(e,1);
    j = CT(e,2);
    % [B] matrix for element-e
    B        = (1/Le(e))*[-1 1];
    % {d} vector for element
    d = [D(i) ; D(j)];
    dudx(i) = B*d;
    dudx(j) = B*d;
end

%.....................................................................

%.....................................................................
% plot_solution.m
%.....................................................................
% Plot the solution
h2 = figure(2);
set(gca,'FontSize',15)
plot(x,D,'-k+',x,dudx,'-or','LineWidth',2)
ylabel('u(x)')
xlabel('x')
grid on
hold on
legend('u', 'dudx')
%.....................................................................
```

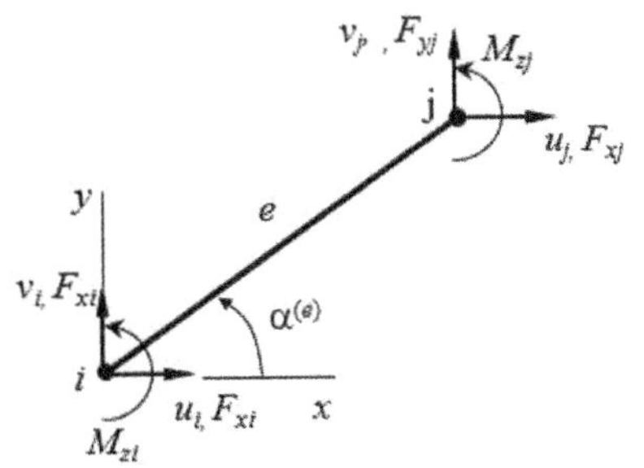

FIGURE B.2 Definition of nodes i and j for an element (e) whose properties are E, A, I, and L.

B.3 Finite element program for a two-dimensional frame

In this section a finite element solver is presented for solving a two-dimensional frame problem, presented in Sections 6.4–6.6. The input data are entered into the program bu using three separate input files.

B.3.1 Data input into the finite element program frame2D

***input.txt*:** this file is used to enter the element connectivity table (CT). This table includes the node numbers for each element, the nodal forces and moments, the material properties, and geometry for each element. The structure of the input file is as follows:

NUMEL	*NNodes*	0	0	0	0	0	0	0	0	0	0	0	0	0
e	*i*	*j*	*Fxi*	*Fyi*	*Mi*	*Fxj*	*Fyj*	*Mj*	*E*	*A*	*I*	*L*	*h*	α
...														
...														

First line of the ***input.txt*** file is used to enter the total number of elements (*NUMEL*) and the total number of nodes (*number_of_nodes*). The rest of this line consists of zeros. Lines following the first line are used to enter the element connectivity table and the element properties. The variables are defined as follows:

i:	low numbered node of the element
j:	high numbered node of the element
Fxi, Fyi, Mi:	loads (consistent and nodally applied) on node-*i*
Fxj, Fyj, Mj:	loads (consistent and nodally applied) on node-*j*
E:	elastic modulus of the element
A:	cross-sectional area of the element
I:	second moment of area of the element
L:	length of the element
h:	thickness of the element
α:	orientation of the element with respect to *x*-axis (Fig. B.2)

***Original_positions.txt*:** is used to enter the original positions of the nodes in the global coordinate system. The file contains the nodal positions of each node in a sequential manner.

xi	yi
...	
...	

***boundary_conditions.txt*:** is used to enter the essential boundary conditions of the problem. The structure of the boundary conditions input file is as follows:

No of EBCs	0	0
Node	*BC_dof*	*BC_magnitude*
...	...	...

No of EBCs: Total number of EBCs that we want to impose.
Node: The node number at which the boundary condition is to be imposed
BC_dof:

1. to impose a boundary condition for u,
2. to impose a boundary condition for v,
3. to impose a boundary condition for θ

BC_magnitude: The magnitude of the imposed boundary condition.

B.3.2 Example

Frame shown in Fig. B.3 is modeled by using five elements. The loading for this example is given as follows: $P_1 = -50$ kN, $M_3 = 10$ kN.m, $q_x = 5$ kN/m, $q_y = -5$ kN/m. The frame has uniform cross section ($h = 10$ cm, $A = 0.01$ m^2, $I = 8.33 \times 10^{-6}$ m^4) and elastic modulus ($E = 72$ GPa).
Element connectivity table and the properties of the elements used in the *input.txt* file are given in Table B.1.

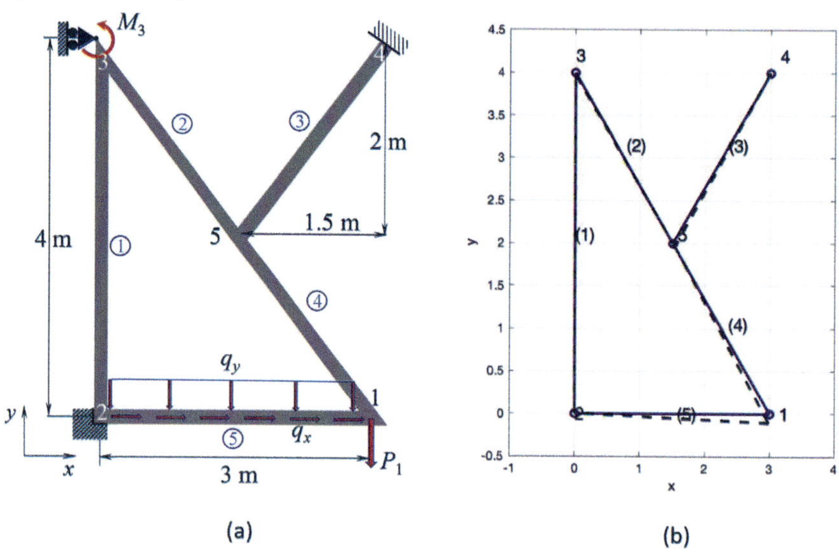

FIGURE B.3 (A) The frame, its boundary conditions, and the loads used as an example in this section. (B) The nodal coordinates, element and node numbers, and the deflected state of the frame (scale factor 100).

TABLE B.1 *input.txt* file structure for the frame shown in Fig. B.3.

NUMEL	i	j	Fx1	Fy1	Mi	Fxj	Fyj	Mj	E	A	I	L	h	alpha
5	5	0	0	0	0	0	0	0	0	0	0	0	0.00	0
1	2	3	0	0	0	0	0	10000	7.20E+10	0.01	8.33E−06	4.00	0.10	90
2	3	5	0	0	0	0	0	0	7.20E+10	0.01	8.33E−06	2.50	0.10	−53.1301
3	4	5	0	0	0	0	0	0	7.20E+10	0.01	8.33E−06	2.50	0.10	−143.1301
4	1	5	0	−50000	0	0	0	0	7.20E+10	0.01	8.33E−06	2.50	0.10	126.87
5	1	2	7500	−7500	−3750	7500	−7500	3750	7.20E+10	0.01	8.33E−06	3.00	0.10	180

TABLE B.2 *original_postions.txt* file structure for the frame shown in Fig. B.3.

xi	yi
3.0	0.0
0.0	0.0
0.0	4.0
3.0	4.0
1.5	2.0

TABLE B.3 *boundary_conditions.txt* file structure for the frame shown in Fig. B.3. The first line indicates a total of seven BCs are entered. The next seven lines indicate the node number, the dof that is being imposed and the magnitude of the imposed boundary condition.

Node	BC-type	BC-magnitude
7	0	0
2	1	0
2	2	0
2	3	0
3	1	0
4	1	0
4	2	0
4	3	0

Nodal positions: can be determined from the diagram given in Fig. B.3B and used in the `original_positions.txt` file (Table B.2).

Boundary conditions of this problem are as follows:

node-2: $u_2 = v_2 = \theta_2 = 0$

node-3: $u_3 = 0$

node-4: $u_4 = v_4 = \theta_4 = 0$

The corresponding `boundary_conditions.txt'` file is given in Table B.3.

Results: The finite element program `2DFrame.m` given in Section B.3.1 is used to solve this problem with the inputs described above. Deformed nodal deflections are plotted in Fig. B.3B with a magnification factor of 10^2. Nodal deflections and stresses are given in Table B.4.

TABLE B.4 The output of the code. The bending stress (sigma_b), the axial stress (sigma_a), and the shear stress (tau) for each element are computed on node-*i* of the element at $z = h/2$. The primary degrees of freedom are found at each node. u_i and v_i are the deflection components in global coordinate system with unit of m. The nodal slope (theta_i) is given in unit of radian. The imposed boundary conditions at nodes 2, 3, and 4 (highlighted) are found to be five orders of magnitude smaller than the other nodal deflections.

Element	sigma_b	sigma_a	tau	Node	u_i	v_i	theta_i
1:	1.21e+07	−6.05e+06	2.27e+05	1:	−1.58e−04	−1.09e−03	−2.44e−03
2:	3.58e+07	7.32e+06	5.02e+05	2:	−7.71e−08	−1.27e−07	−3.12e−09
3:	−3.89e+06	−4.81e+05	−9.86e+04	3:	8.40e−08	−3.36e−04	6.71e−03
4:	−1.34e+07	7.38e+06	−2.18e+05	4:	8.96e−09	4.98e−09	1.37e−09
5:	−9.11e+06	−3.80e+06	−1.03e+05	5:	3.27e−04	−4.09e−04	−7.23e−04

B.3.3 MATLAB-based FEA code: Frame2D

```
%..........................................................................
% frame2D
%
% This is  MATLAB based FEA program used for analyzing a frame with
% axial deformation and bending (EB-beam) capabilities
%
% Dependencies: Read_inputs.m, Assemble_K_and_R.m, Apply_BCs.m,
%               Post_process_stress_strain.m, Post_process_deformed_shape
%
% Sinan Muftu, Northeastern University
%..........................................................................
clear;clc;

Read_inputs
Assemble_K_and_R
Apply_BCs
D = Kg\Rg;
Post_process_stress_strain
Post_process_deformed_shape
%..........................................................................

%..........................................................................
% frame2D: Read_inputs.m
%
% Inputs:
% - input.txt: see text below
% - boundary_conditions.txt: see text below
% - original_position.txt: (x,y) position of nodes
%..........................................................................
% This section reads input.txt:
%
% Line-1:               var1 and var2 on the first line are NUMEL and
%                       number_of_nodes rest of first line are zeros.
% Lines 2 - NUMEL+1:
% node_1, node_2, Fx1, Fy1,M1, Fx2,Fy2, M2, E, A, Ix, L, alpha
%
Input_Vars      = load ('input.txt') ;

NUMEL           = Input_Vars(1,1);
number_of_nodes = Input_Vars(1,2);
```

```
fprintf('Total number of elements : %4i \n',NUMEL);
fprintf('Total number of nodes    : %4i \n',number_of_nodes);

for i=2:NUMEL+1;
    % Connectivity table
    CT(i-1,1)    = Input_Vars(i,2);
    CT(i-1,2)    = Input_Vars(i,3);
    % Consistent nodal loads
    re(i-1,1)     = Input_Vars(i,4);
    re(i-1,2)     = Input_Vars(i,5);
    re(i-1,3)     = Input_Vars(i,6);
    re(i-1,4)     = Input_Vars(i,7);
    re(i-1,5)     = Input_Vars(i,8);
    re(i-1,6)     = Input_Vars(i,9);
    % element properties
    E(i-1)        = Input_Vars(i,10);
    A(i-1,1)      = Input_Vars(i,11);
    Ix(i-1,1)     = Input_Vars(i,12);
    Length(i-1)   = Input_Vars(i,13);
    height(i-1)   = Input_Vars(i,14);
    Alpha(i-1)    = Input_Vars(i,15)/180*pi;
end
%
% End of reading inputs
%.............................................................................
% This section reads the boundary conditions
%
% The order of variables in 'boundary_conditions.txt' are as follows:
%    Line-1:
%     (Total_imposed_BCs) 0 0
%    Lines-2 .. Total_imposed_BCs+1:
%     (Node number) (DOF) (Magnitude)
%
% Note: DOF = 1 --> for restraining u
%       DOF = 2 --> for restraining v
%       DOF = 3 --> for restraining theta
%.............................................................................
Input_BCs = load ('boundary_conditions.txt') ;

Total_imposed_BCs = Input_BCs(1,1);

fprintf('*********************************\n')
fprintf('Total number of imposed BCs : %4i \n',Total_imposed_BCs)
fprintf('BC_Node_Number, BC_dof_Number, BC_Magnitude \n')
fprintf('*********************************\n')

for i=2:Total_imposed_BCs+1;
    BC_Node(i-1)       = Input_BCs(i,1);
    BC_dof(i-1)        = Input_BCs(i,2);
    BC_magnitude(i-1) = Input_BCs(i,3);
    fprintf(' %4i, %4i, %4i, %9.2e \n',i-1,BC_Node(i-1),BC_dof(i-1),...
            BC_magnitude(i-1))    ;
end
%
% End of boundary condition reading
%.............................................................................
% This section reads the nodal positions
%
Nodal_positions = load('original_position.txt');
x_vec = Nodal_positions(:,1);
y_vec = Nodal_positions(:,2);
%
% End of reading nodal positions
```

```
%.........................................................................
%.........................................................................
% frame2D: Assemble_K_and_R.m
% S. Muftu, Northeastern University
%.........................................................................
% Initialize the global stiffness matrix global load vector
dof = 3*number_of_nodes;
Kg  = zeros(dof,dof);
Rg  = zeros(dof,1);
%.........................................................................
% Assemble: the Global stiffness matrix, Kg and Global load vector, Rg
%
% Variables
% NUMEL  : No. of elements
% A      : cross-sectional area
% E      : Elastic modulus
% Length : Element length
% Ix     : Moment of inertia
% Alpha  : Orientation with respect to x axis
% ke     : element stiff matrix
%         SE is a 6x6 matrix capable of
%         - Euler-Bernoulli type bending and
%         - axial deformation
%          Associated dof are {d} = {u1 v1 theta2 u2 v2 theta2}^T
%          This element is oriented at an angle alpha
%          with respect to the global x-axis.
% re     : Element load vector (entered in x-y system, see input.txt)
% Kg     : Global stiffness matrix
% Rg     : Global load vector
% CT     : Element connectivity matrix (see input.txt)
%          CT(e,1): contains global node number for node-1 of the element-M
%          CT(e,2): contains global node number for node-2 of the element-M
%
for e=1:NUMEL
    % Get elem-stiff matrix
    ke = get_element_stiffness(A(e),E(e),Length(e),Ix(e),Alpha(e));
    % Use Global-node numbers to find the Global-dof
    global_dof(3) = 3*CT(e,1);          % theta1
    global_dof(2) = global_dof(3)-1;    % v1
    global_dof(1) = global_dof(2)-1;    % u1
    global_dof(6) = 3*CT(e,2);          % theta2
    global_dof(5) = global_dof(6) - 1;  % v2
    global_dof(4) = global_dof(5) - 1;  % u2
    for i=1:6
        K = global_dof(i);
        Rg(K,1) = Rg(K,1) + re(e,i);     % Assemble the global load vector
        for j=1:6
            L = global_dof(j);
            Kg(K,L) = Kg(K,L) + ke(i,j); % Assemble the global stiffness matrix
        end
    end
end
%..................................................................
%.........................................................................
% frame2D: Post_process_stress_strain.m
% S. Muftu, Northeastern University
%.........................................................................
fid01=fopen('outputs.txt','w'); % open an output file
fprintf(fid01,'Element  sigma_b      sigma_a      tau \n');
fprintf('Element  sigma_b      sigma_a      tau \n');
```

```
%
% Calculate Normal Stresses:
%
max_stress = 0;
max_elem   = 0;
d_local    = [0;0;0;0;0;0];
for e=1:NUMEL
    dof3 = 3*CT(e,1);  d_global(3,1) = D(dof3); % theta1
    dof2 = dof3-1;     d_global(2,1) = D(dof2); % v1
    dof1 = dof2-1;     d_global(1,1) = D(dof1); % u1
    dof6 = 3*CT(e,2);  d_global(6,1) = D(dof6); % theta2
    dof5 = dof6-1;     d_global(5,1) = D(dof5); % v2
    dof4 = dof5-1;     d_global(4,1) = D(dof4); % u2
    alpha= Alpha(e);
    c=cos(alpha);   s=sin(alpha);
    % T= transformation matrix
    T=0;
    T(1,1)=c;  T(1,2)=s;
    T(2,1)=-s; T(2,2)=c;
    T(3,3)=1;
    T(4,4)=c;  T(4,5)=s;
    T(5,4)=-s; T(5,5)=c;
    T(6,6)=1;
    %
    d_local = T*d_global;
    %
    u1 = d_local(1); v1 = d_local(2); theta1 = d_local(3);
    u2 = d_local(4); v2 = d_local(5); theta2 = d_local(6);
    % compute bending moments, normal force and shear force at the
    % LHS of each element (x=0)
    L=Length(e);  x=0;
    % strain_b (compute at the top of the element)
    strain_b(e) = -(height(e)/2)*( (-6/L^2+12*x/L^3)*v1 + ...
        (-4/L+6*x/L^2)*theta1 + (6/L^2-12*x/L^3)*v2 + (-2/L+6*x/L^2)*theta2);
    strain_a(e) = (u2 - u1)/L;
    strain_t(e) = strain_b(e) + strain_a(e);
    %
    Mz1(e)       = E(e)*Ix(e)*strain_b(e)*(-2/height(e)) ;
    Nx(e)        = (A(e)*E(e)/L)*(u2-u1);
    Vy(e)        = E(e)*Ix(e)*( (12/L^3)*v1 + ...
        (6/L^2)*theta1 + (-12/L^3)*v2 + (6/L^2)*theta2 );
    % compute the bending stress, axial stress and shear stress
    % sigma_b is computed at the top of the beam
    sigma_b(e)   = E(e)*strain_b(e);
    sigma_a(e)   = Nx(e)/A(e);
    tau(e)       = 1.5*Vy(e)/A(e);
    fprintf('%3i: %12.2e %12.2e %12.2e \n',e,sigma_b(e),sigma_a(e),tau(e));
    fprintf(fid01,'%3i: %12.2e %12.2e %12.2e \n',e,sigma_b(e),sigma_a(e),tau(e));

end
fprintf('\n'); fprintf(fid01,'\n');
%.......................................................................
%.......................................................................
% frame2D: Post_process_deformed_shape.m
% S. Muftu, Northeastern University
%.......................................................................
%
% Get nodes from D to u, v and theta
%
fprintf('Node     u_i           v_i           theta_i \n');
fprintf(fid01,'Node     u_i           v_i           theta_i \n');

for i=1:number_of_nodes
    u_vec(i) = D(3*i-2);      % Extract ui from D
    v_vec(i) = D(3*i-1);      % Extract vi from D
    theta_vec(i) = D(3*i);    % Extract theta_i from D
```

```
    x_new(i)=x_vec(i)+u_vec(i);
    y_new(i)=y_vec(i)+v_vec(i);
    fprintf('%3i: %12.2e %12.2e %12.2e \n',i,u_vec(i),v_vec(i),theta_vec(i));
    fprintf(fid01,'%3i: %12.2e %12.2e %12.2e \n',i,u_vec(i),v_vec(i),theta_vec(i));

end

%
% Plot the original location of the frame
%
scale_factor = 1.05;

fig01=figure(01);
set(gca,'FontSize',15)
for i=1:NUMEL

    xplot(1) = x_vec(CT(i,1));
    xplot(2) = x_vec(CT(i,2));

    yplot(1) = y_vec(CT(i,1));
    yplot(2) = y_vec(CT(i,2));

    xave = 1.05*(xplot(1)+xplot(2))/2;
    yave = 1.05*(yplot(1)+yplot(2))/2;

    plot(xplot,yplot,'b--o')
    text(xave,yave,num2str(i)); % write element numbers on the plot
    xlabel('x'); ylabel('y')
    grid on
    hold on
end
%
% Plot the deformed location of the frame
%
for i=1:NUMEL

    xplot(1) = x_new(CT(i,1));
    xplot(2) = x_new(CT(i,2));

    yplot(1) = y_new(CT(i,1));
    yplot(2) = y_new(CT(i,2));

    plot( xplot*scale_factor,yplot*scale_factor,'k-')
    hold on
end
fclose(fid01);
print(fig01,'frame_01','-dtiff');
%.....................................................................
```

Appendix C

Finite element analysis with ANSYS

In general, finite element analysis involves the following three stages:

Preprocessing: operations are used to create the geometry, define element type(s), to assign and attribute material properties, and to mesh the geometry.

Solution: operations include defining the loads, applying the boundary conditions, setting the type of analysis, and solving for degrees of freedom.

Postprocessing: operations enable analysis and plotting of the results.

In this appendix we provide a brief introduction to using the commercially available finite element package *ANSYS Mechanical APDL* (ANSYS, Inc., Canonsburg, PA, USA) to solve finite element problems. Interested readers are recommended to consult the *ANSYS Help Manual* which is available through the graphical user interface (GUI) of the program, the *ANSYS APDL Programmer's Guide* [1], and references [2,3] for a more comprehensive use of this package. ANSYS is a versatile software package that enables analysis through one or combination of the following approaches:

GUI-based analysis: parameters of the problem are entered; the solution is obtained; and the solution is postprocessed by using the graphical a GUI. Examples to solving problems by using this approach are presented in Section C.1, and Appendices D–F.

APDL macro-based analysis: parameters of the problem are entered; the solution is obtained; and it is postprocessed by using the ANSYS Parametric Development Language (APDL). Examples of using this approach to solve problems are provided in Sections C.2 and C5.

Hybrid approach: the solution can be obtained by a combination of these two approaches.

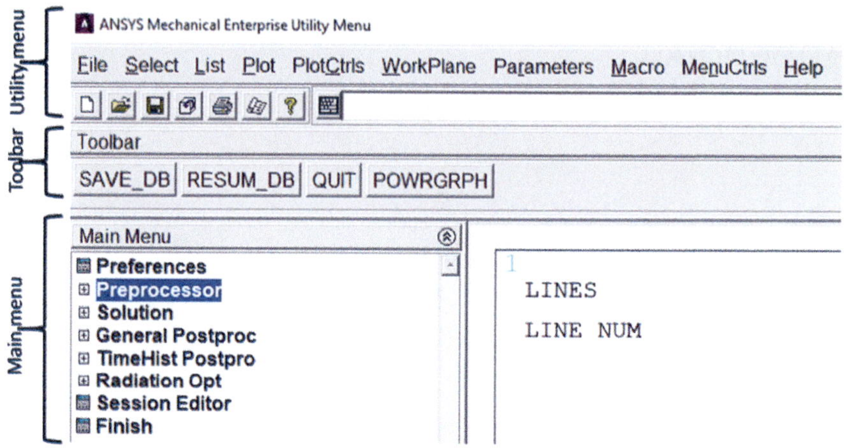

FIGURE C.1 The nomenclature used to address the main menu, the toolbar, and the utility menu in ANSYS. Figure published with the permission of ANSYS, Inc.

Units used in finite element analysis

The user is responsible to specify a consistent unit system such as those given in the following table.

Inputs			Outputs	
Dimensions	*Force*	*Elastic Modulus*	*Stress*	*Deformation*
m	N	N/m^2	N/m^2	m
cm	N	N/cm^2	N/cm^2	cm
mm	N	N/mm^2	N/mm^2	mm

C.1 GUI-based analysis

Upon starting *ANSYS Mechanical APDL*, the user will see three menus (*utility menu, toolbar, main menu*) as shown in Fig. C.1. The *main menu* provides access to the three main stages of finite element analysis: *preprocessing*, *solution*, and *postprocessing*. The first level details of these three main stages are shown in Fig. C.2. References [2,3] are dedicated to detailed use of the menu functions. In this text, we provide an introductory overview of using *ANSYS Mechanical APDL* to help the user get started. Step-by-step descriptions of how to use the menus to solve different example problems are introduced, in Appendices D–F.

In the general, the order of operations involves the following steps.

Preprocessing

Element type

This platform provides analysis capabilities in solid mechanics, heat transfer, fluid mechanics, magnetics, electrostatics, diffusion problems among others. The *Element Type* menu is used to pick the element(s) appropriate for a specific type of analysis.

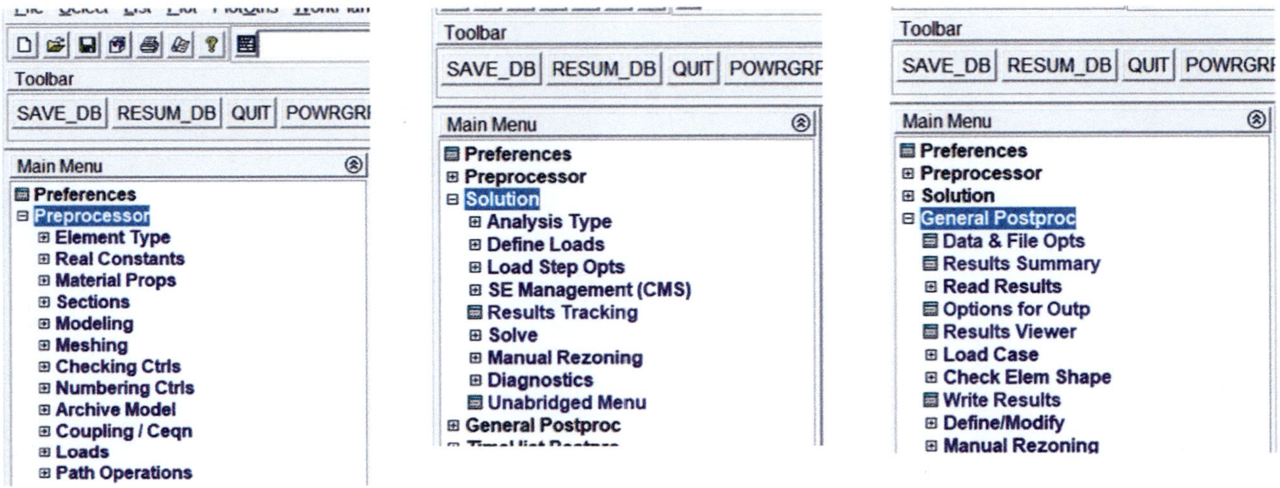

FIGURE C.2 Details of the preprocessor, solution, and general postprocessor menus in *ANSYS Mechanical APDL*. Figure published with the permission of ANSYS, Inc.

Element properties
Depending on the type of element chosen element properties such as thickness, width, area are entered into the program by using one of the *Real Constants* or *Sections* menus.

Material properties
Material properties relevant for the physical system are entered into the program by using the *Material Props* menu. The most common submenu here is *Material Models* which provides choices for structural, thermal, electromagnetic, acoustic, fluid, piezoelectric, thermoelectric, and diffusion model properties to be entered.

Geometry of solution domain
The geometry of the solution domain can be imported from a CAD file or it can be directly created in the ANSYS GUI. The GUI option is invoked from the *Modeling* menu. Geometric entities such as points, lines, area volumes can be entered from the submenu *Create*. If needed, operations such as extrusion, intersection, addition, gluing, division, and partitioning of the geometric entities can be accessed from the submenu *Booleans*.

Meshing
The material and section properties set in the earlier steps can be associated with the subsections (line, area, or volume) of the model by using the submenu *Mesh Attributes*. A practical and powerful meshing tool is provided by the *MeshTool*, which has menu options for setting the element size, element type (e.g., triangular, rectangular) and meshing strategy (free mesh or mapped mesh), and mesh refinement.

Solution

Solution type
The submenu *New Analysis* allows picking the type of analysis (static, modal, harmonic, transient, spectrum, buckling, eigenvalue, and substructuring) to be performed.

Boundary conditions and external loads
The essential boundary conditions and the external loads can be entered by using the submenu *Define Loads*. The submenu *Apply* provides options to apply the boundary conditions, external loads, initial conditions, and more.

Solution
The command to execute the solution is given in the submenu *Solve*.

Postprocessing
Postprocessing operations involve plotting the *Deformed Shape*, obtaining *Contour Plot*(ing), or *List*(ing) *Results* of the primary and the secondary variable at the nodes or for the elements among many other functions.

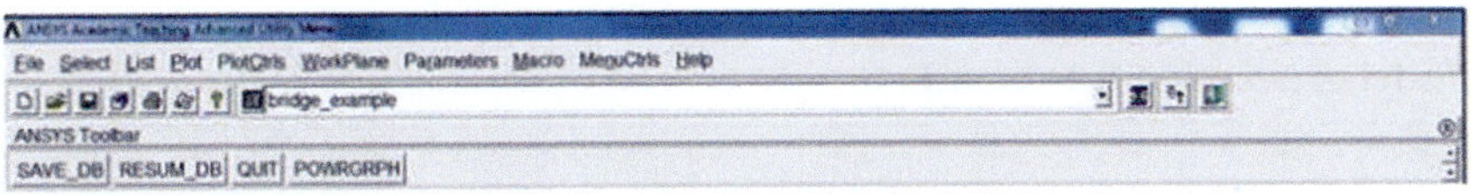

FIGURE C.3 In order to execute the APDL macro, its name is typed into the command line in the utility menu. Figure published with the permission of ANSYS, Inc.

C.2 APDL-based analysis

An APDL macro is a *text file* that contains specific keywords, commands, and programming statements that allows the preprocessing, solution, and postprocessing functions in *ANSYS Mechanical APDL*, without having to use the GUI. Using the APDL macros is particularly useful to allow the user to carryout parametric analyses without having to engage with the GUI.

Creating and running an APDL macro in ANSYS

Open a text file, write your macro (as described below), save it, for example, as "`bridge_example.mac`." Open ANSYS, Click *File*, *Change Directory*, find and select the folder you saved your macro in, click OK. Type, for example, "`bridge_example`" in the command line. Hit enter to run your macro.

Brief description of the APDL commands

Each function that a GUI-user is accustomed to use, has a corresponding APDL command. In fact, the APDL provides a wide range of control to create, execute and analyze a finite element problem. In case the user would like to find out the APDL command that corresponds to a GUI-based operation, it is instructive to note that all commands that are entered into the ANSYS GUI are logged in a *log-file*, which can be accessed through the GUI sequence,

File → List → Log file

In general, an APDL macro is categorized into the same three *stages*

/prep7: model creation preprocessor
/solu: solution processor
/post1: postprocessor

Each stage consists of *commands* that help build the model. Each command is identified by a keyword followed by specific options. For example, the command `K` is used to create keypoints on the geometry and has the following syntax,

```
K, NPT, X, Y, Z
```

where the *control flags* are

`NPT`: Reference number for the keypoint

`X`, `Y`, and `Z`: Keypoint locations in the active coordinate system (can be r, Θ, z if you pick cylindrical coordinate system)

APDL consists of numerous commands. In what follows we follow the list some of the common commands we find useful to create models. The ordering of these commands into the following groups follows reference [2]. The control flags associated with each command can be found in the Help menu of ANSYS, or it can be sought by typing help *command-name* in the command line shown in Fig. C.3

Preprocessor commands
Model generation commands

Command	Description
K	Defines a keypoint
L	Defines a line between two keypoints
A	Defines an area by connecting keypoints
V	Defines a volume through keypoints
LARC	Defines a circular arc
SPLINE	Generate a segmented spline through a series of keypoints
BSPLIN	Generates a single line from a spline fit to a series of Keypoints
BLC4	Creates rectangular area or block volume by corner points
CYL4	Creates rectangular area or cylindrical volume anywhere on the working plane
ADRAG, VDRAG	Generates areas/volumes by dragging a line-pattern/area along a path
VEXT	Generates additional volumes by extruding areas
AADD, VADD	Adds separate areas/volumes to create a single area/volume
LGLUE, AGLUE, VGLUE	Generates new lines/areas/volumes by gluing lines/areas/volumes
LOVLAP, AOVLAP, VOVLAP	Overlaps lines/areas/volumes
CM	Groups geometry items into a component
#DELE	Delete keypoints, lines, areas, volumes, and components `# is one of K, L, A, V, CM`
#PLOT	Plot keypoints, lines, areas, volumes `# is one of K, L, A, V`
#LIST	List keypoints, lines, areas, volumes `# is one of K, L, A, V, CM`

Material properties command

Command	Description
MP	Defines a linear material property as a constant or a function of temperature

Element and meshing related commands

Command	Description
ET	Defines a local element type from the element library
R	Defines the element real constants
N	Defines a node
E	Defines an element by node connectivity
TYPE	Sets the element type attribute pointer
REAL	Sets the element real constant set attribute
MAT	Sets the element material attribute pointer
LESIZE	Specifies the divisions and spacing ratio on unmeshed lines
MSHKEY	Specifies whether free meshing or mapped meshing should be used to mesh a model
#MESH	Mesh lines/areas/volumes `# is one of L, A, V`

(*continued on next page*)

#CLEAR	Deletes nodes and elements associated with selected lines, Areas, or volumes # is one of L, A, V
#DELE	Deletes nodes/elements # is one of N or E
#PLOT	Displays nodes/elements # is one of N or E
#LIST	Lists nodes/elements # is one of N or E

Solution commands

Boundary condition and load commands

D	Defines degrees of freedom constrains at nodes
F	Defines force at the nodes
SF, SFE, SFL, SFA	Defines surface loads on nodes/elements/lines/areas
BF, BFE	Defines body force on a node/element
TUNIF	Defines a uniform temperature to all nodes
LSREAD, LSWRITE	Read/write load and load step option data into the database

Solution commands

SOLVE	Start a solution
LSSOLVE	Reads and solves multiple load steps

Postprocessor commands

FILE	Specifies the data file where results are to be found
SET	Defines the data set to be read from the results
PLDISP	Displays the displaced structure
PLNSOL	Displays solution results as continuous node contours
PLESOL	Displays solution results as discontinuous node contours
PRNSOL	Prints nodal solution results
PRESOL	Prints solution results for elements

Control commands

*GET	Retrieves a value and stores it as a scalar or part of an array parameter
*VWRITE	Writes data to a file in a formatted sequence
*DO, *ENDDO	Defines the beginning/end of a do-loop
*IF, *ELSE, *ELSEIF *ENDIF	Conditional control block commands

Arithmetic and logical operators

+	Addition
-	Subtraction
*	Multiplication
/	Division
**	Exponentiation
<	Less than
>	Greater than
=	Equal to

Session and database commands

`!`	ANSYS ignores any text to the right of this command
`/PREP7`	Enters the model creation preprocessor
`/SOLU`	Enters the solution processor
`/POST1`	Enters the database results postprocessor
`/POST26`	Enters the time-history postprocessor
`FINISH`	Exits normally from a processor
`SAVE`	Saves all current database information
`#SEL`	Select keypoints/lines/areas/volumes/nodes/elements/ `components, where # is one of K, L, A, V, N, E, CM`

References

[1] Anonymous, APDL Programmer's Guide, ANSYS Release1 8.1, Ansys, Inc., Canonsburg, PA, April 2004.

[2] E. Madenci, I. Guven, The Finite Element Method and Applications in Engineering Using ANSYS, Springer, New York, NY, 2006.

[3] M.K. Thompson, J.M Thompson, ANSYS Mechanical APDL for Finite Element Analysis, Butterworth-Heinemann, Cambridge, M, 2017.

Appendix D

ANSYS tutorial: beam and bar elements

D.1 Example: simply supported beam

In this appendix use of *ANSYS Mechanical APDL* is demonstrated by modeling the mechanics of a simply supported beam subjected to uniform loading as an example. The beam, shown in Fig. D.1, is subjected to uniformly distributed load q_y, has elastic modulus E, cross-sectional area $A = b.h$, second moment of area $I_{zz} = b\,h^3/12$, and length L. The mid-point deflection of an Euler–Bernoulli beam is given as [1],

$$v\left(\frac{L}{2}\right) = -\frac{5}{384}\frac{q_y L^4}{EI} \tag{D.1}$$

ANSYS Mechanical APDL provides two types of beam elements:

Euler–Bernoulli beam element (BEAM4)
Timoshenko beam (BEAM188, BEAM 189)

The Timoshenko beam elements are accessible from the GUI and by using APDL, whereas the Euler–Bernoulli beam element is no longer directly available from the GUI but it can be used with APDL. Note that both of these element types are three-dimensional elements with degrees of freedom UX, UY, UZ, ROTX, ROTY, ROTZ that need to be taken into account when the boundary conditions are applied. Deflections predicted by the two theories are not expected to be identical, as the Timoshenko beam theory includes the effects of transverse shear. The predictions of the two theories approach one another if the slenderness ratio (GAL^2/EII) is greater than 30 [2].

In this appendix we provide the step-by-step description of using the GUI, and provide two APDL macros that demonstrate the use of BEAM4 and BEAM188 elements.

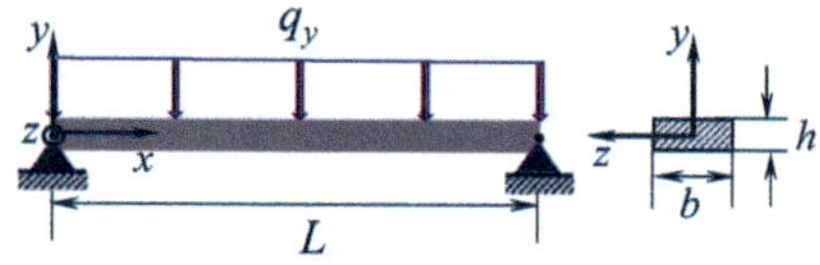

FIGURE D.1 Simply supported beam and its cross section.

TABLE D.1 The width and the height of the beam and the boundary conditions are adjusted as shown in this table, in order to preserve the mechanics of the original problem defined in Fig. D.1, when the loading orientation is changed from –y direction to –z direction.

Loading direction	Beam width (m)	Beam height (m)	Boundary conditions
$-y$	0.1	0.04	UX=UY=UZ=0 ROTX=ROTY=0
$-z$	0.04	0.1	UX=UY=UZ=0 ROTX=ROTZ=0

In what follows we use: $E=100$ GPa, $\nu=0.25$, $b=0.1$ m, $h=0.04$ m, $L=1$ m, $q_y=10$ kN/m.

The slenderness ratio in this case is 5000.

Distributed force: In this problem the distributed load is applied along the $-y$-axis, that is, $\vec{q}=-q_y\hat{j}$. The distributed force can be applied on a three-dimensional ANSYS beam element as a *pressure on a beam* as introduced shortly. The default orientation of this type of pressure is along the $-z$-axis. In what follows we treat this with two different approaches:

GUI-based solution: In Section D.1.1, we interchange the orientations y and z and modify the b and h dimensions of the beam accordingly.[1]

APDL-based solution: In Sections D.1.2 and D.1.3 we preserve the y-orientation, but use the consistent nodal forces that correspond to the distributed force.

Boundary conditions: Note that the beam elements considered in the solution of this problem are three dimensional, but the beam is subjected to a planar loading. While the simple support boundary condition requires restriction of only two displacements (e.g., UX = UY = 0) in a planar problem, here we also need to restrict the UZ and the two appropriate rotational degrees of freedom to properly suppress the rigid body motion.

Table D.1 shows the adjustments we need to make to preserve the original problem when we modify the load orientation from $-y$ to $-z$.

1. Note that it is possible to change the default direction of *pressure on a beam*. But in this introductory treatment we chose to apply the distributed force along the z-axis.

D.1.1 GUI-based solution for a simply supported beam under uniform pressure

Preprocessing:

1. Start ANSYS APDL
2. Utility Menu → File → Change Job Name → Enter "Simply_supported_beam" → OK
3. Preprocessor → Element type → Add/Edit/Delete
   Add → Structural Mass → Beam→2 node 188 → OK (BEAM188)
4. Preprocessor → Sections → Beam → Common Sections →
   ID → Enter "1"
   Name → Enter "Beam"
   B → Enter "0.1" →
   H → Enter "0.04" →
   Nb → Enter "10" →
   Nh → Enter "10" → OK
5. Preprocessor → Material Props → Material Models → Structural → Linear → Elastic → Isotropic
   Enter "1e11" for EX, "0.25" for PRXY → OK → Close the window
6. Utility Menu → Pan-Zoom-Rotate → Pick "Bot"
7. Preprocessor → Modeling → Create → Keypoints → In Active CS
   Enter → Keypoint Number = "1," X = "0," Y = "0," Z = "0" → Apply
   Enter → Keypoint Number = "2," X = "1," Y = "0," Z = "0" → OK
8. Utility Menu → PlotCtrls → Numbering → Check Keypoint Numbers to be On → OK
9. Preprocessor → Modeling → Create → Lines → Straight Line
   Select Keypoint 1 and 2 → OK
10. Preprocessor → Meshing → Mesh Attributes → Picked Lines →
    Select Beam (Lines) → OK
    Select material number "1"
    Element type number "Beam188"
    Element section "1 Beam" → OK
11. Preprocessor → Meshing → Size Cntrls → ManualSize → Lines → Picked Lines → Select Beam (Lines) → OK →
    Enter No. of element divisions = "50" → OK
12. Preprocessor → Meshing → MeshTool → Mesh → Pick ALL
13. Utility Menu → Plot→ Nodes (to view the nodes)
14. ANSYS Toolbar → Save_DB

Solution:

15. Solution → Define Loads → Apply → Structural → Displacement → On Keypoints →
    Enter "1" → Enter

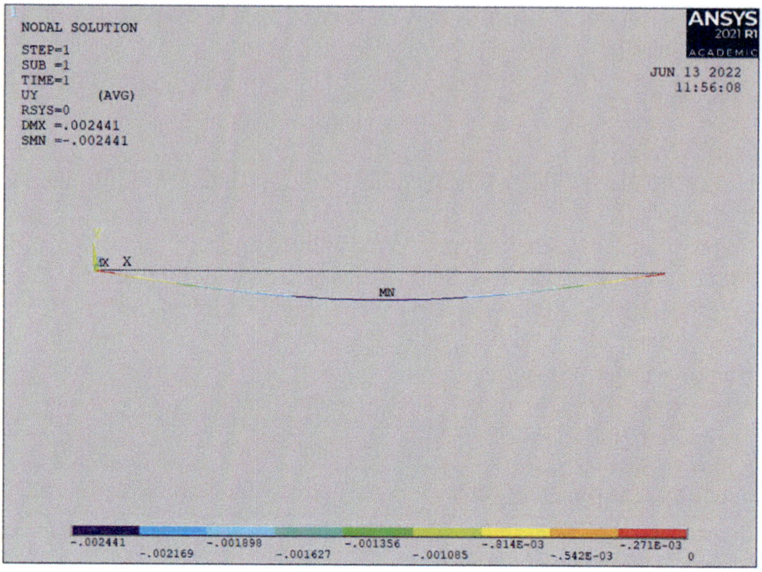

FIGURE D.2 The deformed shape of the simply supported beam subjected to distributed load as prescribed by the PRES variable along the z-axis. Figure published with the permission of ANSYS, Inc.

Enter “2” → OK
Select UX, UY, UZ, ROTX, ROTY and Enter “0” for VALUE → OK

16. Solution → Define Loads → Apply → Structural → Pressure → On Beams → Pick All → OK
    Pressure value at node I, Enter “10,000” → OK
17. Solution → Solve → Current LS → OK
18. Close “the solution is done!” window

Postprocessing:

19. General Postproc → Plot Results → Deformed Shape → OK
20. General Postproc → List Results → Nodal Solution → DOF solution → Select “Z-Component of displacement” → OK →
    You can save this information as a .txt file,
    Command Window Menu → File → Save as
    Maximum displacement is given at the end of the file
21. General Postproc → Contour Plot → Nodal Solu → DOF Solution → Z-Component of displacement → OK

    Maximum displacement occurs (as a negative number) at the center of the beam indicated with blue in the contour plot shown in Fig. D.2.

D.1.2 ANSYS APDL macro for BEAM188

```
!------------------------------------------------------------------------
! Simply supported beam: Timoshenko
!------------------------------------------------------------------------
! Preprocessor
!------------------------------------------------------------------------
/prep7
Modulus = 1e11                    ! Elastic modulus
PRatio  = 0.25                    ! Poisson's ratio
H       = 0.1                     ! Height of beam's cross-section
B       = 0.04                    ! Width of beam's cross-section
Length  = 1                       ! Length of beam
q_load  = 1e4                     ! Distributed force on beam
NELEM   = 50                      ! Number of elements used in analysis
Le      = 1/NELEM                 ! Element length
force   = q_load*Le/2             ! Consistent nodal force
moment  = 0                       ! Consistent nodal moment
k, 1, 0, 0                        ! Position of keypoint-1
k, 2, Length,  0                  ! Position of keypoint-2
l, 1, 2                           ! Create lines
ET,1,BEAM188                      ! Pick element type
MP,EX,1,Modulus                   ! Assign elastic modulus as material-1 property
MP,PRXY,1,PRatio                  ! Assign Poisson's ratio as material-1 property
SECTYPE,1, BEAM, RECT             ! Enter beam cross-section
SECOFFSET, CENT
SECDATA,B,H
LESIZE,1, , ,NELEM, , , , ,1      ! Pick number of elements on Line-1
LMESH, 1                          ! Mesh Line-1
FINISH
!------------------------------------------------------------------------
! Solution
!------------------------------------------------------------------------
/SOLU
D,1, UX, 0                        ! Impose BC UX=0 on Keypoint 1
D,1, UY, 0                        ! Impose BC UY=0 on Keypoint 1
D,1, UZ, 0                        ! Impose BC UZ=0 on Keypoint 1
D,1, ROTX, 0                      ! Impose BC ROTX=0 on Keypoint 1
D,1, ROTY, 0                      ! Impose BC ROTY=0 on Keypoint 1
D,2, UX, 0                        ! Impose BCs on Keypoint 2
D,2, UY, 0
D,2, UZ, 0
D,2, ROTX, 0
D,2, ROTY, 0
F,ALL,FY,-2*force                 ! Apply nodal forces
F,ALL,MZ,moment                   ! Apply nodal forces
SOLVE                             ! Execute solution
FINISH
!------------------------------------------------------------------------
! Postprocessor
!------------------------------------------------------------------------
/POST1
PLDISP,2                          ! Plot displacement component 2
FINISH
!------------------------------------------------------------------------
```

D.1.3 ANSYS APDL macro for BEAM4

```
!----------------------------------------------------------------------------
! Simply supported beam: Euler-Bernoulli
!----------------------------------------------------------------------------
! Preprocessor
!----------------------------------------------------------------------------
/prep7
Modulus = 1                       ! Elastic modulus
PRatio  = 0.5                     ! Poisson's ratio
H       = 0.04                    ! Height of beam's cross-section
B       = 0.1                     ! Width of beam's cross-section
Area    = B*H                     ! Area of the beam
Izz     = B*H**3/12               ! Second moment of area wrt z-axis
Iyy     = H*B**3/12               ! Second moment of area wrt y-axis
Length  = 1                       ! Length of the beam
q_load  = 1e4                     ! Distributed force
NELEM  = 50                       ! Number of elements
Le     = 1/NELEM                  ! Element length
force  = q_load*Le/2              ! Consistent nodal force
moment = 0                        ! Consistent nodal moment
k, 1, 0,  0                       ! Position of keypoint-1
k, 2, Length,  0                  ! Position of keypoint-2
l, 1,  2                          ! Create lines
ET,1,BEAM4                        ! Pick element type
MP,EX,1,Modulus                   ! Assign elastic modulus as material-1 property
MP,PRXY,1,PRatio                  ! Assign Poisson's ratio as material-1 property
R, 1, Area, Izz, Iyy, H, B        ! Enter beam cross-section
LESIZE,1, , ,NELEM, , , , ,1      ! Pick number of element on Line-1
LMESH,1                           ! Mesh Line-1
FINISH
!----------------------------------------------------------------------------
! Solver
!----------------------------------------------------------------------------
/SOLU
D,1, UX, 0                        ! Impose UX=0 as BC on Keypoint1
D,1, UY, 0                        ! Impose UY=0 as BC on Keypoint1
D,1, UZ, 0                        ! Impose UZ=0 as BC on Keypoint1
D,1, ROTX, 0                      ! Impose ROTX=0 as BC on Keypoint1
D,1, ROTY, 0                      ! Impose ROTY=0 as BC on Keypoint1
D,2, UX, 0                        ! Impose BCs on Keypoint 2
D,2, UY, 0
D,2, UZ, 0
D,2, ROTX, 0
D,2, ROTY, 0
F,ALL,FY,-2*force                 ! Apply nodal forces
F,ALL,MZ,moment                   ! Apply nodal moments
SOLVE                             ! Execute solution
FINISH
!----------------------------------------------------------------------------
! Postprocessor
!----------------------------------------------------------------------------
/POST1
PLDISP,2                          ! Plot displacement component 2
FINISH
```

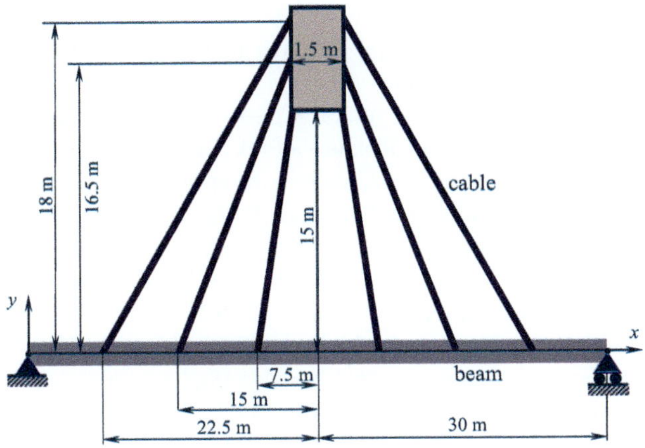

FIGURE D.3 Schematic diagram of the suspended bridge.

D.2 Example: suspended bridge

A planar segment of a bridge, shown in Fig. D.3, consists of a horizontal beam supported by two end supports and six cables. The dimensions of the structure are summarized as follows: the circular cross-sectional area of the cables is 1.27×10^{-2} m^2; the dimensions of the rectangular beams are 0.25×0.08 m^2 ($h \times b$). The elastic modulus and Poisson's ratio for both types of members are 210 GPA and 0.3, respectively. The horizontal members are subjected to a distributed force of 90 kN/m.

Find the maximum deflection and plot the deformed shape of this system. Plot the shear-force and bending moment diagrams of the beam.

D.2.1 GUI-based solution of the suspended bridge problem

Preprocessing:

1. Start ANSYS APDL
2. Utility Menu → File → Change Job Name → Enter "Bridge" → OK
3. Preprocessor → Element type → Add/Edit/Delete
   Add → Structural Mass → Link → 3D finit stn 180 → OK (LINK180)
   Add → Structural Mass → Beam→2 node 188 → OK (BEAM188)
4. Preprocessor → Sections → Beam → Common Sections →
   ID → Enter "1"
   Name → Enter "Beam"
   B → Enter "0.08"
   H → Enter "0.25" → OK
   Preprocessor → Sections → Link → Add →
   Add Link Section with ID → Enter "2" → OK
   Section Name → "Cable"

Link area → "1.27e-2" → OK

5. Preprocessor → Material Props → Material Models → Structural → Linear → Elastic → Isotropic
   Enter "210e9" for EX, "0.3" for PRXY → OK → Close the window
6. Utility Menu → Pan-Zoom-Rotate → Pick "Bot"
7. Preprocessor → Modeling → Create → Keypoints → In Active CS
   Enter → Keypoint Number = "1," X = "0.00," Y = "0," Z = "0" → Apply
   Enter → Keypoint Number = "2," X = "7.50," Y = "0," Z = "0" → Apply
   Enter → Keypoint Number = "3," X = "15.00," Y = "0," Z = "0" → Apply
   Enter → Keypoint Number = "4," X = "22.50," Y = "0," Z = "0" → Apply
   Enter → Keypoint Number = "5," X = "37.50," Y = "0," Z = "0" → Apply
   Enter → Keypoint Number = "6," X = "45.00," Y = "0," Z = "0" → Apply
   Enter → Keypoint Number = "7," X = "52.50," Y = "0," Z = "0" → Apply
   Enter → Keypoint Number = "8," X = "60.00," Y = "0," Z = "0" → Apply
   Enter → Keypoint Number = "9," X = "29.25," Y = "0," Z = "15.0" → Apply
   Enter → Keypoint Number = "10," X = "29.25," Y = "0," Z = "16.5" → Apply
   Enter → Keypoint Number = "11," X = "29.25," Y = "0," Z = "18.0" → Apply
   Enter → Keypoint Number = "12," X = "30.75," Y = "0," Z = "15.0" → Apply
   Enter → Keypoint Number = "13," X = "30.75," Y = "0," Z = "16.5" → Apply
   Enter → Keypoint Number = "14," X = "30.75," Y = "0," Z = "18.0" → OK
8. Utility Menu → PlotCtrls → Numbering → Check Keypoint Numbers to be On → OK
9. Preprocessor → Modeling → Create → Lines → Straight Line
   Select Keypoint 4 and 9
   Select Keypoint 3 and 10
   Select Keypoint 2 and 11
   Select Keypoint 5 and 12
   Select Keypoint 6 and 13
   Select Keypoint 7 and 14
   Select Keypoint 1 and 2
   Select Keypoint 2 and 3
   Select Keypoint 3 and 4
   Select Keypoint 4 and 5
   Select Keypoint 5 and 6
   Select Keypoint 6 and 7
   Select Keypoint 7 and 8 → OK

10. Preprocessor → Meshing → Mesh Attributes → Picked Lines →
    Select Cables (Lines) → OK
    Select material number "1"
    Element type number "Link180" →
    Element section "2 Cable" → APPLY
    Select Beam (Lines) → OK
    Select material number "1"
    Element type number "Beam188"
    Element section "1 Beam" → OK
11. Preprocessor → Meshing → Size Cntrls → ManualSize → Lines → Picked Lines → Select Beam (Lines) → OK →
    Enter Number of element divisions = "6" → APPLY
    Select Cables (Lines) → OK →
    Enter Number of element divisions = "1" → OK
12. Preprocessor → Meshing → MeshTool → Mesh → Pick ALL
13. Utility Menu → Plot→ Nodes (to view the nodes)
14. ANSYS Toolbar → Save_DB

Solution:

15. Solution → Define Loads → Apply → Structural → Displacement → On Keypoints →
    Enter "1" → Enter
    Enter "8" → OK
    Select UX, UY, UZ, ROTX, ROTY and Enter "0" for VALUE → OK
16. Solution → Define Loads → Apply → Structural → Displacement → On Keypoints →
    Enter "9" → Enter
    Enter "10" → Enter
    Enter "11" → Enter
    Enter "12" → Enter
    Enter "13" → Enter
    Enter "14" → Enter
    OK
    Select UX, UY, UZ, ROTX, ROTY and Enter "0" for VALUE → OK
17. Solution → Define Loads → Apply → Structural → Pressure→ On Beams →
    Select All Beam Elements → OK
    Enter "9e4" for Pressure Value at I node → OK
18. Solution → Solve → Current LS → OK
19. Close "the solution is done!" window

Postprocessing:

20. General Postproc → Plot Results → Deformed Shape → OK

21. General Postproc → List Results → Nodal Solution → DOF solution → Select "Z-Component of displacement" → OK →
You can save this information as a .txt file,
Command Window Menu → File → Save as
Maximum displacement is given at the end of the file

22. General Postproc → Contour Plot → Nodal Solu → DOF Solution → Z-Component of displacement → OK
Maximum displacement occurs (as a negative number) at the center of the beam indicated with blue in the contour plot given in Fig. D.4.

23. General Postproc → Element Table → Define Table → Press Add →
Request internal shear force distribution
User label for item → Enter "SF_I"
Results data item → Scroll down and select "SMISC" →
Enter "5" to the right side of the SMISC (SMISC, 5) → Apply →
User label for item → Enter "SF_J"
Results data item → Scroll down and select "SMISC" →
Enter "18" to the right side of the SMISC (SMISC, 18) → Apply →
Request internal bending moment distribution
User label for item → Enter "MY_I"
Results data item → Scroll down and select "SMISC" →
Enter "2" to the right side of the SMISC (SMISC, 2) → Apply →
User label for item → Enter "MY_J"
Results data item → Scroll down and select "SMISC" →
Enter "15" to the right side of the SMISC (SMISC, 15) → Apply →
Plot the shear force and bending moment diagrams for the beam (Fig. D.5)

24. General Postproc → Plot Results → Contour Plot → Line Elem Res →
Element Table Item at node I → Select "SF_I"
Element Table Item at node J → Select "SF_J"

25. General Postproc → Plot Results → Contour Plot → Line Elem Res →
Element Table Item at node I → Select "MY_I"
Element Table Item at node J → Select "MY_J"
List the axial strain in elements

26. General Postproc → Element Table → Define Table → Add →
User label for item → Enter "AXIALSTR"
Results data item → Scroll down and select "LEPEL" →
Enter "1" to the right side of the LEPEL (LEPEL, 1) → OK
General Postproc → List Elem Table → Items to be listed → Pick "AXIALSTR" → OK

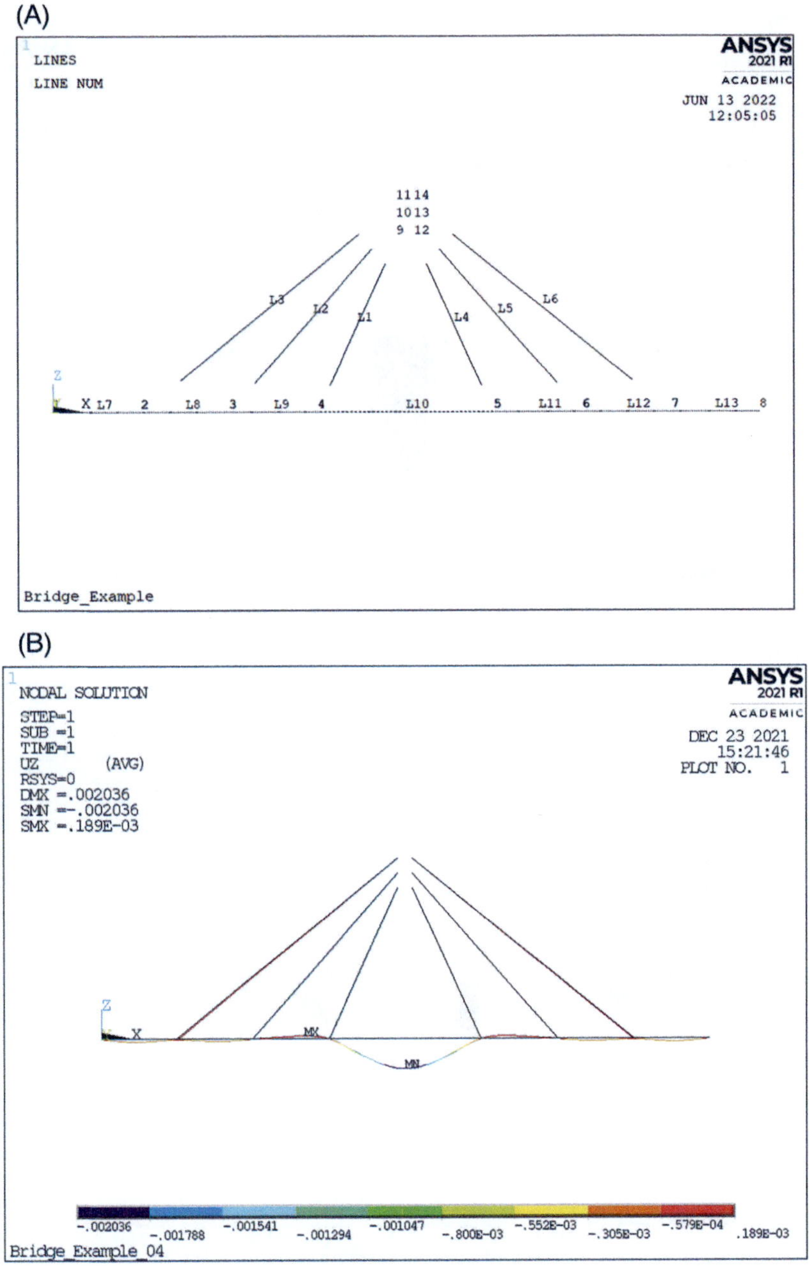

FIGURE D.4 (A) The keypoints and the lines of the bridge model in ANSYS. (B) The deformed shape of the suspended bridge. Figure published with the permission of ANSYS, Inc.

(A)

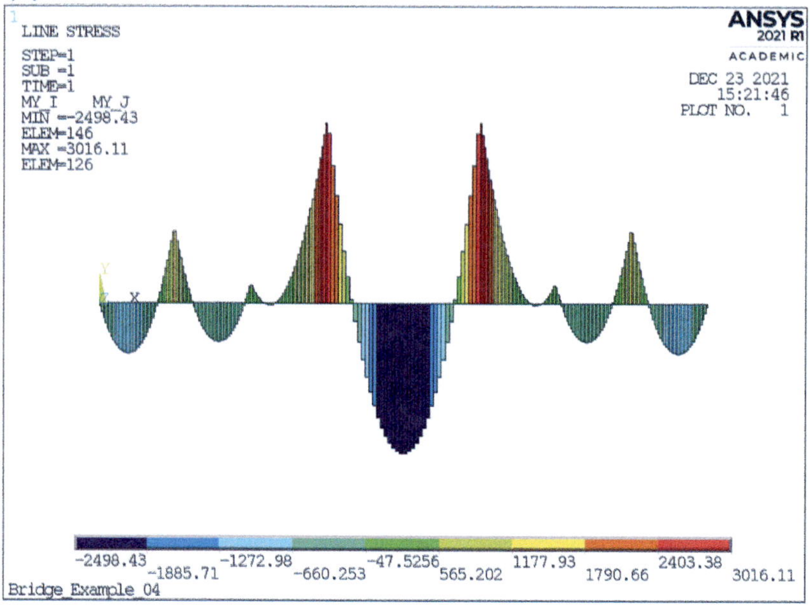

(B)

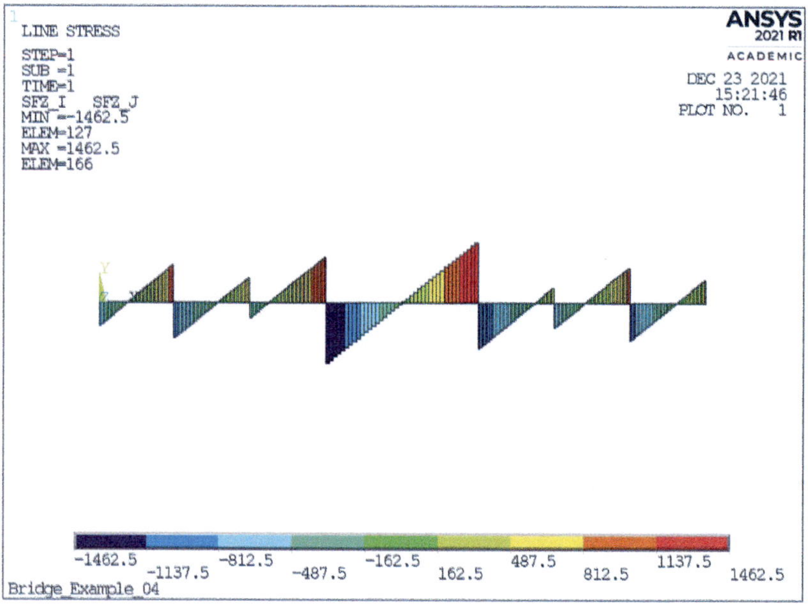

FIGURE D.5 (A) Bending moment and (B) shear force resultant distributions on the horizontal member of the bridge. Figure published with the permission of ANSYS, Inc.

D.2.2 APDL-based solution of the suspended bridge problem

```
!-------------------------------------------------------------------------
! Suspended bridge
!-------------------------------------------------------------------------
/TITLE,Bridge_Example_04            ! Set the title of the job
/FILNAME,Bridge_Example_04_         ! Set the file name
/RGB,INDEX,100,100,100, 0           ! Change the diplay background color
/RGB,INDEX, 80, 80, 80,13
/RGB,INDEX, 60, 60, 60,14
/RGB,INDEX, 0, 0, 0,15
/REPLOT
!-------------------------------------------------------------------------
! Preprocessor
!-------------------------------------------------------------------------
/prep7                              ! Enter the preprocessor
Modulus         = 210e9             ! Modulus of beam
PRatio          = 0.3               ! Poisson's ratio of beam
Modulus_cable = 210e9               ! Modulus of cable
PRatio_cable  = 0.3                 ! Poisson's ratio of cable
A_cable         = 1.27e-2           ! cross-sectional area of cable
B_beam          = 0.08              ! Beam width
h_beam          = 0.25              ! Beam height
q_load          = 9e5               ! Distributed force on beam
NELEM_cable     = 1                 ! Number of elements on cables
NELEM_beam      = 40                ! Number of elements in each beam section
! Create keypoints and lines
k,  1, 0.0,     0, 0                ! Keypoint-1 coordinates
k,  2, 7.5,     0, 0
k,  3, 15.0,    0, 0
k,  4, 22.5,    0, 0
k,  5, 37.5,    0, 0
k,  6, 45.0,    0, 0
k,  7, 52.5,    0, 0
k,  8, 60.0,    0, 0
k,  9, 29.25,   0, 15.0
k, 10, 29.25,   0, 16.5
k, 11, 29.25,   0, 18.0
k, 12, 30.75,   0, 15.0
k, 13, 30.75,   0, 16.5
k, 14, 30.75,   0, 18.0
l, 4,  9                            ! Create line segments for cables
l, 3, 10
l, 2, 11
l, 5, 12
l, 6, 13
l, 7, 14
l, 1,  2                            ! Create line segments for beams
l, 2,  3
l, 3,  4
l, 4,  5
l, 5,  6
l, 6,  7
l, 7,  8
                                    ! Impose BCs on keypoints
DK, 9,UX,0,,1,UY,UZ                 ! Fix cables to the top support
DK,10,UX,0,,1,UY,UZ
DK,11,UX,0,,1,UY,UZ
DK,12,UX,0,,1,UY,UZ
DK,13,UX,0,,1,UY,UZ
DK,14,UX,0,,1,UY,UZ
                                    ! Impose simple-support boundary condtions
```

```
DK,1,UX,0,,1,UY,UZ,ROTX,ROTZ          ! on keypoints 1 and 8
DK,8,UX,0,,1,UY,UZ,ROTX,ROTZ
                                      ! Element types
ET, 1, 188                            ! Set element type-1 as beam188
ET, 2, 180                            ! Set element type-2 as link180
                                      ! Enter material properties
MP, EX,   1, Modulus                  ! Elastic modulus of material type-1
MP, PRXY,1, PRatio                    ! Poisson's ratio of material type-1
MP, EX,   2, Modulus_cable            ! Elastic modulus of material type-2
MP, PRXY,2, PRatio_cable              ! Poisson's ratio of material type-2
                                      ! Enter cross-section information
SECTYPE,   1, BEAM, RECT              ! Set section-1 type to beam
SECOFFSET, CENT
SECDATA,B_beam,H_beam                 ! Enter beam cross-section information
SECTYPE, 2, LINK                      ! Set sction-2 type to link
SECDATA, A_cable                      ! Enter cable cross-section information
                                      !
LESIZE,1,  , ,NELEM_cable             ! Set element size for cable-1
LESIZE,2,  , ,NELEM_cable
LESIZE,3,  , ,NELEM_cable
LESIZE,4,  , ,NELEM_cable
LESIZE,5,  , ,NELEM_cable
LESIZE,6,  , ,NELEM_cable
                                      !
LESIZE, 7,  , ,NELEM_beam             ! Set element size for beam segment
LESIZE, 8,  , ,NELEM_beam
LESIZE, 9,  , ,NELEM_beam
LESIZE,10,  , ,NELEM_beam
LESIZE,11,  , ,NELEM_beam
LESIZE,12,  , ,NELEM_beam
LESIZE,13,  , ,NELEM_beam
                                      ! Meshing the elements
type,2                                ! Assign type-2
secnum,2                              ! Associate secnum-2 with type-2
mat,2                                 ! Associate secnum-2 with material-2
LMESH, 1, 6                           ! Mesh Lines 1 - 6
                                      !
type,1                                ! Assign type-1
secnum,1                              ! Associate secnum-1 with type-1
mat,1                                 ! Associate secnum-1 with material-1
LMESH, 7, 13                          ! Mesh Lines 7 - 13
FINISH                                ! Finish preprocessor
!------------------------------------------------------------------------
! Solver
!------------------------------------------------------------------------
/SOLU                                 ! Enter solution

SFBEAM,ALL,1,PRES,200,,,,,,0          ! Apply surface force for beam

outres,all,all                        ! Write all solution outputs to the database
alls,all                              !

SOLVE                                 ! Execute the solution
FINISH
!------------------------------------------------------------------------
! Postprocessor
!------------------------------------------------------------------------
/POST1                                ! Enter postprocessor
set,last                              ! Read the last data set from the results file
esel,s,type,,1                        ! Select elements of type-1
                                      ! Element Table
ETABLE, SFZ_I, SMISC,  5              ! Request shear force at node-i
```

```
ETABLE, SFZ_J, SMISC, 18          ! Request shear force at node-j
ETABLE, My_I,  SMISC,  2          ! Request bending moment at node-i
ETABLE, My_J,  SMISC, 15          ! Request bending moment at node-j

/show,tiff                        ! Indicate the plot output format
PLLS,MY_I,MY_J,3,0,1              ! Plot moments
/show,close                       ! End plotting
                                  !
/show,tiff
PLLS,SFZ_I,SFZ_J,1,0,1            ! Plot shear froce resultants
/show,close
                                  !
ALLSEL,ALL                        ! Select all entities
/VIEW,1,,-1                       ! Change the viewing angle
/show,tiff
PLNSOL, U,Z, 2,1.0                ! Plot deformed and undeformed shapes
/show,close
FINISH
SAVE                              ! Save the database
```

References

[1] F.P. Beer, R. Johnston, J.T. DeWolf, Mechanics of Materials, eighth ed., McGraw-Hill, New York, NY, 2020.

[2] Anonymous, APDL Programmer's Guide, ANSYS Releas1 8.1, Ansys, Inc., Canonsburg, PA, April 2004.

Appendix E

ANSYS tutorial: two-dimensional linear elastic analysis

Fig. E.1 shows a wrench used to torque a bolt on its left opening, by applying a distributed force 90 N/cm. The edge lengths of the hexagonal regions on the left and right sides of the wrench are 0.7 cm and that at the middle is 0.9 cm. The thickness of the wrench is 0.3 cm. The elastic modulus of the material is 20×10^6 N/cm^2 and the Poisson's ratio is 0.32.

Find the deflected shape and the von Mises stress distribution in the wrench.

In the following a step-by-step description of the finite element solution of this problem by using the GUI of ANSYS is described.

Preprocessing:

1. Start ANSYS APDL
2. Utility Menu → File → Change Job Name → Enter "Wrench" → OK
3. Preprocessor → Element type → Add/Edit/Delete
   Add → Structural Mass → Solid → Quad 8 node 183 → OK
   Options → Element behavior → Select Plane Strs w/thk → OK
   Close
4. Preprocessor → Real Constants → Add/Edit/Delete
   Add → Select Type 1 (Already selected) → OK
   Enter "0.3" for Thickness → OK
   Close
5. Preprocessor → Material Props → Material Models → Structural → Linear → Elastic → Isotropic → EX = "20E6," PRXY = "0.32" → OK
   Click X (top right) to close the window
6. Preprocessor → Modeling → Create → Areas → Circle → Solid Circle
   → X = "0," Y = "0," R = "1.25" → Apply
   → X = "5," Y = "0," R = "1.25" → Apply
   → X = "10," Y = "0," R = "1.25" → OK

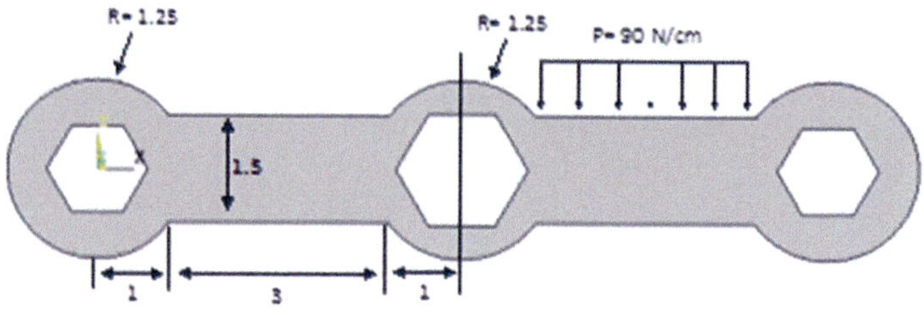

FIGURE E.1 Description of the geometry and loading of a wrench. The units are given in cm.

Preprocessor → Modeling → Create → Areas → Polygon → Hexagon
→ X = "0," Y = "0," R = "0.7," THETA = "120" → Apply
→ X = "5," Y = "0," R = "0.9," THETA = "120" → Apply
→ X = "10," Y = "0," R = "0.7," THETA = "120" → OK
Preprocessor → Modeling → Create → Areas → Rectangle → By Two Corners
→ X = "1," Y = "-0.75," Width = "3," Height = "1.5" → Apply
→ X = "6," Y = "-0.75," Width = "3," Height = "1.5" → OK

7. Preprocessor → Modeling → Operate → Booleans →
Subtract → Areas → Select Left Circle → OK → Select Left Hexagon → OK
Subtract → Areas → Select Middle Circle → OK → Select Middle Hexagon → OK
Subtract → Areas → Select Right Circle → OK → Select Right Hexagon → OK
Preprocessor → Modeling → Operate → Booleans → Add → Areas → Pick All
8. Preprocessor → Meshing → Mesh Tool → Size Controls: Global; Set → Element Edge Length= "0.1" → OK → Mesh → Pick All

Solution:

9. Solution → Define Loads → Apply → Structural → Displacement → On Nodes → Select Nodes at the corners of the left hexagon → OK
DOFs to be constrained → All DOF →→→
Value "0" → OK
10. Solution → Define Loads → Apply → Structural → Pressure → On Lines →
Select Appropriate Line to Apply the Pressure → OK
Enter "90" for VALUE → OK
11. Solution → Solve → Current LS → OK
12. Close "the solution is done!" window

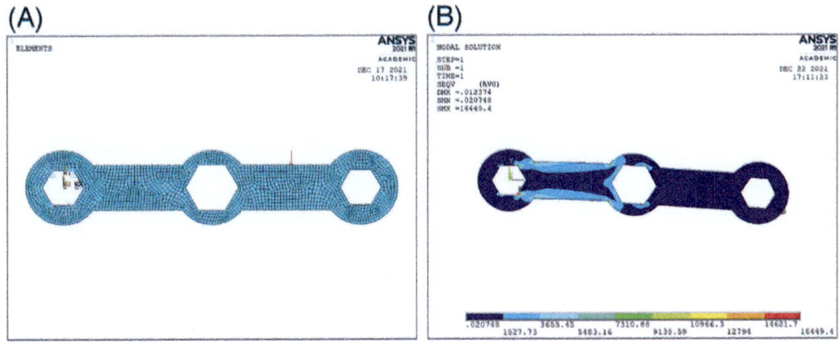

FIGURE E.2 (A) The mesh, boundary conditions, and the external force of the wrench model. (B) The deflected shape and von Mises stress distribution in the wrench. Figure published with the permission of ANSYS, Inc.

Postprocessing:

13. General Postproc → Plot Results → Deformed Shape → OK

14. General Postproc → Plot Results → Contour Plot → Nodal Solution → Select Stress in the first column → Von Mises in the second column → OK

15. ANSYS Toolbar → Save_DB

The mesh, boundary conditions external force, and von Mises stress distribution in the deformed wrench are shown in Fig. E.2.

Appendix F

ANSYS tutorial: thermomechanical deformation

Consider a gas conduit made of two materials (-1 and -2) as shown in the figure. A hot gas at temperature of 75°C flows in this conduit in a direction-oriented out-of-the-plane of the paper. Outside of the conduit, the external air temperature is 20°C. The material properties and necessary heat transfer constants are given in Table F.1.

a. Determine the steady state temperature distribution of the conduit walls.

b. Determine the deflections and internal stress distribution at steady state thermal equilibrium.

Solution

We are going to assume that the material properties are independent of temperature and strain. Therefore, the thermal problem and the structural problem are related, but decoupled. In *ANSYS Mechanical APDL* this problem is solved in two steps. In order to answer part (a), we will model a *steady state heat transfer problem*. Then, we will answer part (b) by reading the results of the thermal solution into the *structural analysis*.

Boundary conditions

Convective heat transfer takes place between conduit and the gases both on the inner and outer surfaces. Recall that convective heat transfer is represented by Newton's law of cooling (Eq. (2.4.3)),

$$f^h = h(T - T_\infty)$$

where q^h is the heat flux through a surface, h is the convection coefficient, T is the wall temperature, and T_∞ is the fluid temperature away from the wall. Inside the conduit T_∞ is the temperature of the gas, and outside the conduit it is the ambient air temperature.

Note that there are two planes of *symmetry* this problem. In *heat transfer analysis*, symmetry is represented by imposing *zero heat flux* through the

TABLE F.1 Parameters of the problem described in Fig. F.1.

	Material-1	Material-2	Inside gases	Outside air
Thermal conductivity, k (W/m K)	0.33	1	–	–
Convection coefficient, h (W/m^2 K)	–	–	65	50
Elastic modulus, E (GPa)	1	20	–	–
Poisson's ratio, ν	0.17	0.15	–	–
Coefficient of thermal expansion, α (m/mK)	5.5E−6	14.5E−6	–	–

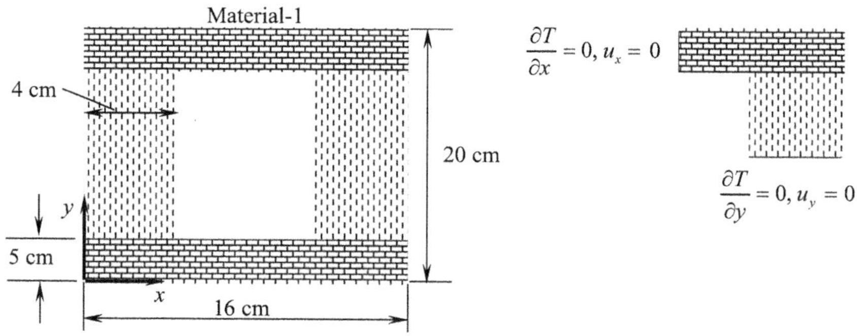

FIGURE F.1 Schematic depiction of a conduit that carries hot gasses inside and exposed to ambient air outside, and the boundary conditions for the symmetry planes.

symmetry plane,

$$f = k\frac{\partial T}{\partial n}$$

where n is the unit outward normal of the symmetry plane. In *structural analysis*, the symmetry condition is imposed by setting the displacement component normal to the symmetry axis to zero as shown in Fig. F.1.

F.1 GUI-based solution of the thermomechanical deformation problem

Heat transfer analysis

1. File → Change Jobname → "thermal-stress"
   File → Change Title → "thermal-stress"
2. Chose element type
   Preprocessor → Element Type → Add/Edit/Delete
   From the Library of Element Types, under Thermal Mass, choose **Solid**, then chose Quad4node 55
3. Set material properties for both materials. This step will be performed twice for each one of material-1 and -2. Note that since we are going to perform

a structural analysis later, it is convenient to set the structural material properties at this step.
Define material-1:
Preprocessor → Material Props → Material Models → Thermal → Conductivity → Isotropic (Kxx = 0.33)
Preprocessor → Material Props → Material Models → Structural → Linear → Elastic → Isotropic (EX = 1E9, PRXY = 0.17)
Preprocessor → Material Props → Material Models → Structural → Thermal Expansion → Secant Coefficient → Isotropic (ALPX = 5.5E−6)
Define material-2:
From the Define Material Model Behavior window
Material → New Model → (Define Material ID 2)
Repeat the above steps for the properties of material-2.

4. Create areas
   Preprocessor→ Modeling → Create → Areas → Rectangle →
   - Create a quarter symmetric model of the problem.
   - Start by creating two areas as shown in Fig. F.1.
   - Remember to convert "cm" to "m" or use consistent material properties.

   These areas need to be *glued* in order for the dof on their common boundaries to be common.
   Preprocessor → Modeling → Operate → Booleans → Glue → Areas
   In the *Glue Areas* window → Pick All → OK
   This glues the two areas.
5. Specify material attributes to areas. Note that this step will be performed once for each area with the proper Material number.
   Preprocessor → Meshing → Mesh Attributes → Picked Areas
   Pick the material-1 area and Apply →
   Area Attributes window → (MAT Material number = 1)
   Repeat for the other area using (MAT Material number = 2)
6. ANSYS Toolbar: SAVE_DB
7. Meshing. Use a rectangular mesh with 0.25 cm elements.
   Preprocessor → Meshing → Size Cntrls → Manual Size→All Areas →
   In *Element Sizes on All Selected Areas* window (SIZE Element edge length = 0.25E−2)
   Preprocessor → Meshing → MeshTool
   - Pick Mesh
   - In *Mesh Areas* window Pick All
   - Apply
8. Apply convection boundary conditions
   On the inside boundary
   Solution → Define Loads → Apply → Thermal → Convection → On lines
   - Pick Line-1 and Line-8 (Fig. F.2) Apply
   - In *Apply CONV on Lines* window
   - Set (VALI Film coefficient = 65)

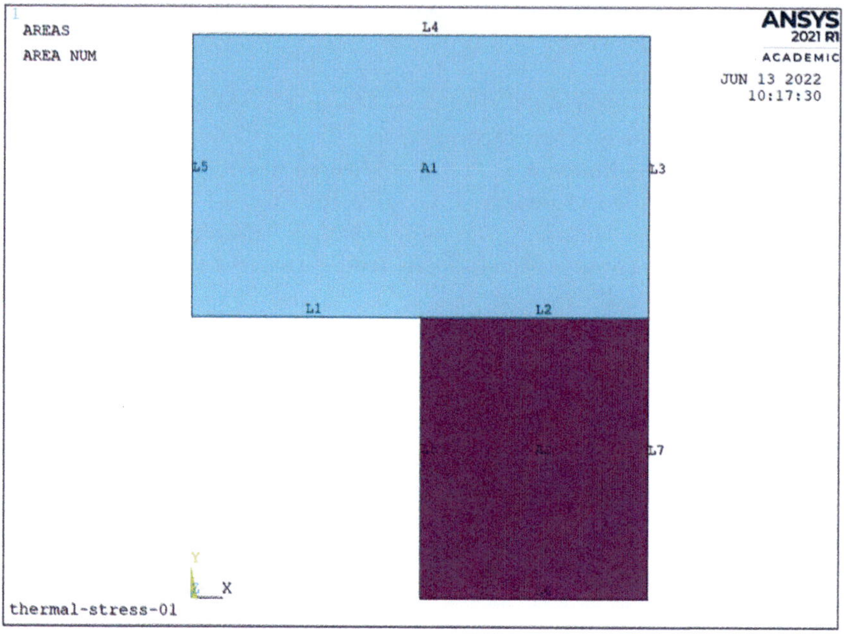

FIGURE F.2 Quarter symmetric model of the problem showing the keypoint and line numbers.

- Set (VAL2I Bulk temperature = 75)
- Apply

On the outside boundary

Solution → Define Loads → Apply → Thermal → Convection → On lines

- Pick Line-3, Line-4, and Line-7 (Fig. F.2) Apply
- In *Apply CONV on Lines* window
- Set (VALI Film coefficient = 50)
- Set (VAL2I Bulk temperature = 20)
- Apply

9. Apply symmetry boundary conditions

Solution → Define Loads → Apply → Thermal → Heat Flux → On lines

- Pick Line-5 and Line-6 (Fig. F.2) Apply
- In *Apply HFLUX on Lines* window
- Set (VALUE Load HFLUX value = 0)
- Apply

10. ANSYS Toolbar: SAVE_DB

11. Obtain solution

Solution → Solve → Current LS

Postprocessing of results

12. Plot temperature contours

General Postproc → Plot Results → Contour Plot → Nodal Solu

- In *Contour Nodal Solution Data* window
- Pick (DOF Solution)
- Pick (Temperature TEMP)
- Apply

13. To save a hard copy of the temperature contours.
    PlotCtrls → Capture Image … → give a file name
14. Plot heat flux q in the conduit.
    - General Postproc → Plot Results → Vector Plot → Predefined
    - In *Vector Plot of Predefined Vectors* window
    - Pick (Flux & gradient)
    - Pick (Thermal Flux TF)
    - Apply
15. ANSYS Toolbar: SAVE_DB
    Note that this step is crucial because it creates the file named `thermal-stress.rtf`, which will be used by the structural analysis. This file contains the nodal temperature values.

Temperature contours $T(x,y)$ and heat flux $q(x,y)$ are provided in Figs. F.3 and F.4, respectively.

Thermal stress analysis

The goal of this part of the analysis is to compute the thermal stresses due to the temperature distribution computed previously. This is accomplished in three steps:

i. Switching the element type from Thermal to Structural
ii. Reading the temperature distribution from the previous step
iii. Carrying out a structural analysis

1. Preprocessor → Element Type → Switch Elem Type →
    - Pick "Thermal to Struc"
      Note: Because we have defined the structural material properties above we do not have to define them here. Otherwise, we would have to assign E, ν, and α here.
2. Read the temperatures calculated at thermal analysis step
    Solution → Define Loads → Apply → Structural → Temperature → From Thermal Analysis →
    - Pick the Fname Name of results file `thermal-stress.rth`
3. Apply structural boundary conditions
    Solution → Define Loads → Apply → Structural → Displacement → On Lines
    - Apply the symmetry boundary conditions on lines L4 and L5
    - Pick L4→Apply→ set UX = 0
    - Pick L5→Apply→ set UY = 0
4. Obtain solution
    Solution → Solve → Current LS

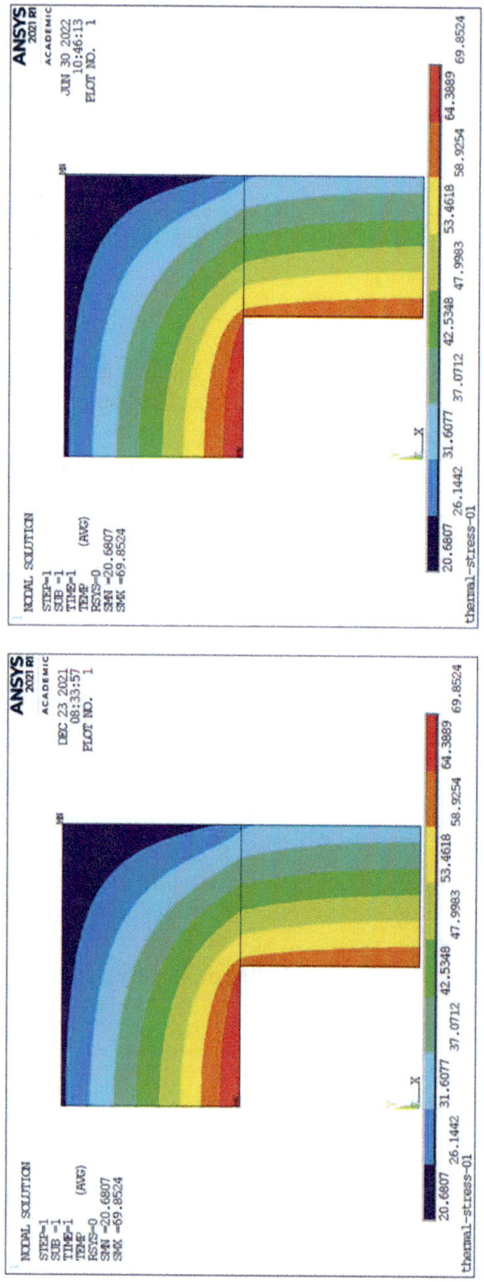

FIGURE F.3 Steady state temperature contours.

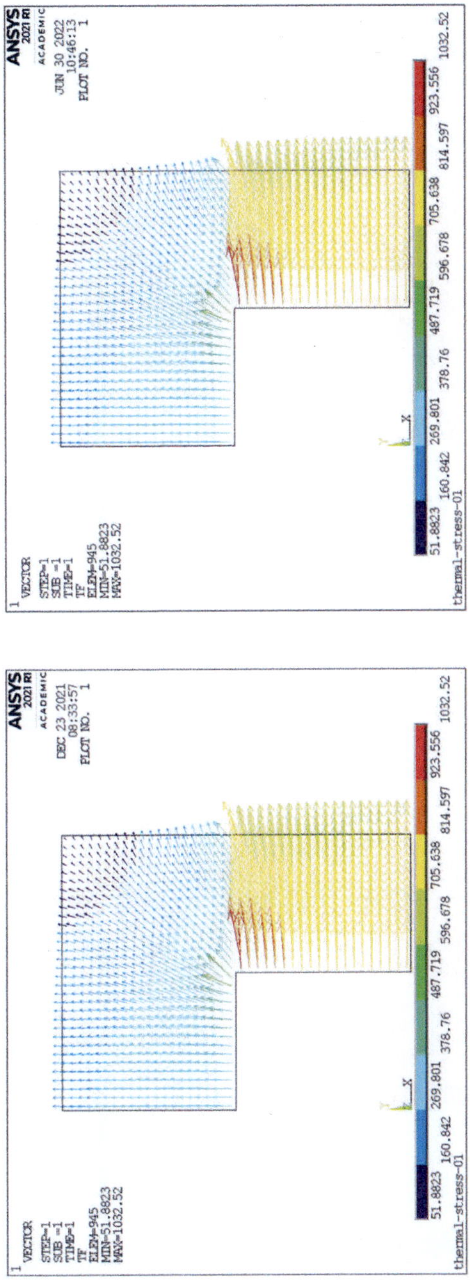

FIGURE F.4 Distribution of the heat flux vector q in the solution domain.

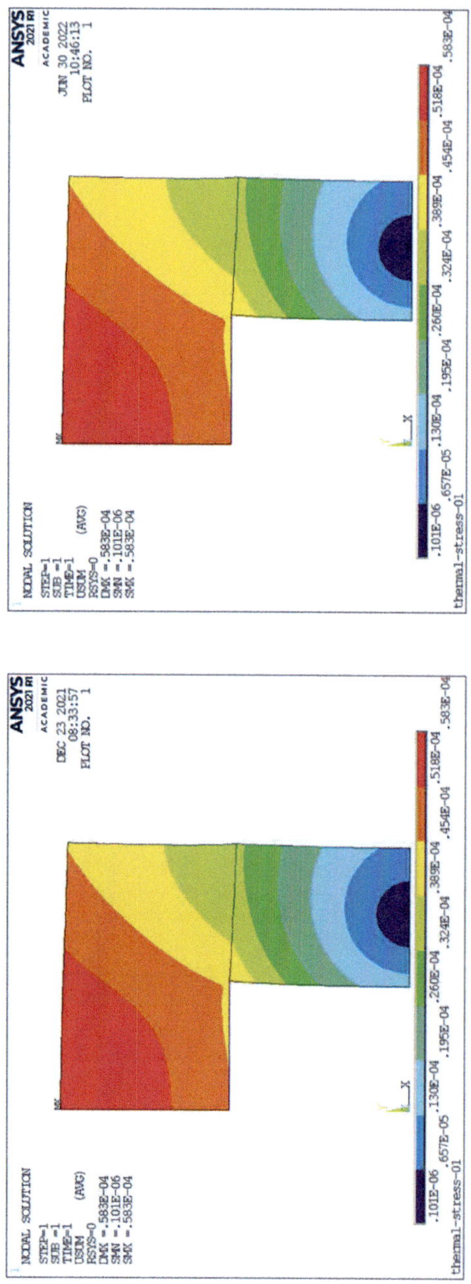

FIGURE F.5 Deformed shape of the conduit due to thermal loading. The contour values indicate the magnitude of the total displacement vector.

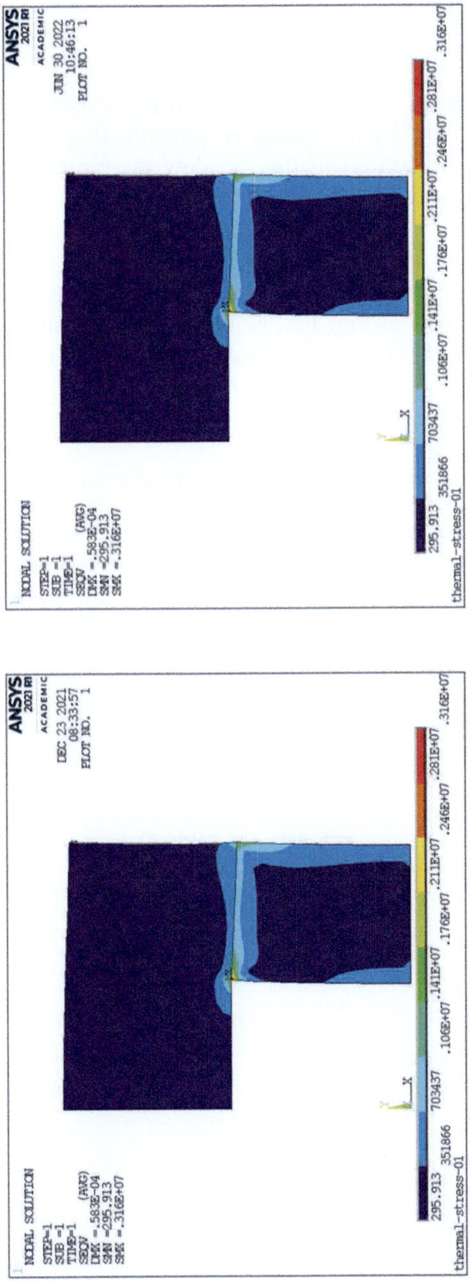

FIGURE F.6 Distribution of the von Mises stress in the conduit.

Postprocessing of results

5. Displacement Solution
   General Postproc → Plot Results → Contour Plot → Nodal Solu → DOF Solution
   - Pick either one of "X-Component of displacement," "Y-component of displacement," or "Displacement vector sum" (Fig. F.4).
   - Note that at the bottom of this window there are various choices for displaying the undeformed structure.
6. Stress Solution
   General Postproc → Plot Results → Contour Plot → Nodal Solu → Stress
   - Pick either one of many stress measures provided. For example, von Mises stress is presented in Fig. F.5. Note that the default contour levels are obscured due to the stress concentration at the inner corner of the junction. You can adjust the contour levels and replot,
   - Top menu: PlotCtrls → Style → Contours → Uniform contours →
     - Pick: User specified
     - Set: VMIN to "0" and VMAX to "900,000" → Apply

Distributions of the heat flux vector, total displacement vector and the von Mises stress are provided in Figs. F.4, F.5 and F.6, respectively.

F.2 APDL-based solution of the thermomechanical deformation problem

```
/TITLE,thermal-stress-01          ! Set the title of the job
/FILNAME,thermal-stress-01,0      ! Set the file name
!......................................................................
! Preprocessor
!......................................................................
/prep7                            ! Enter the preprocessor
! Set problem variables
Modulus_1 = 1e9                   ! Modulus of material-1
Modulus_2 = 20e9                  ! Modulus of material-2
PRatio_1  = 0.17                  ! Poisson's ratio of material-1
PRatio_2  = 0.15                  ! Poisson's ratio of material-2
K1        = 0.33                  ! Thermal conductivity of material-1
K2        = 1.0                   ! Thermal conductivity of material-2
Alpha_1   = 5.5e-6                ! Thermal expansion coefficient of material-1
Alpha_2   =14.5e-6                ! Thermal expansion coefficient of material-1
h_in      = 65                    ! Convection coefficient inside
T_in      = 75                    ! Gas temperature inside
h_out     = 50                    ! Convection coefficient outside
T_out     = 20                    ! Gas temperature outside
! Create keypoints and areas
k, 1,     0,    5e-2              ! Keypoint-1 coordinates
k, 2, 8e-2,    5e-2
k, 3, 8e-2,   10e-2
k, 4,     0,  10e-2
k, 5, 4e-2,    5e-2
k, 6, 4e-2,       0
k, 7, 8e-2,       0
                                  !
A,1,5,2,3,4                       ! Create area-1
A,6,7,2,5                         ! Create area-2
AGLUE,1,2                         ! Glue area-1 and area-2
                                  ! Pick element type
ET,1,PLANE55                      ! Pick element type
                                  ! Set material properties
```

```
MP, KXX, 1, K1                  ! Thermal conductivity of material-1
MP, EX,  1, Modulus_1           ! Elastic modulus of material-1
MP, PRXY,1, PRatio_1            ! Poisson's ratio of material-1
MP, ALPX,1, Alpha_1             ! Thermal expansion coefficient of material-1
MP, KXX, 2, K2                  ! Thermal conductivity of material-2
MP, EX,  2, Modulus_2           ! Elastic modulus of material-2
MP, PRXY,2, PRatio_2            ! Poisson's ratio of material-1
MP, ALPX,2, Alpha_2             ! Thermal expansion coefficient of material-2
! Mesh areas
AESIZE, 1, 0.25e-2              ! Set element size for area-1
AESIZE, 2, 0.25e-2              ! Set element size for area-2
MSHAPE, 0, 2D                   ! Use quadrilateral elements
MSHKEY, 0                       ! Use free meshing
MAT,1                           ! Set material of area-1 to material-1
AMESH,1                         ! Mesh areas 1 and 2
MAT,2                           ! Set material of area-2 to material-2
AMESH,2                         ! Mesh areas 1 and 2
FINISH                          ! Finish preprocessor
!..........................................................................
! Solver for thermal analysis
!..........................................................................
/SOL                            ! Enter the solver
                                ! Apply boundary conditions
SFL,1,CONV, h_in,  ,T_in,       ! Specify surface load (convection) on line 1
SFL,8,CONV, h_in,  ,T_in,       ! Specify surface load (convection) on line 8
SFL,3,CONV, h_out, ,T_out,      ! Specify surface load (convection) on line 3
SFL,4,CONV, h_out, ,T_out,      ! Specify surface load (convection) on line 4

SFL,7,CONV, h_out, ,T_out,      ! Specify surface load (convection) on line 7
SOLVE                           ! Execute solution
FINISH                          ! Finish the solver
                                !
/RGB,INDEX,100,100,100, 0       ! Change the diplay background color
/RGB,INDEX, 80, 80, 80,13
/RGB,INDEX, 60, 60, 60,14
/RGB,INDEX, 0, 0, 0,15
/REPLOT
!..........................................................................
! Postprocessor for thermal results
!..........................................................................
/POST1                          ! Enter the postprocessor
                                !
/SHOW,  TIFF                    ! Indicate the plot output format
PLNSOL, TEMP,, 0                ! Plot nodal temperatures
/SHOW,  close                   ! End plotting
                                !
/SHOW,  TIFF
PLNSOL, TF,SUM, 0               ! Plot thermal flux vector magnitudes
/SHOW,  close
                                !
/SHOW,  TIFF
/VSCALE,1,1,0
PLVECT,TF,,,,VECT,ELEM,ON,0     ! Plot thermal flux vector
/SHOW, close
                                !
SAVE                            ! Save results
FINISH                          ! Finish postprocessor
!..........................................................................
! End of thermal analysis
!..........................................................................
! Preprocessor for structural analysis
!..........................................................................
/PREP7                          ! Enter preprocessor
ETCHG,TTS                       ! Change element type from Thermal to Structural
FINISH                          ! Finish the preprocessor
!..........................................................................
! Solver for structural analysis
!..........................................................................

/SOL                            ! Enter the solver
LDREAD,TEMP,,,,,'thermal-stress-01','rth',' ' ! Read nodal temperature results
DL,5, ,UX,0                     ! Impose boundary condition UX=0 on line 5
DL,6, ,UY,0                     ! Impose boundary condition UY=0 on line 6
SOLVE                           ! Execute solution
FINISH                          ! Finish the solver
```

```
!.............................................................................
! Postprocessor for structural analysis
!.............................................................................
/POST1                          ! Enter the postprocessor
                                !
/SHOW,TIFF
/DSCALE, 1, 100                 ! Set deformation scale to 100
PLNSOL, U, SUM, 0               ! Plot deformation vector magnitude
/SHOW,close
                                !
/SHOW,TIFF
PLNSOL, S,EQV, 0                ! Plot von Mises stress
/SHOW,close
FINISH                          ! Finish the postprocessor
!.............................................................................
! End of deformation analysis
!.............................................................................
SAVE                            ! Save the database
```

Index

Page numbers followed by "*f*" and "*t*" indicate, figures and tables respectively.

A

Adjoint operators, 64
Analytical solutions, 5
ANSYS tutorial
  beam and bar elements, 483
  finite element analysis, 475
  thermomechanical deformation, 503
  two-dimensional linear elastic analysis, 499
APDL, 479
APDL macro-based analysis, 475
Array operations, 451
Array sense, 451

B

Balance laws, 30
  for one-dimensional problem, 30
  for three-dimensional problem, 32
  for two-dimensional problem, 31
Bending stress, 46
Bilinear rectangle, 277
Bilinear rectangular element, 264
  consistent nodal force vector, 267
  element stiffness matrix, 264
Body force, 10
Boundary conditions (BC), 1, 32, 33, 128, 155, 305
  displacement and traction boundaries, 33
  essential, 129, 306
  natural, 128, 305
  numerical implementation, 158
  penalty method, 160
  row and column manipulations, 158
Boundary value problems (BVP), 1, 3, 61, 111
  boundary conditions, 75
  steady state, 3
  weak form, 76
  weak form and functionals, 77
  weighted residual integral, 74

C

Calculus of variations, 65
  Euler–Lagrange equations, 69
  functional derivative, 67
  fundamental lemma, 64
  properties of functionals, 68
  variational derivative, 69
  variation of a functional, 66
Column vector, 448
Computer implementation, 130
  boundary conditions, 131
  element assembly, 130
  element connectivity table, 130
  main-code, 130
Concentrated force, 10
Conduction heat transfer, 50
Conservative systems, potential energy, 36
  elastic body, 37
  linear spring, 36
Consistent nodal force, 140, 145
Consistent nodal load vector, 195
Constant strain triangle (CST) element, 273
Convection heat transfer, 51

D

Damping matrix, 403
Deflection, 113
Deformable solids, initial/boundary value problems, 386
  Euler–Bernoulli beam, 387
  one-dimensional bar, 387
  two-dimensional deformable solid, 386
Dirichlet boundary condition, 2, 61
Discretization, 112, *see also* Finite element formulation
  element connectivity, 112
Distributed force, 10
Divergence theorem, 57, 63

E

Element assembly, 302
Element connectivity table, 130,
see also Computer implementation
Element defects, 275
bilinear rectangle, 277
constant strain triangle element, 275
Element equilibrium equation, 189, 300
boundary value problem, 141, 189
elastic bar, 143
equilibrium equations, 192
for planar bar, 141
problem definition, 141, 189
total potential energy, 142, 191
Element stiffness matrix, 123, 140, 145, 194
constant problem parameters, 123
linearly varying problem, 124
Equation of motion, 421, 429
acceleration-form, 428
average acceleration method, 425
central finite difference approximations, 423
convergence, stability, 431
displacement-form, 428
explicit method, 431
initial conditions, 430
linear and average acceleration methods, 424
α-method, 428
Newmark's method, 426
solution algorithm, 430
stability and consistency, 435
Equilibrium equations, 120
constant problem parameters, 123
linearly varying problem parameters, 124
Essential boundary condition (EBC), 71, 129
Euler–Bernoulli beam
assembly, boundary conditions, and reaction forces, 198
with axial deformation, 197
boundary value problem, 189
C^1-continuous interpolation function, 187
computation of stresses in members, 198
consistent nodal load vector, 195
coordinate transformations, 196
Eigenvalue problem, 389
element equilibrium equation, 189
element stiffness matrix, 194
equilibrium equations, 192
free vibration problem, 391
general beam element, 195
governing equations, 48
kinematics, 45
problem definition, 189
stress–strain relationship, 46
theory, 43
total potential energy, 191
vector transformation, 196
vibration response, 388
Euler–Lagrange equation, 70

F

Finite difference approximations, 348
first derivative, 349
Taylor series expansion, 349
temporal discretization, 348
Finite difference method (FDM), 6
Finite element equations, 147
Finite element formulation
boundary conditions, 128, 129, 305
boundary value problem, 111
computer implementation, 130, 307
constant problem parameters, 123
discretization, 112
element assembly, 125
element equilibrium equations, 294
equilibrium equations, 120
essential boundary conditions, 306
external boundaries, 296
general finite element form, 124
internal elements, 294
Lagrange interpolation functions, 119
linear interpolation, 299
linearly varying problem parameters, 124
linear shape functions, 115
natural boundary conditions, 305
nonconstant coefficient ordinary differential equation, 113
one-dimensional heat transfer equations, 298
one-dimensional interpolation, 114
one-dimensional, steady state, 293
quadratic shape functions, 117
Finite element mesh, 139
Finite element method (FEM), 6
active degrees of freedom, 158
assembly of elements, 151
boundary conditions, 155
computation of stresses and reaction forces, 166
coordinate transformations, 148
one-dimensional-bar element, 149
finite element analysis, 172
formal definition, 156
initial stresses, 166

orthogonal coordinate systems, 148
reaction forces, 167
thermal stresses, 165
Finite element program, 460, 465
Flexible beam, 41
equation of motion of a beam, 43
Euler–Bernoulli beam, 45, 48
Timoshenko beam, 47, 49
Function m-files, 453

G

Galerkin method, 87, 103
Gauss quadrature, 245
coordinate transformation, 245
derivation of second-order, 246
integration of two-dimensional functions, 248
Generalized Hooke's law, 22, 23
for anisotropic materials, 24
initial stress/strain and thermal strain, 25
for isotropic materials, 24
two-dimensional problems, 28
Generalized trapezoidal method
consistent heat capacity matrix, 363
Fourier–von Neumann stability analysis, 362
heat equation discretized, 365
lumped heat capacity matrix, 365
rate of convergence, 366
Global degree of freedom, 140, 152
Global equation of motion, 405
Global force vector, 152
Global stiffness matrix, 140, 152
Green-Gauss theorem, 63
GUI-based analysis, 475, 476

H

Heat flux, 50
Heat transfer, 50, 114
analysis, 504
boundary conditions, 55
conduction, 50
convection, 51
in one-dimensional solid, 53
radiation, 51
three-dimensional solid, 56
Heat transfer equation, direct time integration, 351
Backward difference method, 352
Crank–Nicholson method, 353
Euler method, 351
generalized trapezoidal method, 353
Heat transfer problems
boundary conditions, 315
element assembly, 331
element equilibrium equations, 329
essential boundary conditions and solution, 332, 333
external edges, 324, 326, 327
finite element form, 318
heat equation, 315, 328
heat transfer equation, 316
internal edges, 324
natural boundary conditions, 323
quadrilateral (Q4) element, 320
solution domain and its boundary, 313
Higher order elements, 277
quadratic triangle, 277
Hooke's law, 4, 22

I

Integral formulations and variational methods
calculus of variations, 65
integration by parts, 64
mathematical background, 63
method of weighted residuals, 85
weighted residual integral, 74
Internal forces, 10

J

Jacobian matrix and Jacobian determinant, 230
one-dimensional, for a three-node element, 230
two-dimensional element, 231, 232

L

Lagrange interpolation, 114
Lamé constants, 25
Line plots, 455

M

Mapped isoparametric elements, 238
element stiffness matrix, 241
finite element form, 240
load vector for body forces, 241
strain–displacement operator matrix, 238
surface tractions, 241
Mass matrix, 400
consistent, 400, 401
lumped, 402
MATLAB-based finite element programs, 448
MATLAB-based ODE solver, 462
MATLAB computations, 445
Matrix, 445, 450

Method of weighted residuals (MWR), 85
  Galerkin method, 87
  polynomials, 87
  Rayleigh–Ritz method, 86
Modeling and simulation, 1
  boundary and initial value problems, 1
  boundary value problems, 3

N

Natural boundary condition (NBC), 2, 128
Neutral axis (NA), 45
Newton's law of cooling, 51
Nonessential boundary condition, 2, 61

O

One-dimensional heat transfer, 291
  boundary conditions for, 292

P

Partial differential equations (PDE), 1
Planar bar, element equilibrium equation, 141
  boundary value problem, 141
  elastic bar, 143
  problem definition, 141
  total potential energy, 142
Planar deformation, 18

R

Radiation heat transfer, 51
Rayleigh–Ritz method, 86
Row vector, 447

S

Scatter plots, 456
Script m-files, 453
Self-adjoint operators, 64
Semidiscrete equations, 394
  Euler–Bernoulli beam element, 398
  one-dimensional elastic bar element, 396
  two-dimensional deformable element, 394
Semidiscrete systems, 405
  Eigenvalue problem, 406
  initial excitations, 415
  orthogonality of the eigenvectors, 414
Shear strain, 19
Shear stress, 11
Simulation, 1
Single degree, of freedom systems, 378
  combined response, 384
  free vibrations, 379
  harmonic excitations, 382
  transient vibration, 385
Solution algorithm, 355
  explicit time integration method, 356, 357
  implicit time integration method, 357
Solution methods, 5 9–5 9
Stefan-Boltzmann law, of thermal radiation, 51
Strain energy
  for one-dimensional deformation, 37
  three-dimensional deformation, 39
Strain tensor, 20
Stress tensor, 12
Stress transformations, 12
Structural mechanics, governing equations, 9
  deformation and strain, 17
  external forces, internal forces, and stress, 9
  strain compatibility conditions, 21
  stress transformations, 12
Surface tractions, 10

T

Thermal stress analysis, 508
Three-dimensional deformation, 20
Time integration methods, 358
  generalized trapezoidal algorithm, 361
  semidiscrete first-order equation, 359, 360
Timoshenko beam
  governing equations, 49
  kinematics, 47
  stress–strain relationship, 48
Timoshenko beam theory, 43
Total potential energy
  of deformable body, 41
  of elastic bar, 39
Tractions
  normal and shear components, 15
Transient heat transfer equation, 344
  boundary/initial value problem, 344
  element equilibrium equation, 345
  global boundary conditions, 348
  global equilibrium equation, 348
  heat capacity matrix, 347
Two-dimensional elastic solids
  assembly and boundary conditions, 282
  bilinear rectangular element, 264
  constant strain triangle element, 273
  curvilinear coordinates, 233
  element defects, 275
  element equilibrium equations, 250, 260
  equations of equilibrium, 215
  finite element form, 218
  Gauss quadrature, 245
  general finite element, 216

general quadrilateral and rectangular domains, 229
general treatment of boundary conditions, 253
global equilibrium equations, 252
higher order elements, 277
interpolation across, 221
Jacobian matrix and Jacobian determinant, 230
mapped isoparametric elements, 238
numerical implementation of boundary conditions, 254
outward unit normal, 213
plane quadrilateral elements, 223, 225, 262
plane triangular elements, 228, 263
postprocessing of the solution, 254
solution domain and its boundary, 213
two-dimensional elasticity problem, 259
two-dimensional interpolation functions, 262
two-dimensional mapping, 234
two-dimensional polynomials, 221
two-dimensional shape functions, 223
variational form of the equation, 217
Two-dimensional problems, 27
generalized Hooke's law, 28
plane stress problem, 28, 29

U

Uniaxial deformation, 18

V

von Neumann boundary condition, 2

CPI Antony Rowe
Eastbourne, UK
October 06, 2022